IET MATERIALS, CIRCUITS AND DEVICES SERIES 63

Emerging CMOS Capacitive Sensors for Biomedical Applications

Other volumes in this series:

Emerging CMOS Capacitive Sensors for Biomedical Applications

A multidisciplinary approach

Ebrahim Ghafar-Zadeh and Saghi Forouhi

Institution of Engineering and Technology (IET)

Published by The Institution of Engineering and Technology, London, United Kingdom

The Institution of Engineering and Technology is registered as a Charity in England & Wales (no. 211014) and Scotland (no. SC038698).

The Institution of Engineering and Technology
Michael Faraday House
Six Hills Way, Stevenage
Herts, SG1 2AY, United Kingdom

www.theiet.org

British Library Cataloguing in Publication Data
A catalogue record for this product is available from the British Library

ISBN 978-1-78561-915-1 (hardback)
ISBN 978-1-78561-916-8 (PDF)

Typeset in India by MPS Limited
Printed in the UK by CPI Group (UK) Ltd, Croydon

Contents

About the authors

Ebrahim Ghafar-Zadeh received his B.Sc. and M.Sc. in Electrical Engineering from the KNT University of Technology (Tehran, Iran) and University of Tehran (Tehran, Iran), respectively. He then continued his studies in Polytechnique of Montreal (Montreal, Canada), where he received his Ph.D. degree in Electrical Engineering in 2008. His graduate studies focused on complementary metal-oxide semiconductor (CMOS)-based sensors for lab-on-chip applications. In recognition of his research achievements, he received several fellowship awards including a Postdoctoral Fellowship (PDF) from the Natural Sciences and Engineering Research Council of Canada (NSERC). Then he continued his PDF research studies in Electrical Engineering at the McGill University (Montreal, Canada) and in Bioengineering, at the University of California, Berkeley. As Assistant Professor, in 2013, Ebrahim joined the Department of Electrical Engineering and Computer Science (EECS) in the Lassonde School of Engineering at York University where currently he is Associate Professor, Member of Graduate Programs of Departments of EECS and Biology, and Director of Biologically Inspired Sensors and Actuators (BioSA) research laboratory. His research is aimed at exploring novel integrated sensors and actuators for life science applications. Since 2013, Professor Ghafar-Zadeh has published more than 100 journal and conference articles and trained more than 40 highly qualified personnel (HQP) in the fields of Electrical Engineering and Biology. He is Senior Member of the IEEE and a licensed Professional Engineer in the province of Ontario.

Saghi Forouhi is a postdoctoral researcher at York University (Department of Electrical Engineering and Computer Science (EECS), Biologically Inspired Sensors and Actuators (BioSA) Laboratory), Canada. She received the B.Sc. and M.Sc. degrees in Electrical Engineering from Guilan University, Iran, in 2010 and 2012, respectively. Then, she completed her Ph.D. at Isfahan University of Technology (IUT) in 2019 in an active collaboration between IUT and York University. Her research interests lie in the area of biologically inspired microsystems, CMOS sensors, circuits, and systems.

Table of terminology and definitions

Context	Definition
(3-Aminopropyl)tris (trimethylsiloxy)silane (APTTMS)	A compound with the formula $C_{12}H_{35}NO_3Si_4$
1,4-Phenylene diisothiocyanate (PDITC)	An organic compound with the formula $C_8H_4N_2S_2$ which is a cross-linkers for biological applications
11-Mercapto-1-undecanol	A hydrophilic thiol with the formula $C_{11}H_{24}OS$, which is useful for the production of hydrophilic SAMs
1-Dodecanethiol (DDT)	An alkyl thiol with the formula $C_{12}H_{26}S$ (or $CH_3(CH_2)_{11}SH$) that forms SAM
1-Ethyl-3-(3-dimethylaminopropyl)-carbodiimide (EDC)	A water-soluble carbodiimide usually handled as the hydrochloride with the formula $C_8H_{17}N_3$ generally employed as a carboxyl activating agent for the coupling of primary amines to yield amide bonds
1-Methylimidazole	An aromatic heterocyclic organic compound with the formula $C_4H_6N_2$ (or $CH_3C_3H_3N_2$) useful as a specialty solvent, a base, and as a precursor to some ionic liquids
3-(Aminopropyl)trimethoxysilane (APTMS)	An aminosilane that mainly employed as a silane coupling agent for silanization processes and the surface modification of a variety of nanomaterials
3-[2-(2-Aminoethylamino)ethylamino] propyltrimethoxysilane (AEEPTMS)	A compound with the formula $C_{10}H_{27}N_3O_3Si$ used in chemical synthesis
3-Aminopropyltriethoxysilane (APTES)	An aminosilane with the formula $C_9H_{23}NO_3Si$ used in the process of silanization, the functionalization of surfaces with alkoxysilane molecules, and covalent attaching of organic films to metal oxides such as silica and titania
3T3 cell	A cell line originally established from the primary mouse embryonic fibroblast cells that were cultured by 3T3 protocol (3T3 stands for 3-day transfer, inoculum 3×10^5 cells, and NIH/3T3 is the cell line established from NIH Swiss mouse embryo cultures)
4-(2-Hydroxyethyl)-1-piperazineethane-sulfonic acid (HEPES)	A zwitterionic sulfonic acid buffering agent widely used in cell culture

(Continues)

(*Continued*)

Context	Definition
Acetone (ACE)	An organic compound with the formula C_3H_6O (or $(CH_3)_2CO$) and the simplest and smallest ketone which is highly volatile and flammable liquid
Acetonitrile (MeCN)	The chemical compound with the formula C_2H_3N and the simplest organic nitrile
Active site of an enzyme	The region of an enzyme where substrate molecules bind and undergo a chemical reaction and consists of residues that catalyze a reaction of that catalytic site and amino acid residues that form temporary bonds with the binding site
Adda	A reported alias name for the human gene Add1 (α-adducin) that is a protein in humans encoded by the ADD1 gene
Affibody	Small, robust proteins engineered to bind to a large number of target proteins or peptides with high affinity, imitating monoclonal antibodies, and a member of the family of antibody mimetics
Affimer	Non-antibody binding proteins designed to mimic the molecular recognition characteristics of monoclonal antibodies in different applications that bind to target molecules with similar specificity and affinity to that of antibodies
Affinity binding	The strength of the binding interaction between a single biomolecule (like DNA or protein) to its ligand/binding partner (like inhibitor or drug)
Affinity biosensor	Biosensors that detect a physicochemical change caused by binding analyte molecules to bio-recognition molecules
Alanine	An α-amino acid that is used in the biosynthesis of proteins with the formula $C_3H_7NO_2$
Aldehyde (group)	Any of a class of organic compounds in which a carbon atom shares a single bond with a hydrogen atom, a double bond with an oxygen atom, and a single bond with another atom or a group of atoms
Alpha mouse liver 12 (AML-12) hepatocyte	A cell line established from hepatocytes from a transgenic mice overexpressing transforming growth factor (TGF)α
Amide	An organic functional group characterized by a carbonyl group linked to a nitrogen atom, or a compound including this functional group $CO=NH_2$
Aminosilane	A silicone compound with an amino or diamino functional group and an alkoxy group (formula: H_2NSi)

(*Continues*)

(*Continued*)

Context	Definition
Analyte (chemical species)	A chemical constituent or substance that is of interest in an analytical procedure
Antibody (Ab)	A Y-shaped protein generated mainly by plasma cells used by the immune system to neutralize pathogens like viruses and pathogenic bacteria
Antigen (Ag)	A molecule or molecular structure that can be bound by an antigen-specific antibody or B cell antigen receptor
Aptamer	Oligonucleotide or peptide molecules that bind to a specific target molecule
Aspirin (acetylsalicylic acid)	A medication used to reduce pain, fever, or inflammation with the formula $C_9H_8O_4$
Avidin	A tetrameric biotin-binding protein generated in the oviducts of birds, amphibians, and reptiles and found in the protein content of egg white
Azobisisobutyronitrile (AIBN)	An organic compound with the formula $[(CH_3)_2C]_2N_2$
Bacteria	Microscopic, single-celled organisms that are typically a few micrometers in length and have several shapes, ranging from spheres to rods and spirals
Bacteriophage (phage)	A virus infecting and replicating within archaea and bacteria
Bioelectrogenesis	The generation of electricity by living organisms
Bio-imprinting	A technique to develop dynamic artificial recognition materials that promotes a printing of potential substrates on enzyme structure, inducing a more selective and stable conformation
Biomarker	A characteristic measured as an indicator of pathogenic processes, normal biological processes, or pharmacological responses to a therapeutic intervention
Biomimetic sensor	A technology for collecting data and processing this information in real time, at the high speeds, offered by advanced chips
Biotin	A water-soluble B vitamin that helps to convert certain nutrients into energy and plays an important role in the health of skin, hair, and nails
Biotinylation	The process of covalently attaching biotin to a nucleic acid, protein, or other molecules
Carcinoma	A malignancy developing from epithelial cells that occur when the DNA of a cell is altered or damaged and the cell begins to grow uncontrollably and become malignant
Cardiac-troponin T (cTnT) and troponin I (cTnI)	Cardiac regulatory proteins that control the calcium-mediated interaction between myosin and actin

(Continues)

(Continued)

Context	Definition
Cardiomyocyte (or myocardiocyte)	Cells that make up the cardiac/heart muscle and are primarily involved in the contractile function of the heart that enables the pumping of blood around the body
Cardiotoxicity	The occurrence of heart muscle damage or electrophysiology dysfunction in such a way that the heart becomes weaker and is not efficient in pumping and circulating blood
Catalyst	A substance that enables a chemical reaction to proceed under different condition or at a usually faster rate
$CD4^+$ T-lymphocyte	A type of lymphocyte that helps coordinate the immune response by exciting other immune cells to fight infection
Cell	The basic structural, functional, and biological unit of all known organisms consisting of cytoplasm enclosed within a membrane, which contains many biomolecules such as proteins and nucleic acids
Cell adhesion	The process by which cells interact with each other or with their substratum through specialized molecules of the cell surface
Cell confluence	The percentage of the surface area of a 2D culture that is covered with cells
Cell morphology	An important aspect of the phenotype of a cell that is essential in identifying the shape, structure, form, and size of cells as well as the regulation of cell activities
Cell polarity	Spatial differences in structure, shape, and function within a cell
Cell proliferation	The process by which a cell grows and divides to generate two daughter cells
Cell secretion	A cellular process involving the regulated production and release of intracellular products from cells
Cell seeding	The first stage of cell attachment whose distribution and efficiency can affect the final biological performance of the scaffold
Cell viability	The quantification of the proportion of healthy, live cells within a population
Cerium-oxide	An oxide of cerium with the chemical formula CeO_2
Chemical probe	A small molecule used to study and manipulate a biological system like an organism or a cell by reversibly binding to and altering the function of a biological target within that system
Chemical specificity	The strength of binding between a protein's binding site and specific ligands

(Continues)

(*Continued*)

Context	Definition
Chemical kinetics (reaction kinetics)	The branch of physical chemistry regarding understanding the rates of chemical reactions
Chloroform (CHL)	An organic compound with formula $CHCl_3$ used as a precursor to polytetrafluoroethylene (PTFE) and also a precursor to various refrigerants
Cholesterol	A waxy, fat-like substance found in all the cells in the body to make vitamin D, hormones, and substances and help to digest foods
Collimator	A device which narrows waves or a beam of particles either by aligning the directions of motion in a specific direction or making the spatial cross section of the beam smaller
Colloid	A very tiny and small material dispread uniformly through another substance
Colony-forming unit (cfu)	A unit usually used for the concentration of microorganisms in a sample
Confluence (in cell culture biology)	The percentage of the surface area of a 2D culture covered with cells
Covalent attachment	The covalent attachment includes chemical reactions such as amine, aldehyde, and thiol coupling reactions
CP70 cell	A human ovarian cancer cell line
C-reactive protein (CRP)	A protein made by the liver in response to inflammation
Cross-linking	The process of forming covalent bonds or relatively short sequences of chemical bonds to join two polymer chains together, the use of a probe to link proteins together
Cytometry	The measurement of the characteristics of cells like cell morphology (structure and shape), cell size, cell count, DNA content, cell cycle phase, and the presence or absence of specific proteins on the cell surface or in the cytoplasm
D-(+)-glucose	A monosaccharide which is the main source of energy in the form of adenosine triphosphate (ATP) for living organisms
Deionized water (DI water)	Water that has had almost all of its mineral ions removed, such as anions and cations
Deoxyribonucleic acid (DNA)	DNA is a molecule composed of two polynucleotide chains that form a double helix by coiling around each other. DNA carries genetic information which is coded by four chemical bases: adenine (A), cytosine (C), guanine (G), and thymine (T). DNA base pairs are always A-T and C-G.
Dichloromethane (DCM)	An organochloride compound with the chemical formula CH_2Cl_2

(Continues)

(*Continued*)

Context	Definition
Die singulation	Die singulation, cutting, or dicing is a process of separating a wafer containing multiple identical ICs to individual dies each including one of those ICs
Divinyl sulfone (DS)	A compound with the formula $C_4H_6O_2S$ used as a cross linking agent
DNA hybridization	The generation of a double-stranded nucleic acid (RNA-DNA duplex or a DNA double helix)
DNA probe	Stretches of ssDNA employed for detecting the presence of target nucleic acid sequences that are complementary sequences by hybridization
DNA sequencing	The process of determining the order of nucleic acid sequences (the order of the four bases: A, C, G, and T)
Dopamine (DA)	A hormone and a neurotransmitter with the molecular formula $C_8H_{11}NO_2$ that constitutes about 80% of the catecholamine content in the brain
Drop-cast method	A deposition method more controllable than spray coating for the fabrication of small-area films and coatings that relies on the release of large droplets with controlled sizes and momentum that spread and wet the surface upon impact
E. coli O157:H7	A serotype of the bacteria species *E. coli* and one of the Shiga-like toxin (Stx)-producing types of *E. coli*
Electrogenic	The production of a change in the electrical potential of a cell
Electrophysiology	Investigation of the electrical response characteristics of muscle and nerve
Emulsion polymerization	A kind of radical polymerization in which a monomer (or a mixture of monomers) is polymerized in an aqueous surfactant solution
Enzyme	A protein and a biological catalyst that speeds up the rate of a specific chemical reaction in the cell without being destroyed during the reaction and is used over and over
Enzyme-linked immunosorbent assay (ELISA)	A plate-based assay method for quantifying soluble substances like proteins, peptides, hormones, and antibodies
Epichlorohydrin (ECH)	An organochlorine compound and an epoxide
Epitope (antigenic determinant)	The portion of an antigen that is recognized by the immune system, specifically by antibodies, T cells, or B cells
Epoxy	Any of the basic components or cured end products of epoxy resins

(Continues)

(*Continued*)

Context	Definition
Epoxy resin (polyepoxide)	A class of reactive prepolymers and polymers including epoxide groups that may be reacted (cross-linked or cured) either with themselves through catalytic homopolymerization or with a wide range of co-reactants (hardeners or curatives)
Erythrocyte	A red blood cell that transfers O_2 and CO_2 to/from the body tissues
Escherichia coli (E. coli)	A gram-negative, rod-shaped, coliform bacterium of *Escherichia* that is usually found in the lower intestine of warm-blooded organisms
Escherichia virus T4	A species of bacteriophages that infect *E. coli* bacteria
Ethanol (EtOH)	An organic chemical compound and simple alcohol with the chemical formula C_2H_6O (or C_2H_5OH)
Ethanolamine	An organic chemical compound with the formula C_2H_7NO (or $HOCH_2CH_2NH_2$)
Ethylbenzene (EB)	An organic compound with the formula C_8H_{10} (or $C_6H_5CH_2CH_3$)
Ethylcellulose (EC)	A linear polysaccharide derived from cellulose in which some of the hydroxyl groups on the repeating glucose units are converted into ethyl ether groups
Ethylenediamine pyrocatechol (EDP)	An anisotropic etchant solution for silicon
Extracellular matrix (ECM)	A 3D network of extracellular macromolecules, like enzymes, collagen, and glycoproteins that provide biochemical and structural support to surrounding cells
Faraday's law	The amount of chemical variation occurring at an electrode–electrolyte interface is directly proportional to the current flowing through that interface
Ferrocene	An organometallic compound with the formula $C_{10}H_{10}Fe$ or $(Fe(C_5H_5)_2)$ which is soluble in most organic solvents
Fibroblast	A type of biological cell that synthesizes the collagen and extracellular matrix, produces the stroma for animal tissues, and plays an important role in wound healing
Field-of-view (FoV)	The angle through which the optical devices and sensors can pick up radiation (FoV allows for coverage of an area rather than a single focused point.)
Flow cytometer	A technique for detecting and measuring the chemical and physical characteristics of a population of particles or cells in a sample suspended in a fluid

(Continues)

(*Continued*)

Context	Definition
Fluorescent label (fluorescent tag, fluorescent probe)	A molecule usually comprising a fluorophore that can be chemically attached to a target biomolecule such as an antibody, an amino acid, or a protein and allows for its optical detection
Fluorophore	A fluorescent chemical compound with the ability to absorb light at a particular wavelength and re-emitting at a higher wavelength
Glucokinase (GLK)	An enzyme facilitating phosphorylation of glucose to glucose-6-phosphate
Glucose	A simple sugar with the chemical formula $C_6H_{12}O_6$, A subcategory of carbohydrates and the most abundant monosaccharide
Glucose oxidase (notatin)	An enzyme produced by certain species of fungi and insects that displays antibacterial activity in the presence of oxygen and glucose
Glutamate	A powerful excitatory neurotransmitter released by nerve cells to send signals to other cells that plays an important role in memory and learning under normal conditions
Glutaraldehyde (Glu)	A disinfectant, medication, fixative, and preservative with the formula $C_5H_8O_2$ (or OHC $(CH_2)_3CHO$)
Gold nanoparticle (GNP)	They are small gold particles with a diameter of 1 to 100 nm which, once dispersed in water, are also known as colloidal gold
Granulocyte	A type of white blood cells in the immune system characterized by the presence of specific granules in their cytoplasm
Growth medium (culture medium)	A solid, liquid, or semi-solid containing nutrients designed to support the growth of cells or microorganisms
H1299 cell line	A human cell lung carcinoma cell line established from a lymph node
Hek293 cell line	An embryonic kidney immortalized cell line
HeLa cell	An immortal cell line belonging to a strain that has been continuously cultured since 1951
Heparin	An anticoagulant that prevents the formation of blood clots
Hepatitis B surface antigen (HBsAg)	The surface antigen of the hepatitis B virus (HBV) that indicates current hepatitis B infection
Hepatocyte	The liver cells that perform endocrine, metabolic, and secretory functions
Hexane	A straight-chain alkane with six carbon atoms with the molecular formula C_6H_{14}
Horseradish peroxidase (HRP)	A metalloenzyme with many isoforms found in the roots of horseradish that catalyzes the oxidation of various organic substrates by hydrogen peroxide

(*Continues*)

(*Continued*)

Context	Definition
Human chorionic gonadotropin (hCG)	A hormone produced by trophoblast cells that are surrounding a growing embryo, which eventually forms the placenta after implantation and can be used for the maternal recognition of pregnancy
Human epidermal growth factor receptor 4 (Her4)	A prognostic marker of breast cancer
Hydrogel	A network of cross-linked polymer chains that are hydrophilic
Hydrogen peroxide	A peroxide with the formula H_2O_2 which is used as an oxidizer, antiseptic, and bleaching agent
Hydrophilicity	The physical property of a molecule that tends to be dissolved by water and is attracted to water molecules
Hydrophobicity	The physical property of a molecule that is not dissolved by water and is repelled from water molecules
Hydroxyapatite (HAP)	A naturally occurring mineral form of calcium apatite and a major constituent of hard tissues like teeth and bones with the formula $Ca_{10}(PO_4)_6(OH)_2$
Immunoglobulin G (IgG)	The main type of antibody found in blood that has two antigen binding sites and helps to prevent infections
Immunology	The study of the immune system
Immunosensor	The sensors that use antibodies as BRE and a transducer that converts the antibody–antigen binding event to a measurable physical signal
In vitro	The studies performed with cells, microorganisms, or biological molecules in a controlled environment outside their normal biological context
In vivo	The studies performed on whole living organisms or cells, as opposed to in a laboratory method that is done on a partial or dead organism
Indium tin oxide (ITO)	A widely used transparent conducting oxide composed of three different elements of oxygen, tin, and indium in varying proportions
Interferon-γ	A dimerized soluble cytokine and the only member of interferons type II that is used to test for tuberculosis
Interleukin 6 (IL6)	An endogenous chemical that is active in B cell maturation and in inflammation
Kapton	A polyimide film which remains stable across a wide range of temperatures, from $-269\ °C$ to $+400\ °C$
Lag period	The period of time between the introduction of a microorganism like cells into a culture medium and the time it starts increasing exponentially
L-alanine	L-isomer or L-enantiomer of alanine

(Continues)

(*Continued*)

Context	Definition
Layer-by-layer (LBL) deposition	A thin film fabrication method by depositing alternating layers of oppositely charged materials
L-serine	L-isomer or L-enantiomer of serine
Magnetic nanoparticle (MNP)	A class of nanoparticle, composed of a chemical component as well as a magnetic material like metals that can be manipulated using magnetic fields
MDA-MB-231 breast cancer cell	An epithelial, human breast cancer cell line
Metabolite	Any substance generated during metabolism (digestion or other bodily chemical processes)
Methanol (MeOH) or methyl alcohol	A chemical with the formula CH_4O (or CH_3OH)
Microcontact printing	A type of soft lithography that employs the relief patterns on a master PDMS stamp to create patterns of SAMs of ink on the surface of a substrate through conformal contact like in the case of nano-transfer printing
Microcystin	A class of toxins made by certain freshwater blue-green algae
Moiety	A fragment of a molecule, especially one that is composed of an identifiable unit
Molecularly-imprinted polymer (MIP)	An artificially synthesized affinity polymer formed by using the molecular imprinting method that has binding cavities complementary to the chosen template molecules in size, shape, and functionality
Monoclonal antibody	Antibodies that come from a single cell lineage are produced from cloning a unique white blood cell and have monovalent affinity, binding only to the same epitope
Morphology	The study of the shapes and arrangement of parts of organisms, to determine their function, their development, and how they may have been shaped by evolution
Mouse embryonic stem cells (mESC)	The cells derived from the inner cell mass of the developing the grade one embryo (blastocyst1)
Nano-imprint lithography (NIL)	A simple nanolithography process for making nanometer-scale patterns that creates patterns by mechanical deformation of imprint resist and subsequent processes
Neurochemical	A small organic molecule or peptide involved in neural activity
Neurotransmitter	Chemical substances released by neurons to transmit a message from a nerve cell across the synapse to a target cell and stimulate other neurons or muscle or gland cells
Neutrophil	The most abundant type of granulocytes in mammals and vital components of the immune system

(*Continues*)

(*Continued*)

Context	Definition
N-hydroxysuccinimide (NHS)	An organic compound with the formula $C_4H_5NO_3$ (or $(CH_2CO)_2NOH$) used as a reagent for preparing active esters in peptide synthesis
Nitrilotriacetic acid (NTA)	Aminopolycarboxylic acid with the formula $N(CH_2CO_2H)_3$
Nonpolar solvent	Nonpolar solvents include bonds between atoms with similar electronegativities, like hydrogen and carbon. They have low dielectric constants and are not miscible with water
Normal saline	A mixture of water and salt that contains 0.9% salt
Nuclear magnetic resonance (NMR)	A physical phenomenon used for determining the structure of organic molecules in which nuclei in a strong constant magnetic field are perturbed by a weak oscillating magnetic field and respond by generating an electromagnetic signal with a frequency characteristic of the magnetic field at the nucleus
Nucleic acid	The main information-carrying molecules of the cell that determine the inherited characteristics of every living thing by directing the process of protein synthesis and their two main classes are DNA and RNA
Oligomer	A molecule composed of a few identical repeating units that could be derived, conceptually, from copies of a smaller molecule (its monomer)
Oligonucleotide	A short single-stranded fragment of nucleic acid (like RNA or DNA), or similar fragments of analogs of nucleic acids (like Morpholinos or peptide nucleic acid)
Organic molecule	A type of molecule normally found in living systems
Organic solvent	Carbon-based substances capable of dispersing or dissolving one or more other substances
Oxygen plasma	Any plasma treatment carried out while introducing oxygen to the plasma chamber
Parasite	An organism living in or on a host organism that gets its food from or at the expense of its host
Paraxylylene	An aromatic hydrocarbon with the formula C_8H_{10} (or $C_6H_4(CH_3)_2$) and one of the three isomers of dimethylbenzene known collectively as xylenes
Parylene (poly paraxylylene)	A polymer obtained by polymerization of *para*-xylylene (*p*-Xylene). The most used variety is Parylene C

(*Continues*)

(Continued)

Context	Definition
Pathogen (infectious agent)	A biological agent, especially a virus, bacterium, or other microorganisms, that causes illness or disease to its host
Peptide	The amino acids that are the building blocks of certain proteins required by the skin, such as elastin and collagen
Perovskite	Any material with a crystal structure similar to the mineral adopted by many oxides that have the chemical formula ABO_3 like $SrTiO_3$ which is a cubic perovskite
Petroleum jelly	A semi-solid mixture of hydrocarbons
Phosphate-buffered saline (PBS)	A water-based salt solution commonly used in biological research to maintain a constant pH
Phosphorylation (of a molecule)	The attachment of a phosphoryl group
Physical adsorption (physisorption)	A process in which the electronic structure of the molecule or atom is perturbed upon adsorption
Plasmodium falciparum (P. falciparum)	The deadliest species of *Plasmodium* and a unicellular protozoan parasite of humans that causes malaria
Polar solvent	These solvents have large dipole moments. They contain bonds between atoms with very different electronegativities, such as hydrogen and oxygen
Poly(3,4-ethylenedioxythiophene) (PEDOT)	A conducting polymer based on 3,4-ethylenedioxythiophene (EDOT) which is an organosulfur compound with the formula $C_6H_6O_2S$ (or $C_2H_4O_2C_4H_2S$)
Poly(allylamine) hydrochloride (PAH)	A cationic and biodegradable polyelectrolyte prepared by the polymerization of allylamine
Poly(amidoamine) (PAMAM) dendrimer (4 generation)	A compacting agent to bring in the conformational variations in the DNA molecule
Poly(dimethylsiloxane) (PDMS)	The most widely used silicon-based organic polymer belonging to a group of polymeric organosilicon compounds with the formula $(C_2H_6OSi)_n$
Poly(epichlorohydrin) (PECH)	A compound with the formula C_3H_5ClO
Poly(methyl methacrylate) (PMMA) or acrylic	A transparent thermoplastic with the formula $(C_5O_2H_8)_n$
Poly(styrenesulfonate)	A strong anionic polyelectrolyte
Poly(vinylidene difluoride) (PVDF)	A highly nonreactive thermoplastic fluoropolymer generated by the polymerization of vinylidene difluoride
Polycarbonate	A group of thermoplastic polymers containing carbonate groups
Polyclonal antibody	Antibodies that are secreted by different B cell lineages and are a collection of immunoglobulin molecules that react against a specific antigen, each identifying a different epitope

(Continues)

(Continued)

Context	Definition
Polyelectrolyte	Polymers whose repeating units produce an electrolyte group and are used in the formation of new types of materials known as PEMs using an LBL deposition technique
Polyether urethane	A type of polyurethane which is an elastomer with excellent hydrolytic resistance
Polyethylene terephthalate (PET)	The most common thermoplastic polymer resin of the polyester family with the formula $(C_{10}H_8O_4)_n$
Polyimide (PI)	A polymer of imide monomers with high heat resistance
Polystyrene	A synthetic aromatic hydrocarbon polymer with the formula $(C_8H_8)_n$ made from the monomer known as styrene
Polyurethane	A polymer comprising organic units joined by carbamate (urethane) links which has two main types: polyester and polyether (polyester urethane and polyether urethane are elastomers)
Potential of hydrogen (pH)	A scale employed to specify the basicity or acidity of an aqueous solution
Propylene glycol monoether acetate (PGMEA)	A photoresist solvent for the application of surface adherents on silicon wafers
Protein	Essential nutrients for the human body and one of the building blocks of body tissue that can also serve as a fuel source
Quantum dot (Qdot)	Man-made nanoscale crystals that can transport electrons (When UV light hits these artificial semiconductor nanoparticles, they can emit light of various colors.)
Receptor	Protein molecules on the target cell surface or its inside that receive a chemical signal
Ribonucleic acid (RNA)	A polymeric molecule essential in coding, decoding, regulation, and expression of genes which is similar to DNA, but unlike DNA, it is single-stranded
Saccharide	An organic compound including a sugar or sugars
Salmonella	A genus of gram-negative rod-shaped bacteria of Enterobacteriaceae
Self-assembled monolayer (SAM)	A molecule thick layer of material that bonds to a surface in an ordered way as a result of physical or chemical forces during a deposition process
Serine	An α-amino acid that is used in the biosynthesis of proteins with the formula $C_3H_7NO_3$
Serum	The fluid and solute component of blood which does not play a role in clotting and does not contain white blood cells, red blood cells, platelets, or clotting factors

(Continues)

(*Continued*)

Context	Definition
Signaling molecule (often called ligand)	Molecules that bind specifically to other molecules (like receptors) and the messages carried by them are often relayed through a chain of chemical messengers inside the cells
Silane	An inorganic compound with chemical formula SiH_4
Single-stranded DNA (ssDNA)	Each of the two strands of DNA molecules
SKBR3 cells	A human breast cancer cell line
Soft lithography	A family of techniques for making or replicating structures using molds, conformable photomasks, and elastomeric stamps
Solid-phase synthesis	A technique in which molecules are covalently bound on a solid support material and synthesized step-by-step in a single reaction vessel employing selective protecting group chemistry
Sonication	The process of applying sound energy to agitate particles in a liquid sample
Spacer DNA	A region of non-coding DNA between genes
Staphylococcus aureus (S. aureus)	A gram-positive, round-shaped bacterium and a member of the Firmicutes that is usually found on the skin and in the upper respiratory tract
Staphylococcus epidermidis (S. epidermidis)	A gram-positive bacterium, and one of the species belonging to the genus *Staphylococcus*
Streptavidin	A tetrameric protein derived from the bacterium *Streptomyces avidini* that shows extraordinary affinity for biotin
Strontium titanate	An oxide of titanium and strontium with the chemical formula $SrTiO_3$
SU8	An epoxy-based negative photoresist
Surface grafting (in polymer chemistry)	The addition of polymer chains onto a surface
Surface plasmon resonance (SPR)	A phenomenon useful for measuring adsorption of material onto the surface of metal nanoparticles or onto planar metal surfaces, where the conduction electrons in the metal surface are stimulated by photons of incident light with a certain angle of incidence and propagate parallel to the metal surface
Tetrabutylammonium hexafluorophosphate (TBAPF$_6$)	A type of salt with the formula NBu_4PF_6
Tetramethyl ammonium hydroxide and water (TMAHW)	An anisotropic silicon etchant useful for silicon micromachining as an alternative to the more commonly etchants like EDP
Thiol	The sulfur analogue of alcohols containing a sulfur atom in place of the oxygen atom in the hydroxyl group of an alcohol
Thrombin	An enzyme that acts both as an anticoagulant and a procoagulant for regulating hemostasis and maintaining blood coagulation

(*Continues*)

(*Continued*)

Context	Definition
Toluene (TOL)	An aromatic hydrocarbon
Transferrin	The glycoproteins found in vertebrates which bind to and mediate the transport of iron through blood plasma
Triethylamine	The chemical compound with the formula $C_6H_{15}N$ (or $N(CH_2CH_3)_3$)
Tumor necrosis factor (TNF)	A cell-signaling protein (cytokine) involved in systemic inflammation and one of the cytokines that make up the acute phase reaction
Tyramine	A compound generated by the breakdown of an amino acid called tyrosine that helps to regulate blood pressure
Vascular endothelial growth factor (VEGF)	A signal protein generated by cells stimulating the formation of blood vessels
Yeast cell	Single-celled microorganisms categorized as members of the fungus kingdom
Zika virus	A mosquito-borne flavivirus
α-Adducin	α-Adducin is a protein that in humans is encoded by the ADD1 gene
β-Mercaptoethanol	A chemical compound with the formula C_2H_6OS (or $HOCH_2CH_2SH$)

List of frequently used acronyms and abbreviations

Explanation	Acronym/ abbreviation
(3-Aminopropyl) tris (trimethylsiloxy) silane	APTTMS
(3-Glycidoxypropyl)trimethoxysilane	GOPTS
1,4-Phenylene diisothiocyanate	PDITC
11-Mercapto-1-undecanol	MCU
11-Mercaptoundecanonic acid	MUA
1-Dodecanethiol	DDT
1-Ethyl-3-(3-dimethylaminopropyl)-carbodiimide	EDC
2-(N-morpholino)ethanesulfonic acid	MES
3-(Aminopropyl) trimethoxysilane	APTMS
3-[2-(2-Aminoethylamino) ethylamino] propyltrimethoxysilane	AEEPTMS
3-Aminopropyltriethoxysilane	APTES
4-(2-Hydroxyethyl)-1-piperazineethanesulfonic acid	HEPES
6-Mercapto-1-hexanol	MCH
Acetone	ACE
Acetonitrile	MeCN
Acrylamide	AAm
Action potential	AP
Alpha mouse liver 12	AML-12
Alternating current	AC
Aluminum	Al
Aminophenylboronic acid	APBA
Aminopropyldimethylmethoxysilane	APDMMS
Analog-to-digital converter	ADC
Antibody	Ab
Antibody-conjugated magnetic bead	MB-Ab
Application-specific integrated circuit	ASIC
Atomic layer deposition	ALD
Austria Mikro System	AMS
Azobisisobutyronitrile	AIBN
Back-end-of-line	BEOL
Benzocyclobutene	BCB
Biological recognition element	BRE
Biological/chemical phenomenon	BCP
Capacitance-to-digital converter	CDC
Capacitance-to-frequency converter	CFC

(Continues)

(*Continued*)

Explanation	Acronym/abbreviation
Capacitance-to-voltage converter	CVC
Carbon nanotube	CNT
Carboxyphenylboronic acid	CPBA
Cardiac-troponin T	cTnT
Cerium	Ce
Charge injection-induced error-free	CIEF
Charge-based capacitance measurement	CBCM
Chemical vapor deposition	CVD
Chloroform	CHL
Chromium	Cr
Classical swine fever virus	CSFV
CMOS image sensor	CIS
Colony-forming units	cfu
Common-mode feedback	CMFB
Common mode	CM
Complementary metal-oxide semiconductor	CMOS
Computer-aided design	CAD
Continuous time	CT
Copper	Cu
Correlated double sampling	CDS
C-reactive protein	CRP
Current-controlled oscillator	CCO
Current-to-voltage (converter)	I/V
Cyclic voltammetry	CV
Data acquisition	DAQ
Deep reactive ion etching	DRIE
Deionized water	DI water
Deoxyribonucleic acid	DNA
Design rule check	DRC
Diacrylate bisphenol A	DABA
Dichloromethane	DCM
Dielectrophoresis	DEP
Digital-to-analog converter	DAC
Digital-to-capacitance converter	DCC
Direct current	DC
Direct-write fabrication process	DWFP
Divinyl sulfone	DS
Dopamine	DA
Dry-film resist	DFR
Electric double layer	EDL
Electrochemical deposition	ECD
Electrochemical impedance spectroscopy	EIS
Electroless nickel immersion gold	ENIG
Enzyme-linked immunosorbent assay	ELISA
Electrical signal	ES
Escherichia coli	*E. coli*
Ethanol	EtOH

(*Continues*)

(*Continued*)

Explanation	Acronym/abbreviation
Ethylbenzene	EB
Ethylcellulose	EC
Ethylene glycol dimethacrylate	EGDMA
Ethylenediamine pyrocatechol	EDP
Ethylenediaminetetraacetic acid	EDTA
Extracellular matrix	ECM
Field-of-view	FoV
Field-programmable gate array	FPGA
Fluorescence energy transfer	FRET
Frequency response analysis	FRA
Gadolinium	Gd
Glucokinase	GLK
Glutaraldehyde	Glu
Gold	Au
Gold nanoparticle	GNP
Hanks' balanced salt solution	HBSS
Hepatitis B surface antigen	HBsAg
Hepatitis B virus	HBV
Horseradish peroxidase	HRP
Human chorionic gonadotropin	hCG
Human epidermal growth factor receptor 4	Her4
Human immunodeficiency virus	HIV
Hydrogen fluoride	HF
Hydroxyapatite	HAP
Immunoglobulin G	IgG
Indium tin oxide	ITO
Injection-locked oscillator	ILO
Inner Helmholtz plane	IHP
In-phase (signal)	I
Input dynamic range	IDR
Input/output	I/O
Integrated circuit	IC
Interdigitated electrode	IDE
Interdigitated microelectrode array	IDMA
Interleukin 6	IL6
Intermediate frequency	IF
Ion-selective field-effect transistor	ISFET
Iron oxide	Fe_3O_4
Isopropyl alcohol	IPA
Isotropic conductive adhesive	ICA
Laboratory-on-chip	LoC
Laser-induced forward transfer	LIFT
Layer-by-layer	LBL
Least-significant bit	LSB
Lens-free Ultra-wide-field Cell monitoring Array platform based on Shadow imaging	LUCAS
Light-emitting diode	LED

(Continues)

(*Continued*)

Explanation	Acronym/ abbreviation
Limit of detection (or detection limit)	LoD
Linear feedback shift register	LFSR
Localize backside etching	LBE
Low-pass filter	LPF
Low-temperature co-fired ceramic	LTCC
Low-temperature oxide	LTO
Luria–Bertani medium	LB medium
Mach–Zehnder interferometer	MZI
Magnetic bead	MB
Magnetic nanoparticle	MNP
Metal-oxide-metal	MOM
Metal-oxide-semiconductor field-effect transistor	MOSFET
Methacrylic acid	MAA
Methanol	MeOH
Microelectrode array	MEA
Microelectromechanical system	MEMS
Micro-nuclear magnetic resonance	μNMR
Micro-total analysis system	μTAS
Molecularly imprinted polymer	MIP
Mouse embryonic stem cells	mESC
Multi-labs-on-a-single chip	MLoC
Mycobacterium avium Complex	MAC
N-(3-aminopropyl)methacrylamide hydrochloride	NAPMA
N,N'-methylenebis(acrylamide)	MBAm
N,N-diethyldithiocarbamic acid benzyl ester	INIFERTER
Nano-imprint lithography	NIL
Nanoparticle	NP
Nanorod	NR
N-ethyldiisopropylamine	EIPA
N-hydroxysuccinimide	NHS
Nickel	Ni
Nitrilotriacetic acid	NTA
Non-return-to-zero	NRZ
N-type metal-oxide-semiconductor	NMOS
Nuclear magnetic resonance	NMR
Operational amplifier	Op-amp
Otter Helmholtz plane	OHP
Oxide/nitride/oxide	ONO
Palladium	Pd
Pentaerythritol tetrakis(3-mercaptopropionate)	PETMP
Personal computer	PC
Phase-locked loop	PLL
Phosphate-buffered saline	PBS
Physical vapor deposition	PVD
Plasma-enhanced atomic layer deposition	PEALD
Plasma-enhanced chemical vapor deposition	PECVD
Plasmodium falciparum	*P. falciparum*

(*Continues*)

(*Continued*)

Explanation	Acronym/ abbreviation
Platinum	Pt
Point-of-care	PoC
Poly(3,4-ethylenedioxythiophene) doped with Poly(styrenesulfonate)	PEDOT:PSS
poly(allylamine) hydrochloride	PAH
Poly(amidoamine)	PAMAM
Poly(cyanopropylmethylsiloxane)	PCM
Poly(dimethylsiloxane)	PDMS
Poly(epichlorohydrin)	PECH
Poly(ethylene glycol)	PEG
Poly(ethylene glycol) 400-dimethacrylate	PEG-400 DMA
Poly(ethyleneimine)	PEI
Poly(methyl methacrylate)	PMMA
Poly(sodium styrene sulfonate)	PSS
Poly(vinylidene difluoride)	PVDF
Polyelectrolyte multilayer	PEM
Polyether urethane	PEUT
Polyethylene terephthalate	PET
Polyimide	PI
Porous silicon	PSi
Potential of hydrogen	pH
Printed circuit board	PCB
Propylene glycol monoether acetate	PGMEA
Prostate-specific antigen	PSA
Proton nuclear magnetic resonance	H NMR
P-type metal-oxide-semiconductor	PMOS
Quadrature (signal)	Q
Quadrature voltage-controlled oscillator	QVCO
Quantum dot	Qdot
Radio frequency	RF
Radio frequency identification	RFID
Reactive ion etch	RIE
Reference electrode	RE
Ribonucleic acid	RNA
Ring oscillator	RO
Self-assembled monolayer	SAM
Single-sensor evaluation module	SEM
Sigma-Delta	$\Sigma\Delta$
Signal-to-noise ratio	SNR
Silicon	Si
Silicon dioxide	SiO_2
Silicon nitride	Si_3N_4
Silver	Ag
Single-stranded DNA	ssDNA
Single-walled carbon nanotube	SWNT
Sodium carboxymethyl cellulose	Na-CMC
Sodium chloride	NaCl
Species	spp.

(Continues)

(*Continued*)

Explanation	Acronym/ abbreviation
Spermidine/spermine N1 acetyltransferase	SSAT
Spin-on-dielectrics	SOD
Spin-on-glasses	SOG
Staphylococcus aureus	*S. aureus*
Staphylococcus epidermidis	*S. epidermidis*
Stereolithography	SLA
Strontium titanate perovskite	$SrTiO_3$
Successive-Approximation Register	SAR
Surface plasmon resonance	SPR
Switched capacitor	SC
Taiwan Semiconductor Manufacturing Company	TSMC
Tantalum	Ta
Tape automated bonding	TAB
Tetrabutylammonium hexafluorophosphate	$TBAPF_6$
Tetrahydrocannabinol	THC
Tetramethyl ammonium hydroxide and water	TMAHW
The primary antibody	Ab1
The secondary antibody	Ab2
Three-dimensional	3D
Titanium nitride	TiN
Titanium-tungsten	TiW
Toluene	TOL
Transimpedance amplifier	TIA
Trimethylolpropane trimethacrylate	TRIM
Tris(2-carboxyethyl)phosphine	TCEP
Troponin I	cTnI
Tryptic soy broth	TSB
Tumor necrosis factor	TNF
Two-dimensional	2D
Two-polysilicon-four-metal	2P4M
Ultraviolet	UV
User interface software	UIS
Variable gain amplifier	VGA
Vascular endothelial growth factor	VEGF
Voltage-controlled oscillator	VCO
Voltage-to-current (converter)	V/I
Working electrode	WE
Zero-resistance ammeter	ZRA
Zinc	Zn

Parameters

Parameter	Definition	Page number
μ_p	The whole mobility coefficient	77
A_{amp}	The gain of the amplifier	92
A_c	The overlapping area of the electrodes	37
A_{CCO}	The gain of the CCO	96
A_I	The gain of current mirror	77
a_{IDE}	The width of the stripped area of IDE	39
A_{inv}	The gain of the inverter	56
A_{TIA}	The gain of the TIA	63
A_u	The unit value of the integrating parts	94
C	Capacitance	37
C_{12}	The parasitic capacitances between two combs of IDE named E_1 and E_2	163
C_{1s}	The parasitic capacitances between one comb of IDE named E_1 and the substrate	163
C_{2s}	The parasitic capacitances between one comb of IDE named E_2 and the substrate	163
C_{bio}	A combination of the surface functionalization layer, BREs, grafting molecules, binding of analytes, and any contribution of the Helmholtz layer	36
C_{cell}	Cell layer capacitance	37
C_{dl}	EDL capacitance	35
C_E	The spreading capacitance of the electrode to the electrolyte	100
C_f	Fringe capacitance	37
C_{fb}	Feedback capacitor	54
C_{GC}	Diffuse layer capacitance in Gouy–Chapman model	35
C_{gm}	The total of all capacitive couplings of the growth medium above the cell layer to all the neighboring conductors	38
C_{gs}	Gate-source capacitance	71
C_H	Helmholtz capacitance	35
C_I	Interfacial capacitance	36
C_i	The capacitance of the node N_i	54
C_{ins}	The capacitance of the insulating layer	36
C_{int}	Integrating capacitor	81
$C_{int\pm}$	The integrating capacitances used in a differential mode	88
C_{inter}	The dielectric interconnect capacitance	40
C_{os}	Offsetting capacitor	72

(Continues)

(*Continued*)

Parameter	Definition	Page number
C_{ox}	The gate capacitance per unit area	77
C_p	Parasitic capacitance	56
C_{pi}	The parasitic capacitances of the node N_i	53
C_R	Reference capacitance	54
C_S	Sensing capacitance	53
C_{sensed}	Sensed capacitance	37
C_{sol}	The electrolyte solution capacitance	40
$C_{standing}$	Standing capacitance	72
C_{stray}	Stray capacitance	40
C_{surf}	Surface capacitance	30
C_{tot}	Total capacitor	36
d_c	The distance between the plates of the electrodes	37
d_{OHP}	The distance of the OHP from the metal electrode	35
e_n	Current error	85
f_0	Initial frequency	59
f_{CCO}	The output frequency of CCO	96
F_{CCO}	The average of the output pulse frequencies of CCO	96
f_{int}	Integrating frequency	88
f_{Out}	Output frequency	57
f_{ref}	Reference frequency	61
f_S	Sampling frequency	67
I_B	Current reference	57
i_{CCO}	The input current of CCO	94
I_{D0}	A process-dependent parameter in the subthreshold region	86
I_{inj}	The current of the injection source or injection current	90
I_{osc}	The current of the oscillator	62
i_R	Reference current	67
I_R	The average of the reference current	67
i_S	Sensing current	67
I_S	The average of the sensing current	67
J_0	Zeroth Bessel function of the first kind	39
k_B	Boltzmann constant	36
L_D	Debye length	35
L_i	The length of the transistor M_i	78
l_{IDE}	The length of the fingers of IDE	38
M	The number of RO stages	58
n	The subthreshold slope factor	86
N	The number of the cycles of sampling time	92
N_{Count}	Total number of pulses counted by the counter	96
N_{div}	Division ratio	61
n_{IDE}	The numbers of the fingers of IDE	38
Q	Total charge	99
q	Electronic charge	36
Q_{LC}	Quality factor of LC tank	62
q_n	Output pulse of the comparator	85
R_E	The spreading resistance of the electrode to the electrolyte	100
R_{leak}	Insulator resistance	28

(*Continues*)

(Continued)

Parameter	Definition	Page number
R_P	Potentiometer	84
R_R	Reference resistance	92
R_S	Sensing resistance	92
R_{sol}	The electrolyte solution resistance	28
$S_{\Delta V/\Delta C}$	The sensitivity of the output voltage versus the capacitance changes	83
T	Temperature	36
t	Time	84
t_{IDE}	The thickness of IDE	38
T_m	The interval of unit	95
T_{osc}	Period of oscillator	62
T_S	Sampling period	67
V_0	The potential at the electrode	36
V_A	The amplitude of a pulse signal	55
V_{Am}	The voltage dropped over the ammeter	67
V_C	Control voltage	61
V_{cm}	Common-mode voltage	81
V_{dd} and V_{ss}	Dual supply voltages	53
V_{exc}	Excitation voltage signal	63
V_i	The voltage of the node N_i	63
V_I	In-phase component of voltage	63
V_{N1}	Switching threshold voltage	56
V_{off}	Offset voltage	84
$V_{off\pm}$	The offset voltages in a differential mode	88
V_{osc}	Certain threshold voltage of oscillator	58
V_{Out}	Output voltage	53
$V_{Out\pm}$	Differential output voltages	88
V_Q	Quadrature component of voltage	63
V_{ref}	Reference voltage	54
V_{sgi}	The source-gate voltage of the transistor M_i	77
V_T	Thermal voltage	36
V_{th}	Threshold voltage	57
V_{thp}	The threshold voltage of P-channel transistors	77
V_{tog}	Toggling voltage	95
W_i	The width of the transistor M_i	78
w_{IDE}	The width of the fingers of IDE	39
$w_{ss}(x,y)$	The displacement of the flexible electrode of surface stress-based device	41
Y_S	Sensing admittance	63
z	The valence of the ion	36
Z_S	sensing impedance	63
ΔC	Capacitance variation	59
ΔC_0	The mismatch of the sensing and the reference capacitors	73
$\Delta C_{tot}(f_0)/C_{tot}$	The relative shift in the total capacitance with respect to the C_{tot}	60
Δf	Frequency shift	59
$\Delta f/f_0$	Relative frequency shift	59
Δi	The difference between sensing and reference current	82

(Continues)

(*Continued*)

Parameter	Definition	Page number
$\Delta i_{\pm}$	The differences between the sensing and the reference currents in a differential mode	88
ΔR	Resistance variation	92
$\Delta v/\Delta t_0$	The slope of the time-variant voltage	65
ΔV_{off}	The difference between the offset voltages of a differential mode	89
ε	Permittivity	38
ε_0	The vacuum permittivity	37
ε_r	The relative permittivity of the dielectric material	37
λ_{IDE}	IDE spatial wavelength or the periodicity of the fingers	39
σ_n	Noise	22
τ	Time constant	84
τ_d	Delay of comparator	57
ω	Angular frequency	63
ω_0	Free-running angular frequency	62
ω_{inj}	The angular frequency of injection source	62

Chapter 1

Introduction

Complementary metal-oxide semiconductor (CMOS)-based capacitive biosensors are among the most widespread types of CMOS sensors that have been developed for various biosensing applications. Charge-based capacitance measurement (CBCM) as a highly accurate and scalable capacitive sensing technique has attracted much attention in the field of life sciences. This book gives an overview of different parts of capacitive biosensors, notably core-CBCM ones, as well as their applications.

1.1 CMOS sensors

Emerging and development of CMOS technologies have opened up new horizons on life sciences and demonstrated great advantages for the design and implementation of a single small chip featuring millions of sensors such as Ion Torrent™ deoxyribonucleic acid (DNA) sequencer [1] which is the third-generation DNA sequencing system including an array of millions of ion-selective field-effect transistors (ISFETs) using standard 0.35 CMOS process [2].

CMOS foundries, with huge investment, have paved the way for efficient mass production of various platforms comprising microelectronic devices whose costs are low enough to be affordable for the end users. The rising trend in their scale of integration, predicted based on Moore's law, allows for the monolithic integration of several active elements like ISFETs and microelectrodes, signal conditioning and processing circuits on a single chip, and making highly dense systems with high signal-to-noise ratios (SNRs) due to the reduced external noises and parasites. The distinct cost, accessibility and reliability of CMOS technology are the reasons why their replacement with other technologies would not be economical. Additionally, this technology has been developed for providing some electrical characteristics such as high-speed, low-noise, or high-frequency electrical circuits which can also be useful for biochemical applications. So, it is more economical to adapt the new biochemical platforms with standard CMOS technology rather than introducing a new technology dedicated to such applications.

CMOS technology offers the advantage of implementing an array of a large number of sensors and their associated electronic circuitry to make a single laboratory-on-chip (LoC) device to speed up the time-consuming biological analysis, such as cancer detection [3,4], neurochemical detection [5], DNA analysis [6], and

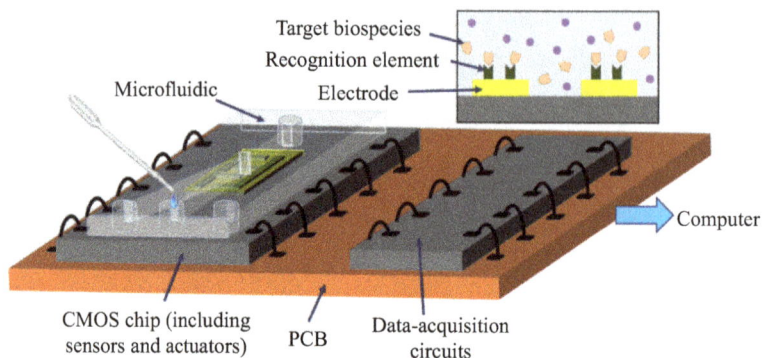

Figure 1.1   Conceptual view of a simple hybrid CMOS-microfluidics system for LoC applications

continuous glucose monitoring [7]. LoC technology has revolutionized biochemical assays by miniaturization and automation of the equipment of traditional laboratories. The main problem in designing CMOS biosensors is that they should not be in direct contact with analytes due to the biochemical contamination and imposed uncertainty and instability to microelectronic characteristics. Thanks to developments in both CMOS and microfluidic technology, hybrid CMOS-microfluidic techniques have emerged to overcome this obstacle. Hybrid CMOS-microfluidic LoC platforms (as shown in Figure 1.1) are being developed to prepare, control, and sense biochemical samples and perform high throughput screening with only a tiny amount of them. CMOS sensors are the critical elements of LoC systems as seen in Figure 1.1. While microfluidics help to direct or even prepare the fluidic sample toward sensing/actuating sites, the microelectronics provide actuators, sensors, and the required readout circuits. Actuators stimulate and manipulate bioparticles in a fluidic sample by using mechanical [8], magnetic [9], or electrical [10] forces, whereas sensors measure the physical changes (like optical [11], magnetic [12], or capacitive [13]) sensed by their transducers.

To date, there is a vast amount of literature on the CMOS sensors developed for LoC applications using optical [11,14], magnetic [12,15], potentiometric [16,17], impedimetric [18,19], capacitive [13,20,21] techniques using opto-diodes, microcoils, ISFETs, and microelectrodes as transducers, respectively (see Figure 1.2).

In affinity-based biosensors, the physiochemical transducer is incorporated with a biological recognition element (BRE) or probe such as a DNA probe [22], antibody [23], or bacteriophage [24] and undergo physiochemical variations as a result of selective bonding between the BRE and target biospecies of biological/chemical sample. So, the challenge of detecting the target in the sample will be transferred to detecting an electrical signal [25]. Finally, a readout instrument or circuit measures the characteristics of the transducer and amplifies, processes, and converts them to readable signals like voltage, current, and digital data.

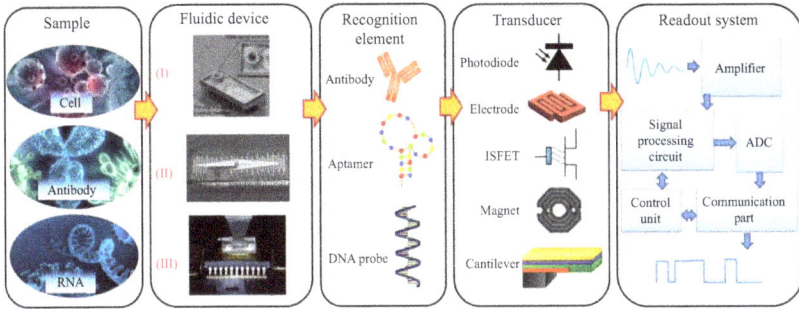

Figure 1.2   An illustration of a CMOS biosensor (the photos (I), (II), and (III) of the integrated CMOS-microfluidic systems shown in fluidic block are extracted from [26], [27], and [28], respectively)

1.1.1   Optical

Optical detection methods especially fluorescence imaging using fluorescent labels attached to the target are among the most popular methods in laboratories [29]. Most of these methods suffer from complex fluorescent labeling processes, high-cost, time-consuming and bulky instruments like optical scanner system comprising light sources, lenses, and cameras [29]. Table 1.1 shows some of the recent advances of CMOS optical biosensors.

In fluorescence-based techniques, the analyte is stimulated by an excitation light and the frequency change of the radiation emission is monitored [38]. The presence of specific targets might be transduced by using an added dye or directly by a change of reaction without any label. Fluorescence energy transfer is another approach based on the generation of a unique fluorescence signal by one of two paired fluorophores with overlapped emission and excitation wavelengths within a small distance (about few angstroms) from each other which is stimulated by the excitation of the pairing one [38,39]. A transducer like a photodetector is used to convert the emitted signal to electrical signals. Photodetectors can be implemented by embedded PN-junctions in the standard CMOS technology (Figure 1.3(a)). To integrate fluorescence modules into a single CMOS chip, it is noticeable that the optical signals should not be blocked by the metal layers above the photodetector. Also, if the intensity of the excitation signal is stronger than the fluorescent signal, the photodetector might be saturated and causes system malfunction [39]. Using optical filters is one of the solutions. For instance, Jang *et al.* [40] employed a thin-film long-pass optical filter atop the chip and also a fiber-optical faceplate to direct the signal. In another effort, Ta-chien *et al.* [41] proposed to use time-resolved fluorescence detection without using any external filters. Hong *et al.* [42] could fully integrate a CMOS fluorescence biosensor by utilizing an on-chip nanophotonic filter. The same group [36] reported an integrated optical sensor that did not require any post-processing and external collimators, lenses, or optical filters (Figure 1.3(b)). However, it required an off-chip light source like a continuous-wave diode laser or a

Table 1.1   A comparison of some CMOS optical biosensors

Application	Target	Label	Array #	Ref.
SPR for detection of pathogens	*E. coli* and *S. aureus*	–	–	[30]
Lens-free on-chip microscopy using a fiber-optic array	*P. falciparum*	–	23	[31]
Parallel on-chip monitoring of various cell types	Blood cells, NIH/3T3 fibroblasts, murine embryonic stem cells, AML-12 hepatocytes	–	4	[32]
High-throughput lens-free imaging and characterization of a heterogeneous cell solution on a chip	Red blood cells, fibro-blasts, hepatocytes, mESC and polystyrene microbead, yeast cells	–	–	[33]
Lens-free imaging for PoC testing	CD4$^+$ T-lymphocyte	–	–	[34]
Lens-free holographic imaging for on-chip cytometry and diagnostics	Heterogeneous solutions of red blood cells, *E. coli*, yeast cells, and various sized microparticles	–	–	[35]
In vitro and *in vivo* fluorescence biosensing	–	Qdot	–	[36]
Identifying microscopic disease during cancer removal	Breast and prostate cancer cells	Qdot	32 × 32	[14,37]
Total serum cholesterol quantification	Cholesterol	Active enzyme	16 × 16	[11]

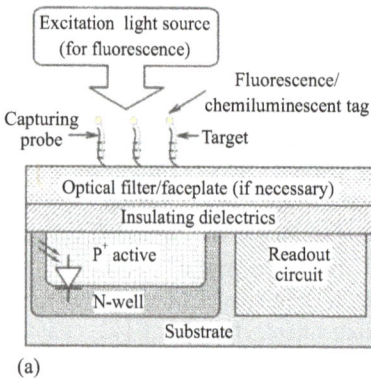

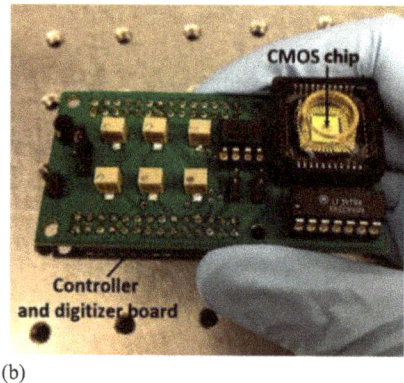

Figure 1.3   (a) Fluorescence/chemiluminescence imaging; (b) fully integrated fluorescence-based biosensor with nanoplasmonic filters [36]

light-emitting diode (LED); 48 zeptomoles of streptavidin-coated quantum dot (Qdot)-based fluorophores could be detected on the chip surface.

In another effort, Papageorgiou *et al.* [37] presented an angle-selective fluorescence imager comprising 32 × 32 pixels for blur reduction. They fabricated micro-collimators above each photodiode and designed it for near-field tumor cell imaging and used a custom optical filter. The same group [14] improved the performance of their image sensor for imaging prostate and breast cancer cells. In another work, Al-Rawhani *et al.* [11] reported a colorimetric sensing platform including off-chip LED and a CMOS photodiode array for total cholesterol quantification in blood serum which worked based on the color change of enzyme assays.

In chemiluminescence, the chemiluminescent tags avert the external light source to excite the chemical reaction and prevent the saturation of the photodiode. These sensors can use photodetectors to detect the emitted chemiluminescent signals and do not need optical filters. Also, the distance from the tag to the surface of the photodetector can be shortened that paves the way for supercritical angle luminescence and efficient signal detection [39,43].

Some lens-free techniques such as contact imaging, lens-free holographic microscopy, and shadow imaging have been developed to eliminate bulky and expensive optics like lenses and mirrors. In contact imaging, the object close to the focal plane is imaged with the resolution in the order of the pixel size. On-chip shadow imaging and holographic microscopy take advantage of the shadows of small objects [44]. Shadow imaging is based on the diffraction patterns that are generated when optically semitransparent micro-objects (like cells) located on top of an optoelectronic image sensor are illuminated by an incoherent or partially coherent light source. In this approach, there is no need for the magnification of the original optical plane where the targets are located and it can fully monitor the whole active imaging region [45]. On the other hand, lens-free holographic devices take advantage of digital refocusing of object waves to image an object plane regardless of optical diffraction and computer algorithms are used to reconstruct the original image from the recorded hologram [46].

CMOS lens-free imaging technology is reported for different biomedical applications like cardiotoxicity screening [47], real-time monitoring of cell growth [48], and analyzing the diffraction signature of different cells [33]. For instance, as mentioned in Table 1.1, a high-resolution lens-less microscope was proposed by Bishara *et al.* [31] for the real-time monitoring of malaria parasite infection (*Plasmodium falciparum* (*P. falciparum*)) in thin blood smears. This imaging system consists of 23 LEDs, a CMOS sensor, and a color filter. Utilizing a holographic technique with an improved numerical algorithm helped to achieve sub-pixel resolution (<1 μm).

Lens-free Ultra-wide-field Cell monitoring Array platform based on Shadow imaging (LUCAS) is one of the significant techniques in the field of CMOS cell imaging. Figure 1.4(a) shows a shadow imaging method. This technique does not require a laser as a light source and even an LED can be utilized for illumination. In comparison to lens-based microscopy, the LUCAS-based technique has a large

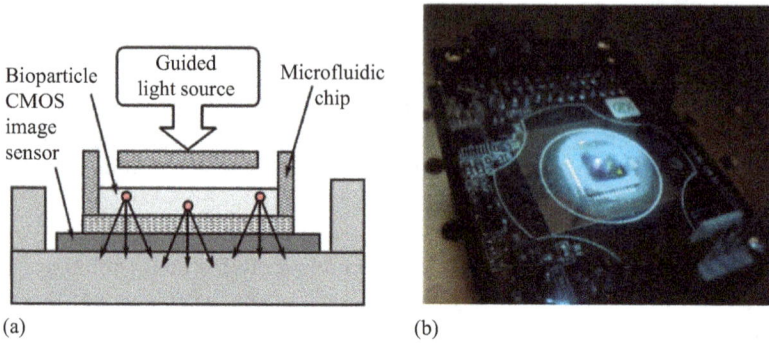

Figure 1.4   (a) Shadow imaging method; (b) experimental apparatus of the holographic-LUCAS platform used in [35]

field-of-view (FoV) and does not require scanning stages and lenses [49]. For instance, Ozcan *et al.* [32] reported a high-throughput array using a LUCAS technique that does not need microscope imaging and fluorescent tagging. The cells are exposed to a regular incoherent white-light, and after a diffraction-limited propagation, the sensor array records the shadow pattern of each cell. This technique was only applicable to homogenous cell solutions. The same group [33] improved this platform by a custom-developed decision algorithm enabling to identify the particle location in three-dimensional (3D) space and characterize various cell types within a heterogeneous solution. Additionally, they improved the sample volume (of ~4 mL) and the depth-of-field (of ~4 mm). In another effort, Seo *et al.* [35] proposed a lens-free and LUCAS-based holographic microscope for on-chip cytometry of heterogeneous solutions of red blood cells, *Escherichia coli* (*E. coli*), yeast cells, and various sized microparticles (Figure 1.4(b)). In this work, a high-resolution CMOS image sensor (CIS) array recorded a two-dimensional (2D) holographic diffraction pattern of each cell or microparticle. In contrast to other LUCAS-based techniques, this system utilized spatially coherent illumination instead of incoherent light which resulted in an improvement in the signature differences among different cell types, the signature uniformity of the cell, the SNR, and resolution of the system. In another work, Moon *et al.* [34] used a shadow imaging technique and proposed a microfluidic-integrated platform for automatically counting and enumerating target $CD4^+$ T-lymphocytes from whole blood for human immunodeficiency virus (HIV) point-of-care (PoC) testing (see Table 1.1). The immobilized anti-CD4 antibody on one side of the microfluidic chip helped to capture rare cells selectively and the imaging platform detects the captured cells rapidly.

There are also some other optical techniques using CISs. For example, interferometer-based biosensors measure the change of the refractive index by the propagation of light through an optical waveguide. For example, Blanco *et al.* [50] integrated a 3D microfluidic network with a wafer containing Mach–Zehnder interferometer (MZI) nano-photonic biosensor devices for label-free biochemical detection (Figure 1.5(a)). A typical MZI transducer comprises two branches called

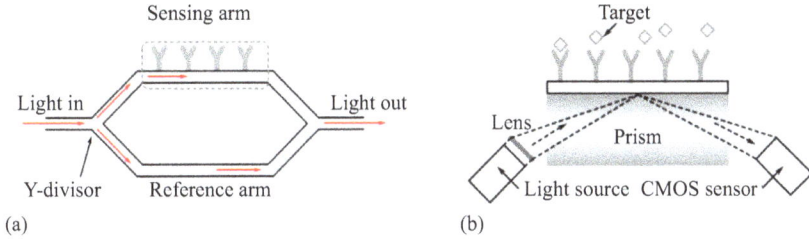

Figure 1.5   (a) Principle of an MZI biosensor; (b) principle of an SPR platform using a prism

sensing and reference arms with two Y-junctions for dividing the guiding light into these two branches and recombination of them. The interference of the two branches with each other leads to the generation of a sinusoidal variation related to the refractive index changes of the surrounding medium.

In another work, Tokel *et al.* [30] used a CMOS sensor in a microfluidic-integrated surface plasmon resonance (SPR) to detect *E. coli* and *Staphylococcus aureus* (*S. aureus*) (see Table 1.1; Figure 1.5(b)). In Figure 1.5(b), a cylindrical lens is illuminated by an LED and collimates the light onto a rectangular prism. The CMOS sensor captures the reflected light and the image will be processed by a computer.

1.1.2   Magnetic

Magnetic biosensors work based on measuring the variations of the magnetic field created by the magnetic particles bound to target biospecies. Magnetic-based platforms have the advantages of high sensitivity because there is no virtual background due to the weak magnetic properties of most biological samples. In these sensors, transducers and samples are not in direct contact because the magnetic field can penetrate the insulating layers. The same magnetic labels can also be used for magnetic manipulation for sample preparation [51]. For example, Murali *et al.* [12] reported a CMOS flow cytometer chip using a spiral transformer. This sensor was embedded into a microfluidic channel and used for the detection of breast cancer cells tagged with magnetic labels (see Table 1.2). Figure 1.6(a) and (b) shows the schematic of the magnetic sensor, magnetic labeling of SKBR3 cells, and the fabricated platform.

Aytur *et al.* [52] proposed a CMOS-based magnetic bead bioassay platform using immunological recognition to detect antigen of purified mouse immunoglobulin G (IgG) and human anti-dengue virus IgG in clinical serum samples. The sensor chip included a 1024-element array of Hall sensors bonded to a printed circuit board (PCB). A gold (Au)-plated well on the PCB with a small hole at the bottom allowed the sensor surface array to be exposed to sample fluid. The biosensor was placed into the gap of a custom electromagnet core which could be operated in either a direct current (DC) washing mode or an alternating current (AC)

Table 1.2   A comparison of some CMOS magnetic biosensors

Sensor type	Application	Target	Label	Array #	Ref.
Magnetic	Flow cytometry	Breast cancer cells	Magnetic labels	2	[12]
Hall sensor	Infectious disease diagnosis	Mouse IgG and human anti-dengue IgG	Magnetic bead	1024	[52]
Magnetoresistive immunosensor	Detection of *E. coli* O157: H7 in food and clinical samples	*E. coli* O157:H7	Magnetic microspheres	4	[53]
Frequency-shift magnetic diagnostic sensor	Nucleic acid or protein assays	DNA oligomer, interferon-γ protein	Magnetic label	48	[51]
Magnetic	Medical diagnostics, food pathogen detection, and water analysis	Mouse IgG	Magnetic bead	21	[15]
NMR	Biomolecular sensing	Avidin, hCG, and human bladder cancer cells	Biotin-coated MPs[1], anti-hCG, and bladder cancer cell surface markers	1	[54]
NMR	High-resolution H NMR spectroscopy and relaxometry	DI water, EtOH, L-alanine, EB[2], aspirin, L-serine, D-(+)-glucose	–	–	[55]
NMR	PoC applications	Avidin and CuSO$_4$	MNP[3]	–	[56]

[1]Magnetic particles
[2]Ethylbenzene
[3]Magnetic nano-particles

measurement mode. In the AC measurement mode, an excitation signal led the magnetic beads to produce local magnetic fields detectable by the sensor. In the DC washing mode, the electromagnet can help to remove specifically bound beads from the sensor surface. In another effort, Mujika *et al.* [53] proposed a microfluidic-integrated platform including a magnetoresistive immunosensor for the detection of *E. coli* O157:H7 in food and clinical samples. A giant magnetoresistive multilayer structure was used as a sensing film and silicon nitride (Si_3N_4) was utilized as the sensor surface coating for immobilization of antibodies to react with target antigens. This sensor can measure the small magnetic field variations arising from the presence of superparamagnetic beads bound to the antigens.

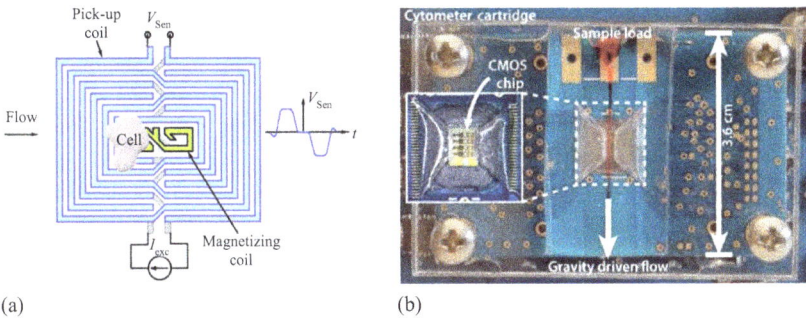

Figure 1.6 (a) Schematic of the magnetic sensor used by Murali et al.; (b) the flow cytometer proposed in [12]

CMOS frequency-shift-based magnetic resonant sensing is one of the magnetic methods reported for biosensing applications including inductor-capacitor (LC) resonators. The decrease in the oscillation frequency caused by the presence of a magnetic bead that is close to the inductor of the resonator can be electronically measured by counting the number of cycles over a fixed interval. Pai *et al.* [51] reported such a magnetic sensor for the detection of antigen and nucleic acid whose sensitivity was increased by correlated double sampling (see Table 1.2). They also proposed a magnetic freezing technique that neutralizes the effect of magnetic beads on the sensor by saturating the magnetization of the beads with a small permanent magnet. By this technique, there is no need for a baseline measurement before the experimental assay and also the SNR is enhanced. Moreover, all the measurements can be done after the biological assay and the measurement time is reduced. The entire functionality of the sensor is integrated onto a reader PCB and a disposable open-well cartridge combines all sensor sites in a single reaction well without the need for microfluidic structure and pumps.

In another effort, Zheng *et al.* [15] proposed an immunoassay platform comprising microfluidics, micro-coil sensor array, a CMOS application-specific integrated circuit to detect the sensing signals, and a microcontroller-based interface circuit for data acquisition (DAQ) and achieved a high-sensitivity sensor comparable to traditional optical enzyme-linked immunosorbent assay (ELISA).

Several magnetic techniques have also been developed for cellular analysis. Nuclear magnetic resonance (NMR) is a method useful for the detection and molecular analysis of cells in which a weak oscillating magnetic field perturbs nuclei in a strong constant field. The electromagnetic response signal produced by the nuclei has the frequency characteristic of the magnetic field at the nucleus. For instance, Sun *et al.* [54] presented two miniature NMR systems and a partially integrated CMOS radio frequency (RF) transceiver for the detection of the weak NMR signal as a result of the used small-sized magnet. The first NMR system could be held in the palm of a hand and weighed 0.1 kg. The second one included an on-chip NMR coil and the transceiver designed for the palm system. This system is useful for the detection of human chorionic gonadotropin (hCG), human bladder cancer cells, and avidin (see Table 1.2). In another work, Ha *et al.* [55] reported a highly integrated NMR spectrometer chip combined with a compact permanent magnet which was useful for high-resolution proton NMR (H NMR) spectroscopy and relaxometry of different biological, organic, and drug compound molecules. Lei *et al.* [56] also reported a micro-NMR (μNMR) transceiver for automated biochemical assays and integrated a 2D multielectrode digital microfluidic device inside a portable magnet (see Table 1.2).

1.1.3   Electrochemical

Electrochemical methods can be used in label-free biosensors without adding labels like magnetic beads or fluorescent materials to target biospecies. Labeled methods suffer from larger detection time, cost, and complexity whereas label-free electrochemical sensors offer the advantages of simplicity and real-time measurement as well as cost and volume reduction. Moreover, their scalability allows them to be easily adapted to miniaturized and integrated equipment. However, their sensitivity is still under investigation [57] and it seems labeled sensing has a better performance to achieve ultra-low (sub-pM) concentration of biospecies in a complex sample like serum [58].

In these sensors, a physiochemical transducer, like an electrode, is coated with a biointerface layer which is usually composed of immobilized BREs. These variations change the electrical characteristics of sensing transducer like dielectric constant (permittivity), electrical charge, or conductivity. Finally, a readout instrument or circuit measures and converts these characteristics to readable signals.

Based on the approach used for the measurement of these characteristics, electrochemical biosensors can be classified into amperometric, potentiometric, impedimetric, conductometric, and capacitive sensors. Table 1.3 compares various CMOS electrochemical biosensors that have been reported for sensing different types of biospecies.

In voltammetric approaches (Figure 1.7(a)), a voltage is applied to two electrodes in contact with the fluidic sample and linearly swept in time. For example, in the cyclic voltammetry (CV) approach, the potential of the working electrode (WE) is ramped linearly versus time. Then, the transducer current is measured by an interface circuit. An amperometric approach is similar to voltammetric techniques, except that it measures the current generated during the reactions at a constant DC voltage [69]. For example, a two-electrode time-based potentiostat is presented by Massicotte *et al.* [5] for multi-neurotransmitter detection. In other efforts, Ghoreishizadeh *et al.* [59] reported an amperometric sensor for high-sensitivity glucose measurement. In another work, Niitsu *et al.* [60] proposed two types of electroless plated microelectrode arrays (MEAs) and an amperometry circuit for direct bacteria and HeLa cell counting. For noise reduction, the same group [61] implemented a current integrator in conjunction with these bacterial-sized MEAs and developed their sensor to a large-scale (1024 × 1024) array. In another work,

Table 1.3   A comparison of some CMOS electrochemical biosensors

Sensor type	Application	Target	Array #	Ref.
Amperometric	Glucose detection	H_2O_2	1	[59]
Amperometric	Selective detection of neurochemicals	DA[1], glutamate	2	[5]
Amperometric	Direct counting of cells	Bacteria, HeLa cell	4×4, 16×16	[60]
Amperometric	High-sensitivity bacterial counting	Microbeads with the same size as bacteria	1024×1024	[61]
Amperometric and potentio-metric	Analysis and measurement of chemical solutions	MeCN, TBAPF$_6$[2], ferrocene	2 single sensors	[62]
Potentiometric (ISFET)	pH detection	The pH of DI water, running water, normal saline	10	[17]
Potentiometric (ISFET)	Rapid screening of foodborne bacteria	*E. coli*	512×128	[16]
Potentiometric (ISFET)	pH detection	pH variations induced by the cell population	12	[63]
	Monitor the electrical activity of the neuron	The electrical and metabolic activity of a neuronal cell population	40	
Potentiometric (ISFET)	Extracellular imaging of hydrogen-ion activity of cell cultures	Hydrogen-ion activity	16×16	[64]
Impedimetric	A protein-based biosensor for continuous glucose monitoring	Glucose	4	[65]
Impedimetric	Impedance spectroscopy and imaging of biological cells	mESC	59,760	[18]
Impedimetric	DNA hybridization detection	Zika virus oligonucleotides	16×20	[19]
Impedimetric and CV	Neurotransmitter detection	DA	32 and 28	[66]
Conductometric	Growth monitoring and sensing of bacteria	*E. coli*	8	[67]
Capacitive	Bacteria growth monitoring	*E. coli*	3	[68]
Capacitive	Real-time measurement of cell proliferation	Human ovarian cancer cells	4×4	[13]
Capacitive	Detection of single bacterial cell	*S. epidermidis*	16×16	[20]
Capacitive	Cell growth monitoring	Human lung carcinoma cell	8×8	[21]

[1]Dopamine
[2]Tetrabutylammonium hexafluorophosphate

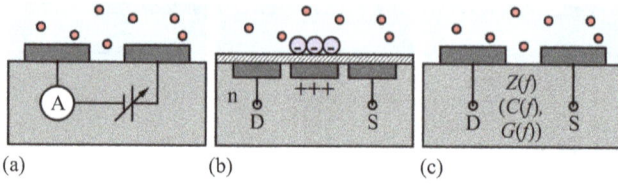

*Figure 1.7   Interfacing techniques for (a) voltammetric sensors, (b) p-channel
ISFET, and (c) impedimetric (or capacitive, or conductometric) sensing*

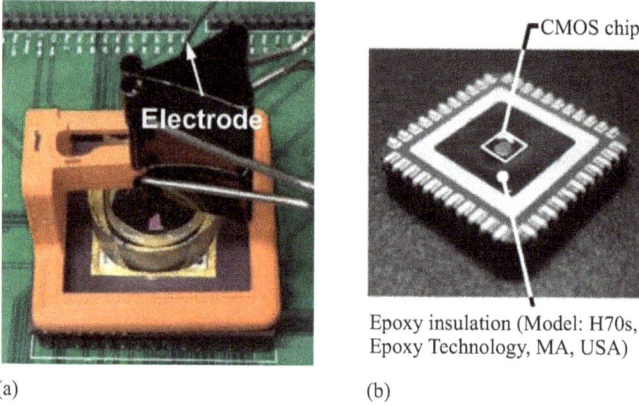

*Figure 1.8   (a) The packaged ISFET chip with electrode reported in [16]; (b) the
packaged EIS biosensor proposed in [70]*

Milgrew *et al.* [64] used ISFET in a 16 × 16 proton camera array for direct
extracellular imaging of the hydrogen-ion activity of *in vitro* cell cultures.

In potentiometric methods, ions or free electrons in biological samples induce
an electrical potential across the dielectric material. ISFET is a suitable element
compatible with CMOS technology which can be utilized in potentiometric measurements [6]. Figure 1.7(b) illustrates a p-channel ISFET. For example, hybrid
amperometric and potentiometric sensors were proposed by Giagkoulovit *et al.* [62]
for electroanalytical applications. ISFET has been reported for a variety of biochemical sensing applications like pH (potential of hydrogen) detection of food-
borne bacteria screening [16] and water [17]. Figure 1.8(a) shows the packaged
ISFET sensor chip with the electrode used in [16]. Martinoia *et al.* [63] proposed
two different microsystems for bioelectrochemical measurements of cell populations. The first system consists of 12 ISFETs for the detection of small pH variations induced by the cell population. The second system monitors the electrical
activity of neurons with an array of 20 Au microelectrodes and 40 ISFETs.

Impedimetric [71], conductometric [67], and capacitive [13] (Figure 1.7(c))
sensors process the frequency-dependent parameters of biological samples like
impedance, conductance, and capacitance, respectively [69].

Manickam *et al.* [70] took advantage of electrochemical impedance spectroscopy (EIS) for label-free detection of various biological analytes, such as DNA and proteins (see Figure 1.8(b)). Ghafar-Zadeh *et al.* [65] proposed an impedimetric biosensor platform for glucose biosensing and used genetically engineered glucokinase (GLK) as receptor proteins. In another effort, a CMOS chip is reported by Vijay *et al.* [18] for EIS and electrophysiology recordings on 59,760 electrodes for monitoring biological cells such as mouse embryonic stem cells (mESC). The integration of a large number of electrodes in the silicon (Si) area is one of the important challenges of small CMOS biosensors. So, some researchers have tried to address this challenge. For example, Hsu *et al.* [19] developed a polar-mode impedance measurement technique by designing a digital readout circuitry occupying a small footprint and used it for the measurement of hybridization of Zika virus oligonucleotides.

Some multifunctional systems are also reported in the literature. For example, Dragas *et al.* [66] presented a system comprising 32 current recording units, 28 CV units, and 32 impedance measurement units for neurotransmitter detection (see Table 1.3). Additionally, this system included 2048 action potential (AP) recording units, 16 dual-mode stimulation units, and 32 local-field-potential recording units.

Yao *et al.* [67] used a conductometric approach for bacteriophage-based bacteria detection. Since no biochemical reaction is involved in bacteriophage-based bacteria detection and no electrical charge is created, the amperometric method is not suitable. On the other hand, impedimetric methods are complex. Conductometric methods can be the alternatives that only measure the resistance of the analyte between reference electrode (RE) and WE and can be performed with DC stimulation. Yao *et al.* used T4 bacteriophages as BREs to detect *E. coli* (see Table 1.3).

Ghafar-Zadeh *et al.* [68] reported a capacitive readout architecture technique to monitor the growth of bacteria in the Luria–Bertani medium. In another effort, Couniot *et al.* [20] reported a 16×16 capacitive biosensor whose readout interface converted capacitance to a voltage based on charge sharing. This sensor is capable of real-time detection of *Staphylococcus epidermidis* (*S. epidermidis*) with a sensitivity of 2.18 mV/bacterial cell. Nabovati *et al.* [21] and Senevirathna *et al.* [13] also reported CMOS capacitive biosensors for the detection of cell attachment and growth of human lung carcinoma cell line as well as real-time cell viability measurements, respectively.

1.2 CMOS capacitive sensor

Affinity-based biosensors can be divided into labeled and label-free sensors. In labeled sensors [11,12,14,15,36,56], some labels like fluorescent markers, Qdots, magnetic beads, and active enzymes are utilized to confirm the reactions between the biointerface layer and target biospecies. These sensors usually need costly, time-consuming assays as well as bulky equipment which is a limiting factor to miniaturize these sensing systems. Also, labeling might change the molecular structure of target cells or molecules (especially for protein targets) preventing real-time measurement in some applications like studying the kinetics of the reactions and dominant physical processes [57].

As aforementioned, electrochemical sensors pave the way for label-free and real-time measurement as well as low complexity. Among these label-free affinity-based sensors, capacitive biosensors have the benefit of low-temperature dependency, scalability, and low-noise properties. They have been employed for widespread applications such as bacteria detection [20], real-time measurements of cell viability and proliferation [13,21], as well as DNA detection [22].

It is also important to distinguish between faradaic and non-faradaic biosensors. In a faradaic process, charge is transferred across the electrode–solution interface governed by Faraday's law. For instance, in faradaic EIS, redox species is alternatively reduced and oxidized by the transfer of an electron from and to the metal electrode. By contrast, in a non-faradaic processes like adsorption or desorption, the charges are distributed on the surface through physical processes without using a chemical mechanism like breaking and forming chemical bonds [72]. The term capacitive biosensor is usually dedicated to a sensor based on a non-faradaic scheme which is usually assessed at a single frequency [25].

1.2.1   LoC-based versus MEMS-based capacitive sensors

There are some differences between the capacitive sensors used for LoC applications and the ones reported for microelectromechanical systems (MEMS) like the systems for measuring vibration, acceleration, or pressure.

As seen in Figure 1.9(a) and (b), both LoC- or MEMS-based capacitive sensors require sensing electrodes connected to an interface readout circuit. But, as depicted in Figure 1.9(a) and (c), in MEMS-based applications, the capacitive electrodes are used as an off-chip device wire-bonded or flip-chip bonded to the CMOS chip. On the contrary, in LoC applications, the sensing electrodes are fabricated above the CMOS chip as illustrated in Figure 1.9(b) and (d) [73].

MEMS-based capacitors usually need long and continuous measurement time as well as a built-in self-calibration module to correct the accumulating errors during the whole measurement time, whereas LoC-based capacitive sensors often measure the subtraction of sensing capacitances in the presence and absence of biochemical sample for only a short time (see Figure 1.10). So, if a time-variant random error does not change significantly in the short interval before and after applying a sample, the subtraction of the sensing capacitances will be constant and does not suffer from accumulative errors over a long duration of time [76].

In many MEMS applications such as vibration monitoring, the capacitance variation should be measured over a very short period. So, they require low-noise and relatively high-speed sensors. But, LoC-based capacitive sensors (see Figure 1.11) have enough time for the measurement before and after applying biochemical samples and measure the average of capacitances over a long duration of time. So, a low-speed circuitry can meet their requirements [76].

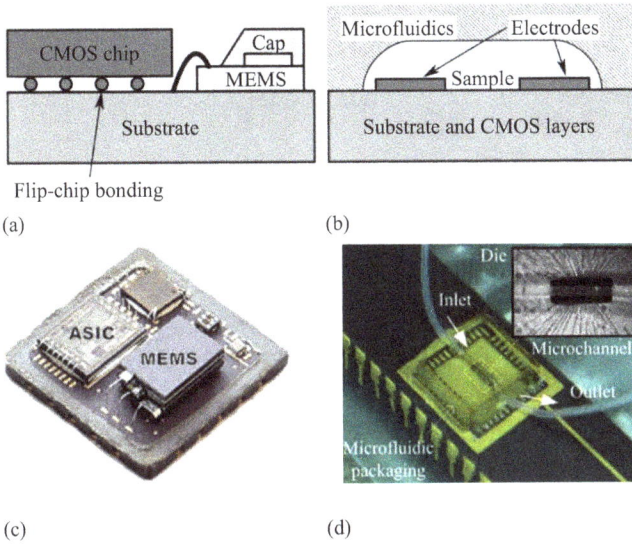

Figure 1.9   *Conceptual view of (a) wire-bonded electrodes used for MEMS applications, (b) on-chip electrodes used for LoC applications, (c) a decapsulated Colyibrys MS9000-series accelerometer [74], and (d) a microfluidic packaging of sensor chip (based on direct-write fabrication process (explained in Section 5.2.3)) [75]*

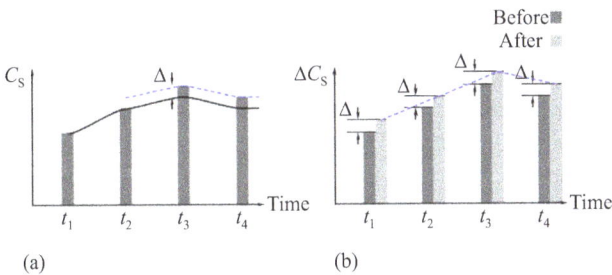

Figure 1.10   *The effect of offset error on (a) MEMS-based capacitive sensors and (b) LoC-based capacitive sensors*

In contrast to MEMS-based capacitive sensors that are protected from direct contact with dust, fluids, or even air, LoC-based capacitive sensors are always exposed to aqueous samples and their characteristics can be affected by the presence of non-bonded bioparticles or other remnants [76].

On the other hand, an encapsulated MEMS-based capacitive sensor in a vacuum chamber senses a pure capacitance variation between electrodes, while

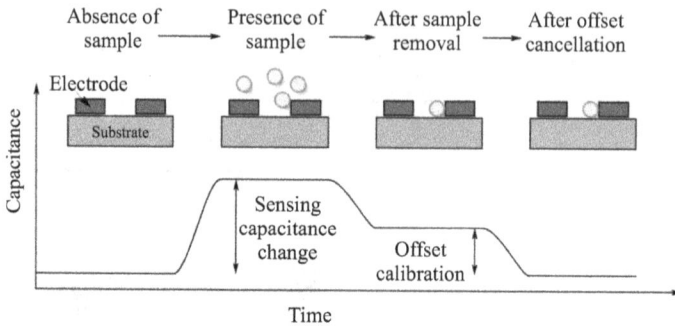

Figure 1.11   An illustration of offset cancellation of LoC-based capacitive sensors

biochemical samples above the electrodes of LoC-based capacitive sensors are modeled with capacitance and resistance in series and the sensor should be able of extracting the sensing capacitance component.

1.3   Organization of this book

A CMOS-based system for capacitive biosensing application includes different parts and can be characterized by various parameters. Chapter 2 is dedicated to introduce some metrics and methods for the design and implementation of such systems.

Capacitive biosensors require capacitive transducers or sensing elements like electrodes. Well-known transducers for capacitive biosensors are interdigitated electrodes (IDEs) and floating electrode interfaces. IDEs, especially coplanar ones, are the most commonly used transducers in integrated capacitive biosensors. They work based on variations of the dielectric constant of the biochemical sample and can be easily implemented in CMOS technology [21,68,77,78]. Floating electrode interfaces are sensitive to the thickness and/or permittivity of the dielectric of the sample. Additionally, surface stress-based transducers like cantilever and capacitive micromembrane have received much attention over recent years, which work based on the variation of the distance between the plates of a sensing capacitor [57]. But, to the best of our knowledge, the latter has not been adapted with standard CMOS technology yet. The capacitive transducers and the materials and techniques useful for the fabrication of them in CMOS technology will be reviewed in Chapter 3.

An intermediate layer is also needed in between the surface of sensing transducers and the biochemical sample which is sensitive to biochemical variations and differs based on the applications. The applications can be divided into three categories: (1) organic solvent sensing, (2) nonselective cell monitoring and toxicity test, and (3) selective sensing. In nonselective cell monitoring and toxicity test, the intermediate layer is created by its nature. For instance, in cell monitoring or cell growth and proliferation, a biofilm is formed above the electrodes and the contact between cells and the passivation layers on top of the electrodes changes surface charge [79] or dielectric property [80] that can affect the value of the capacitor under measurement [76]. In organic solvent sensors [81], the electrical conductivity of solvents affects the sensing capacitance [76,82]. In selective sensing, the surface of the sensing transducer is coated with BREs or receptors to react with target biospecies with an appropriate selectivity and sensitivity and consequently to change the properties of the transducers. Antibodies [83–87], enzymes [88,89], nucleic acids [90–93], and whole cells [94,95] are among the well-known BREs used in electrochemical assays. Furthermore, recent years have seen increasing growth in artificial BREs such as aptamers [96,97], phages [98,99], affimers [100,101], and molecularly imprinted polymers (MIPs) [102–104]. Different biological applications of CMOS and non-CMOS capacitive sensors will be reviewed in Chapter 6.

The signals sensed by the capacitive transducer should be amplified, processed, and converted to readable electrical signals using a readout interface circuit. Commonly investigated types of capacitive interface circuits are switched capacitor (SC) circuits [4], charge-sensitive amplifiers [105–107], charge sharing approaches [20,79], capacitance-to-frequency converters (CFCs) based on ring oscillators (ROs) [13,108–114] or relaxation oscillators [115–117], triangular waveform techniques [118,119], lock-in detection technique [120–130], voltage-controlled oscillator-based circuits [131–140], and CBCM method [21,22,68,73,77,141–144]. Among all these techniques, the latter is attracting an increasing interest due to its high sensitivity and low complexity. In LoC applications, the offset capacitance of the transducers can affect considerably the measurement because CMOS process tolerances make this offset capacitance different from chip to chip and it is difficult to measure it. CBCM method as a sensitive and differential sensing approach seems a suitable method for the characterization of sensing transducers of CMOS-based LoCs [76]. These striking features have encouraged researchers to use this method in high throughput biosensors in which several small biochemical signals must be sensed in parallel. CBCM is a useful approach for the measurement of static capacitances and even dynamic capacitances atop the chip [76]. A growing body of literature has used the CBCM method for several LoC applications such as organic solvent monitoring [145], cell viability and growth monitoring [21,142], bioparticle sensing [146], and DNA detection [22]. In addition to LoC applications, some other applications like the characterization of nonlinear and bias-dependent capacitances [147–150] and measurement of metal-oxide-metal capacitance matching [151–153], interconnection capacitances [154–162], leaky capacitors [163], plasma process-induced charging damage [164], and the mobility of devices with various channel widths [165] are reported in the literature. So, after a brief overview of different interface circuits of capacitive sensors in Chapter 4, we will review the core-CBCM capacitive biosensors more comprehensively in this chapter.

A key problem with much of the literature [27,166–174] to hybrid CMOS-microfluidic capacitive biosensors is that the material and methods used for the fabrication of microfluidics and packaging must be biocompatible and reliable. Si, glass, and polymers like poly(dimethylsiloxane), poly(methyl methacrylate), and SU8 are the most reported materials for the fabrication of microfluidics that can be formed by techniques like rapid prototyping, soft lithography, direct-write fabrication process (DWFP), and other techniques. These techniques will be reviewed in Chapter 5.

After a review of the biological/chemical applications of capacitive biosensors in Chapter 6, current technologies and the future work will be discussed in Chapter 7.

Chapter 2

Design, implementation, and characterization of CMOS capacitive biosensors

The main goal of this book is to provide the reader with the required skills and knowledge to design and implement a complementary metal-oxide semiconductor (CMOS) capacitive biosensing system. Figure 2.1(a) shows the main parts of this system including CMOS chip, microfluidics, biological recognition element (BRE), and data acquisition (DAQ). As seen in Figure 2.1(b), the biological or chemical partials or related reactions (so-called biological/chemical phenomenon (BCP)) are transduced into electrical signals (ESs) such as a dielectric change. A high-precision interface circuitry is used to detect minute changes of these ESs and convert them to digital for monitoring and further signal processing in the computer. Before discussing the design strategies for the development of such a system, let us briefly discuss the design metrics. This section will be continued with the methods.

2.1 Design metrics

There are various metrics for the design and implementation of custom-made and specific capacitive sensors that characterize key aspects of biosensing. As seen in Figure 2.2, the main metrics are applicability, reusability, integration and minia-turization capability, complexity, power consumption, biocompatibility, biost-ability and lifetime, selectivity, multiplexing, noise immunity, linearity, limit of detection (LoD), resolution, input dynamic range (IDR), sensitivity, detection time, and reproducibility. Each one of these metrics will be described in the following subsections. As will be discussed in the following sections, some of these features are closely related to each other.

2.1.1 Applicability

The first step toward designing a capacitive biosensor is investigating whether it is useful for the desired application or not. This type of electrochemical biosensors is useful for various applications including chemical solvent monitoring, cell growth monitoring, toxicity test, biomarker detection, and various nucleic acid-based and protein-based assays. Different applications of capacitive biosensors (CMOS and non-CMOS) will be reviewed in Chapter 6.

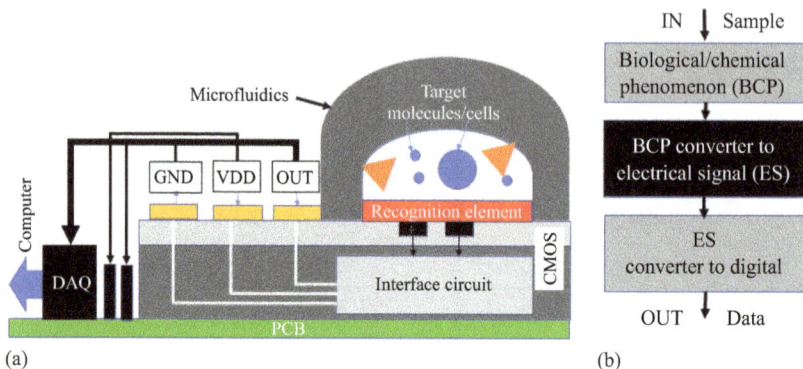

Figure 2.1   Design of CMOS: (a) capacitive biosensor illustration of a CMOS capacitive sensing system including BRE, electrodes (or transducer), interface circuit, microfluidics, DAQ, and PCB; (b) main blocks of a CMOS capacitive biosensor

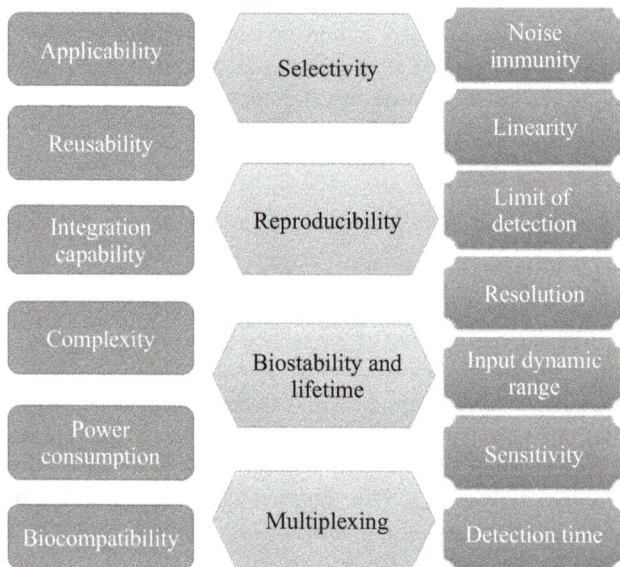

Figure 2.2   Various design metrics

2.1.2   Noise immunity

Noise means all unwanted statistically fluctuating signals and disturbances which are superimposed on the output signal. Electrical and biological noise are the main sources of noise in a capacitive biosensor which can cause uncertainty in the output signal of the biosensor. Electrical noise is generated by the electrical device,

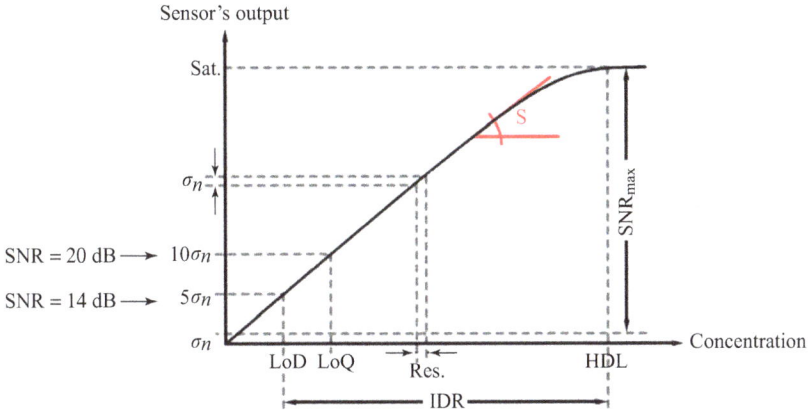

Figure 2.3   Illustration of a response curve showing the sensor output versus target substance concentration, highlighting LoD, LoQ, HDL, IDR, resolution (Res.), the noise floor (σ_n), the saturation level (Sat.), the maximal SNR (SNR_{max}), and the sensitivity (S)

whereas biological noise is due to biological phenomena such as the stochastic mass-transfer processes within the reaction chamber, or probabilistic molecular-level bindings within the sensing sites. The dynamical behavior of electrons and photons within solid-state matter can be characterized by statistical mechanical methods. But quantum chemical and biological systems are considerably more complicated than quantum electronic systems and little works have been reported for modeling the biological noise [175].

LoD, sensitivity, and resolution (explained in Sections 2.1.4, 2.1.7, and 2.1.5, respectively) of a biosensor are limited by the noise of the system. Considering the noise is the measure of the signal generated from the sum of all the noise sources within the system, signal-to-noise ratio (SNR) can serve as a metric for estimating whether a substance signal is actually a demonstration of the presence of the substance in the sample.

One of the metrics which is usually used, especially for the optimization of the sensor parameters, is maximum SNR demonstrated by SNR_{max} in Figure 2.3. SNR_{max} is defined as the ratio between noise floor, σ_n, and saturation level (Sat.) at the sensor's output which is usually expressed (in dB) as follows [176]:

$$SNR_{max} = 20\log_{10}\frac{Sat.}{\sigma_n} \tag{2.1}$$

2.1.3 Linearity

When a sensor is linear, it means its transfer function (input–output) can be represented by a straight line. The nonlinearity of a sensor affects the operational dynamic range (explained in Section 2.1.6) of the sensor. *R*-squared (R^2) is a statistical measure showing how much a sensor response is linear and how close the output signal is to a fitted straight regression line. For example, if $R^2 = 0.9995$, it means that 99.95% of the output results can be explained by a linear model for the input values.

2.1.4   Limit of detection

In analytical chemistry, LoD, which is sometimes mistaken with analytical sensitivity, is the minimum detectable amount or concentration of a target substance that can be reliably distinguished from the absence of that substance [176]. Unfortunately, there is no universal method for determining LoD [25]. One approach is measuring the smallest concentration of the target at which the sensor response to a dilution series can be clearly distinguishable from the response to a blank solution. But the sensor response to a blank solution is not necessarily zero. Thus, the LoD might be overestimated. Another technique is to calculate LoD based on the standard deviation of the blank response. In this approach, LoD is defined as three times of the standard deviation of the blank response [25]. In Figure 2.3, LoD is defined as a concentration at which the output signal is 5 times larger than the noise floor, σ_n (SNR = 10 dB) [176]. There are several concepts derived from LoD, one of which is limit of quantization (LoQ) that is the minimum detectable concentration corresponding to an output ES larger than σ_n by a factor 10 or SNR = 20 dB [176].

LoD is almost always measured in the absence of confounding nontarget species. But real-world samples (like clinical tests) are rarely such clean and are usually a mixture of target/nontarget species which is challenging. Moreover, the achievable LoD, fundamentally, depends on the strength of the probe–target interaction and noise floor [25]. The variation between sensors is another factor affecting the measured LoD which can be alleviated by calibration of each sensor. However, it can be cumbersome for a large number of experiments. In many practical situations, lack of sensor reproducibility (see Section 2.1.10) and nonspecific binding (explained in Section 2.1.9) dictate LoD.

2.1.5   Resolution

As shown in Figure 2.3, resolution (Res.) is the minimum detectable change in target concentration which is associated with the sensor noise, σ_n, which is due to systematic and irreducible noise [25,176]. In other words, it is the output uncertainty due to noise divided by the slope of the sensor response curve.

2.1.6   Input dynamic range

Generally, IDR is the range from the smallest to the largest values of a certain quantity (concentration, capacitance, etc.) that can be measured by a sensor.

As depicted in Figure 2.3, IDR can be considered as the range of detectable concentration from LoD (see Section 2.1.4) to the highest detectable level (HDL) which is usually expressed (in dB) as follows [176]:

$$IDR = 20\log_{10}\frac{HDL}{LoD} \tag{2.2}$$

The output signal corresponding to the HDL is its saturation level (Sat.). This definition of IDR provides a quantification of the target concentration [25,176].

If the response curve is the sensor output signal versus input capacitance variations, IDR can be defined as the range between the smallest and the largest measurable physical input (capacitance in this case) in such a way that the response curve for the input capacitances in this range is linear.

2.1.7 Sensitivity

Sensitivity (*S*), as depicted in Figure 2.3, is the slope of the response curve which means the change of the output signal per unit variation of the target substance concentration [176].

If the response curve is the sensor output signal versus input capacitance variations, an electrical sensitivity can be defined as the change of the sensor's output signal (like voltage and frequency) per unit variation of the physical input (capacitance in this case) and, according to the type of the output signal, it can be expressed in V/F, Hz/F, etc. [176].

2.1.8 Detection time

Detection time is the analysis time required to achieve a stable output starting from the initial collection of the biological sample including the time needed for sample conditioning, pre-concentration as well as the response time of the electrical transducer itself [176]. When the capacitive sensor is used for end-point measurement mode, detection time can be defined as the time required to reach 90% of steady-state response (Figure 2.4) [177].

2.1.9 Selectivity

In selective sensing applications (which will be reviewed in Section 6.3), selectivity is a figure of merit demonstrating the ability of the sensor to discriminate a particular target analyte (such as a target bacteria) in a complex mixture without interferences from other background components (like viruses, other bacteria and proteins). This metric plays a key role in real samples and is one of the most

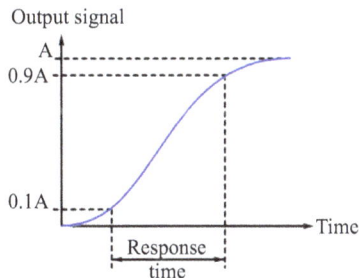

Figure 2.4   Response time

challenging features to be provided for real-world applications because the target concentration might be much less than nontarget species. For example, for detecting prostate-specific antigen (PSA), as a biomarker of prostate cancer, in blood serum, 2 ng mL^{-1} of PSA must be discriminated from 70 mg mL^{-1} total protein content in blood serum. So, a biosensor that is capable of detecting 1 ng mL^{-1} PSA in saline, but shows even 1 ppm response to blood proteins, would not be appropriate for clinical tests unless some compensation techniques (like depletion of interfering proteins) are applied. This shows the trade-off between sample preparation complexity and selectivity requirements [25].

This ability can be provided by various BREs like antibodies, enzymes, bacteriophages, DNA, and peptides. In surface-based methods, BREs cover the surface of the transducer. Background noise or nonspecific bindings due to the imperfect coverage and selectivity is one of the drawbacks of these methods. As illustrated in Figure 2.5, nonspecific bindings prevent target–probe binding and can cause false positives. However, different approaches have been reported to ameliorate this issue. For example, exposing the sensor to a solution containing blocking agents such as bovine serum albumin which covers the unoccupied regions, or antifouling agents like polyethylene glycol which prevents or retards fouling, can reduce nonspecific bindings and prevent target depletion. Differential measurement using a reference sensor with no target binding is another approach to hopefully remove common-mode signals due to extraneous factors including nonspecific binding. If both reference and working sensors react similarly to nonspecific binding, differential measurement can improve both sensitivity and selectivity of the system. However, the nonspecific component of the sensor response is usually subtracted out imperfectly. Another approach is washing the surface before readout which is useful for an end-point measurement, not a real-time approach, and also is not necessary for homogeneous assays [25].

Additionally, low reproducibility (see Section 2.1.10), low stability under flow (see Section 2.1.16), and low robustness against the forces of the fluidic flow are some other problems of surface-based technique to form the BRE layer. To avoid BREs covering the entire microfluidic channels rather than only sensing sites and, consequently, decreasing the sensitivity, it is recommended to functionalize the

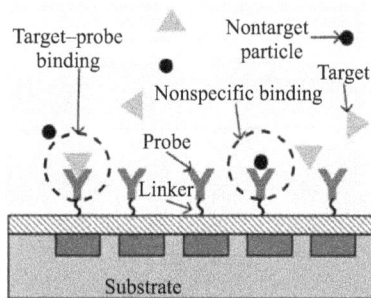

Figure 2.5   Target–probe binding and nonspecific binding

sensor before the encapsulation. It is noteworthy that the fluidic package should be sealed because BREs can be denatured by the annealing step. Patterning spots with specific BREs by a spotter is a technique appropriate for multiplexed biosensors utilizing the microarray technology.

Other techniques of creating the BRE layer are grafting BREs like enzymes or antibodies through physical adsorption (physisorption) of self-assembled monolayers (SAMs) (cross-linkers). In physical adsorption, BREs are directly incubated on the naked surface. Robustness to wash, antibody denaturation, poor coverage, uniformity, and reproducibility are some of the features of this method [176]. By the other technique, SAMs can provide a covalent link with chemical terminations of BREs. For example, bacteriophages, as BREs, can be covalently immobilized on an electrode surface by chemical processing of the surface using 1-ethyl-3-(3-dimethy-laminopropyl)-carbodiimide (EDC). Then, bacteriophages can capture and digest target bacteria in a selective way [178]. Layer-by-layer (LBL) assembly [179] and polymer coating [180] are also other techniques of surface-based methods.

2.1.10   Reproducibility

Some factors like the random binding of biological species on random positioned BREs at the sensor surface can cause sensor-to-sensor variability [175]. Reproducibility denotes the ability of similar biosensors to produce identical responses for similar experimental conditions.

2.1.11   Multiplexing

Multiplexing is the capability of the sensor to detect several biospecies within a sample by employing a single biochip and steering ESs. This feature is typically provided by a biosensor microarray whose sensing area is divided into several sensing sites which are differently functionalized. This way, cost and sample volume per data point will be significantly reduced. These merits make microarray biosensors suitable for point-of-care (PoC) applications and high-throughput screening [176].

As shown in Figure 2.6, cross-reactivity (binding targets to multiple probes or vice versa) makes it complicated to detect some targets like protein in a multiplex way, especially in the real world. But, for clinical applications, detecting several biomarkers can provide far more information than a single biomarker. Thus, regardless of readout setup, the main restriction on multiplexing is due to cross-reactivity of the affinity step for most label-free biosensors.

2.1.12   Reusability

Reusability of a platform is a challenging factor that allows to use a platform up to several times without the degradation of its performance. Most affinity-based bio-sensors are typically single-use and the biosensor will be discarded after a single use to avoid technical failure and contamination. However, to reduce the analysis cost, it is preferred to develop biosensors that can be regenerated and reused after washing procedures up to several times [25,176].

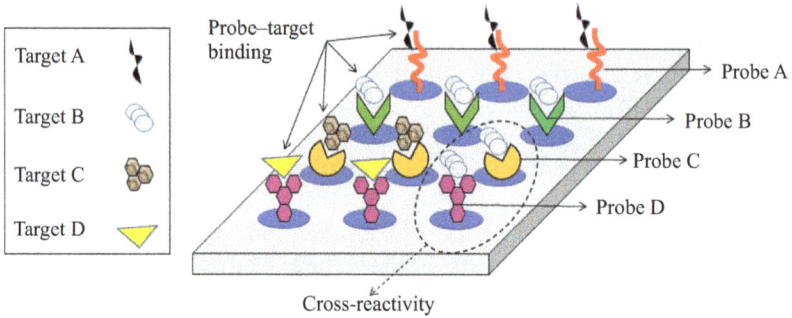

Figure 2.6   Illustration of a microarray, probe–target binding, and cross-reactivity

2.1.13   Miniaturization and integration capabilities

Miniaturized biosensors are necessary for many applications like PoC and laboratory-on-chip (LoC) platforms. These biosensors require a portable integrated system including transducers, readout circuits, signal processing units along with wireless modules and microcontrollers. Miniaturization and integration capabilities have important effects on the size of the sensor as well as its cost and portability. In addition, miniaturized systems can increase throughput and automate the assays by consuming tiny volumes of samples (like blood) and also reagents (like antibodies) and, consequently, help to the simplicity, easiness, and cost of biological/chemical tests. CMOS technology can enhance both integration and miniaturization capability and offers dramatic cost reduction for mass production [57,176].

2.1.14   Complexity

The complexity of a biosensor can affect the cost, noise, sensitivity, and selectivity of the sensor. Some parts of the complexity of the system stem from electrical parts. Another part might be due to the sample complexity like chemical properties (permittivity, ionic strength, ion composition, etc.), the matrix physical nature (solid, liquid, or semisolid), parasitic and microbial compositions (proteins, red blood cells, etc.) which can cause challenges for the selectivity of the sensor (see Section 2.1.9) and restrict the biosensing performance [176].

2.1.15   Power consumption

One of the limitations of portable and implantable biosensing applications is the biochip power consumption which is generally dominated by the transducer and readout interface [176]. For these applications, the power should be provided wirelessly (especially for implantable devices) or by using a battery. This imposes some restrictions that the sensor must be designed accordingly.

2.1.16   Biostability and lifetime

Biostability and lifetime are essential for long-term applications where the sensor must be in touch with fluidic analyte during hours and days. This feature is closely related to the stability and robustness under flow. BREs should be robust and stable under the flow of the fluidics in microfluidic applications and probable intense washing procedures so that they can keep biological reaction kinetics identical and withstand biofouling. In addition, electrochemical corrosion of the surface material comprising the biointerface layer and BREs must be much slower than the operating time of the sensor [175].

2.1.17   Biocompatibility

This feature denotes the ability of a biosensor to keep fluidic analyte in contact with the electrical parts without malfunction of the circuit due to the leakage of the fluidic sample or altering the biological sample due to the release of pollutants by the sensor itself. Especially for *in vivo* applications, inappropriate biocompatibility can lead to immunogenicity, thrombogenicity, mutagenicity, carcinogenicity, and toxicity [175].

2.2   Design, implementation, and test steps

In this book, we discuss various methods related to the development of capacitive biosensors. As seen in Figures 2.7 and 2.8, different steps must be taken to design, implement and test such platforms which are related to different parts of a CMOS capacitive biosensor system. Sample preparation, BRE coating, modeling of the biological/chemical interface, readout circuit design, post-processing procedures for implementation of CMOS-based electrodes, and also microfluidic packaging are the main steps for designing. Generally, before implementing electrodes and on-chip/off-chip readout circuit and DAQ, their behavior must be simulated and analyzed by circuit and multiphysics simulation software. The next step would be the fabrication and packaging of the chip and implementation of DAQ system and experimental setup for desired chemical and biological tests.

Figure 2.8 illustrates the order of these steps. First of all, we should determine the desired application and investigate the applicability of capacitive biosensors for the application. If these sensors are appropriate, the design metrics (which were discussed in Section 2.1) required for that specific application must be defined. Then, the CMOS-based elements including the transducer (or electrodes) and readout circuit as well as DAQ system and microfluidics are designed and simulated. In case of validity of the simulation results, CMOS chip, printed circuit board (PCB) and microfluidics will be fabricated. After designing an experimental setup, first, the electrical behavior of the CMOS chip is characterized. If the electrical part of the system works correctly, the whole system is tested by a low-complexity chemical assay. In case of validity, biological tests will be carried out in the next step. After passing the previous steps, the system would be ready for clinical tests

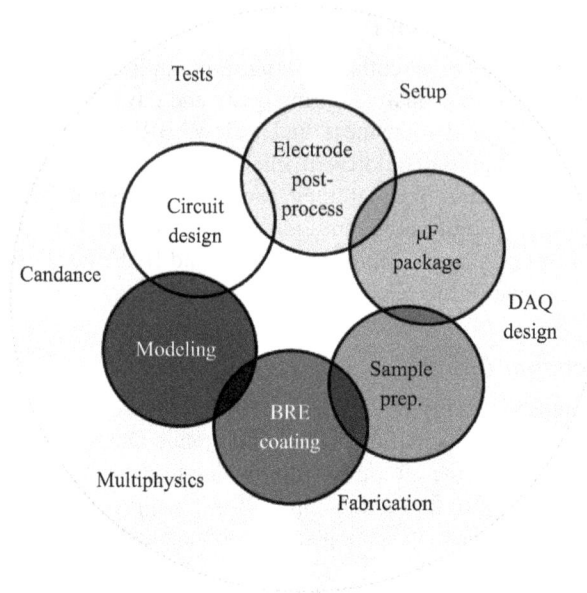

Figure 2.7   Design, implementation, and test steps of CMOS capacitive sensors

with real-world samples. Various factors are affecting the clinical test results arising from the complexity of the samples in the real world. So, several tests are required to validate a system for mass production. Because of the limited number of packaged chips available in the research phase, the clinical trials must be repeated at least three times.

Each of the above-mentioned steps and their challenges are elaborated in the following sections.

2.2.1   Electrode design and biointerface layer

Size, shape, and type of the electrode (or transducer) can affect the sensitivity of the sensor. Although small electrodes are more appropriate for high-throughput screening and have smaller offset capacitance, they can cause undesirable effects on the sensitivity, LoD, and settling time of the sensor. The materials used for electrodes are closely associated with the biointerface layer (or BRE layer) required for the desired sensing applications. Functionalization of the sensor surface depends on the type of the materials used in the surface as well as the application and target particle.

There is not a unique model for different biointerface layers and modeling biological materials can be complicated and different from each other. But Figure 2.9 illustrates a common resistor-capacitor (RC) model used for non-faradaic interfaces, where R_{Sol} is the resistance of the solution, C_{surf} denotes the surface capacitance, and R_{leak} is resistive path usually due to an insulator on the

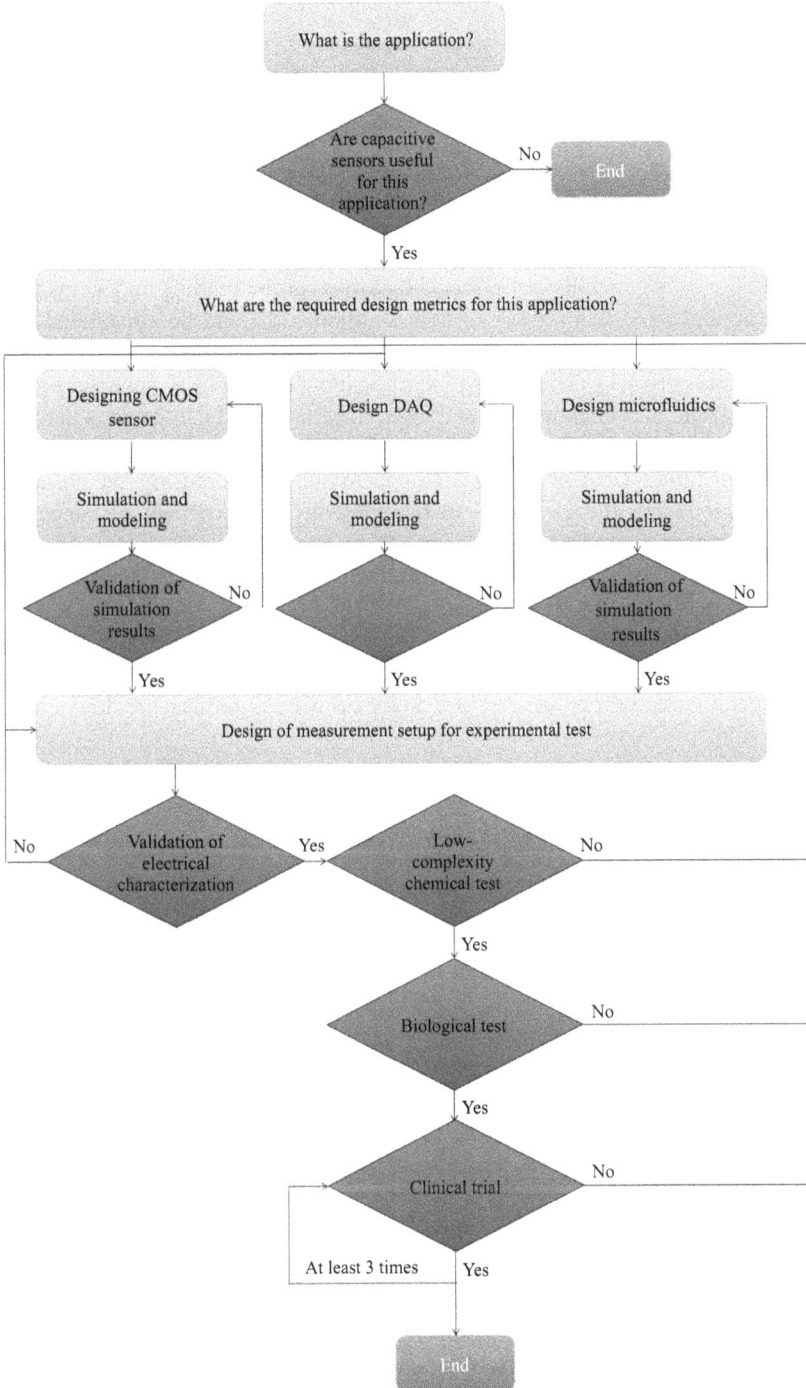

Figure 2.8   Flowchart of design, implementation, and test steps

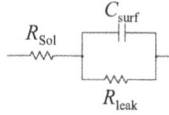

Figure 2.9   Common circuit models for non-faradaic interface

surface [25]. R_{Sol} is generally not affected by target–probe binding and stems from the finite conductance of the ions in bulk solution. C_{surf} can be considered as a model for a series combination of surface modification and double-layer capacitances (see Section 3.1). The surface modification capacitance depends on the dielectric constant and thickness of the biointerface layer and can be considered as a parallel-plate capacitor. If the insulator is ideal, R_{leak} would be infinite. But, in practice, it is finite. Decreasing C_{surf} (for example, by increasing the insulator thickness or reducing the electrode area) allows the measurement of capacitive behavior at higher frequencies. Reducing R_{Sol} (for example, by increasing salt concentration) affects high-frequency impedance. Decreasing R_{leak} makes it difficult to measure C_{surf} at low frequencies.

Decreasing the sensor area changes the values of C_{surf}, R_{Sol}, and R_{leak} and reduces the absolute number of immobilized BREs. The absolute change upon probe–target binding will be smaller and might be difficult to measure due to noise and alike [25].

Modeling and simulation of electrodes along with the biointerface layer require multiphysics simulation software such as COMSOL and ANSYS. However, these software tools still do not combine electrical, biological, and fluidic effects efficiently [181].

2.2.2   Circuit design, modeling, and fabrication

Modeling and simulation of integrated readout circuits are holistic methodologies for the automation and streamlining the design and verification flow for the fabrication of multi-die heterogeneous systems. This step can be taken by using the software tools dedicated to designing integrated circuits such as Cadence Virtuoso. This way, the performance of the circuit can be optimized and many error sources can be omitted before fabrication. Designing the circuit must be based on the design metrics defined for the desired application. Different types of circuits reported for capacitive biosensors will be reviewed in Chapter 4.

2.2.3   Sample preparation

For portable and implantable devices, it is required to automate sample preparation and manipulation steps including centrifugation, droplet confinement, immuno-magnetic separation, applying electrokinetic force, and dielectrophoresis (DEP) [176]. However, in current studies, laboratory-based equipment is still required for sample preparation.

2.2.4   Microfluidics

The simplest approach employed in rapid prototyping methods (which will be elaborated in Section 5.2.1) to keep the analyte in contact with the sensing sites is using reservoir tanks including preferably a closable lid to prevent evaporation and the subsequent variation of the sample concentration. But there is a growing need to develop microfluidics and adapt them with integrated biosensors for full automation of delivering and also manipulation of small-volume analytes. As shown in Figure 2.10, continuous-flow microfluidics and digital microfluidics (or droplet-based microfluidics) are two main types of microfluidics. Continuous-flow microfluidics deal with handling the continuous flow of a fluidic sample through microchannels which include pumping, mixing, and separation of biospecies. But digital microfluidics are used for manipulating discrete droplets which can be categorized as droplet-based and liquid-marble-based microfluidics [182]. Continuous-flow microfluidics are suitable for high-throughput applications, but they usually need bulky external liquid delivery and optical microscopy for characterization. Digital microfluidics is the solution for the problem of the relatively large sample volume as well as bulky external systems.

2.2.5   Experimental setup and test

After design and fabrication of the chip, DAQ system, and microfluidics, the whole system along with the signal processing devices such as microcontrollers or power

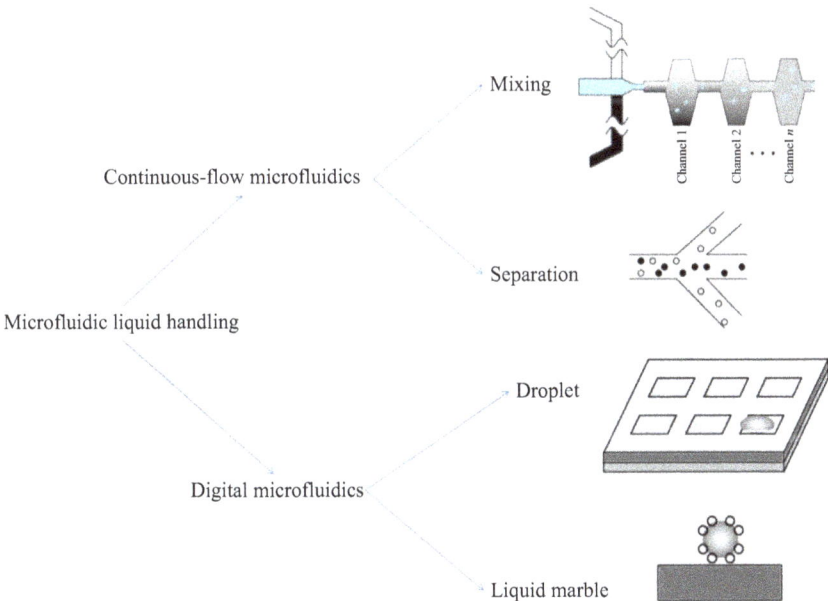

Figure 2.10   Microfluidic liquid handling

supply must be prepared. For portable and implantable devices, such equipment should also be supported by miniaturized devices. However, in the primary tests, the platform is designed and tested by reliable devices and even other laboratory equipment such as microscopes and commercialized biosensors alike are usually used to compare and verify the responses. Figure 2.11 shows an experimental setup as an example which includes field-programmable gate array (FPGA) platform (clock generator and decoder), microprobe, and power supply, as reported in [183].

2.2.6 Calibration

Various errors can deviate the sensor's response from an accurate measurement result. These errors might stem from the noise of the system, environmental factors (like temperature), remnants, device-to-device variations, etc. So, there is a need for a calibration strategy to adjust the sensor to function as accurately as possible. Calibration techniques are usually based on comparing the values measured by the designed sensor with those of a calibration standard of known accuracy. For example, the results achieved by standard laboratory-based instruments and chemical analytes with known dielectric values can be used to prepare a calibration curve or lookup table. Data analysis and in-circuit trimming are among the most popular techniques reported in the literature [13,142].

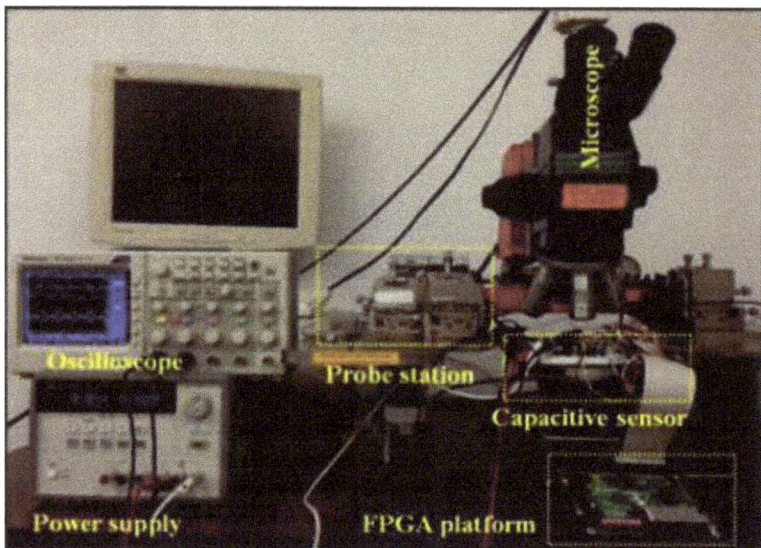

Figure 2.11   An experimental setup including oscilloscope, FPGA platform (clock generator and decoder), microprobe, and power supply, reported in [183]

2.3 Summary

In this chapter, different parts of a capacitive biosensing system were reviewed. To give an overview of the important features that must be considered before designing such a system, some of the main design metrics are also introduced in this chapter. Then, different steps for design, implementation, and test are outlined. It is noteworthy that, in practice, there are many factors that should be taken into account such as temperature and other environmental factors, disposability of the sensor, and diffusion of ions from the sample solution into the chip.

Chapter 3

Microelectrodes

Since the electrodes used in biochemical sensors are in contact with the liquid analyte, first we describe the related principles of such capacitive transducers in this chapter. Then, after the introduction of different configurations of these electrodes, various techniques reported for their fabrication in complementary metal-oxide semiconductor (CMOS) technology will be reviewed.

3.1 Electrode–solution interfaces

In electrode–solution interfaces, a working electrode (WE), usually passivated by a thin insulating layer, is functionalized as one of the plates of the capacitor and is immersed in an electrolyte solution, through an auxiliary electrode of a two-electrode configuration, as the other plate [184]. Under a polarization voltage, an electric double layer (EDL) is formed at the WE–solution interface characterized by an EDL capacitance (C_{dl}) (see Figure 3.1). This capacitor is composed of two layers of ions with opposite polarities at the surface of the metal electrode and in the electrolyte solution separated by a single layer of solvent molecules adhered to the surface of the WE as the dielectric of the capacitor.

Solvent molecules and sometimes specifically adsorbed ions or molecules in the solution side make Helmholtz or Stern layer. The nearest solvated ions to the metal electrode create the outer Helmholtz plane (OHP) and the specifically adsorbed ions make the inner Helmholtz plane (IHP) [185]. The Helmholtz capacitance, C_H, is defined as [186]:

$$C_H = \frac{\varepsilon_0 \varepsilon_r}{d_{OHP}}$$ (3.1)

where d_{OHP} stands for the distance of the OHP from the metal electrode. The distribution of the nonspecifically adsorbed ions also creates a diffuse layer (proposed by Gouy [187] and Chapman [188]) extending from the OHP into the bulk of the solution. The concentration of electrolyte in the solution, and consequently its conductivity, affects the thickness of the diffuse layer which is represented by Debye length (L_D) and can be modeled by a capacitor denoted by C_{GC} [186]:

$$C_{GC} = \frac{\varepsilon_0 \varepsilon_r}{L_D} \cosh\left(\frac{zV_0}{2V_T}\right)$$ (3.2)

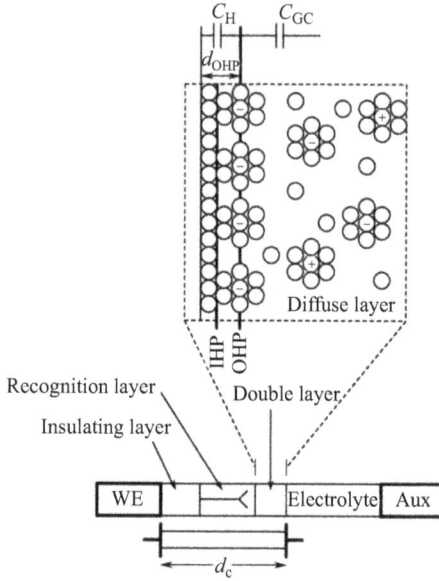

Figure 3.1   Electrode–electrolyte interface

where V_T is thermal voltage ($V_T = k_B T/q$, k_B is the Boltzmann constant, T defines temperature, and q is an electronic charge) and V_0 is the potential at the electrode. z denotes the valence of the ion. Interfacial capacitance, C_I, is given by the series combination of Helmholtz (Stern) layer and diffuse layer:

$$\frac{1}{C_I} = \frac{1}{C_H} + \frac{1}{C_{GC}} \tag{3.3}$$

Non-passivated electrodes result in lower impedance than passivated ones. However, the surface of the non-passivated electrodes is usually coated by a thin chemical or biological layer, considered as a semi-passivated layer, to reduce the faradaic current between the electrolyte solution and the electrode. The capacitance of the insulating layer is identified by C_{ins}.

The thickness of the EDL is determined by the thickness of the IHP, OHP, and diffuse layers. After modification of the electrode surface with an insulating layer and with a recognition layer, the distance between the plates will be increased. After specific probe–target binding, the distance between the plates will become greater. So, the electrode–solution interface behaves like a capacitor. Consider C_{bio} stands for a combination of the surface functionalization layer, biological recognition elements (BREs), grafting molecules, binding of analytes, and any contribution of the Helmholtz layer. Then, the total capacitor, C_{tot}, can be represented by three capacitors, C_{ins}, C_{bio}, and C_{GC}, in series:

$$\frac{1}{C_{tot}} = \frac{1}{C_{ins}} + \frac{1}{C_{bio}} + \frac{1}{C_{GC}} \tag{3.4}$$

The binding between BREs and target biospecies changes C_{bio} and consequently C_{tot}. The smallest capacitance causes the most change in C_{tot}. So, the insulating layer must have a high dielectric constant and low thickness resulting in large values for C_{ins} [57].

3.2 Capacitive transducers and their models

In a simple structure, the capacitor of a capacitive sensor consists of two parallel plates as electrodes with a dielectric material in between them and the value of the capacitance is obtained by

$$C = \varepsilon_0 \varepsilon_r \frac{A_c}{d_c} \tag{3.5}$$

where ε_0 and ε_r are the vacuum permittivity and the relative permittivity of the dielectric material, respectively. A_c identifies the overlapping area of the electrodes and d_c denotes the distance between the plates. Variations of each one of the parameters ε_r, d_c, and A_c change the capacitance value. To date, the latter has not been reported for biochemical sensing [57].

Generally, capacitive devices used for biochemical sensing can be categorized into three general groups: (1) floating electrodes, (2) interdigitated electrodes (IDEs), and (3) surface stress-based devices.

3.2.1 Floating electrodes

In floating electrodes, an electrode–solution interface is formed at the surface of a WE immersed in the liquid sample and the reference electrode (RE) is ideally assumed to be grounded in an infinite distance. In CMOS technology, the top metal layer is used to implement the WEs and a grounded wire forms the RE which is connected through the bonding pad. For example, Prakash *et al.* [189] utilized this configuration for living cell monitoring and tracking cell proliferation. Figure 3.2(a)–(d) illustrates such a floating electrode with a cell exposed to a weak low-frequency electric field. Being an ionic solution, the cell-culture medium acts as a conducting layer above the electrode. This being the case, the cells play the role of an insulation layer between the two conducting materials (ionic medium and the metal electrode) and the exposure of low-frequency electric fields leads their membranes to become polarized as shown in Figure 3.2(a)–(c). When a cell suspension comes into contact with the electrode, cell–substrate interaction goes through three phases including sedimentation, adhesion, and proliferation of cells. Figure 3.2(d) shows cell adhesion and proliferation phases. The sensed capacitance, C_{sensed}, can be expressed by

$$\frac{1}{C_{sensed}} = \frac{1}{C_{cell} + C_f} + \frac{1}{C_{ins}} + \frac{1}{C_I} + \frac{1}{C_{gm}} \tag{3.6}$$

where C_{cell} and C_{ins} stand for cell layer capacitance and insulation layer capacitance, respectively. C_I is the interfacial capacitance that can be modeled by the Gouy–Chapman–Stern theory [190]. C_f is fringe capacitance arising from the lateral

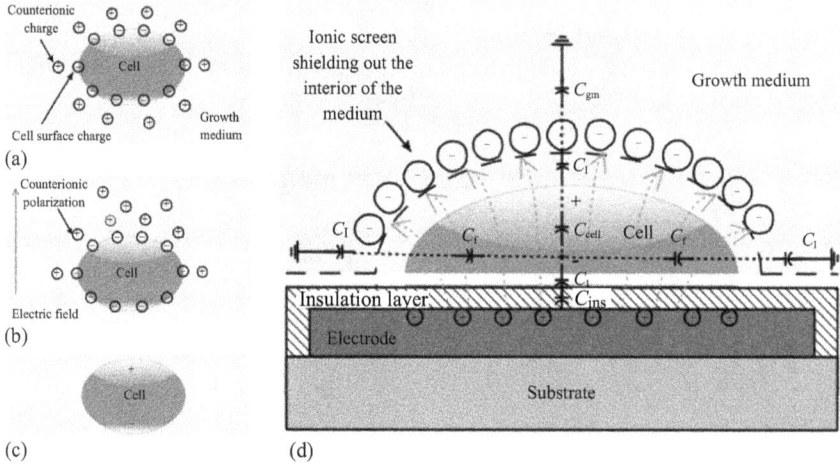

Figure 3.2   (a) Insulation cell suspended in a conductive growth medium; (b) induced polarization on exposure to an electric field; (c) induced cell dipole; (d) floating capacitor and a cell exposed to an electric field (in cell adhesion, and proliferation phases)

coupling of the electrode with neighboring metal lines. As depicted in Figure 3.2(d), an ionic screen is produced by the growth medium shielding out its interior from the sensing field. During adhesion and proliferation, the cells are inside the ionic screen and exposed to the sensing field. The bulk of the growth medium can be considered as an ideal ionic conductor that is electrically neutral under equilibrium conditions. C_{gm} is the total of all capacitive couplings of this conductor above the cell layer to all the neighboring conductors (voltage biases, power supply, and ground metal resting at DC potentials) except the sensing electrodes. Given that the adherence of healthy cells to the electrodes is more tightly than dead or unhealthy cells, the healthier cells create higher values of sensed capacitance due to the stronger capacitive coupling between them and underlying electrodes [189].

3.2.2   Interdigitated electrodes

The operation of IDE is based on the changes of dielectric constant (see Figure 3.3(a)). IDEs can be functionalized with probe molecules that can react with target molecules and monitor the reactions as a capacitance or impedance change (see Figure 3.3(b)).

If the electrodes are thick and edge effects can be neglected, the capacitance of an IDE (see Figure 3.3(a)) can be simplified as (3.7) [57]:

$$C = n_{IDE}\varepsilon \frac{l_{IDE}t_{IDE}}{d_{IDE}} \tag{3.7}$$

where n_{IDE}, t_{IDE}, and l_{IDE} are the numbers of fingers, the thickness, and the length of IDE, respectively, and ε denotes the permittivity of the coated material and d_{IDE}

stands for the distance between electrodes. Den Otter *et al.* [191] approximated the capacitance of an IDE by assuming the negligible thickness of the fingers in comparison to the other dimensions, a large number of parallel strips with uniform width (w_{IDE}), and the space in between two of which (d_{IDE}). Considering λ_{IDE} is IDE spatial wavelength or the periodicity of the fingers and is defined as $\lambda_{IDE} = 2$ ($w_{IDE} + d_{IDE}$), they proved that

$$C = l_{IDE} a_{IDE} \frac{8}{\pi \lambda_{IDE}} \varepsilon \sum_{n=1}^{\infty} \left[\frac{1}{2n_{IDE} - 1} J_0^2 \left(\frac{(2n_{IDE} - 1)\pi d_{IDE}}{\lambda_{IDE}} \right) \right] \tag{3.8}$$

where J_0 is zeroth Bessel function of the first kind and a_{IDE} identifies the width of the stripped area. The locations of charges (between or on the strips) are not discriminated in this approximation. This equation shows that although the capacitance of IDE does not change linearly with the variations of its area, there is a linear relationship between the capacitance and the permittivity of the dielectric.

The smaller sizes of IDEs in comparison to traditional electrodes have benefits like higher sensitivity, consuming tiny amounts of samples and consequently compactness, simplicity, and portability. However, they are limited in the production of the short and reproducible distance between the fingers [192]. The performance of an IDE-based sensor is optimized when the features of IDE are comparable with the size of target biospecies [193]. For the cases where IDEs are nanoscaled, more complicated modules and more expensive production facilities are required. Gerwen *et al.* [194] have proven that the electric field between the electrodes is restricted in a layer thinner than λ_{IDE}. Some analytical models are developed for the characterization of the corresponding capacitance of IDEs and their frequency response [195–199]. Igreja *et al.* [196,197,200] have shown that the capacitance of the IDE saturates at a layer thickness of $\lambda_{IDE}/2$ above the surface. Since analytical evaluation of IDE capacitance is complex, some numerical approximations like numerical Finite Element models [201] have also been developed to do so. In order to simulate and analyze the behavior of IDEs different software such as COMSOL, Sonnet, etc., can be used. An example is explained briefly in Appendix A.

In addition to planar structures of IDE [91], some three-dimensional (3D) IDEs [202,203] are also introduced for capacitive and impedimetric biosensing in which insulating barriers are used between adjacent digits of the electrodes. An analysis of impedance for both planar and 3D structures is presented in [204] indicating that planar IDE provides lower impedance but a higher sensitivity in impedance change and 3D structures offer only a higher interdigit surface area and more space for molecular immobilization. But the utilized model in this study did not consider the surface-charge-governed properties [205–207]. In [208], another comparison between planar and 3D structures is presented using a model taking into account these properties. The surface resistance causes both IDE structures to show a nonlinear behavior in low-ionic strength solution. Bäcker *et al.* [208] assembled a charged polyelectrolyte layer onto the barrier surface that affected the surface resistance. In this study, 3D IDEs showed higher surface-charge sensitivity than the planar IDEs with the same fingerprint.

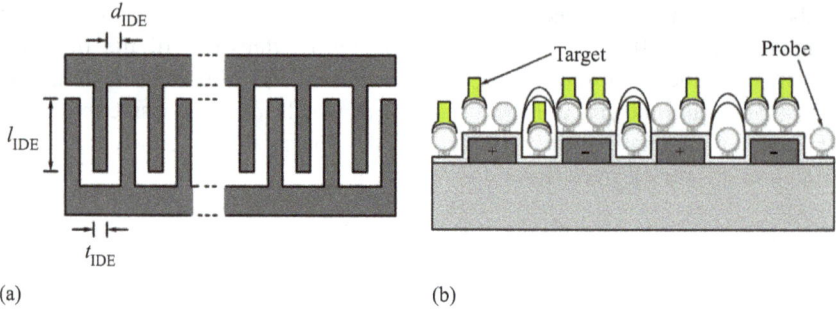

(a)   (b)

Figure 3.3   (a) Top-view of an IDE and (b) cross section of a functionalized IDE with probe–target binding

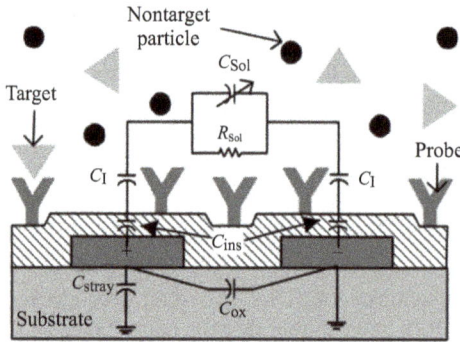

Figure 3.4   A coplanar IDE, immobilized probes, and target biospecies

IDEs as the most popular coplanar electrodes composed of two planar electrodes can be fabricated on the topmost metal layer in CMOS technology. A wide range of applications has been reported for them such as the detection of neurotransmitter dopamine (DA) [115], bacteria growth monitoring [68], and drug test [28].

Figure 3.4 shows a simple electrical model of a coplanar IDE in which C_{stray}, C_{inter}, C_I, R_{sol}, and C_{sol} denote the stray capacitance of the WE terminal, the dielectric interconnect capacitance, the C_{dl} created in the electrode–solution interface, as well as the electrolyte solution resistance and capacitance, respectively.

3.2.3   Surface stress-based devices

These devices are attracting more interest in recent years. Surface stress-based devices, such as cantilevers and capacitive membranes, work based on altering the distance between the plates. They include a rigid and a flexible electrode. The flexible plate is functionalized with probe molecules. When the probes interact with the appropriate target biospecies, they induce surface stress changing the deflection

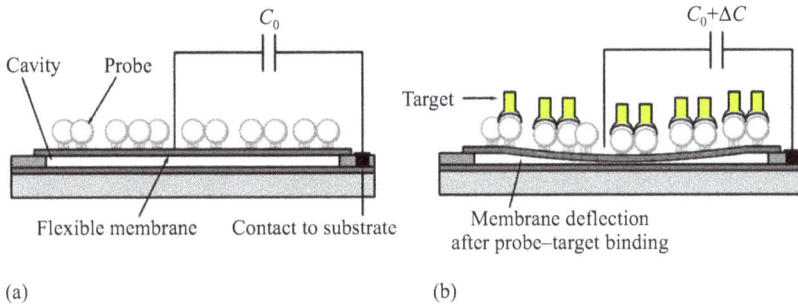

Figure 3.5 *The operation of the surface stress-based devices (capacitive membrane): (a) before binding; (b) after probe–target binding (adapted from [57])*

of the flexible structure and the distance between the two plates of the capacitor (Figure 3.5(a) and (b)). If we show the displacement of the flexible electrode with $w_{ss}(x,y)$, (3.9) can be modified as [57]

$$C = \varepsilon_0 \varepsilon_r \iint\limits_{A_c} \frac{1}{(d_c - w_{ss}(x,y))} \qquad (3.9)$$

Micro-cantilevers show high sensitivity. Differently functionalized cantilevers can be parallelized in an array structure. Usually, optical or piezoresistive approaches are utilized to detect cantilever bending. But optical techniques require costly and bulky setups and it is difficult to use them in opaque liquids like blood. On the other hand, piezoresistive approaches suffer from less sensitivity and temperature dependency. Capacitive sensing can provide highly sensitive detection. However, the electrolyte solutions cause faradaic currents between the plates of cantilever biosensors, make capacitive detection infeasible, and reduce the signal-to-noise ratio (SNR). So, cantilevers are usually used for air-based and gas sensors and require microelectromechanical system processes [209,210].

Using a membrane instead of a cantilever and sealing the capacitor plates from the electrolyte solution provides reliable detection. Furthermore, the sensitivity of the micromembrane capacitive biosensors is comparable with cantilever sensors using optical detection [184,211,212]. The membrane can be made of a conductive material, like highly doped Si, or might be constructed based on a thin conductive layer on an insulator, like Si_3N_4, silicon dioxide (SiO_2), or polymers such as Parylene, poly(methyl methacrylate) (PMMA), and poly(dimethylsiloxane) (PDMS) [184].

3.3 CMOS-based integrated electrodes

This section is a review of the techniques used for the implementation of sensing electrodes for different applications. These techniques employ standard CMOS and post-CMOS processing structures. Aluminum (Al) and Si impurities are the main

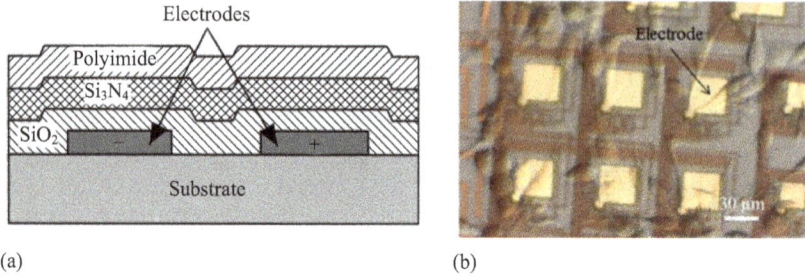

(a)   (b)

Figure 3.6   (a) Cross section of a typical passivated electrode (IDE); (b) the
floating electrodes fabricated by passivated topmost metal reported
for tracking cancer cell proliferation [189]

materials constituting the topmost metal layer of standard CMOS technology. For instance, the topmost metal layer of 0.18 μm process consists of Al with 1% Si [213]. As shown in Figure 3.6(a), some passivation layers made of SiO_2, Si_3N_4, and poly-imide (PI) are stacked on the topmost metal layer. Table 3.1 compares different electrode structures for various applications which can be divided into eight general categories: (1) passivated electrodes; (2) high-sensitivity passivated electrodes; (3) quasi IDEs; (4) Al/Al_2O_3 electrodes; (5) polymer-coated electrodes; (6) Au-coated electrodes; (7) platinum (Pt)-coated electrodes; (8) titanium nitride (TiN) electrodes.

3.3.1   Passivated electrodes

Figure 3.6(a) illustrates a typical passivated electrode structure in which the pas-sivation layers of standard CMOS technology are not removed from the metal layers of this technology. Detection of neurotransmitter DA [115] and cell growth monitoring [79,189,220,221] are some of the applications reported for this sensing structure. A microphotograph of this type of sensing electrodes is depicted in Figure 3.6(b) which is used for tracking the proliferation of cells by Prakash *et al.* [189]. According to Table 3.1, they used three different sizes of 20 μm × 20 μm, 30 μm × 30 μm, and 40 μm × 40 μm in their experiments.

Lu *et al.* [115,222] used the Al layer (Metal 3) of two-polysilicon-four-metal (2P4M) 0.35-μm CMOS technology as the electrode material and employed the inter-metal SiO_2 on Metal 3 for the immobilization of the probe molecules required for the detection of neurotransmitter DA (see Table 3.1). The top passivation layers of this technology (Si_3N_4 and SiO_2) were removed by dry etching (see Appendix B, Section B.1.2). To improve the sensitivity, it is possible to thin down the remaining oxide thickness by the optional dielectric reactive etching process.

The equivalent parasitic capacitance of the coating passivation layer is in series with the sensing capacitance. So, a thinner passivation layer and consequently a larger parasitic capacitance can lead to a higher sensitivity of total capacitance measurement. However, a custom process [79] is required to reduce the thickness of the passivation layers. In some applications where specific bioparticles and BREs should be immo-bilized above the electrodes, passivated electrodes cannot meet the needs.

Table 3.1 CMOS-based electrodes used for capacitive and impedance sensing

Tech.	Application	Target biospecies	Sensing element material	Electrode size (μm^2)	Array #	Ref.
0.5-μm 2P3M	Cell proliferation monitoring	Breast cancer cells	Al/1Pass[1]	20 × 20, 30 × 30, 40 × 40	28	[189]
0.35-μm 2P4M	Detection of the neurotransmitter DA	CPBA	Al/1Pass	20 × 27, 30 × 39, 50 × 67, 100 × 135, 150 × 203	5×5	[115]
0.25 μm	Bacteria detection	S. epidermidis	Al/Al₂O₃	220 × 230	1	[109]
0.25 μm	Detection of single bacterial cell	S. epidermidis	Al/Al₂O₃	14 × 16	16 × 16	[20]
0.18-μm 1P6M	Rapid identification of bacteria	E. coli	Al/Al₂O₃	~ 35 × 35	2	[214]
0.18 μm	Bacteria growth monitoring	E. coli	Op-fin[2]	100 × 100	3	[68]
0.35 μm	Pathogenic bacteria detection	Pathogenic bacteria	Q-IDE[3]	100.3 × 102.8	30	[78]
0.35 μm	Cell growth monitoring	Human lung carcinoma cell	Al/PDMS	50 × 50	8 × 8	[21]
0.35 μm	Cell culture monitoring and drug test	H1299 and Hek293 cells	Al/Polyelectrolyte	50 × 50	8 × 8	[28]
0.5 μm	Detection of DNA hybridization	DNA	Au	200 × 200	8 × 16	[117]
0.35 μm	EIS	DNA, proteins	Au	40 × 40	10 × 10	[70]
90 nm	High-frequency impedance spectroscopy	Insulating and conducting particles	AuCu	0.025	256 × 256	[144]
0.18-μm 2P6M	Stimulation, impedance measurement, and fast CV for neurotransmitter detection	Neurotransmitter	Pt	3 × 7.5	332 × 180	[66]
0.35 μm	Intracellular electrophysiological imaging	Neonatal rat ventricular cardiomyocytes	Pt	-	32 × 32	[215]
0.6 μm	Extracellular recording and stimulation of electrogenic cells	Chicken dorsal root ganglion neurons	Pt	~ 24.5	11,011	[216]
0.35 μm	Recording and stimulation of electrogenic cells in vitro	Rat cortical neurons	Pt	9.3 × 5.4	26,400	[217]
0.18 μm	Impedance spectroscopy and electrophysiological imaging of cells	Cardiac cells, brain slices	Pt	1 × 1, 2 × 2, 4 × 4, 8 × 8	59,760	[218]
0.13 μm	Electrical recording and impedance imaging for neuro-electrophysiological studies	Primary hippocampal neurons	TiN	11 × 11, 7 × 7, 4.5 × 4.5, 2.5 × 3.5	16,384	[219]

[1]Topmost metal with one passivation layer
[2]Passivated IDE with a window in between the fingers
[3]Quasi IDE

3.3.2   High-sensitivity passivated electrodes

To boost the electric field in between the conductors of electrodes and consequently enhance the sensitivity, the passivation layer in between the conductors of electrodes can be removed by the pad-etch process as shown in Figure 3.7(a). However, the length, width, and thickness of electrodes are less than the pad. So, design rule check violations are required for their simulation in Cadence software. Since the electric field generated in the space between parallel electrodes are larger than the electric field formed above them, increasing the height of the sensing electrode by using a thick topmost metal layer will improve the performance of the sensor. Ghafar-Zadeh *et al.* used such a structure by removing the passivation layer in between the fingers of IDEs [75,223] as seen in Figure 3.7(b). It is worth mentioning that the mask must be matched with the gap between the fingers. The lack of any stop layer above the substrate in these electrodes led to a little bit of erosion in the substrate and consequently deepening the gap in Figure 3.7(b). These electrodes were fabricated in the size of 100 μm × 100 μm by using 0.18-μm CMOS technology (see Table 3.1). This group took advantage of this structure for chemical solvent detection [224] and bacteria growth monitoring [68].

3.3.3   Quasi IDE

Quasi IDEs are implemented in two topmost metal layers as seen in Figure 3.8(a) without the passivation layers in such a way that the beneath electrode is utilized as the reference and the top electrodes play the role of WEs. The pad-etch process is used to remove the passivation layers on IDEs. This structure [78,225–227] leads to strengthen the electric field around the electrode and enhances the sensitivity. In some literature such as [226,228,229], this structure has been used as a CMOS capacitive gas sensor as shown in Figure 3.8(b). A polymeric thin-film covering the surface of the chip is usually used for gas sensing applications. For example, a quasi IDE coated with a polymer is reported by Kummer *et al.* [225] for the detection of volatile organics in ambient air. In another effort, Tashtoush *et al.* [78] used this structure for pathogenic bacteria detection by coating the sensing area with bacteriophage as the BRE (see Table 3.1) and covering unexposed regions with epoxy.

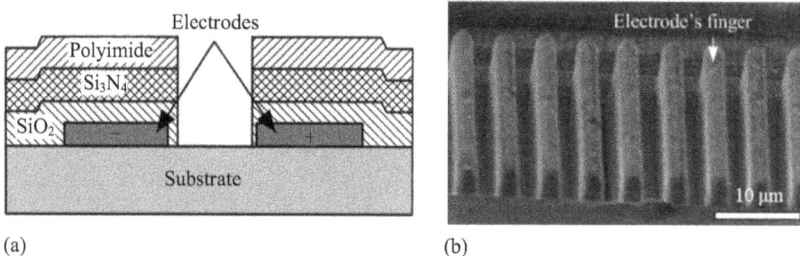

(a)   (b)

Figure 3.7   (a) Cross section of a passivated IDE with a window in between the fingers; (b) the IDE utilized in [223] in which the passivation layers between the fingers are removed

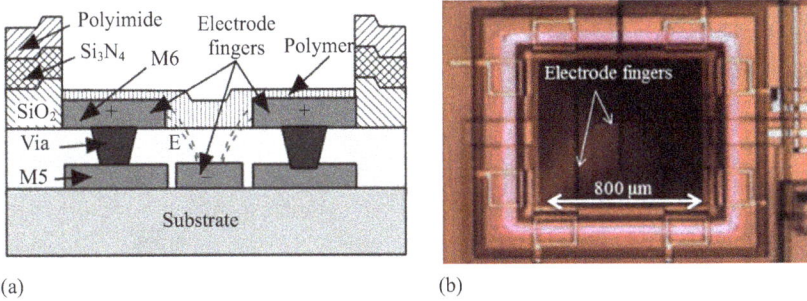

(a)   (b)

Figure 3.8   (a) Cross section of a quasi IDE; (b) a polymer-coated quasi-interdigitated sensing capacitor [229] for gas analysis

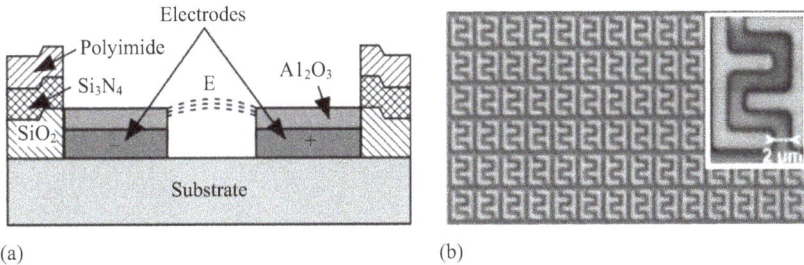

(a)   (b)

Figure 3.9   (a) Cross section of a bare electrode; (b) post-processed Al/Al₂O₃ IDE array utilized in [20]

3.3.4   Al/Al₂O₃ electrodes

Al is very prone to oxidization. Although removing the passivation layers using a standard pad-etch mask in CMOS processes leads the Al electrode to be exposed to the sample directly, Al_2O_3 can be formed natively above the bare Al electrodes in exposure to air (as shown in Figure 3.9(a)) [183]. The Al_2O_3 layer acts as an insulating layer protecting the electrodes against the biological and chemical environment and provides a suitable surface for functionalization purposes [230]. Such bare electrodes are reported for various applications like bacteria growth [214], DNA detection [230], or chemical solvent monitoring [183]. As above-mentioned, an option to create a sensing area in between the electrodes is using thick topmost metal layers and also creating a large space between the electrodes by choosing a pad-etch mask [76]. For example, Nikkhoo *et al.* [214] only removed the passivation layer above the topmost metal layer and implemented two electrodes with a size of approximately 35 μm × 35 μm and a trench in between them for bacteria detection.

Al_2O_3 can also be deliberately coated above non-passivated Al electrodes by post-CMOS processes [20] (see Figure 3.9(b)). The electrodes used by Couniot *et al.* [20,109] which were also reported for bacteria detection (see Table 3.1) were not covered with SiO_2 and Si_3N_4 passivation layers, but to protect the used topmost metal layer from corrosion, a thin passivation layer of Al_2O_3 was post-processed and deposited by plasma-enhanced atomic layer deposition (PEALD) on it (see Appendix B, Section B.3.3.1).

3.3.5   Polymer-coated electrodes

Usually, sensing electrodes are coated with polymeric membranes (as depicted in Figure 3.10(a)) to provide higher sensitivity for gaseous samples [226,228,229]. Additionally, a layer of polymer helps to improve biocompatibility as reported by Nabovati *et al.* [21]. They spin-coated (see Appendix B, Section B.6) a PDMS layer on top of bare Al electrodes and used them for cell growth screening as shown in Figure 3.10(b). In addition to polymers, the same group [28] also reported the advantage of polyelectrolyte multilayer (PEM) as an intermediate layer to improve cell adhesion for applications like cell growth monitoring and drug test. They used a layer-by-layer (LBL) coating technique to deposit PEM on Al electrodes and showed the functionality of these electrodes in life science applications. As mentioned in Table 3.1, an 8×8 array of these electrodes was implemented in both works, [21] and [28]. It is worth mentioning that some bioactive coating approaches based on PEMs are also useful for applications like glucose detection [231,232] and drug delivery [233].

3.3.6   Au-coated electrodes

Au, as a good electron conductor and a biocompatible, stable, inert, and noble metal [76,234], is among the most widely used materials for different biosensing applications such as anti-Rubella [235] and DNA detection [117,236,237]. Furthermore, Au allows self-assembled monolayer (SAM) to be formed as a linker between electrodes [76] and different BREs like glucose oxidase [238] and DNA oligonucleotides [239] helping the immobilization of these BREs on Au electrodes. However, the fabrication of Au electrodes above the CMOS chip requires post-CMOS processing steps. Au can be deposited on different materials such as Pt, copper (Cu), palladium (Pd), and Al [240] or nickel (Ni) [241]. An interlayer like chromium (Cr), zinc (Zn), or Ni is usually needed to improve the Au adhesion on Al (see Figure 3.11(a)) [234].

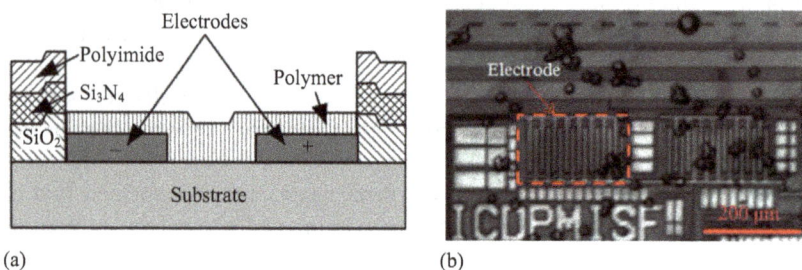

(a)   (b)

Figure 3.10   *(a) Cross section of a typical polymer-coated electrode; (b) the polymer-coated electrodes utilized in [21] for cellular analysis*

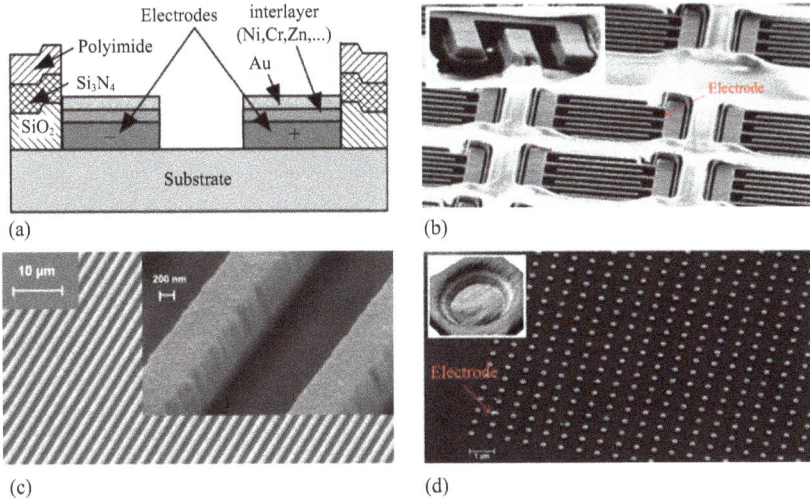

Figure 3.11 (a) Cross section of a typical Au-coated electrode; (b) the nano-well structure and the chip reported in [235]; (c) Au IDEs for DNA chip used in [117]; (d) array of Au nanoelectrodes with a zoom-in image of a single electrode reported by Widdershoven et al. [144,242]

Several techniques are reported to deposit Au on Al electrodes in CMOS technology like photolithography and electroless deposition [235,240,243]. In the photolithographic deposition of Au as shown in Figure 3.12, standard photo-lithography steps are used to deposit Au. In this technique, the positions of Au electrodes are patterned by a photomask on a photoresist as a sacrificial layer. Then, an interlayer of Ni or Cr is deposited using e-beam or thermal evaporation [227,234] (see Appendix B, Section B.5.2). In the next step, Au is deposited using sputtering or e-beam techniques (see Appendix B, Section B.5). Finally, the liftoff process (see Appendix B, Section B.2) helps to remove the excess Au on unre-quired regions [234]. Guiducci *et al.* [237] used this technique for capacitive DNA sensing and employed tantalum (Ta) as the interlayer between Au and Al (see Figure 3.11(c)). According to Table 3.1, this group created an array of 128 Au electrodes for the same application [117]. The alignment of a photomask with the fabricated electrodes is the main difficulty of this technique.

Electroless deposition, usually known as electroless nickel immersion gold (ENIG), is an alternative to tackle the problem mentioned for photolithography. As shown in Figure 3.13, after cleaning, rinsing and washing, and drying the chip, a three-step process is employed to deposit Au on Al. In the first step, the chip is placed in a Zn plating solution, and consequently, Al ions are replaced by Zn ions. In the next step, the immersion of the chip in an Ni plating solution causes the replacement of the Zn layer with Ni layer. Finally, the chip is immersed in Au resulting in adhesion of the Au atoms to Ni-plated regions [234]. In another effort, a two-step electroless deposition process is presented in [240] in which no interlayer

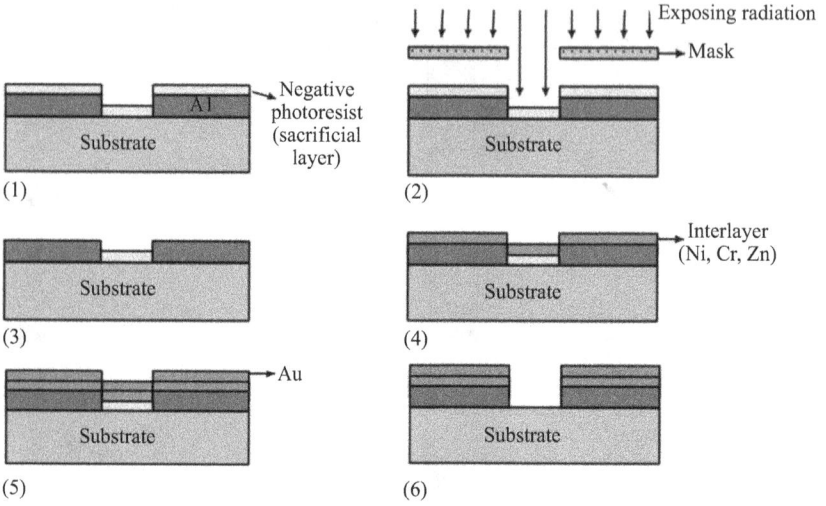

Figure 3.12   The process of photolithographic deposition of Au

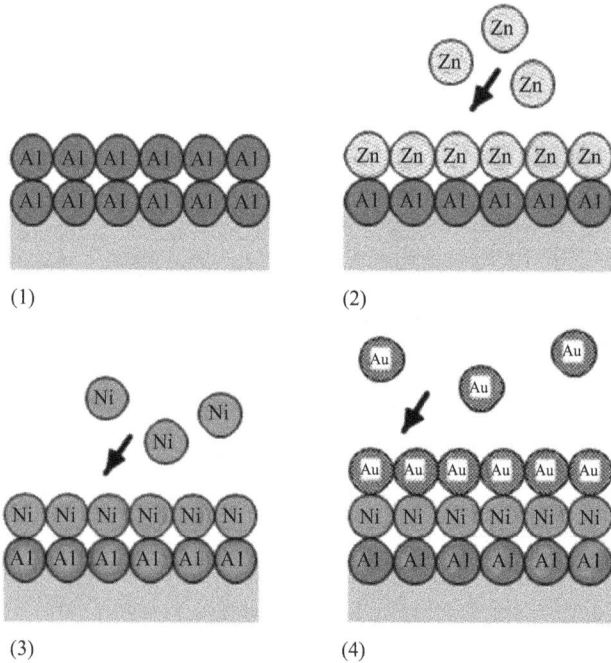

Figure 3.13   The process of ENIG

is used and only a small droplet of electroless solution can be utilized to perform it. In this method, after the treatment of the Al surface with Atomex solution, the Au can be deposited by galvanic displacement. In the second step, Catagold 2A solution is used for electroless deposition. Cheap self-aligned process and the possibility of making trench structures are the promising features of electroless plating making it suitable for the implementation of bacteria-sized Au microelectrodes. Using the trench structure improves sensitivity due to the minimization of the current flow from the peripheral part of the microelectrode which is unaffected by the target. Additionally, this structure allows for decreased pitches due to the mitigation of the expansion of the diffusion layer [60,244]. Sun *et al.* [235] used ENIG plating procedure to fabricate Au IDEs for a high-density 4096-pixel electrochemical biosensor array. As shown in Figure 3.11(b), 3D trenches are formed between the two combs by removing the passivation layer which are useful as nano-wells for cell culture. Manickam *et al.* [70] also implemented biocompatible Au microelectrodes using ENIG for the detection of DNA and proteins.

In another work, Widdershoven *et al.* [144,242,245] took advantage of nano-electrodes for high-frequency impedance spectroscopy and imaging. Figure 3.11(d) shows a single nanoelectrode and an array of nanoelectrodes reported by this group. These nanoelectrodes are connected to a vertical pillar composed of a stack of four Cu islands (vias) fabricated during CMOS process. The Cu of the fourth via is alloyed with the Au deposited on the chip surface and created Au/Cu nanoelectrodes. As mentioned in Table 3.1, the smallest size of electrodes is reported by this group which is utilized in a large array (256×256) in comparison with the other works.

3.3.7 Pt-coated electrodes

Pt is also an inert metal like Au, but not that much stable against oxidization. Additionally, the fabrication of these electrodes needs post-CMOS processing. Usually, Pt is deposited and patterned by photolithography, and titanium–tungsten (TiW) layers are used as an adhesion layer between Al and Pt [234] (see Figure 3.14(a)). Pt electrodes are reported for some applications like stimulation, impedance measurement, and recording of electrogenic cells such as cardiomyocytes or neurons [66,217,246,247]. In these works [66,217,247], Al electrodes are covered with Pt by performing a simple two-mask post-processing procedure. The electrode is shifted and the underlying Al is sealed with the electrode metal and a passivation stack. The electrode metal comprising TiW, an adhesion promoter, and Pt is sputter-deposited and structured through a liftoff process (see Appendix B, Section B.2). Then, an alternating Si_3N_4/SiO_2 passivation stack is deposited by PECVD (see Appendix B, Section B.3.2). A mixed frequency PECVD process matches the stress with that of the underlying Si_3N_4 of the CMOS process. Finally, reactive ion etch (RIE) (see Appendix B, Section B.1.2.1) is utilized to etch openings in the nitride (the passivation layer) and defines the shape and the size of the electrodes. As seen in Table 3.1, Dragas *et al.* [66] used 332×180 of such microelectrodes with the size of $3 \ \mu m \times 7.5 \ \mu m$ in their system occupying a large active area of $4.48 \ mm \times 2.43 \ mm$ (Figure 3.14(b)). Similarly, Frey *et al.* [216]

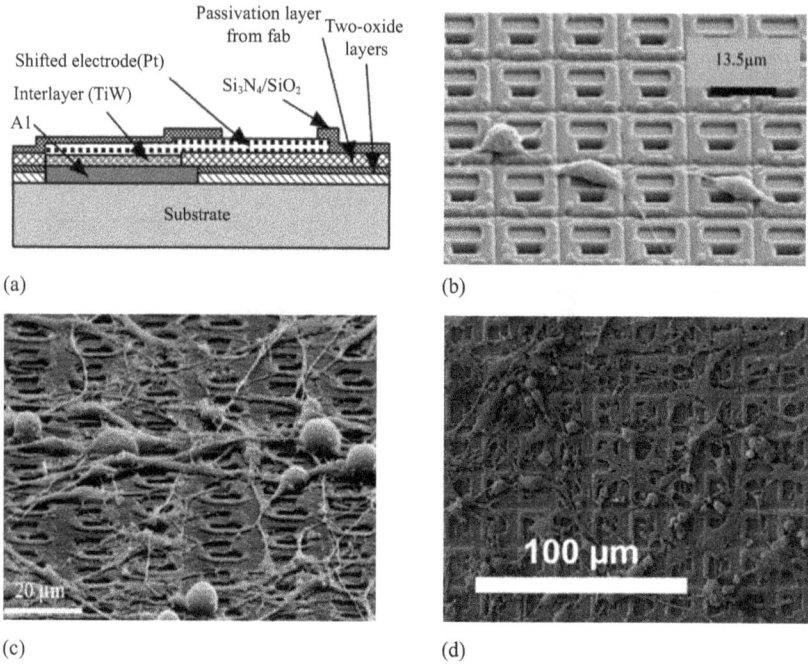

(a)   (b)

(c)   (d)

Figure 3.14   (a) *Cross section of a shifted Pt electrode (in floating configuration);*
    (b) *Pt electrodes with cultured rat cortical neurons on top of them*
    [66]; (c) electrodes plated with chicken dorsal root ganglion neurons
    [216]; (d) the post-processed Pt electrodes, plated with rat cortical
    neurons, reported in [217]

presented an 11,011 array of such Pt electrodes along with 126 channels for extracellular bidirectional communication with electrogenic cells (Figure 3.14(c)). The same group [217] employed such a post-processed shifted-electrode as shown in Figure 3.14(d) which is plated with rat cortical neurons. As seen in Table 3.1, a 1024-channel MEA with 26,400 of such Pt electrodes was implemented for recording and stimulation of electrogenic cells. Viswam *et al.* [218] used a similar approach to fabricate 59,760-array of Pt-microelectrodes densely packed within a 4.5 mm × 2.5 mm sensing region.

3.3.8   TiN electrodes

TiN electrodes (see Figure 3.15(a)) are other kinds of electrodes reported in some literature for biosensing applications [219,248]. Lopez *et al.* [248] reported an active MEA chip including 16,384 TiN electrodes which were fabricated on top of the Si substrate in a 0.13-μm Al CMOS technology using a six-metal-layer Al back-end-of-line (BEOL) stack (see Figure 3.15(b)). In the first step, the top of the last BEOL metal layer (M6) is passivated by the deposition of a layer of SiO_2 using

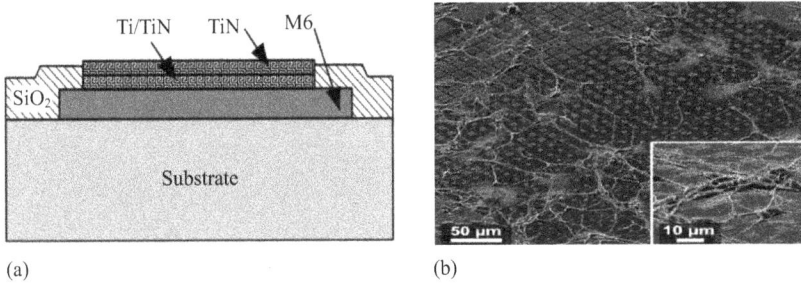

Figure 3.15 *(a) Cross section of a TiN electrode adapted (in floating configuration); (b) cultured neuron on a single active area including TiN electrodes [219]*

PECVD (see Appendix B, Section B.3.2). A second SiO_2 layer is also deposited to achieve an appropriate passivation thickness. In the next step, to reach the M6 contacts, RIE (see Appendix B, Section B.1.2.1) is used to open via holes in the passivation layer. Then, a Ti/TiN seed layer is deposited and covered by a TiN layer using reactive physical vapor deposition and RIE is employed to pattern this stack. This group [219] showed the functionality of their chip for neuro-electrophysiological studies. As seen in Table 3.1, four different sizes of these electrodes were presented on the chip surface, ranging from 2.5 μm × 3.5 μm up to 11 μm × 11 μm.

3.4 Summary

This chapter discussed the different techniques utilized for the implementation of sensing electrodes above CMOS chips. The appropriate configuration of the electrodes should be selected based on the employed biosensing method. The simplicity of the design and fabrication of these on-chip configurations is their important advantage. But choosing the minimum dimensions or the materials exposed to analyte are some of their limitations. Although nanoscale CMOS processes have paved the way for the implementation of large arrays of nanoscale electrodes, a compromise between the minimum features of the sensing electrodes, the speed of sensor, and the applied voltage or current should be taken into consideration.

Chapter 4

CMOS interface circuits of capacitive biosensors

After a brief review of different types of complementary metal-oxide semiconductor (CMOS) capacitive interface circuits in this chapter, the principle of charge-based capacitance measurement (CBCM) will be introduced and the development process of this kind of capacitive biosensors for life science applications will be outlined.

4.1 Different CMOS interface circuits of capacitive biosensors

There have been various types of CMOS capacitive interface circuits which can be categorized into eight groups: (1) charge-sharing method; (2) charge-sensitive amplifier-based and switched capacitor (SC) circuits; (3) capacitance-to-frequency converter (CFC) using comparator-based relaxation oscillators; (4) ring oscillator (RO)-based CFC; (5) voltage-controlled oscillator (VCO)-based sensors; (6) lock-in detection; (7) triangular voltage analysis; (8) CBCM.

4.1.1 Charge-sharing method

The charge-sharing method is a simple technique. In the circuit shown in Figure 4.1 [79,189], charging and discharging the capacitance of the two nodes N_1 and N_2 are controlled by switches M_1–M_3.

Whether the circuit works in the reset or evaluation phase depends on the applied controlling clock pulses. During the reset phase, the capacitances of nodes N_1 and N_2 are charged to V_{dd} and V_{ss}, respectively. During the evaluation phase, the charge is redistributed between the capacitances of nodes N_1 and N_2. It can be proved that the output voltage is equal to

$$V_{Out} = \frac{(C_{p1} + C_S)V_{dd} + C_{p2}V_{ss}}{C_{p1} + C_{p2} + C_S} \tag{4.1}$$

which is a function of the sensing capacitance (C_S). In this equation, C_{p1} and C_{p2} denote the parasitic capacitances of the nodes N_1 and N_2, respectively.

As another example, the charge-sharing technique proposed by Couniot et al. [20] is depicted in Figure 4.2. This figure shows one pixel of their sensor which was used in a 16×16 array for detection of a single bacterial cell. The controlling voltages impose four different phases to this circuit: initialization, reset, integration, and readout.

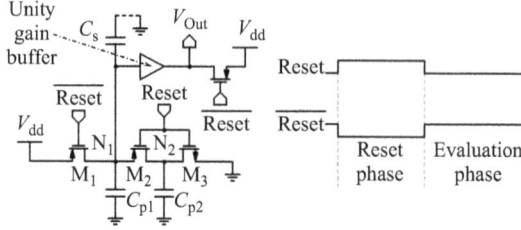

Figure 4.1   The charge-sharing structure utilized by Prakash et al.

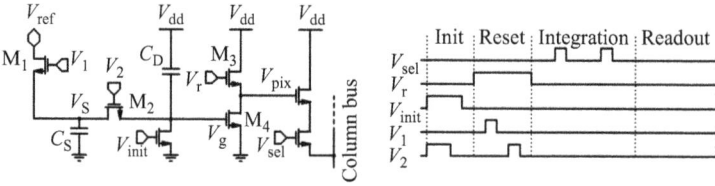

Figure 4.2   The charge-sharing technique utilized by Couniot et al.

The voltages V_g and V_S are grounded during the initialization phase. In the reset phase, switch M_3 sets V_{pix} to its reset value. First, C_S is loaded to V_{ref} via M_1, and then enabling M_2 results in the charge distribution between C_S and the fixed capacitance $C_D + C_g$ in which C_g is the capacitance of the node V_g. So, V_g is a function of C_S:

$$V_g = \frac{C_S V_{ss}}{C_D + C_g + C_S} \tag{4.2}$$

During the integration phase, V_g does not change from this value. When M_4 is placed in the subthreshold, V_g is amplified exponentially. In this phase, M_4 discharges V_{pix} progressively. The values of V_{pix} at the start and the end of the integration period are stored in column amplifiers (which are not shown in Figure 4.2). During the readout phase, these values are sent to an output stage. Finally, an off-chip analog-to-digital converter (ADC) converts the output analog voltages to digital. In comparison to [79,189], this circuit offers the advantages of more linear response, additional subthreshold gain stage, and the calibration capability with V_{ref}.

4.1.2   Charge-sensitive amplifier-based and SC techniques

In addition to the charge-sharing principle, the function of charge-sensitive amplifier-based sensors and SC circuits are also based on charge redistribution. A simple structure of a capacitive sensor employing a charge-sensitive amplifier is depicted in Figure 4.3(a) in which a voltage pulse (Φ) is applied to C_S [105–107]. When the pulse ends up, the charge of C_S transfers to a reference capacitor (C_R). Then, an integrator circuit comprising an amplifier and a feedback capacitor (C_{fb}) integrates the voltage across C_R. The reset switch prevents the saturation of the amplifier. It can be verified that the output voltage of the amplifier is proportional

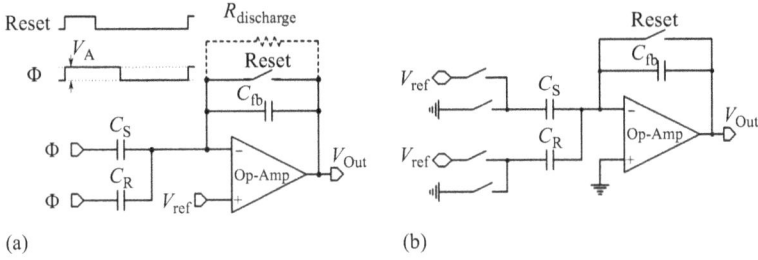

Figure 4.3   (a) A charge-sensitive amplifier; (b) a switched capacitor circuit

to the difference between C_S and C_R [106]. In other words, C_S is compared with the parallel C_R and after the completion of the integration, the output voltage of the integrator is calculated by

$$V_{Out} = V_A \frac{C_S - C_R}{C_{fb}} \tag{4.3}$$

where V_A stands for the amplitude of the Φ signal. The period of the reset signal is the same as the period of Φ but its duty cycle is lower. When the reset signal is high, C_{fb} is discharged. If the non-inverting input of the amplifier is connected to V_{ref} instead of ground, the V_{Out} is increased by V_{ref}.

SC capacitive sensors can be considered as charge-sensitive amplifiers in which the voltage pulse input is replaced by DC voltage input and several switches to provide the charging and discharging phases of C_S and C_R [106] (see Figure 4.3(b)). It is worth mentioning that the low impedance input of the amplifier can make it easy to cancel parallel parasitic capacitances.

A charge-sensitive amplifier is presented by De Venuto *et al.* [105] in which C_R is not used, and as depicted with a dotted line in Figure 4.3(a), the feedback capacitor is discharged via a parallel resistor ($R_{discharge}$) instead of a reset switch [106]. The varying current that discharges C_{fb} leads to V_{Out} of the form

$$V_{Out} = -\frac{Q}{C_{fb}} \exp\left(-\frac{t}{\tau}\right) \tag{4.4}$$

where $\tau = C_{fb} R_{discharge}$ and Q is the charge integrated by C_{fb} during each period of Φ. In [105], a technique by using Schottky diodes was presented to maintain the discharge current stable and increase the input dynamic range (IDR) of the charge-sensitive amplifier-based capacitive sensor.

The sawtooth output signal of the amplifier can be frequency modulated utilizing an amplifier and a counter. But if the output signal is a voltage level, an ADC [249,250] like successive sigma-delta ADC ($\Sigma\Delta$ ADC), successive-approximation register (SAR) ADC, pipeline ADC, and alike [106] can convert it to a digital representation of C_S. SC circuits can be easily adapted to various types of ADCs [4,251,252]. These sensors might directly convert C_S to digital or first, convert C_S to voltage and then digitize the voltage.

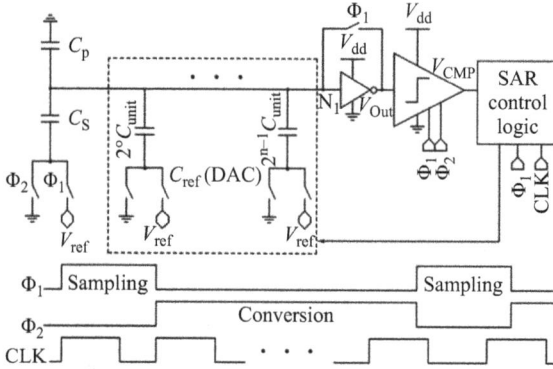

Figure 4.4   An SAR capacitance-to-digital converter (CDC) based on SC techniques

For example, Alhoshany *et al.* [4] proposed an SC sensor adapted to SAR ADC which is illustrated in Figure 4.4. This sensor includes a binary-weighted digital-to-analog converter (DAC), a single-ended comparator for cascaded amplification, and control SAR logic in addition to an amplifier with feedback loop. Two clock phases, Φ_1 and Φ_2, and a SAR clock (CLK) determine the operation phase of the circuit which might be sampling or conversion phase.

During sampling phase, Φ_1 is high and a unity-gain amplifier is formed by turning on the feedback switch and helps to charge the binary-weighted DAC, C_S, and parasitic capacitance, C_p, by using the switching threshold voltage (V_{N1}). While the bottom plates of the binary-weighted DAC are grounded by the switches, the bottom plate of C_S is connected to V_{ref}. To reduce the average power consumption, the comparator is turned off. During the conversion phase, Φ_1 becomes low, the feedback switch is turned off, and an open-loop amplifier is formed. Depending on the *n*-bit SAR logic output, the bottom plates of the binary-weighted DAC might be either connected to ground or V_{ref}. The comparator is turned on for cascaded amplification. The DAC is employed to compare the capacitive input with the digital output. It can be proved that the output voltage of the amplifier is given by

$$V_{Out} = V_{N1} + \Delta V_{Out} = V_{N1} + \frac{(C_S - C_{ref,ON})V_{ref}}{C_S + C_{ref} + C_p}A_{inv} \tag{4.5}$$

where the second term shows the output voltage changes. A_{inv} denotes the finite gain of the inverter. The output voltage of the comparator is obtained by

$$V_{CMP} = \begin{cases} 0, \Delta V_{Out} < 0 \\ 1, \Delta V_{Out} > 0 \end{cases} \tag{4.6}$$

Changing $C_{ref,ON}$ by the successive-approximation algorithm will be continued until $C_{ref,ON}$ matches C_S with an error less than one least-significant bit. The sign of $(C_S - C_{ref,ON})$ determines V_{CMP} regardless of the absolute value of ΔV_{Out}. Thus, the sensor is not sensitive to the variations of C_p and V_{ref}.

The switches of SC circuits can add some problems like charge injection and offsets to the circuits. The most famous techniques to mitigate these imperfections in SC circuits are auto-zeroing and double sampling which will result in higher power consumption and more complexity of the whole circuit.

4.1.3   CFC using a comparator-based relaxation oscillator

In these sensors, which are composed of a relaxation oscillator [115–117], C_S changes the output frequency. In a simple circuit like what is depicted in Figure 4.5, a current reference (I_B) charges and discharges C_S and a comparator compare the voltage of the capacitor with a threshold voltage (V_{th}). When this voltage reaches the V_{th} of the comparator, the switch controlled by the output signal of the comparator resets the voltage of the capacitor to zero. It can be proved that the output pulse frequency (f_{Out}) is inversely proportional to C_S [115]:

$$f_{Out} = \left(\frac{C_S V_{th}}{I_B} + \tau_d \right)^{-1} \tag{4.7}$$

where τ_d denotes the negligible delay created by the comparator.

In another configuration, a comparator with hysteresis (like Schmitt trigger) can be used to control the switching of two equal reference currents, which are source and sink, and charge and discharge C_S, respectively. Figure 4.6 shows a similar idea [117] by using a single reference current whose direction is reversed by switching. Also, switching is used to create two symmetric threshold voltages in two phases

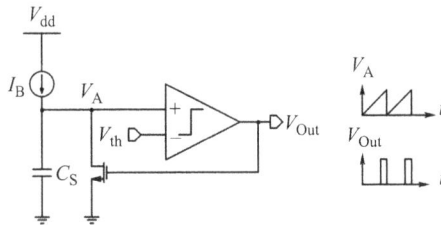

Figure 4.5   Comparator-based relaxation oscillator technique

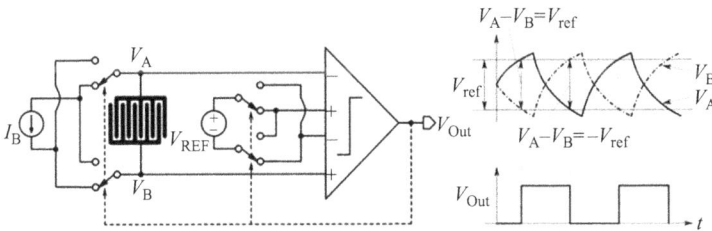

Figure 4.6   A capacitive biosensor utilizing comparator-based relaxation oscillator

with a comparator working with a single threshold voltage. Switching to push or pull mode leads to the transient voltage waveforms on the two plates of the sensing electrode, V_A and V_B, as shown in Figure 4.6. When $V_A - V_B$ becomes higher than V_{ref} or lower than $-V_{ref}$, the direction of the reference voltage and current are reverted simultaneously through the switches. The time constant of these waveforms is dominated by the capacitive component of the electrode–solution impedance. Considering solution resistance is negligible, it can be verified that the frequency of the generated squared waveform is inversely proportional to C_S:

$$f_{Out} = \left(2RC \ ln \left(1 / \left(1 - \frac{V_{ref}}{I_B R} \right) \right) \right)^{-1} \tag{4.8}$$

where R and C are the contribution of the resistance and capacitance of both electrode and solution. If the frequency is not too low, the following equation is given by the first-order Taylor approximation:

$$f_{Out} = \frac{I_B}{2V_{ref} C} \tag{4.9}$$

The output frequency can be measured by means of a counter. This circuit is reported for the detection of DNA hybridization in an 8×16 array.

Comparator-based relaxation oscillators require accurate comparators and current references. Also, the output frequency of these sensors changes with parasitic capacitances and the values of some temperature- and process-variable parameters of the circuit such as transistor transconductance or a resistor.

4.1.4   RO-based CFC

C_S can be converted to frequency by using ROs [13,108–114]. Figure 4.7 shows such a capacitive sensor in which C_S is included as a load of one delay cell of a three-stage RO. The delay of each cell depends on the charge of its load capacitor via the current I_B to a certain threshold voltage V_{osc}. So, the period T_{osc} of an M-stage RO will be equal to [112]:

$$T_{osc} = 2 \left[\frac{C_S V_{osc}}{I_B} + (M - 1) \frac{C_R V_{osc}}{I_B} \right] \tag{4.10}$$

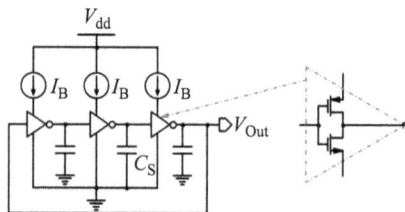

Figure 4.7   An RO-based structure

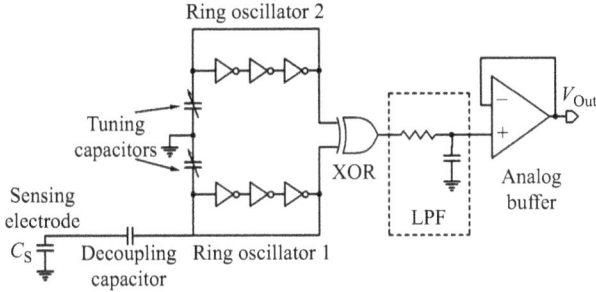

Figure 4.8   A differential RO-based sensor

where C_R identifies the reference capacitance, and $C_S = C_R + \Delta C$ is the sensing capacitance under measurement (ΔC shows the capacitance changes).

A differential RO-based sensor with two separate ROs is reported by Mohammad *et al.* [110]. As depicted in Figure 4.8, the frequency difference of the two ROs is detected by an Exclusive-OR (XOR) gate and a low-pass filter (LPF). Ring oscillator 1 is connected to the sensing electrode in a microfluidic channel. The XOR gate compares the oscillation frequency of ring oscillator 1 and ring oscillator 2 and generates a signal with the fundamental frequency of the ring oscillators which is pulse-width modulated at the difference frequency. Then, the LPF removes the high-frequency components and an amplifier buffers the intermediate frequency (IF) signal. Then a frequency counter determines the difference frequency. Two variable capacitors were used to tune the sensor output IF frequency.

It is noticeable that supply voltage, process, temperature, device sizes, and alike can affect the output frequency of ROs. However, all of these CFCs including comparator-based relaxation oscillator CFC and RO-based CFCs provide the advantage of generating semi-digital output signals which can be simply converted to digital data by a counter without ADCs and decrease power consumption and complexity.

4.1.5   VCO-based sensors

These oscillator-based capacitive sensors are usually reported for dielectric spectroscopy and complex permittivity sensing in mega-to-gigahertz frequency ranges [132–140]. As seen in Figure 4.9(a), these sensors include a sensing VCO and a frequency detector to identify a frequency shift, Δf.

Figure 4.10 shows a typical inductor-capacitor (LC) VCO-based capacitive sensor in which C_S is inserted in the LC tank of an LC VCO [132,135]. The sample induces electric field perturbations leading to a frequency shift in the VCO. The relative frequency shift ($\Delta f/f_0$) of the LC tank with respect to LC oscillator's frequency before exposure (f_0) can be expressed as follows [140]:

$$1 + \frac{\Delta f}{f_0} = \frac{1}{\sqrt{1 + \Delta C_{tot}(f_0)/C_{tot}}} \tag{4.11}$$

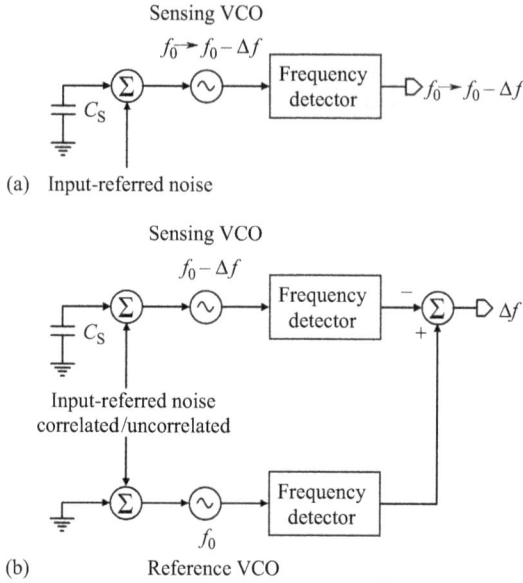

Figure 4.9   (a) A single VCO-based sensor; (b) a VCO-based sensor with a referencing and sensing VCO

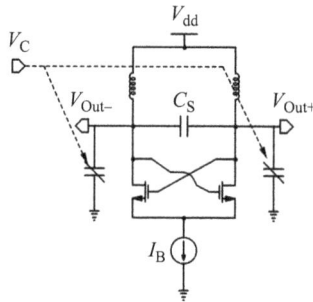

Figure 4.10   A typical LC VCO-based capacitive sensor

where C_{tot} is the total capacitance including C_S, parasitic capacitors, and varactors. $\Delta C_{tot}(f_0)/C_{tot}$ is the relative shift in the total capacitance with respect to C_{tot}.

The sensitivity of single VCO-based sensors to low-frequency temperature and environmental variations is an important disadvantage of these circuits. An approach to improve the sensor performance is utilizing a reference and a sensing VCO and differentiating their output frequencies as depicted in Figure 4.9(b). One of the shortcomings of this technique is that the two VCOs should be as identical as possible. The two VCOs can be activated periodically in such a way that only one of them operates at time to avoid VCO frequency pulling and to filter the correlated low-frequency noise between the two VCOs [135]. A frequency counter is a common frequency detector, but it requires long measurement times. Moreover, the open-loop system including the VCO leads to absolute oscillator frequency drift which is an obstacle for a precise characterization of the sample.

To measure the capacitance at a certain frequency and provide a considerably faster measurement time, a phase-locked loop (PLL) circuit as a closed-loop frequency detector circuit can be used to tune the frequency shift and keep it stable. For a fixed reference frequency, f_{ref}, and division ratio, N_{div}, any change in the free-running frequency of the VCO is translated into the variation of the control voltage, V_C. The DC tuning voltage of the PLL loop represents C_S that can be read by an ADC [132,133].

For example, Figure 4.11 shows the block diagram of a dielectric sensor readout circuitry based on an ADC and a frequency synthesizer loop for permittivity detection [132]. The VCO is placed inside a frequency synthesizer loop consisting of a phase and frequency detector, a frequency divider with the frequency division ratio of N_{div}, a loop filter, and a charge pump. The VCO control voltage, V_C, is adjusted by the synthesizer loop in such a way that the output frequency of the VCO, f_{Out}, is equal to N_{div} times of the reference frequency, f_{ref} (or $f_{Out} = N_{div} f_{ref}$). If N_{div} and f_{ref} are constant, f_{Out} remains constant. So, the variation of the C_S is converted to a change in V_C to compensate it. The voltage drop across the C_1 of the loop filter shown in Figure 4.11 changes according to V_C and is utilized as an input to the ADC for further compensation and processing. Then, the ADC digitizes V_C to be used to control through either a variable N_{div} or a variable f_{ref}. The accuracy of the frequency shift detection might be limited by the resolution of ADC. In another effort, Elhadidy *et al.* [140] reported an RO-based PLL whose output is semi-digital.

Some other sensors [137–139] use quadrature VCO (QVCO) to improve the SNR and accuracy in which LC tanks are parts of injection-locked oscillators (ILOs). Figure 4.12 illustrates a single sensing channel of a dual-channel interferometer-based capacitive sensor using an ILO for high-speed flow cytometry [137] which includes a

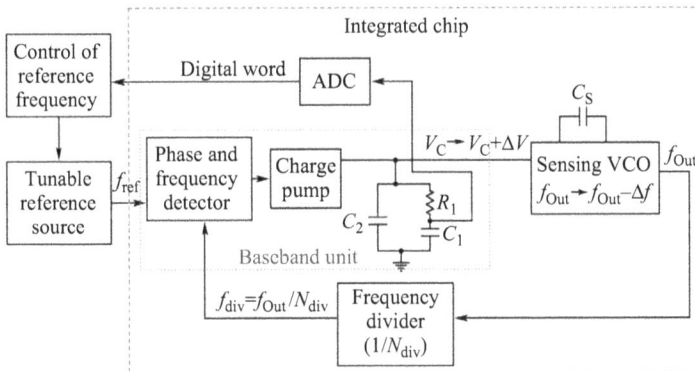

Figure 4.11   A VCO-based sensor using a PLL and an ADC as a frequency detector

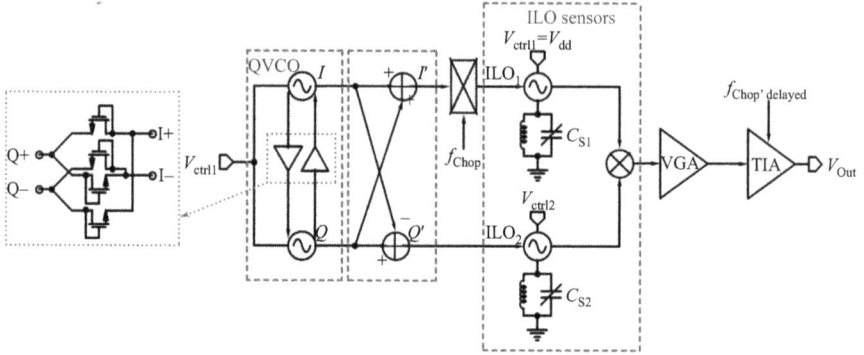

Figure 4.12   A single sensing channel of a dual-channel interferometer-based capacitive sensor using an ILO

VCO, a mixer as phase detector, two ILOs, and the signal conditioning analog front end. The injection source frequency, ω_{inj}, deviates from the free-running frequency of the frequency-locked oscillator, ω_0. By maintaining Barkhausen criteria, the phase shift between the input and output is expressed by [137]:

$$\sin\theta \approx \frac{2Q_{LC}}{\omega_0}\frac{I_{osc}}{I_{inj}}\left(\omega_0 - \omega_{inj}\right) \tag{4.12}$$

where Q_{LC} is the quality factor of LC tank. I_{osc} and I_{inj} stand for the currents of the oscillator and injection source. The ILO can be considered as a phase amplifier preceding the phase detector with a gain controlled through I_{osc}/I_{inj}. The injection lock of the sensing oscillators is done by a diode-coupled QVCO. I and Q signals are summed and subtracted with each other before injection to maximize mixer gain by the elimination of the quadrature phase error to the first order. A variable gain amplifier (VGA) interfaces the mixer followed by a baseband amplifier and a transimpedance amplifier (TIA) performs demodulation. Chopping is used in one of the signal injection paths to minimize the effect of flicker noise.

One of the problems of this technique is the need for an expensive tunable frequency source [132]. The required inductor for LC tank in LC VCOs occupies a large die area. To avoid substrate or electromagnetic coupling of inductors, the inductors are suggested to be separated by a large area. To raise the sensitivity of frequency deviation-based sensors for microwave permittivity sensing, Nagulapalli *et al.* [253] utilized relaxation oscillators instead of VCO.

4.1.6   Lock-in detection

Figure 4.13 shows the principle of this approach in which a specific signal modulates the capacitance under measurement. Low-frequency noise and offset are also modulated to a high frequency. After voltage-to-current (V/I) or current-to-voltage (I/V) conversion (as required) as well as amplification, the signal is down-converted. Synchronous demodulation and LPF can help to retrieve the original

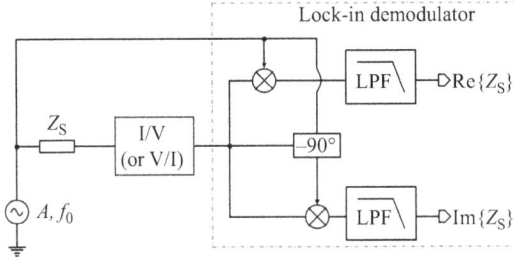

Figure 4.13   The block diagram of a simple lock-in architecture

signal. The real and imaginary parts of sensing impedance (Z_S) can be extracted by a quadrature demodulator. This technique is commonly utilized for impedance spectroscopy [66,70,254,255]. It is also reported for ultra-wideband dielectric spectroscopy and complex permittivity sensing in some literature [124,125,130].

The first fully integrated impedance-based biosensor in CMOS technology was implemented by Manickam *et al.* [70] which comprised a 10×10 array of on-chip electrodes (see Figure 4.14). Each pixel of this array consisted of a coherent demodulator including two mixers and a low-noise TIA. The current of the sensing impedance, Z_S, is amplified by the TIA. The two mixers multiply the output of the TIA by orthogonal sinusoidal signals (in-phase (I) and quadrature (Q) signals) at the angular frequency ω. The amplitude and phase of the sensing admittance, $Y_S(\omega) = Z_S(\omega)^{-1}$ can be obtained by the DC values of the output voltages of these two mixers, V_I and V_Q:

$$|Y_S(\omega)| = \frac{\sqrt{V_I^2 + V_Q^2}}{A_{TIA} \cdot |V_{exc}(\omega)|} \tag{4.13}$$

$$\angle Y_S(\omega) = \tan^{-1}\left(\frac{V_Q}{V_I}\right) \tag{4.14}$$

where A_{TIA} denotes the gain of the TIA and $V_{exc}(\omega)$ identifies a sinusoidal excitation voltage signal.

The lock-in technique reported by Ciccarella *et al.* [120,128] for airborne particle detection could provide a resolution better than 100 zeptofarad (zF). However, some limitations such as synchronization and mismatch errors as well as high complexity are added to the circuit. Figure 4.15 illustrates the schematic of each channel of the 32-channel sensor proposed by this group consisting of a charge preamplifier with adjustable high-pass filtering, a first-order g_m-C tunable LPF, square-wave mixer, and a digital network. The capacitive noise is only 65 zF rms, which makes the sensor suitable for counting single micrometer-sized airborne particles.

Two sets of IDEs, C_{up} and C_{dw}, share a common electrode connected to a charge preamplifier whereas the other two sides are driven with the counter-phased square wave voltages (with amplitude V_{Ref}) generated by two on-chip clocks, Φ_1 and Φ_2.

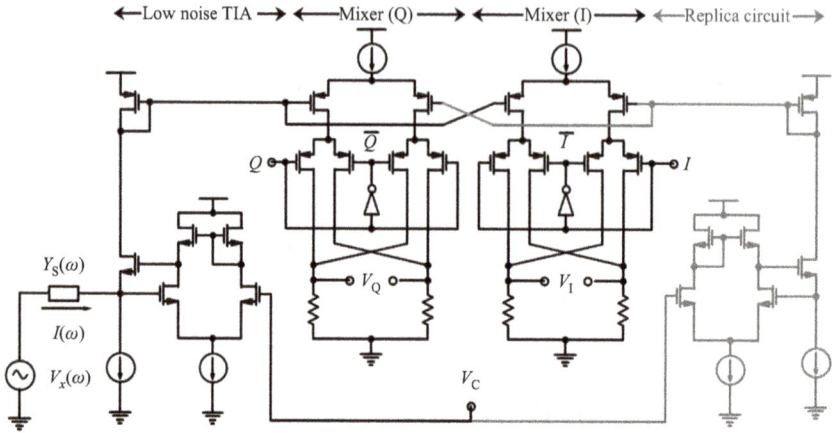

Figure 4.14   The entire pixel of an EIS biosensor (excluding bias circuits)

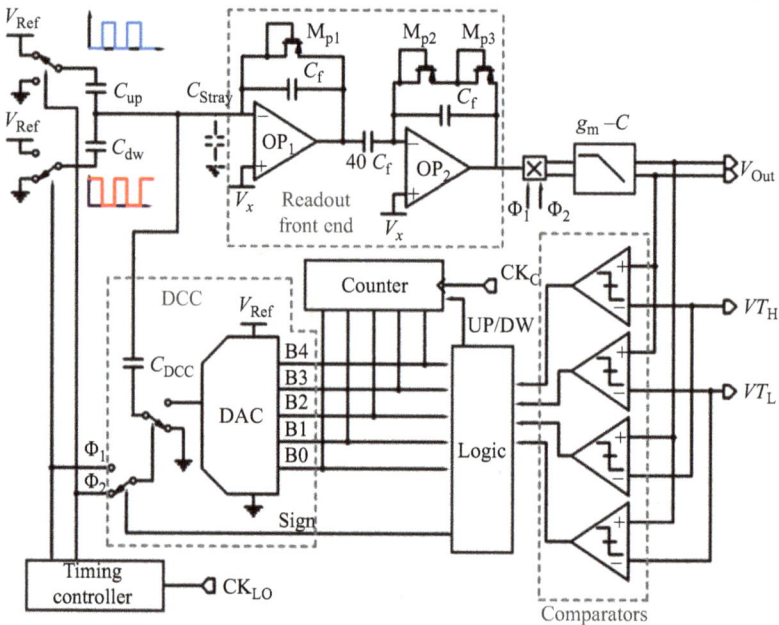

Figure 4.15   Each channel of the 32-channel sensor CMOS capacitive sensor for counting single micrometer-sized airborne particles

V_{Ref} is set to be equal to V_{dd} by employing a rail-to-rail square-wave as excitation voltage instead of a sinusoidal curve to maximize the charge injected in the pre-amplifier. The charge preamplifier with a small feedback capacitor, C_f, read the current of the sensing capacitor. In an ideal case, it measures the difference in charge on one of the two IDEs and the differential structure allows a high gain for the input amplifier by canceling out the large capacitance of the sets of IDEs and rejects the noise of V_{Ref} and capacitance fluctuations due to environmental factors like humidity and temperature variations. A non-return-to-zero passive mixer and the low-pass g_m-C filter perform the lock-in demodulation. A voltage amplifier between the charge preamplifier and the lock-in demodulator helps to limit the noise of the g_m-C filter and a digital-to-capacitance converter was used to maintain a balanced condition of the differential input capacitance and cope with the saturation of the front-end amplifier.

4.1.7 Triangular voltage analysis

A triangular voltage signal can be created utilizing a pulse current stimulator (Figure 4.16) or be used to stimulate the biochemical sample (Figure 4.17). This approach can help to measure both capacitance and resistance of the sample independently.

In the first scenario like what is shown in Figure 4.16, the slope of each rising or falling segment of the output voltage ($\Delta v/\Delta t_0$) can be used to calculate C_S through the equation $C_S = (I_0 \cdot \Delta t_0)/\Delta v$. Druart *et al.* [119] used a current-controlled oscillator (CCO) for charging and discharging C_S as seen in Figure 4.16. With this self-oscillating system, there is no need to stimulate the circuit by a variable signal source.

In the second scenario like what is depicted in Figure 4.17, a triangular voltage waveform is applied to the sensing electrode whose slope identifies the amplitude

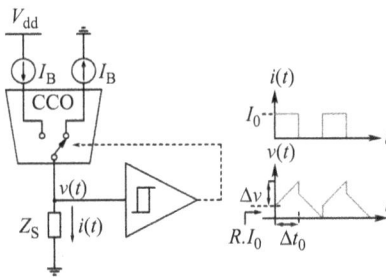

Figure 4.16   The first scenario with an output triangular voltage waveform

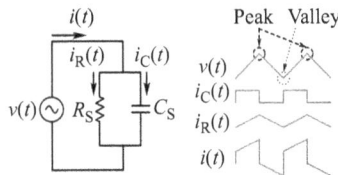

Figure 4.17   The second scenario with an input triangular voltage waveform

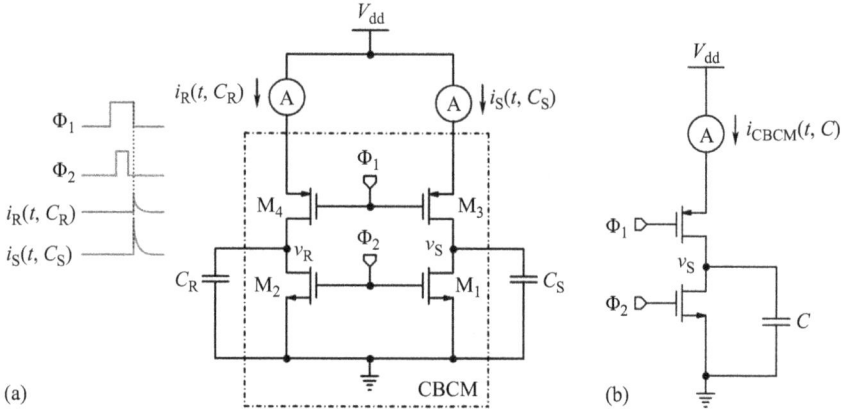

Figure 4.18   (a) The core of CBCM method; (b) single-electrode core of CBCM

of the square current flowing through C_S. This capacitance is only determined in the "valley" and "peak" regions of the input triangular voltage waveform where its second derivative is nonzero. Measurements have to be done after the output current reaches steady-state value [118]. Lee *et al.* [118] designed an SC circuit to process the output current which was utilized in a 16×8 array for detection of DNA hybridization.

In comparison to lock-in detection, triangular voltage analysis helps to measure both components of the impedance under measurement directly from the output current or voltage of the sensor without measuring the amplitude and phase shift of the sensing impedance [118].

4.1.8   Charge-based capacitance measurement

The basic core of the CBCM method is depicted in Figure 4.18. This circuit consists of a sensing (C_S) and a reference (C_R) capacitor along with a pair of pseudo-inverters and ammeters. The sensing capacitor is covered by an intermediate layer sensitive to target biospecies and is exposed to the biochemical sample under test. The reference capacitor might be either left unconnected to electrodes or connected to an identical electrode that is not exposed to the sample or not covered by the sensitive intermediate layer. So, differently from C_S, C_R is insensitive to the sample under test. Consequently, any change arising from the sample, ΔC, causes a difference between C_S and C_R ($\Delta C = C_S - C_R$) which can be measured by this circuit. The two pseudo-inverters form the core of the CBCM circuit. Each pseudo-inverter comprises a pull-up and a pull-down switch to charge or discharge these capacitors, respectively. In Figure 4.18, P-type metal-oxide-semiconductor (PMOS) transistors (M_3 and M_4) and N-type metal-oxide-semiconductor (NMOS) transistors (M_1 and M_2) are used as pull-up and pull-down switches, respectively. These switches are controlled by two clock pulses, Φ_1 and Φ_2. The nonoverlapped form of these pulses prevents short-circuit current [155].

When Φ_1 and Φ_2 become low, both capacitors, C_S and C_R, are charged via pull-up switches (M_3 and M_4) up to a known voltage, for instance, $V_{dd} - V_{Am}$ in which V_{dd} and V_{Am} denote the supply voltage and the voltage dropped over the ammeters, respectively. Whenever these clock pulses go high, the capacitors are discharged via pull-down switches (M_1 and M_2). As shown in Figure 4.18, the currents of the two branches of this circuit, $i_S(C_S,t)$ and $i_R(C_R,t)$, are time-dependent sharp exponential currents. The instantaneous currents of the two branches can be obtained by

$$i_S(t) = C_S \frac{dv_S(t)}{dt} \tag{4.15}$$

$$i_R(t) = C_R \frac{dv_R(t)}{dt} \tag{4.16}$$

where $v_S(t)$ and $v_R(t)$ denote the instantaneous voltages across C_S and C_R, respectively. The averages of these currents throughout one period of Φ_1 and Φ_2 are proportional to the related capacitances and can be expressed as follows:

$$
\begin{aligned}
I_S &= \frac{1}{T_s} \int_0^{T_s} C_S \frac{dv_S(t)}{dt} dt \\
&= \frac{C_S}{T_s} \int_0^{V_{dd}-V_{Am}} dv_S \\
&= \frac{C_S(V_{dd} - V_{Am})}{T_s}
\end{aligned} \tag{4.17}
$$

$$
\begin{aligned}
I_R &= \frac{1}{T_s} \int_0^{T_s} C_R \frac{dv_R(t)}{dt} dt \\
&= \frac{C_R}{T_s} \int_0^{V_{dd}-V_{Am}} dv_s \\
&= \frac{C_R(V_{dd} - V_{Am})}{T_s}
\end{aligned} \tag{4.18}
$$

where I_S and I_R stand for the average of currents $i_S(C_s,t)$ and $i_R(C_R,t)$, respectively. T_S is the period of Φ_1 and Φ_2 which is the inverse of their frequency, f_S ($T_S = 1/f_S$). So, the difference between C_S and C_R ($\Delta C = C_S - C_R$) can be given by the subtraction of the two averaged currents:

$$I_S - I_R = f_S(V_{dd} - V_{Am})\Delta C \tag{4.19}$$

So, ΔC can be extracted from C_S with a high accuracy. This method can also be performed by half of this circuit (like Figure 4.18(b)) comprising one electrode and one pseudo-inverter but the measurements must be done in two steps. In the first step, the electrode is not exposed to the biochemical sample and the output current is only affected by parasitic capacitances of electrode and switches. In the second step, the biochemical sample changes the capacitance under the measurement from its initial value [76,256]. As we will see in the following sections, in recent advances of core-CBCM capacitive sensors, integrated readout circuits are realized to process CBCM currents instead of ammeters and off-chip elements.

4.1.9   Comparison of different capacitive sensors

Table 4.1 compares the IDR, power, area, the number of arrays, and the sensitivity of the various techniques discussed in this section. As seen in Table 4.1, a very high resolution down to 0.065 aF is achieved by the lock-in technique used in [120], but the power consumption is high and the IDR is limited to 1 fF. On the other hand, the combination of triangular voltage analysis and SC technique proposed in [118] for output processing resulted in a resolution of about 1 pF. The accuracy of this sensor is affected by the accuracy of the stimulating or the generated triangular voltage as well as the approach of processing the output signal.

As seen in Table 4.1, the above-mentioned CMOS-based techniques are reported to design CFCs, capacitance-to-voltage converters (CVCs), and capacitance-to-digital converters (CDCs). The RO-based sensors reported in [116] and [109] show resolutions of 2.5 and 10 fF, respectively. But using ROs in the CFCs proposed in [13,108,112] provided higher sensitivities and finer resolutions. In [112] and [13], resolutions about tens of attofarads (aF) with a reasonably wide IDR about tens of femtofarads (fF) are achieved using ROs. The comparator-based relaxation oscillator reported in [115] which is used in a 5 × 5 array demonstrated an IDR between 12 and 700 fF. The 8 × 16 array presented in [117] is the largest array of CFCs in this table which provided an IDR between 330 pF and 10 nF.

Charge sharing, charge-sensitive amplifier, and SC-based methods are charge redistribution-based techniques that have demonstrated roughly high sensitivities in some cases. Despite the low resolution (5 fF) of the 28-array sensors working based on the charge-sharing principle which is reported in [189], the charge-sharing-based sensor proposed by Couniot *et al.* [20] showed a resolution of about 450 aF and its low complexity led to a reasonable low power consumption about 29 μW and a total area of about 0.1 mm^2. The charge-sensitive amplifiers reported in [107] provided a higher resolution of about 21 aF and an IDR of about 0.42 fF. Although this circuit was used in a large array of 320 × 320 sensors, its outputs are not digital and need an external ADC to convert the outputs to digital. The resolution provided by the SC technique proposed in [4] which was adapted to SAR ADC (about 4.5 fF) was not as fine as the other charge redistribution-based methods, but it consumed the lowest power in comparison to the other sensors reported in Table 4.1 which was totally about 2.1 μW for a 9 array.

Since biological signals are very small, accuracy is one of the most significant factors in designing capacitive biosensors for life science applications. On the other hand, the low complexity of circuits paves the way for high-throughput measurements. Although RO-based CFC and CBCM are among popular techniques for array

Table 4.1 CMOS interface circuits of capacitive biosensors

Power (V_{dd})	Area (mm^2)	IDR	Sensitivity	Resolution	Array #	Digital/ Analog	D/ ND	Method	Tech.	Application	Ref.
29 μW (2.5 V)	0.1 (TA)	0.45–57 fF	55 mV/fF	450 aF	16 × 16	Analog	ND	CVC by ChS	0.25 μm	Detection of single bacterial cell	[20]
NA (3 V)	2.25 (TA), 0.16 (EA)	NA	NA	5 fF	28	Analog	ND	CVC by ChS	0.5 μm	Cell proliferation monitoring	[189]
NA (3.3 V)	NA	0.42 fF	345 mV/fF	21 aF	320 × 320	Analog	D	CVC by CSA	0.35 μm	Localization of bioparticles	[107]
2.1 μW (0.9 V / 1 V)	0.06 (AA)	16.137 pF	NA	4.5 fF	9	Digital	ND	SC+ SAR	0.18 μm	Cancer biomarker detection	[4]
NA	NA	12–700 fF	21 kHz/fF @ 4.9 MHz @ 123 fF	NA	5 × 5	Digital	ND	CFC by Rel.O	0.35 μm	Detection of neuro-transmitter DA	[115]
NA	28.8 (TA)	330 pF–10 nF	23 Hz/pF @ 7.5 kHz @ 330 pF	NA	8 × 16	Digital	D	CFC by Rel.O	0.5 μm	Detection of DNA	[117]
NA	2.25 (TA), 1 (AA)	±100 fF	235 mV/fF	10 aF	3 × 4	Digital	D	CFC by RO	0.13 μm	Cell volume growth monitoring	[112]
8 mW (3.3 V)	9 (TA), 0.0144 (EA)	12 fF	590 kHz/fF	14.4 aF	4 × 4	Digital	ND	CFC by RO	0.35 μm	Real-time measurement of cell proliferation	[13]
NA	4 (TA)	NA	570 kHz/fF@ 1.37 GHz @ 500 fF	14 aF	1	Digital	D	CFC by RO	0.35 μm	Single-cell analysis	[108]
NA	NA	NA	223 kHz/fF @ 5.2 MHz @ 23 fF	2.5 fF	8 × 8	Digital	ND	CFC by RO	0.35 μm	Detection of cells and biomolecules (using magnetic bead detection)	[116]
29 mW (2.5 V)	0.05 (AA), 0.0506 (EA)	NA	11 kHz/fF @ 254 MHz @ 17.5 fF	10 fF	1	Digital	ND	CFC by RO	0.25 μm	Bacteria detection	[109]

(Continues)

Table 4.1 (Continued)

Power (V_{dd})	Area (mm²)	IDR	Sensitivity	Resolution	Array #	Digital/ Analog	D/ ND	Method	Tech.	Application	Ref.
NA (−1.2 to 1.3 V)	1.8 (TA)	NA	10 fF/mV	1 fF	3 × 3	Analog	ND	CVC by CBCM	0.18 μm	Detection of nano- and microparticles	[257]
165 μW (3 V)	3	25 fF	200 mV/fF	15 aF	6 × 6	Analog	D	CVC by CBCM	0.5 μm	Cell monitoring	[142]
NA (1.8 V)	2 (TA)	~ 2.7 fF	250 mV/fF	10 aF	3	Digital	ND	CBCM+ΣΔ ADC	0.18 μm	Bacteria growth monitoring	[68]
NA (3.3 V)	9	10 fF	350 mV/fF	10 aF	8 × 8	Digital	D	CBCM+ΣΔ ADC	0.35 m	Cell growth monitoring	[21]
15 mW (1.2 V)	2,525,253 (TA), (Radius of NE = 90 nm)	4 fF	NA	1 aF	256 × 256	Digital	ND	CBCM+ADC	90 nm	High-frequency impedance spectroscopy and imaging	[144]
103 μW (1.8 V)	0.0645 (TA), 0.007224 (AA), 0.0236 (EA)	70 fF	138 pulses/fF	873 aF ^{NP} 10 aF ^{P}	1	Digital	ND	CBCM+CCO	0.18 μm	—	[73]
NA	10 (TA)	10 pF–10 nF	NA	1 pF	16 × 8	Digital	ND	TVA+ SC+ ADC	0.35 μm	Detection of DNA hybridization	[118]
84 mW (3.3 V)	6 (TA), 1.15 (EA)	Less than 1 fF	NA	0.065 aF	32	Digital	D	Lock-in + ADC	0.35 μm	Airborne particle detector	[120]

ChS, charge sharing; ND, non-differential; TA, total area; EA, electrode(s) area; CSA, charge-sensitive amplifier; D, differential; AA, active area; Rel.O, relaxation oscillator; RO, ring oscillator; NE, nanoelectrodes; NP, without pre-distortion; P, with pre-distortion; TVA, triangular voltage analysis; NA, not available.

structures due to their lower complexity, the accuracy of RO-based sensors is usually not as much as the CBCM method and it is vulnerable to supply voltage, process and temperature. According to Table 4.1, CBCM offers the advantage of high accuracy in the range of attofarad as well as low power consumption even in the range of microwatt. So, it seems to be a suitable candidate for such applications. The rest of this chapter is dedicated to the development procedure of core-CBCM circuits.

4.2 Some nonidealities of CBCM

CBCM circuits might cause some errors in the results due to leakage current [258], charge injection [258–261], noise [262], and mismatch [81,142,262].

Subthreshold leakage and junction leakage are the main sources of leakage current. In addition to switching mode, standby mode (with no switching) arising from unselected devices can create leakage current [263].

Charge injection is another nonideality which is more important for submicron CMOS technologies (lower than 90 nm). When the clock Φ_1 goes high (Figure 4.18 (a)), a current is injected through the gate-source capacitance (C_{gs}) of the PMOS switches onto the measurement bus [258]. Figure 4.19 illustrates the origin of this error. The current injection follows a relation like $I = C dv/dt$, where dv/dt is the edge rate of Φ_1.

The noise coupling from other devices is another factor that might affect the CBCM performance through supply voltage and substrate. Guard ring and distinct supply voltages for mixed-signal circuits are some conventional approaches to mitigate the noise coupling [262].

Process and device mismatches such as the mismatches of switches and the transistors used in readout circuits and parasitic capacitances cause the two branches of the circuit not to be completely balanced. Furthermore, remnants of bio-particles on the electrodes, temperature variations, and some other unpredictable impacts can create an offset current so that for $\Delta C = 0$, i_S and i_R are not equal. This offset should be calibrated (according to the structure of the readout circuit) and the variability of the sensor should be mitigated. Especially in the voltage-mode

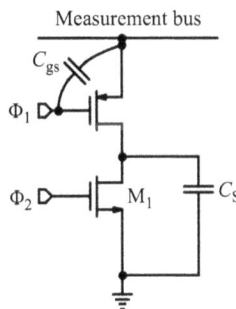

Figure 4.19 Current injection through C_{gs}

circuits [81,142] in low-supply voltage technologies, the offset current is converted to an offset voltage that can significantly limit the dynamic range and resolution of the circuit. Using larger devices can reduce the impact of mismatches, but in larger arrays of the sensor, this approach increases the chip footprint.

Regarding parasitic capacitances of core-CBCM sensors, two capacitors can be used to model these parasitics including stray capacitances from the readout circuit (C_{stray}) and standing capacitance from electrodes and fluidic environment ($C_{standing}$). Sensing capacitance includes C_{stray}, $C_{standing}$, and sensed capacitance ($C_S = C_{stray} + C_{standing} + C_{sensed}$). As aforementioned, the node of the reference capacitance may be connected to an electrode or not. If it is left unconnected (as shown in Figure 4.20(a)), the capacitance in the reference node is equal to C_{stray} and as a result $\Delta C = C_{standing} + C_{sensed}$. Since $C_{standing}$ will limit the dynamic range of the sensor, some techniques are proposed to compensate $C_{standing}$. One technique is replicating the circuit in such a way that both nodes of sensing and reference capacitance are attached to identical electrodes. So, we will have $C_R = C_{stray} + C_{standing}$ and $\Delta C = C_{sensed}$ [77] (as shown in Figure 4.20(b)). Another way is the attachment of an offsetting capacitor (C_{os}) (using interlayer capacitances) to the reference node ($C_R = C_{stray} + C_{os}$) so that $C_{os} > C_{standing}$ and $\Delta C = C_{standing} + C_{sensed} - C_{os}$. Thus, a negative capacitive offset will be obtained that should be in the IDR of the sensor [264].

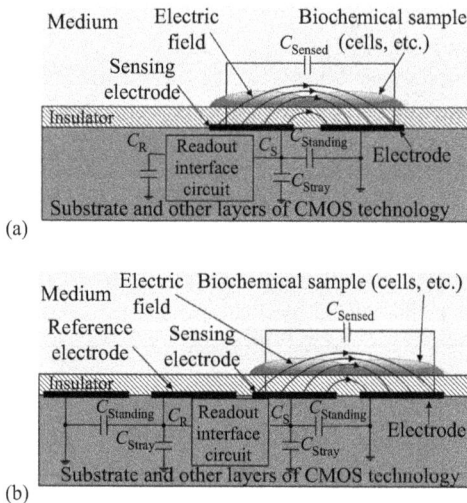

Figure 4.20   *Simplified diagram of parasitic capacitances of electrodes where $C_S = C_{stray} + C_{standing} + C_{sensed}$, but (a) the node of reference capacitance is unconnected to electrode and $C_R = C_{stray}$ (pseudo-differential method); (b) both reference and sensing capacitances are connected to identical electrodes and $C_R = C_{stray} + C_{standing}$ (differential method)*

Some approaches are proposed in the literature for the modification of the basic CBCM block. Although they are proposed for industrial applications, they might be useful for life science applications too. Bach *et al.* [258] proposed to use thick-oxide transistors and high sampling frequency (to increase the measured current) to reduce the impact of the gate leakage current. However, the leakage current between the reference and the sensing branch is constant, and consequently, the difference between I_S and I_R and the measured capacitance will be correct. They also proposed to reduce the gate-source capacitance (by minimizing the transistor width) and the edge rate of the clock Φ_1 (by increasing the transistor length) to reduce charge injection.

Vendrame *et al.* [261] utilized transmission gates to mitigate charge injection effects. Ning *et al.* [263] improved CBCM by using a transmission gate scheme and proposing a leakage suppresser. Charge injection-induced error-free (CIEF) CBCM circuits are also presented in [259,260] suppressing this error. Furthermore, in [265], a self-differential CBCM method is provided to improve CIEF CBCM which can operate at a very high frequency.

In [262], an on-chip self-timed control scheme circuitry is proposed for industrial applications operating in gigahertz ranges. It comprises a pseudo-inverter, two transmission gates connected to C_S and C_R, and a nonoverlapping signal generator. Thick-oxide transistors are also used to reduce the gate leakage current. Utilizing only a single pseudo-inverter (instead of two) results in the elimination of charge-injection errors through two pass measurements and also minimizes transistor mismatches. Also, Si calibration is performed to reduce mismatch. It means that first C_S with empty load and C_R are employed to measure and calculate the mismatch parameter, $\Delta C_0 = C_S - C_R$. To mitigate the noise coupling from the substrate, guard rings are used to surround the CBCM block. Yu *et al.* [266] proposed an on-chip signal generator for CBCM measurement circuits to reduce the mismatch between the two clock signals Φ_1 and Φ_2. There are also some calibration techniques which will be reviewed in the following subsections.

4.3 Core-CBCM interface circuits

The first report on the CBCM method was carried out in 1996 to investigate the crosstalk of integrated circuits [155]. In this work, DC ammeters were used to measure the charging and discharging current of the CBCM circuit. Many attempts have been made to develop CBCM, especially for life science applications, and integrate the basic block along with the other required parts like the readout interface circuit and the electrodes on a single chip which will be reviewed in this section. Table 4.2 summarizes the information of some of the important core-CBCM capacitive sensors reviewed in this paper. As can be seen in Table 4.2, a resolution about tens of attofarads can be obtained by the CBCM method. Its low complexity, low power and a small occupied area have paved the way to achieve large arrays of core-CBCM sensors like the 246 × 256 array proposed in [144] which was designed for high-frequency impedance spectroscopy.

Table 4.2 *Comparison of different CMOS-based core-CBCM readout interface circuits*

Power	Area (µm²)	Array #	Single-ended (SE)/fully differential (FD)	Output type	Frequency	Dynamic range of ΔC (fF)	Sensitivity	Capacitance resolution	Supply voltage (V)	Tech.	Application	Ref.
–	6.272×10^5	4	SE	A	13 Hz to 800 kHz	2	1 V/fF	10 aF	5	0.8 µm	Particle detection	[77]
–	1.8×10^6	3×3	SE	D	–	–	0.1 mV/fF	1 fF	$-1.2 \leq V_{dd} \leq 1.3$	0.18 µm	Detection of nano- and micro particles	[257]
–	2×10^6	3	SE	D	1 kHz	~2.7	250 mV/fF	10 aF	1.8	0.18 µm	Bacteria growth monitoring	[68]
165 µW	$*1.45 \times 10^2$	6×6	FD	A	1 kHz	25	200 mV/fF	15 aF	±3	0.5 µm	Cell monitoring	[142]
580 µW	10^4	1	FD	D	150 kHz	10	350 mV/fF	10 aF	±3.3	0.35 µm	Chemical solvent detection	[183]
1.5×10^4 µW	2.5×10^{12}	256×256	SE	D	1–70 MHz	4	–	1aF	1.2	90 nm	High-frequency impedance spectroscopy	[144]
–	450×300	4×4	FD	A	1 kHz	10 to 10^3	3 mV/fF	10 fF	±2.5	1.2 µm	DNA hybridization	[267]
1.09 mW	–	1	SE	D	200 Hz	–	0.38 mV/fF	100 aF	3.3	0.35 µm	femtomolar detection of bacterial pathogens	[78]
1.7 mW	9×10^6	1	SE	A	~167 Hz	100 aF	29.3 mV/fF	41 fF	5	0.8 µm	Detection of deadly bacteria such as E. coli and Salmonella	[24]
103 µW	6.45×10^4	1	SE	D	100 kHz	NP 873 aF, P 10 aF	138 pulses/fF	70 fF	1.8	0.18 µm	–	[73]

A, analog; SE, single-ended; D, digital; FD, fully differential; *One sensor pixel without electrodes and sensor evaluation modules; NP, without pre-distortion; P, with pre-distortion.

4.3.1 Core-CBCM capacitive biosensors using discrete components

Stagni and Guiducci *et al.* [22,237,256,268,269] used the CBCM method for DNA hybridization detection. In their seminal papers, this group showed the functionality of this method for DNA detection by a (non-integrated) prototype setup including discrete components [256,268]. They used Au electrodes (12 mm in diameter) in a simple two-electrode electrochemical system (without an RE) forming an electrode–solution interface (see Figure 4.21). Their new methodology was a replacement for conventional three-electrode setup cells which also require an RE and a potentiostat. The two-electrode system is more suitable for integration on a single chip. In Figure 4.21, the WE is functionalized to capture single-stranded DNA (ssDNA). The capacitance of this electrode–solution interface is affected by the physicochemical characteristics of the interface which is measured in two steps, before and after exposure to ssDNA. The utilized capacitance-to-current transducer works based on the CBCM method and the average current flowing through the functionalized electrode is measured by a multimeter.

As a continuation of this work, the same group proposed a microcontroller-based system instead of all laboratory instruments [269]. The microcontroller identifies the maximum useable frequency and controls a DAC to generate the required input pulse waveforms. The output current of the electrode is converted to a voltage using a resistor and an operational amplifier (Op-amp) amplifies the signals. Then, an ADC samples and sends them to the microcontroller which is programmed to average them at different frequencies and calculate the equivalent capacitance using the slope of the curve of average voltage versus frequency.

This group put the first step toward fully integrated capacitive biosensor for DNA detection by integrating Au microelectrodes with a silicon CMOS standard process [237]. Since the ionic aqueous condition can add a resistive path to the sensing capacitor and cause an imperfect behavior [256], they show the potentiality of using the CBCM method in ionic conditions by a qualitative study in [237]. The same group proposed a fully electronic DNA sensor chip [22] in which the CBCM

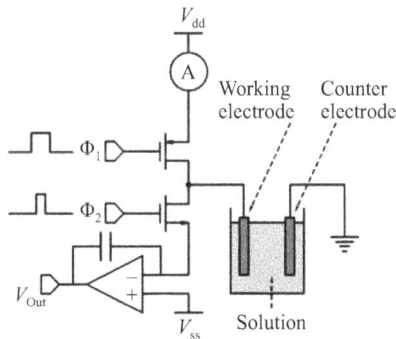

Figure 4.21   The two-electrode measurement setup

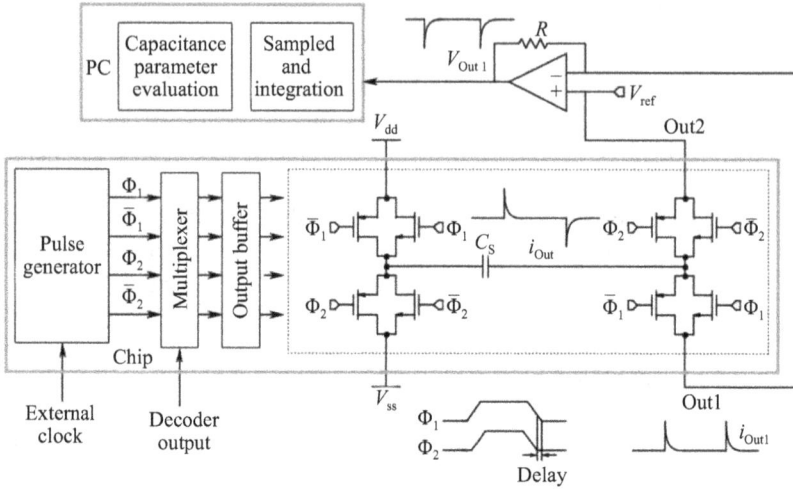

Figure 4.22   The circuit associated with each cell of the array of capacitive biosensors

structure and pairs of Au electrodes are integrated with CMOS technology, but the readout circuit is implemented off-chip. Figure 4.22 shows a single cell of the array of the capacitive sensor used in [22]. The output transient current of the chip is converted to a voltage signal by an external I/V converter circuit and the result is sampled and processed (using trapezoidal integration method) by a personal computer to give the capacitance value. Carrara *et al.* used the same measurement system and improved it by new immobilization strategies for DNA detection, immunosensing, and cancer biomarkers detection [23,270,271].

IDEs are suitable electrodes for replacing three-electrode structures with two-electrode setup in integrated sensors. Each IDE consists of two parts in which one part is grounded and the other part plays the role of a WE. Sampson *et al.* [272] used IDE and the basic principle of CBCM with one pseudo-inverter for the detection of paraffinophilic Mycobacterium avium complex (MAC) (a dangerous pathogen). They deposited paraffin wax over Cr/Au interdigitated array microelectrodes which were designed in conventional and square spiral-type. Bacteria consume this dielectric layer and change the sensing capacitance. Dual N- and P-channel pairs of metal-oxide-semiconductor field-effect transistor (MOSFET) integrated chip were used to achieve measurement setup. The operation of the circuit was calibrated through a comparison of the measured values of commercial capacitors.

4.3.2   Current mirror integrated with CBCM structure

A simple current mirror like (M_3 and M_4) in Figure 4.23 can be employed instead of an off-chip ammeter to pave the way for on-chip measurement of the CBCM current. The CBCM current is sensed and amplified by the current mirror with a gain equal to A_I. When Φ_1 is low and C_S is charging rapidly, the drain current of M_4 (i'_S) is A_I times

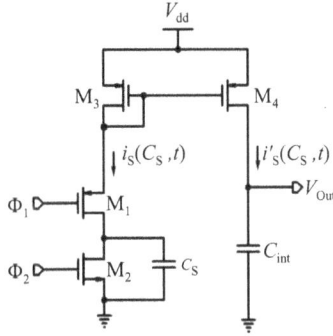

Figure 4.23   On-chip core-CBCM CVC using a current mirror

the drain current of M_3 (i_S) meaning $i'_S = A_1 \cdot i_S$. By discounting the on/off resistances of the switches, M_1 and M_2, the relation between them can be expressed as follows

$$i'_S = A_1 C_S \frac{dV_S}{dt} = A_1 \mu_p C_{ox} \left(V_{sg3,4} - V_{thp}\right)^2 \tag{4.20}$$

where μ_p and C_{ox} stand for the whole mobility coefficient and the gate capacitance per unit area, respectively. $V_{sg3,4}$ and V_{thp} are the source-gate voltage and the threshold voltage of M_3 and M_4, respectively.

By considering $V_{sg3,4} = V_{dd} - V_S$ (where V_S is the source voltage) and assuming $V_S = 0$ at $t = 0$ while Φ_1 and Φ_2 are high and C_S is discharging, V_S can be obtained by

$$V_S = \left(V_{dd} - V_{thp}\right) - \frac{\left(V_{dd} - V_{thp}\right)C_S}{\mu_p C_{ox}\left(V_{dd} - V_{thp}\right)t + C_S} \tag{4.21}$$

Combining the above equations, we deduce that i'_S which is a function of C_S and time, t, can be given by

$$i'_S(C_S, t) \approx \frac{\left[\left(V_{dd} - V_{thp}\right)C_S\right]^2 \mu_p C_{ox} A_1}{\left[\mu_p C_{ox}\left(V_{dd} - V_{thp}\right)t + C_S\right]^2} \tag{4.22}$$

By using equation (4.23) and integrating the current $i'_S(C_S, t)$ with an integrating capacitor, C_{int}, the output voltage, V_{Out}, for $t \gg 0$, can be ideally calculated and expressed as equation (4.24).

$$i'_S = C_{int} \frac{dV_{Out}}{dt} \tag{4.23}$$

$$V_{Out} = \frac{C_S}{C_{int}} \frac{W_4 L_3}{W_3 L_4} \left(V_{dd} - V_{thp}\right) \tag{4.24}$$

where W_3 and W_4 are the width of M$_3$ and M$_4$ and L_3 and L_4 are the length of these transistors, respectively. Since $C_S = \Delta C + C_R$ and $\Delta C \gg C_R$, the dynamic range is limited [76]. To extract capacitance changes, ΔC, from C_S and measure ΔC, the signals related to C_S and C_R (current, voltage, frequency, etc.) can be subtracted before or after averaging like the differential voltage- and differential current-based techniques discussed in the following sections.

4.3.3 Non-differential CVCs

To fully integrate core-CBCM capacitive sensors, their sharp exponential current signal (see Figure 4.18(a)) is usually averaged by an integrator circuit or a capacitor. In this way, the current signal is converted to a voltage signal proportional to the averaged current and consequently the related capacitance under measurement. York *et al.* [141] proposed a capacitance tomography system based on an adapted CBCM, more commonly known as charge transfer (see Figure 4.24). In this circuit, the charging current of an unknown capacitor is mirrored to a known integrating capacitor. The gain MOSFET works in an active region and two integrating capacitors are utilized to increase the measurement range without the reduction of sensitivity. Six programmable offset capacitors (C_{os}) are used in parallel in this circuit for three reasons: (1) reduction of the measurement range and consequently improvement of the measurement sensitivity; (2) providing the biasing current required to turn on the transistors for the first stage current mirror; (3) providing discrete test capacitors in the range of femtofarad for initial testing. As shown in Figure 4.24, a virtual ground node can be obtained using a single transistor in a negative feedback loop. This feedback path helps to increase the circuit stray immunity and reduce charge injection effects from the switching transistors. On the other hand, the transistors of the first stage operate in weak inversion to achieve a high transconductance and increase the sensitivity (80 mV/fF in this case).

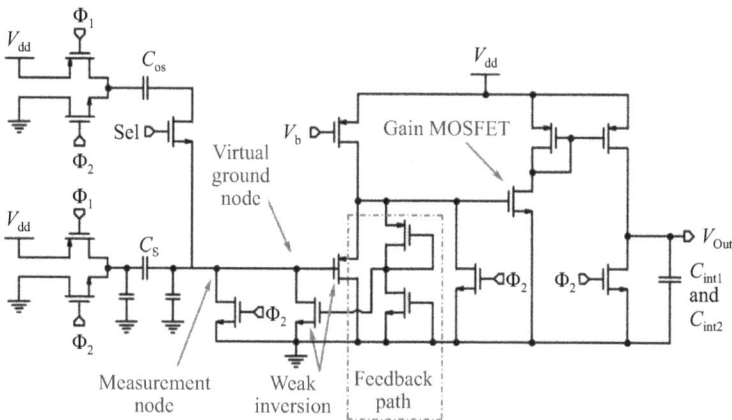

Figure 4.24   Core-CBCM capacitance tomography system

4.3.4 The CVCs based on differential voltage

In differential voltage-based approaches, the currents of the basic CBCM core are averaged by an integrator immediately after extraction from the basic core.

In an approach reported in [77,273], at first, the currents i_S and i_R are averaged separately using two individual integrating capacitors. Then, the voltages of the two integrating capacitors are amplified and subtracted by a differential amplifier as shown in Figure 4.25. In this approach, more voltages across integrating capacitors sacrifice the resolution and dynamic range of the sensor for higher sensitivity. This is because higher input voltages push the differential amplifier to its nonlinear region. However, it has a great resolution of about 10 aF. York and Evans *et al.* used this capacitive sensor for particle detection in [77,274] (see Table 4.2). This group utilized an IDE consisting of Al fingers on the surface of the Si chip. Standing capacitance and the capacitance changes in the presence of small particles were determined using 3D finite-element modeling in [274].

Tanskanen *et al.* [273] used the same idea along with novel current–voltage mirrors for microelectromechanical systems (MEMS) applications (see Figure 4.26). Each of these current–voltage mirrors consist of a two-transistor configuration of a simple current mirror (M_6, M_7) (or (M_{12}, M_{13})) to mirror their input and output currents and a four-transistor configuration (M_5, M_8, M_9, and M_{10}) (or (M_{11}, M_{14}, M_{15}, and M_{16})) to mirror their input and output voltages. A virtual short between their input and output caused by the transistors (M_9, M_{10}) (or (M_{15}, M_{16})) helps to their equalization. Since $V_{sg10} = 0$ (or $V_{sg16} = 0$), the subthreshold current flows through (M_9, M_{10}) (or (M_{15}, M_{16})) and the source-gate voltage of M_9 (or M_{15}) is virtually short, $V_{sg9} = 0$ (or $V_{sg15} = 0$). So, it can mirror both their input and output voltages and currents. Utilizing these new current–voltage mirrors, CBCM currents are mirrored more accurately at the expense of diminishing the output voltages due to the limited output voltage swing. The sensitivity of this circuit in 0.35 µm CMOS technology is decreased from 47.73 to 28.7 mV/fF after replacing the simple current mirrors with the new one.

Yusof *et al.* [143] proposed a fully differential CVC based on CBCM for the measurement of electric double-layer (EDL) capacitances at the electrode–solution

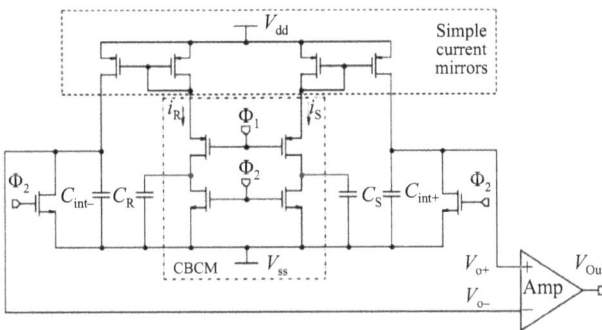

Figure 4.25   A core-CBCM CVC where CBCM currents are integrated before subtraction

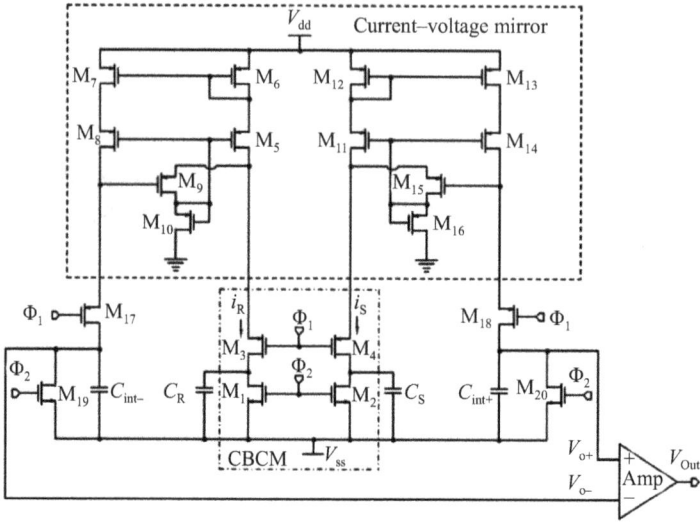

Figure 4.26   *The capacitive sensor with a new current–voltage mirror for MEMS applications*

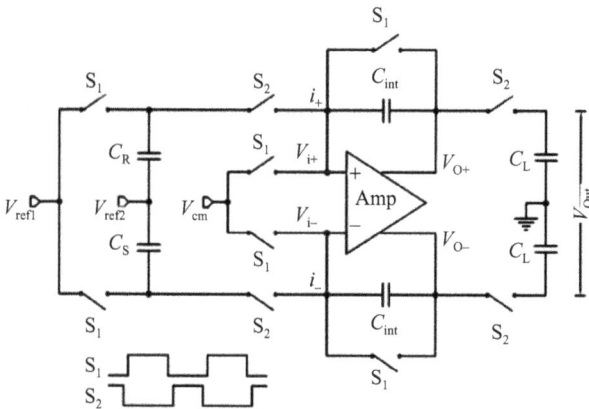

Figure 4.27   *The capacitive sensor proposed by Yusof et al.*

interface. Figure 4.27 shows the proposed circuit. To reduce the charge injection and clock feedthrough effects, transmission gates are used as switches. If S_1 and S_2 are ideal switches and only non-faradaic impedance is effective, the average of the transient current over sensing and reference capacitors, C_S and C_R, after applying voltage steps to the switches is roughly equal to

$$I_{S,R} \approx f_S C_{S,R} (V_{ref1} - V_{cm}) \tag{4.25}$$

where V_{ref1} and V_{ref2} stand for two different reference voltages and V_{cm} which is equal to $(V_{dd}+V_{ss})/2$ identifies a reference for common-mode (CM) voltage (see Figure 4.27). These transient currents are averaged by integrating capacitors, C_{int}. Then, a differential amplifier, with a gain equal to A_{amp}, amplifies the integrated voltages. The differential output voltage at a stable state is obtained by

$$V_{Out} = \frac{\Delta C(V_{ref1} - V_{cm})A_{amp}}{C_{int}(1 - A_{amp})} \tag{4.26}$$

where C_{int} is the value of integrating capacitor. The functionality of this sensor is characterized in sodium chloride (NaCl) solutions with an ionic concentration between 0.1 mM and 1 M over a wide range frequency. Additionally, DNA hybridization in a phosphate-buffered solution resulted in a 20% change in capacitance (10 fF).

Kim *et al.* [275] used this sensor along with a $\Sigma\Delta$ ADC, radio frequency identification circuit and an on-chip spiral inductor tag antenna to implement a wireless biosensor. Nakazato [276,277] used this core-CBCM sensor in a multi-modal sensor array along with amperometric and potentiometric sensors for DNA detection. The same group presented a 4×4 array of this sensor in [267] as seen in Table 4.1. In the array structure, the differential CBCM cell is placed in a matrix, and the Op-amp and SC integrator are shared between cells. N-type and P-type array-clock controllers are also designed to select the array cells.

4.3.5 The single-ended circuits based on differential current

To mitigate the problem of limited dynamic range in differential voltage mode, another approach is presented by Ghafar-Zadeh *et al.* [278] as shown in Figure 4.28. The proposed idea is the subtraction of CBCM currents before averaging. First, M_1 and M_4 are on and M_2 and M_3 are off and discharging currents are subtracted, Δi, in the output node, V_{Out}. Then, M_1 and M_4 are turned on and M_2 and M_3 are turned off. So, no current flows through the output node. The differential current, Δi, can be integrated by an integrating capacitor or an integrator circuit.

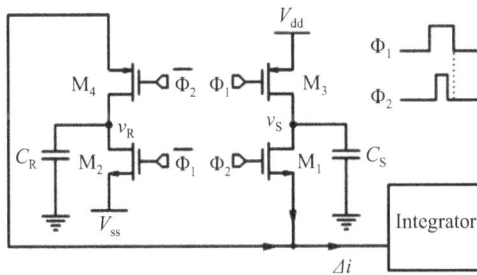

Figure 4.28 A core-CBCM capacitive sensor where CBCM currents are subtracted before integration

The mismatch of PMOS and NMOS transistors are the main cause of the error in this circuit.

Ghafar-Zadeh *et al.* [81] developed their circuit as shown in Figure 4.29(a). In this technique, the currents i_S and i_R are subtracted and amplified by current mirrors, first ($\Delta i = i_S - i_R$). Then, the differential current is injected into an integrating capacitor. The simulation results of this single-ended core-CBCM CVC are discussed in Appendix C.

In [81], the response of this circuit is analyzed for low-conductive and ionically high-conductive solutions atop the electrodes of the biosensor. Figure 4.30 shows the equivalent circuit of the parasitics generated on top of the electrodes where R_{sol} is the resistivity of the analyte and C_{ins} and C_1 denote the capacitance of the passivation layer and the analyte capacitance. C_R is the same for high-conductive and low-conductive solutions. In case of a low-conductive analyte, R_{sol} is very large

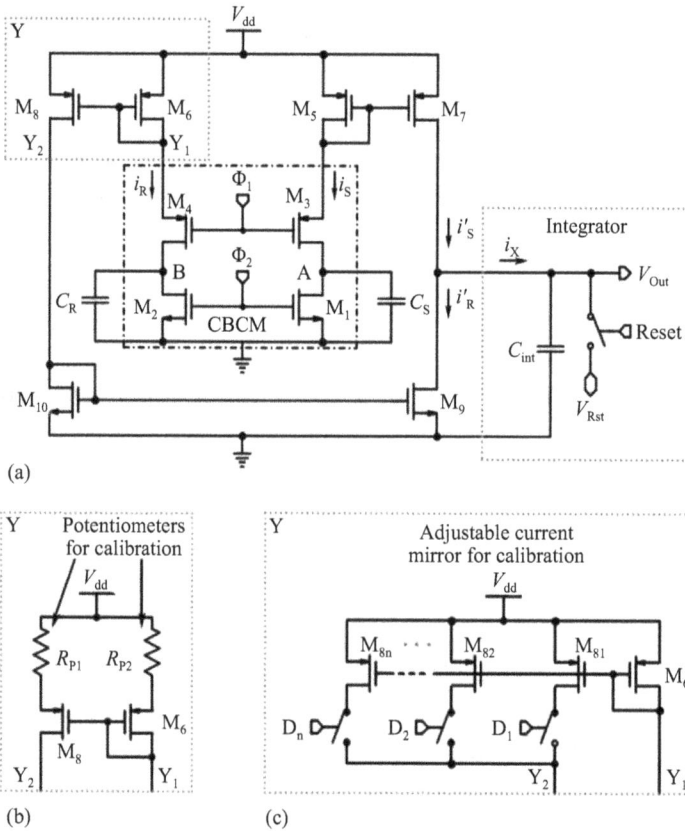

Figure 4.29   *(a) A core-CBCM CVC; (b) potentiometers used for calibration (instead of the circuit shown in the rectangle X illustrated in (a)); (c) adjustable current mirror used for calibration (instead of X)*

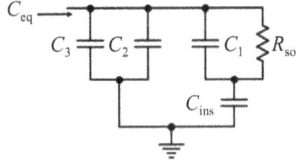

Figure 4.30 The equivalent circuit of the parasitics generated on the electrodes

resulting in $C_{eq} = C_1 \| C_2 \| C_3$, where $C_{ins} \gg C_1$. For ionically conductive solution, R_{sol} is negligible and $C_{eq} = C_2 \| C_3 \| C_{ins}$ [81].

For nonconductive solution, $C_S = C_1 + C_2 + C_3$ and the output voltage is [81]:

$$V_{Out} \approx \frac{C_{T1}}{C_{int}} A_{I1} \left(V_{dd} - V_{thp} \right) - \frac{C_{T2}}{C_{int}} A_{I2} \left(V_{dd} - V_{thp} \right) + V_{off} \tag{4.27}$$

where V_{off} is output offset voltage.

The on/off resistances of switches M_1–M_4 are discounted. C_{T1} and C_{T2} are equivalent capacitances from points A and B in Figure 4.29 and $C_{T1} = C_A + C_S$ and $C_{T2} = C_B + C_R$, where C_A and C_B denote the parasitic capacitances created by M_1–M_4. A_{I1} and A_{I2} are the gain of the current mirrors which are equal to [81]:

$$A_{I1} = \frac{(W/L)_7}{(W/L)_5} \tag{4.28}$$

$$A_{I2} = \frac{(W/L)_8}{(W/L)_6} \times \frac{(W/L)_9}{(W/L)_{10}} \tag{4.29}$$

Equation (4.27) can be rewritten as follows [81]:

$$V_{Out} \approx \frac{C_{T1} - C_{T2}}{C_{int}} A_{I1} \left(V_{dd} - V_{thp} \right) + V'_{off} \tag{4.30}$$

where V'_{off} stands for the equivalent output offset voltage which is equal to [81]:

$$V'_{off} = \frac{C_{T2}}{C_{int}} (A_{I1} - A_{I2}) \left(V_{dd} - V_{thp} \right) + V_{off} \tag{4.31}$$

By assuming $C_A = C_B$, the sensitivity of the sensor with respect to $\Delta C = C_{T1} - C_{T2}$ can be given by [81]:

$$S_{\Delta V / \Delta C} = \frac{\Delta V_{Out}}{\Delta C} = \frac{A_{I1} \left(V_{dd} - V_{thp} \right)}{C_{int}} \tag{4.32}$$

where ΔV_{Out} denote the variation of output voltage. Higher A_{I1} and lower C_{int} result in higher sensitivity. However, C_{int} cannot be less than the parasitic capacitance at node A. A multiple stage amplifier can provide higher A_{I1} and higher sensitivity.

The larger electrodes create larger ΔC, but an array of small sensing electrodes is preferable because it decreases the incorporated parasitic capacitances.

In high-conductive solutions, V_{Out} for $t > 0$ can be obtained by [81]:

$$V_{\text{Out}} = \frac{C_{\text{ins}}}{C_{\text{int}}} A_{11} \left(V_{\text{dd}} - V_{\text{thp}} \right) \left[1 - \frac{C_1}{C_1 + C_{\text{ins}}} \exp \left(\frac{-t}{R_{\text{sol}}(C_1 + C_{\text{ins}})} \right) \right] + V_{\text{off}}''$$

(4.33)

where V_{off}'' is equal to [81]:

$$V_{\text{off}}'' = \frac{\left(V_{\text{dd}} - V_{\text{thp}} \right)}{C_{\text{int}}} \left[(C_2 + C_3 + C_A)A_{11} - (C_R + C_B)A_{12} \right] + V_{\text{off}}$$

(4.34)

The time constant $\tau = R_{\text{sol}}C_{\text{ins}}$ is smaller than the sampling time and $C_{\text{ins}} \gg C_S$. In these fluids, for low-frequency clock pulses ($t \gg 0$) [81],

$$V_{\text{Out}} \approx \frac{C_{\text{ins}}}{C_{\text{int}}} A_{11} \left(V_{\text{dd}} - V_{\text{thp}} \right) + V_{\text{off}}''$$

(4.35)

So, by applying low-frequency clock pulses, R_{sol} does not change V_{Out} and the output voltage will be independent of C_S. Thus, the sensor can be employed for the detection of similar conductive fluids [81]. With appropriate circuitry and differential measurement, C_S can be sensed in series with C_{ins} [68,145,279].

If two different analytes are directed toward the sensing and reference capacitors, a differential measurement can be used for the comparison of the analytes.

To calibrate the offset voltage, V_{off}, they proposed to utilize two potentiometers, R_{P1} and R_{P2} (see Figure 4.29(b)), and adjustable current gain (see Figure 4.29(c)) in addition to the current mirrors. Since the relation between variations of V_{Out} and the changes in value of the potentiometers is nonlinear, only a coarse adjustment can be performed in R_{P1} or R_{P2}. So, the digital calibration circuitry is used to adjust current gain in order to decline the offset voltage and achieve a high resolution.

Corresponding to each digital input signals $D_1, \ldots, D_n$ depicted in Figure 4.29(c), I_R is generated through M_{81-8n} with the following aspect ratios:

$$\left(\frac{W}{L} \right)_6 = 2^1 \left(\frac{W}{L} \right)_{81} = 2^2 \left(\frac{W}{L} \right)_{82} = \ldots = 2^n \left(\frac{W}{L} \right)_{8n}$$

(4.36)

where $(W/L)_{Mi}$ is the aspect ratio of the transistor M_i ($i = 6$ and 81, 82, $\ldots$, $8n$). The digital inputs, $D_1, \ldots, D_n$, can be produced by an on-chip or off-chip programmable unit. The residual offset after offset cancelation is automatically removed by the subtractions of two subsequent measurements before and after analyte injections.

If the sensitivity of the sensor shown in Figure 4.29 is adjusted for about 250 mV/fF in 0.18 μm technology and 1.8 V power supply voltage, an IDR of about 2.7 fF and a resolution of about 10 aF will be obtained (see Table 4.2) [68].

To step toward an integrated LoC, Ghafar-Zadeh *et al.* used the direct-write fabrication process (DWFP) technique in [172] to fabricate polymer-based microfluidics above the CMOS chip and fluidic packaging (see Section 5.2.3). The simple circuit proposed in [81] was integrated with microfluidics to prove the concept. In [223], the feasibility and the compatibility of this new procedure to CMOS-based applications and its functionality for on-chip measurement of the parasitic capacitance are demonstrated using a polyelectrolyte hydrogel and dichloromethane as an ionically conductive and a nonconductive analyte, respectively.

As a continuation of this work, Ghafar-Zadeh *et al.* proposed a low-complexity $\Sigma\Delta$ modulator adapted to the differential circuit shown in Figure 4.29 in [145] converting the voltage across the integrating capacitor to digital data. Figure 4.31 (a) shows this CDC using a voltage comparator and also illustrates the adjustable current mirrors used for calibration. Generally, $\Sigma\Delta$ modulators have the merits of high resolution, simple architecture, low power consumption, small size, and easy implementation. But differently from the conventional $\Sigma\Delta$ modulators, the input voltage of the modulator proposed by Ghafar-Zadeh *et al.* is a step waveform instead of a ramp.

Figure 4.31(b) demonstrates a simple model for this $\Sigma\Delta$ modulator. The integrated voltage on C_{int} is compared with the reference voltage, V_{ref}, and the output pulse of the comparator, q_n, is generated and applied to the switch S_z resulting in the generation of $i_z(C_R, t)$ by the transistor M_z. i_z changes the DC voltage of V_{Out} by charging C_{int}. The interactions corresponding to C_S lead V_{Out} to a value more than V_{ref} so that the switch S_z is closed and the value of V_{Out} is declined to a level lower than V_{ref} as seen in Figure 4.31(c). The generated i_S in each period of Φ_1 and Φ_2 increments up the voltage on C_{int} that plays the role of a delay integrator [76].

In Figure 4.31(a), the switch S_z and the current mirror (M_6, M_z) constitute a 1-bit DAC for this first-order $\Sigma\Delta$ ADC that converts the digital input to i_z. Considering i'_S and i'_R as the amplified currents of i_S and i_R, respectively, a current error is generated which is equal to $e_n = i'_S - (i'_R + i_z)$. This current error is converted to a voltage error and accumulated to the capacitor output voltage. The quantization is performed by a track and latch voltage comparator followed by an RS flip-flop. Since $i'_S - i'_R$ is a linear function of C_S, the digital output of the comparator, q_n, is also proportional to C_S.

The viability of this hybrid CMOS-microfluidics sensor employing the $\Sigma\Delta$ modulator and DWFP-based microfluidics is examined by using polyelectrolyte solutions and chemical solvents with known dielectric constants in [75,145,280] and for bacteria growth monitoring in [68,279]. Additionally, a new algorithm for decoding the data of a first-order $\Sigma\Delta$ modulator is proposed by this group [281] which is useful for LoC applications.

Miled *et al.* [282] proposed to use the subthreshold conduction of the current mirror transistors (which amplifies the CBCM current) instead of the saturation region to increase the sensitivity and reduce the power consumption of the CBCM method.

If the current mirror transistors, for example, M_5 and M_7 operate in the saturation region, the sensitivity of the drain current, $i_{5,7}$ to their gate-source

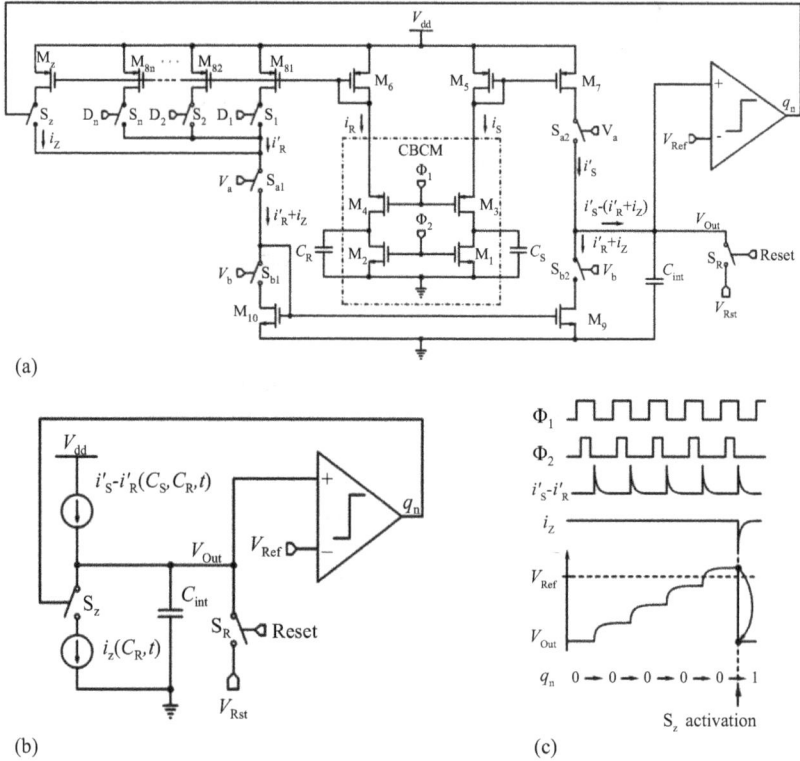

(a)

(b)   (c)

Figure 4.31   (a) Core-CBCM $\Sigma\Delta$ capacitive sensor circuit; (b) simple model of the
$\Sigma\Delta$ modulator; (c) applied clock pulses, differential current, $\Sigma\Delta$
current, output voltage, and the bitstream of the sensor

voltage, $V_{sg5,7}$ is linear:

$$\frac{di_5}{dv_{sg5}} = \mu_p C_{ox} \frac{W_5}{L_5} \left(V_{sg5} - V_{thp} \right) \tag{4.37}$$

M_5 and M_7 in Figure 4.29 are intentionally utilized at the subthreshold region
meaning their gate-source voltage is lower than their threshold voltage ($V_{sg5,7} <$
V_{thp}). Therefore, their drain current is more sensitive to the variation of V_{sg5}. The
sensitivity of i_5 to V_{sg5} in subthreshold region is expressed by

$$\frac{di_5}{dv_{sg5}} = \frac{W_5}{L_5} I_{D0} \frac{V_{sg5}}{nV_T} \exp\left(\frac{V_{sg5}}{nV_T}\right) \tag{4.38}$$

where I_{D0} denotes a process-dependent parameter and n stands for the subthreshold
slope factor. By comparing (4.37) and (4.38), it is obvious that the drain current of
the transistor operating in the subthreshold region is more sensitive to its source-
gate voltage because of the exponential term in (4.38). So, any few changes in C_S

results in a considerable current variation with small power consumption. However, the behavior of the transistors in the subthreshold region is not well known. Moreover, it results in a limited dynamic range. A simple circuit with one current mirror was simulated to test this idea.

The same group proposed a dielectrophoresis device taking advantage of electric field for micro- and nanoparticle manipulation which was integrated with an array of electrodes and used an adapted version of the circuit shown in Figure 4.29 whose current mirror transistors work in the subthreshold region [257,283,284]. A 3×3 array of capacitors is implemented to evaluate this circuit that can be activated line by line and measure three different capacitances at the same time (see Figure 4.32 and Table 4.1).

Each one of the command lines En_0, En_1, and En_2 shown in Figure 4.32 activates one line including three different capacitive sensor cells and make it possible to take measurement of the three capacitances at different microchannel locations simultaneously. A sensing current circuit connected to each column of the array detects any current variation in the two branches of the sensor cells of that column. The output differential current is converted to a voltage by an integration circuit and then digitized by a $\Sigma\Delta$ ADC.

Yao *et al.* [24] utilized a similar circuit as shown in Figure 4.33 for the detection of deadly bacteria such as *Salmonella* and *E. coli* (see Table 4.1). The resistors R_{P1} and R_{P2} in Figure 4.33 are used for balancing offset, and the source follower consisting of M_{11} and M_{12} drives the load capacitor as an LPF.

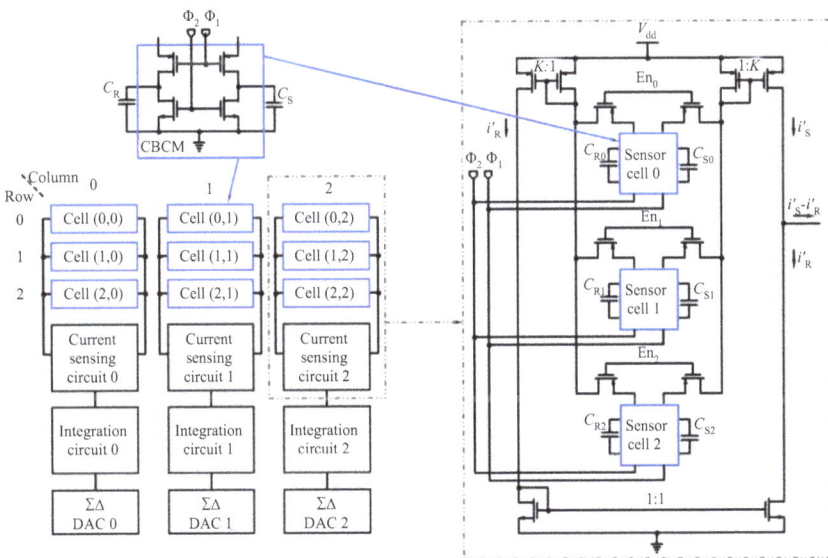

Figure 4.32 A 3 × 3 sensor array configuration with the capability of simultaneous evaluation of three different capacitances

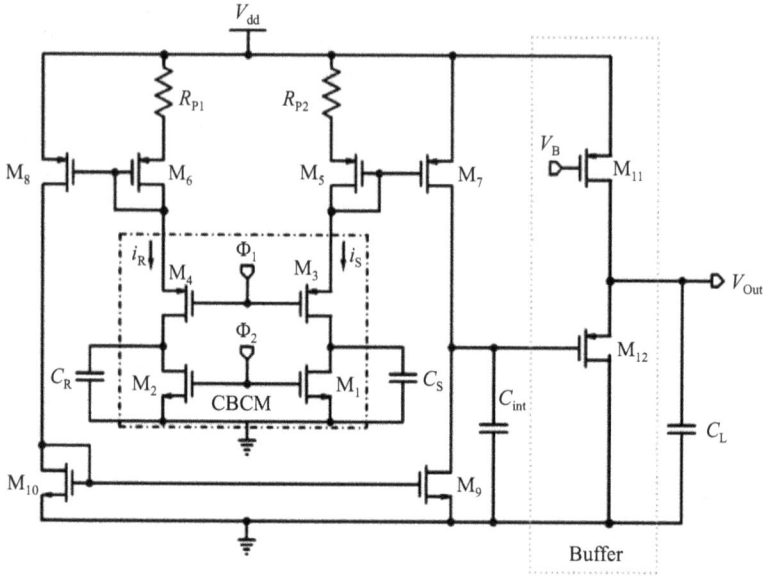

Figure 4.33   The core-CBCM capacitive sensor used by Yao et al.

4.3.6   The fully differential circuits based on differential current

A fully differential rail-to-rail CVC is proposed in [142,264] for on-chip cell monitoring (Figure 4.34). Some problems of single-ended structures due to parasitic capacitances and CM noise are solved in the fully differential circuit and the dynamic range of the capacitance measurement is doubled.

Consider the integrating capacitors are equal to each other ($C_{int+} = C_{int-} = C_{int}$), $\Delta i_+ = i_S - i_R$, $\Delta i_- = i_R - i_S$, and both currents Δi_+ and Δi_- are amplified with a gain of $A_I = A_{I1} = A_{I2}$ using current mirrors. Then, the output voltages ($V_{Out\pm}$) of this circuit are calculated by [264]

$$V_{Out\pm} = \pm \frac{A_I \Delta C (V_{dd} - V_{ss} - |V_{thp}|) f_S}{C_{int} f_{int}} + V_{off\pm} + V_{cm} \qquad (4.39)$$

where V_{dd} and V_{ss} are dual supply voltages and V_{cm} denotes the CM voltage at the output nodes. $V_{off\pm}$ stands for the offset voltage after the integration of Δi_+ and Δi_-. f_{int} is integrating frequency and in this work, it is equal to the sampling frequency ($f_{int} = f_S$). In other words, integration is performed only during a single period of Φ_1 and Φ_2. After subtraction of these two voltages, ΔV_{Out} is equal to [264]

$$\begin{aligned} \Delta V_{Out} &= V_{Out+} - V_{Out-} \\ &= \frac{2A_I \Delta C (V_{dd} - V_{ss} - |V_{thp}|) f_S}{C_{int} f_{int}} + \Delta V_{off} \end{aligned} \qquad (4.40)$$

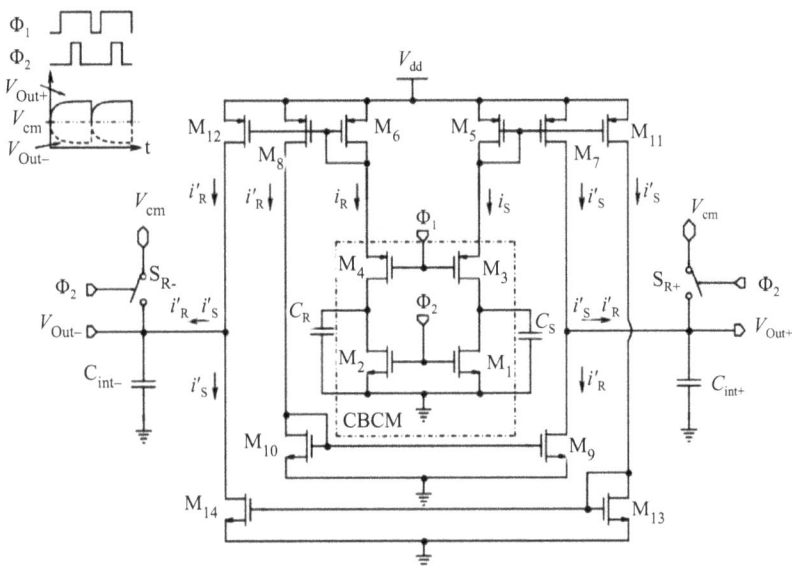

Figure 4.34   A fully differential core-CBCM capacitive sensor

where $\Delta V_{off} = V_{off+} - V_{off-}$, which is zero under ideal conditions. But practically, the mismatches in the transistors of both sides of the circuit lead to $\Delta V_{off} \neq 0$.

In [142,264], the current bus shielding used in the fully differential capacitance sensor array protects the large array of the sensor from noise and junction leakage and simplifies calibration because there is no need to calibrate pixels individually. Figure 4.35(a) and (b) illustrates the horizontal view of the shielded current bus and a pixel of the fully differential capacitance sensor array with a column-parallel architecture based on the bus, respectively. As seen in these figures, the current bus lines are the extension of the nodes N_+ and N_- which makes it possible to share a common single-sensor evaluation module (SEM) consisting of the current mirrors, subtractors, integration capacitors, as well as output buffers with a column of CBCM pixel units. A larger area metal shield fabricated in a lower metal layer can isolate the bus line from the substrate. In addition to the improvement of the immunity of the current bus to substrate noise, the effects of source-to-bulk junction leakage and bus-to-shield capacitance are canceled out by employing the buffer amplifiers and driving the shield line and the individual n-wells of the PMOS switches in each CBCM pixel with a potential which tracks the bus line potential.

Additionally, in [142,264], floating gate transistors are utilized in the current mirrors to cancel the voltage offset in the output of a fully differential architecture. A combination of electron injection and tunneling is also employed in this calibration technique. Long-term retention, programming capability, and standard CMOS fabrication are some of the merits of using floating-gate transistors. The transistors M_7 and M_{12} in the current subtractors of the circuit depicted in Figure 4.34 are replaced by

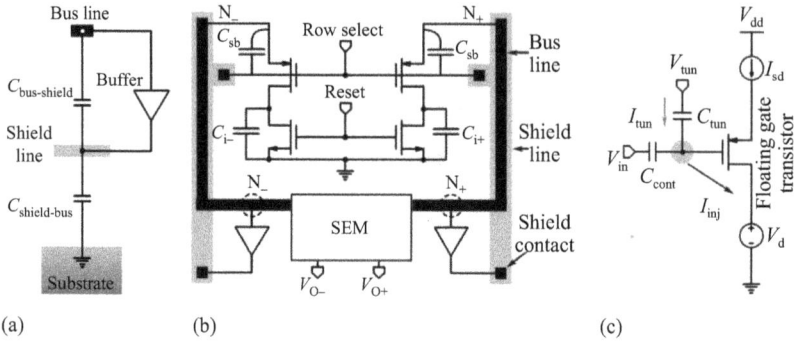

Figure 4.35   *(a) Horizontal view of the shielded current bus; (b) a pixel of a fully differential capacitance sensor array with a column-parallel architecture based on a shielded current routing bus; (c) conceptual view of a floating-gate transistor*

floating-gate transistors. So, the gate offset voltage can be modified and the corresponding offset current can be canceled. Figure 4.35(c) shows the conceptual view of the floating-gate transistor whose gate is isolated by a high-quality insulator (SiO_2) providing a nonvolatile charge storage node at this node. The poly 1 and 2 of the CMOS process are used for the floating gate and the control gate, respectively, that together form the capacitor C_{cont}. C_{tun} depicted in Figure 4.35(c) is a PMOS capacitor whose gate is connected to the floating gate, and the source, drain, and bulk which are connected to the tunneling terminal. Impact-ionized hot electron injection and Fowler–Nordheim tunneling are two different mechanisms employed for programming the floating-gate transistor. In the former approach, a negative shift in the gate offset voltage is generated by imposing a high source-to-drain voltage on the transistor and consequently the generation of electrons at the drain edge of the drain-to-channel depletion region via hot-hole impact ionization. The source-to-drain voltage and the source-to-drain current control the rate of injection current, I_{inj}. In the latter mechanism, Fowler–Nordheim tunneling, which is an electric field-assisted tunneling, by imposing a very high electric field on C_{tun} generates a positive shift in the gate offset voltage.

As seen in Table 4.1, using the supply voltage of $\pm3V$, Prakash *et al.* [142] achieved a dynamic range of tens of femtofarads with attofarad resolution which is suitable for on-chip cell monitoring. The same group discussed three different array architectures of the sensor, mismatch compensation technique, and electrode configurations in [285].

A fully differential CDC is proposed in [183] which is realized by a fully differential CVC and a $\Sigma\Delta$ modulator specifically adapted to this structure (see Figure 4.36(a)–(c)). This $\Sigma\Delta$ modulator is an extended version of the single-ended one proposed by Ghafar-Zadeh *et al.* [145].

Figure 4.36(a) shows this CDC which is composed of a comparator, a quantizer, and a DAC. The voltages of the integrating capacitors C_{int+} and C_{int-} which

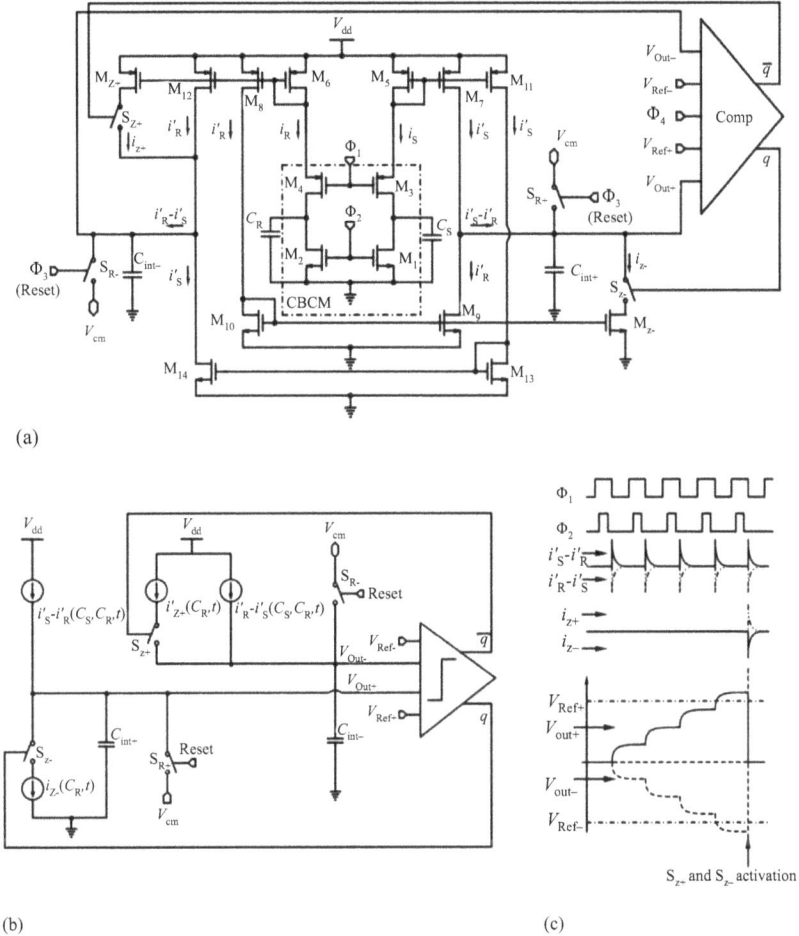

(a)

(b)   (c)

Figure 4.36   (a) A fully differential core-CBCM ΣΔ CDC; (b) a ΣΔ modulator
              adapted to a fully differential core-CBCM capacitive sensor; (c) the
              related signal diagrams

realize a summing node and LPF in the first-order ΣΔ modulator are compared with
the reference voltages in each quantizer's sampling period. If $V_{Out+} - V_{Out-} >
V_{Ref+} - V_{Ref-}$, the output signals generated by the comparator will activate the
switches S_{z+} and S_{z-} to inject i_{z+} and i_{z-} which are dependent on C_R to the nodes
V_{Out-} and V_{Out+}, respectively. As a result, the current i_{z-} will be subtracted from
the positive current $i'_S - i'_R$ and the current i_{z+} will be added to the negative
current $i'_R - i'_S$. S_{z+} and S_{z-} and the current sources of i_{z+} and i_{z-} (M_{z+} and M_{z-})
play the role of a DAC in the first-order ΣΔ modulator.

The simple model of this modulator is depicted in Figure 4.36(b) and the
related signal diagrams are illustrated in Figure 4.36(c). To increase the sensitivity

of the fully differential structure, the CBCM currents are integrated over N cycles of the periods of Φ_1 and Φ_2 ($f_S = N f_{int}$). So, the sensitivity will be equal to

$$S_{\frac{dV_{Out}}{dC_s}} = \frac{2A_1\Delta C\left(V_{dd} - V_{ss} - |V_{thp}|\right)N}{C_{int}} \tag{4.41}$$

This sensor is characterized by Nabovati *et al.* [183] for chemical solvent detection. As mention in Table 4.1, by integration over 10 cycles of sampling time ($N = 10$), a sensitivity of about 350 mV/fF is obtained and the dynamic range is 10 fF in 0.35 µm CMOS technology. As a continuation of this work, the same CDC is used in an 8 × 8 array for high-throughput cell growth screening in [21]. The same group [286] employed this capacitive sensor along with a pH sensor to develop a multifunctional biosensor.

Jun-Rui *et al.* used a fully differential core-CBCM sensor in [287] like the one shown in Figure 4.34 for low-energy biomarker detection, but the simple current mirrors are replaced by cascade current mirrors. This type of current mirror helps to achieve a high enough output impedance that does not affect the output current and both the steady-state and transient current can be studied correctly. In this work, the potentiality of applying the CBCM method in ionic conditions has been studied quantitatively. The binding of biomarkers onto functionalized Au IDEs changes both real and imaginary components of the sensing impedance. If a parallel resistance–capacitance (RC) model for sensing and reference electrodes ($R_S\|C_S$ and $R_R\|C_R$, respectively), two variables are defined as $\Delta R = R_S - R_R$ and $\Delta C = C_S - C_R$. The behavior of this circuit is studied in two states: (1) $\Delta R = 0$ and sweeping ΔC; (2) $\Delta C = 0$ and sweeping ΔR. ΔC affects the transient response of the circuit and the peak value of the output voltage, V_{Out}, in the first case and ΔR changes the final value of V_{Out} in the second case. So, the peak and the final values of the transient response can help to extract real and imaginary parts of the sensing impedance.

To achieve high sensitivities, the output voltages of the fully differential CVC presented in [142] are fed to a differential voltage amplifier in [288] (Figure 4.37). The sensitivity of this sensor is increased to 1.42 V/fF by the gain of the amplifier, A_{amp} (see Appendix C). It means that the sensitivity of this sensor is equal to

$$S_{\frac{dV_{Out}}{dC_s}} = \frac{2A_{amp}A_1\Delta C\left(V_{dd} - V_{ss} - |V_{thp}|\right)N}{C_{int}} \tag{4.42}$$

Although this circuit has a great resolution, the restricted output voltage swing limits considerably the output dynamic range (about 300 aF). Furthermore, it is proposed to use a common-mode feedback circuit to fix the output CM voltage of the first fully differential stage on half of the supply voltage. This results in more symmetrical responses at the two differential output nodes of the first stage (V_{Out+} and V_{Out-}) and prevents the overshooting of these voltages.

4.3.7   Core-CBCM CFC

Some sensors utilize a combination of the CBCM method and CFC. Figure 4.38 illustrates the circuit proposed in [289] in which the sharp exponential currents of

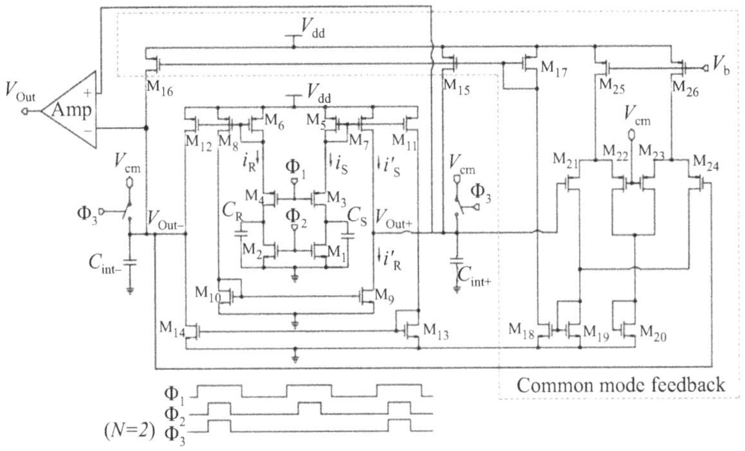

Figure 4.37   *Differential core-CBCM capacitive sensor (V_{cm} is the required CM voltage equal to $V_{dd}/2$)*

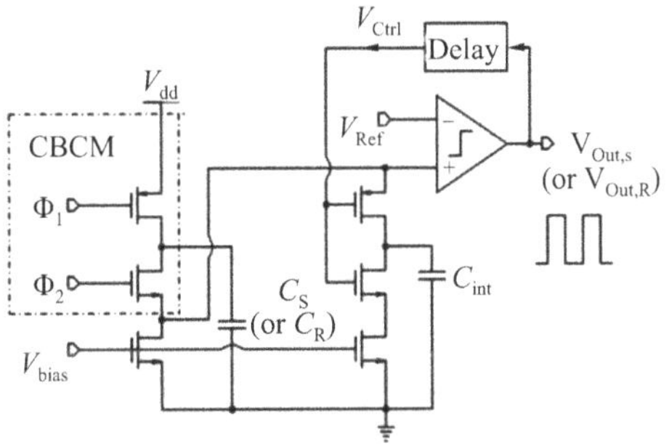

Figure 4.38   *A combination of CBCM method and CFC.*

each branch of the CBCM block are averaged using an integrating capacitor in the analog domain and then this voltage is converted to frequency. Two similar structures are used for each of the sensing and reference capacitors to convert the related capacitance to frequency. The subtraction of their outputs is proportional to ΔC. This circuit requires a large integrating capacitor to achieve high accuracy and its IDR is from femto-to picofarad.

A similar approach is proposed in [290] utilizing a combination of the CBCM method and CFC (Figure 4.39). It utilizes an integrator and a Schmitt trigger. The CBCM current is mirrored and accumulated by the integrator and results in a tri-angular signal. The Schmitt trigger converts it to a square signal and fed it to the digital block containing a simple counter. The counter compares the period of this sensing signal with a reference signal obtained from the symmetrical circuit. Whenever the reference signal is counted up to a known number (2^8), the counting

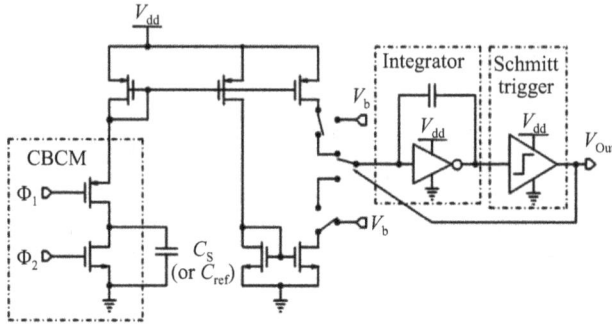

Figure 4.39   The capacitive sensor proposed by Yamane et al.

of the sensing signal is stopped. The number of the counter at this time reveals C_S/C_R. But this sensor suffers from low resolution.

Tashtoush [78,291] used the same core-CBCM sensor proposed in [24] as well as a comparator-based signal processing circuit in a multi-labs-on-a-single-chip (MLoC) system (see Table 4.1). Without the signal processing circuit, the output can be read as a voltage signal. The signal processing circuit converts it to a frequency. In addition to the CBCM-based capacitive sensor used for femtomolar detection of bacterial pathogens, the MLoC consists of an optical CMOS biosensor, electrochemical spectroscopy biosensor, and CMOS micro-coil incorporated with interdigitated microelectrode array and microfluidic.

Prior core-CBCM sensors have limitations in dynamic range because of working in voltage mode. In all of these circuits, the output current of the CBCM block is converted to voltage even before digitization. Integration in the analog domain results in this effect, because the supply voltage restricts the swing of the integrating capacitor voltage. Current-mode circuits are appropriate candidates for low-supply voltage CMOS technologies. So, a core-CBCM CFC working in the current-mode is presented in [292] overcoming this limitation. Figure 4.40(a) shows a simple circuit to show the proposed concept. In this approach, the amplified currents of the CBCM core are fed to a CCO after subtraction. One of the most important features of CCO is its capability to follow the variations of the sharp exponential currents of the CBCM block in such a way that it can modulate them to pulse frequency. A counter is employed to average the output frequency because the average of the differential current is proportional to ΔC. In other words, in this sensor, differently from the previous works, the required averaging operation is done part by part using the CCO in the analog domain and then the counter accumulates the total integrated parts (that each one has a unit value equal to A_u), in the digital domain.

Figure 4.40(b) illustrates a basic design of a CCO only to elaborate the performance of the circuit shown in Figure 4.40(a). The input current of the CCO (i_{CCO}) alternatively charges the nodes V_1 and V_2 of the CCO depicted in Figure 4.40(b) resulting in the variation of the output voltages V_3 and V_4 which are the inversions of

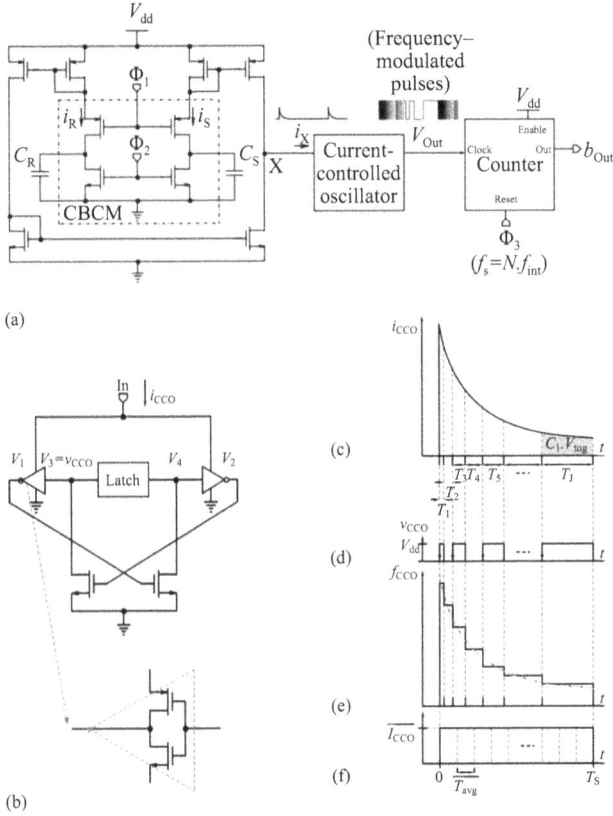

Figure 4.40   *(a) Current mode core-CBCM capacitive sensor; (b) a basic design of a CCO; (c) CCO input current; (d) CCO output voltage; (e) frequency of the output pulses of the CCO and its envelope; (f) CCO-averaged input current based on the mean value theorem (by exaggeration in the sizes of the unit blocks)*

each other due to the latch. The output frequency is determined by the intervals (T_m) during which V_1 and V_2 reach a toggling voltage (V_{tog}) and the state of the latch is changed. So, the integration of i_{CCO} during each interval is calculated as follows

$$\int_{\left(\sum_{z=1}^{m} T_z\right) - T_m}^{\sum_{z=1}^{m} T_z} i_{CCO}(t, C_S, C_R) dt = C_{t1} V_{tog}, \, m = 1, 2, \ldots, J \qquad (4.43)$$

where C_{t1} denotes the total capacitance at nodes V_1 and V_2 for a symmetrical CCO which is a small capacitance composed of the junction capacitances of the transistors. Any small changes in i_{CCO} leads to significant variation in the output frequency of the CCO. If $C_{t1} V_{tog}$ is roughly constant, the smaller amplitude of i_{CCO} results in longer interval (T_m) and vice versa. Therefore, as seen in Figure 4.40(c) by exaggeration, the total area under $i_{CCO}(t)$ during a certain time is divided into very small blocks, each with an area of $C_{t1}V_{tog}$ that we call them unit blocks here. Assume J unit blocks are obtained during the sampling time ($T_S = 1/f_S$) (see Figure 4.40(c)–(f)). As depicted in Figure 4.40(d), after each T_m ($m = 1, 2, \ldots, J$), the state of the output voltage (v_{CCO}) changes high to low or conversely low to high. The integration of i_{CCO} during the total integration time $T_{int} = 1/f_{int}$ is obtained by

$$\int_0^{T_{int}} i_{CCO}(t, C_S, C_R)dt = N\sum_{m=1}^{J}\left(\int_{\left(\sum_{z=1}^{m}T_z\right)-T_m}^{\sum_{z=1}^{m}T_z} i_{CCO}(t, C_S, C_R)dt\right) \quad (4.44)$$

$$= NJC_{t1}V_{tog}$$

These blocks can be counted by a counter. So, the integration process is continued and completed in the digital domain. If the rising edges of the output pulses are considered, the total number of pulses counted by the counter (N_{Count}) during T_{int} is equal to $NJ/2$. So, it can be easily proved that

$$N_{Count} = \frac{NA_1(V_{dd} - V_{thp})\Delta C}{2C_{t1}V_{tog}} \quad (4.45)$$

where N is an even number. If V_{tog} is constant, N_{Count} is linearly proportional to ΔC.

As seen in Figure 4.40(e), the CCO does not convert i_{CCO} to frequency in every moment of T_{int}, but the output frequency (f_{CCO}) can follow the envelope of the discrete variable frequency of the output pulses. The time-variant frequency of the CCO can be given by

$$f_{CCO}(t, C_S, C_R) = A_{CCO}i_{CCO}(t, C_S, C_R) \quad (4.46)$$

where A_{CCO} stands for the gain of the CCO. On the other hand, it can be easily verified that

$$i_X(t, C_S, C_R) = A_1\Delta C\frac{dv}{dt} \quad (4.47)$$

Combining (4.46) and (4.47) and averaging the results over T_S, we obtain the average of the variable output pulse frequencies, F_{CCO}, over T_S as follows

$$F_{CCO}(\Delta C) = \frac{A_{CCO}A_1\Delta C(V_{dd} - V_{thp})}{T_S} \quad (4.48)$$

It can be easily proved that averaging over the total integration time (T_{int}) results in

$$F_{CCO}(\Delta C)T_{int} = A_{CCO}NA_1\Delta C(V_{dd} - V_{thp}) \quad (4.49)$$

According to Figure 4.40(d), each unit block generates a half period of each pulse. Using the mean value theorem, it can be shown that the counter number (N_{Count}) is equal to

$$N_{Count} = \int_0^{T_{int}} f_{CCO}(t, C_S, C_R)\,dt = F_{CCO}(\Delta C)T_{int} = \frac{NJ}{2} \qquad (4.50)$$

So, the sensitivity of the sensor is given by

$$S_{\frac{d(Numb)}{d(\Delta C)}} = A_{CCO}NA_1\left(V_{dd} - V_{thp}\right) \qquad (4.51)$$

So, three parameters including the sensitivity of the CCO (A_{CCO}), the gain of the current mirrors (A_1), and the number of the cycles of Φ_1 and Φ_2 (N) affect the sensitivity of the capacitive sensor among which the latter is off-chip controllable. Increasing the gain of the current mirror (A_1) results in more sensitivity but the IDR of the CCO and the size of the counter can limit the IDR. Raising the number of cycles (N) leads to a longer measurement time.

To recapitulate, this circuit integrates the CBCM current signals in two steps. First, the current is integrated by the CCO over the time of each unit block in the analog domain. In the next step, the number of unit blocks is counted by the counter and the integration is completed in the digital domain. Each unit block generates an individual output frequency. Finally, a string of the output pulses with discrete frequencies are produced that follows the envelope of the instantaneous frequency modulated current response.

The original circuit used in [73] is calibrated by a fast up-down counter and is designed in such a way that it can work in five phases: initialization, down counting, saving, up counting, and cycling. In this regard, a linear feedback shift register (LFSR) is adapted to count bidirectionally and also work as a serial-input serial-output shift register. A digital buffer is also required before the counter to mitigate the loading effect. The integration over N cycles of the sampling period ($f_s = N \cdot f_{int}$) increases the sensitivity of the sensor. The nonlinearity of the CCO is the main cause of the limited resolution of this sensor. To improve its resolution, the pre-distortion technique is proposed. In this circuit, the IDR of the CCO and the size of the counter are the limiting factors of the IDR of the sensor. But, as mentioned in Table 4.1, this method results in a dynamic range of about 70 fF in 0.18 μm CMOS technology which is high in comparison to previous core-CBCM sensors. Some additional explanations and the simulation results of Figure 4.40(a) are outlined in Appendix D.

4.3.8   Core-CBCM capacitance sensor with nanoelectrodes

While a growing body of literature has studied single and array of nanoelectrodes in voltammetric or amperometric electrochemical assays, the nanoscale is received only limited attention in electrochemical impedance spectroscopy (EIS). In the absence of faradaic contribution and in the typical frequency of impedance spectroscopy which is low (about 10 kHz), the EDL has enough time to be rearranged. The instantaneous electric field will encounter an exponential decay from the electrode surface whose length is equal to the equilibrium Debye screening length L_D. So, the target analyte can only be sensed when the separation distance between the target analyte and the electrode is on the order of L_D or less (a few nanometers). In other words, a target analyte beyond the EDL formed by screening salt ions, even at a short distance from the sensor, cannot be sensed at low frequencies. But, in some applications, it is desirable to probe beyond EDL under physiological salt conditions [144,245,293].

One solution is increasing the frequency of the AC excitation. If it is sufficiently increased, ions of the solution will not be rearranged so fast that the formed EDL perturbs the electric field. Thus, the electric field penetrates radially from the electrode to the solution at a distance on the order of the smallest electrode dimension and also the sensitivity to small adsorbates within the EDL is decreased. This offers high selectivity and sensitivity to, for instance, supra- or macromolecular entities using nanoelectrodes. Probing beyond the EDL occurs when the modulation frequency is above the salt-concentration-dependent cutoff frequency which is equal to

$$f_{\text{cutoff}} = \frac{1}{2\pi R_{\text{sol}}(C_{\text{dl}} + C_{\text{sol}})} \tag{4.52}$$

where R_{sol} and C_{sol} stand for the solution resistance and capacitance, respectively. Thus, much high enough frequency is required when nanoelectrodes are used. But stray capacitances limit the frequency to less than 1 MHz. One solution is increasing the L_D by lowering the electrolyte ionic strength under physiological conditions, but it might change or even denature the biomolecule properties. Another solution is the parallel connection of a large number of nanoelectrodes to increase the total admittance with respect to the stray capacitances. In this way, a small volume can be probed beyond EDL at the expense of missing individual detection at each electrode [245,293]. Integration of the array of nanoelectrodes and the required readout circuits can minimize the background capacitance and facilitates the high-frequency operation at physiological salt concentrations.

Widdershoven *et al.* [242] used the CBCM method for capacitive measurement of a densely packed and individually addressable array of 256×256 nanoelectrodes at high frequencies (up to 200 MHz). Additionally, an on-chip temperature sensor allows for analyzing temperature-induced variations. The same group [293] used their massively parallel biosensor for high-frequency impedance spectroscopy and real-time imaging of the dynamic variations of BEAS, THP1, and MCF7 cancer cells in growth medium and achieved an attofarad resolution on the sub-micrometer scale. Their nano-capacitor biosensor can image microparticles and living cells under physiological salt conditions. This group extended their biosensing platform with microfluidics and calibration systems in [144] (see Table 4.1). Capacitance spectra can be recorded by calibrated multifrequency measurements and reveal new information about the analyte.

Figure 4.41(a) illustrates the core-CBCM readout circuit and Figure 4.41(c) shows the array architecture of the biosensor proposed by Widdershoven *et al.* [144,245].

In this biosensor, all the switches of the CBCM core are made of NMOS transistors and charge and discharge the sensing capacitor to V_1 and V_2, respectively.

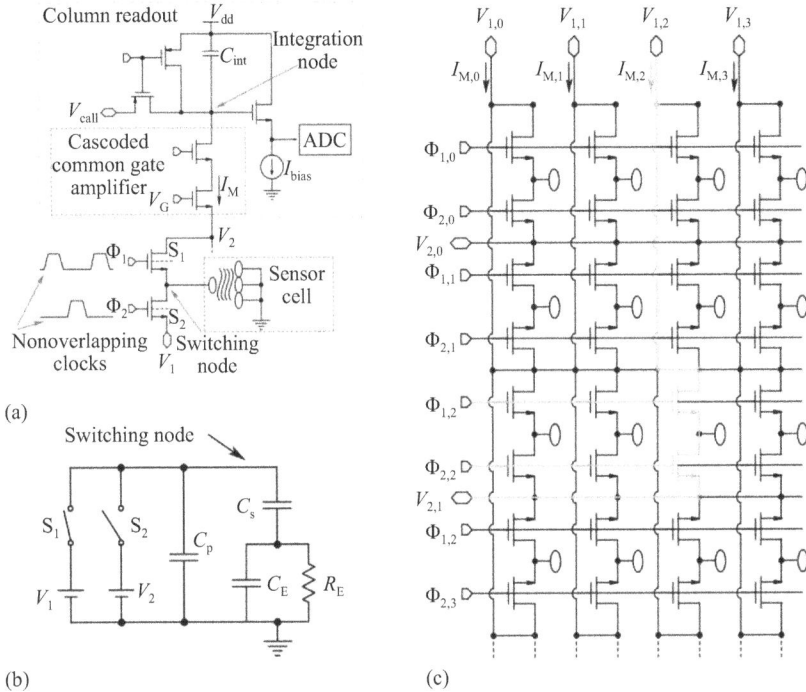

Figure 4.41   (a) Array structure and (b) the equivalent circuit of the switching model; (c) readout circuit of core CBCM biosensor used for real-time imaging and impedance spectroscopy

After N cycles of Φ_1 and Φ_2, the total charge transferred by one of these sensor cells is obtained by

$$Q = N(V_1 - V_2)(C_S + C_p) \tag{4.53}$$

where C_p stands for the parasitic capacitance of the nanoelectrode. After sensing the analyte, changing the capacitance result in a change in Q:

$$\Delta Q = N(V_1 - V_2)\Delta C \tag{4.54}$$

The circuit does not produce flicker noise, but the transistor reset noise of the $2N$ switching steps will be added to the signal which is equal to

$$\sigma_n^2 = 2Nk_B T(C_S + C_p) \tag{4.55}$$

where k_B identifies Boltzmann's constant and T denotes the absolute temperature. So, the intrinsic SNR is obtained by

$$\left(\frac{\Delta Q}{\sigma_n}\right)^2 = \frac{N(V_1 - V_2)^2 \Delta C^2}{2k_B T(C_S + C_p)} \tag{4.56}$$

The small dimensions of nanoelectrodes increase ΔC and decrease $C_S + C_p$ results in less noise. So, SNR can be maximized. Making small C_S needs non-CMOS technologies like electron-beam lithography, but regarding small C_p, sub-micrometer dense integration in modern CMOS technology is required [245].

As shown in Figure 4.41(c), one row of the array architecture can be measured simultaneously by applying Φ_1 and Φ_2 of Figure 4.41(a). The nonselected nano-electrodes will be disabled by the low potential on their respective Φ_1 and isolated from the column lines connected to V_1. High voltages of Φ_2 of the nonselected nanoelectrodes connect them to AC grounded discharge lines. In this way, the non-selected nanoelectrodes are merged and play the role of a large reconfigurable on-chip counter electrode which provides a low-impedance AC return path from the electro-lyte to the chip and prevents unwanted electrochemical modifications of the electrode surfaces. Figure 4.41(b) illustrates the simplified equivalent circuit of the proposed switching model where C_p denotes the parasitic capacitance, C_{surf} is the surface capacitance (a self-assembled monolayer (SAM), if present, in series with EDL), C_E and R_E, respectively, stand for the spreading capacitance and resistance of the elec-trode to the electrolyte. In low-conductive solutions and/or at high frequencies, the C_{dl} is less dominant. But, in ionically conductive solutions and/or at low frequencies, the C_{dl} is dominant which is dependent on the effective area of the electrode.

The voltage V_1 is controlled by the cascoded common-gate amplifier. As illu-strated in Figure 4.41(a), the column current is averaged by parasitic capacitance and passes to the integrating capacitor C_{int}. After integration, an ADC samples the vol-tage at the node V_{int}. Then, the calibration voltage V_{cal} is sampled and passed to the integration node by the ADC. Next, the ADC converts the difference between the two samples. Eight on-chip 13-bit ADCs are used for digitization of V_{int} in parallel. Each ADC processes 32 columns consecutively using time-division multiplexing.

4.4   Summary

This chapter gave an overview of various CMOS capacitive interface circuits including charge redistribution techniques like charge-sharing method, charge-sensitive amplifier-based and SC circuits as well as CFCs using comparator-based relaxation oscillators and ROs, VCO-based sensors, lock-in detection, triangular voltage analysis, and CBCM method dedicated to biosensing applications.

Among these techniques, lock-in detection, RO-based CFC, and CBCM have shown the privilege of high resolution. However, the complexity of lock-in detection usually results in high power consumption and large occupied area of the circuit. On the other hand, RO-based CFCs usually suffer from their sensitivity to temperature, device sizes, process, and supply voltage. Due to the accuracy and low-complexity of core-CBCM biosensors, these sensors were selected to be elaborated in this chapter. So, the development process of core-CBCM circuits was then discussed.

Although CMOS capacitive biosensors have attracted great attention, further efforts are still required to be made to develop generic capacitive biosensors with a large array of sensing electrodes functionalized appropriately for various biological applications.

Chapter 5

Microfluidic packaging

This chapter is dedicated to the microfluidic packaging techniques which are compatible with integrated circuit (IC) technology as well as the aqueous environment. Although all of these techniques are not only for capacitive biosensors, they might be suitable candidates for the packaging of this kind of biosensor.

5.1 Materials and challenges

At the first glance, it seems that if a droplet of the liquid sample is dropped accurately on top of the sensing part of the complementary metal-oxide semiconductor (CMOS) sensor in such a way that the other parts like bonding wires are passivated, the sample can be in direct contact with the sensor without any microfluidics. But, according to the application and the properties of the materials used in the surface such as hydrophobicity and hydrophilicity, dispersion and/or evaporation of the droplet might be occurred. So, there is a need for fluidic devices to contain and direct the sample through the CMOS sensor. Generally, three approaches are reported to do so which are illustrated in Figure 5.1(a)–(c): (1) a big chamber like a petri dish or well with a small hole to give access to an open-top CMOS sensor (Figure 5.1(a)), (b) single-channel microfluidics directing the liquid through one channel toward the sensor (Figure 5.1(b)); (c) multichannel microfluidics directing the liquid through multiple channels toward different sensing parts arranged in an array structure (Figure 5.1(c)). The latter is useful for high-throughput experiments but it is more complex than the other two configurations.

Primarily, thermal and chemically stable materials like Si and glass were used for microfluidic fabrication. But they usually require complicated and costly techniques. After the advent of polymers like SU8, poly(methyl methacrylate) (PMMA) and poly(dimethylsiloxane) (PDMS), they are attracting widespread interest due to their biocompatibility. Among them, PDMS is the most widely used polymer because of its transparency, compatibility with rapid-prototyping, low cost and biocompatibility [294]. But several challenges exist in the development of hybrid CMOS-microfluidic devices such as

1. Biocompatibility of the materials: According to the application, the material used in packaging should be compatible with the liquid sample and can protect the sample without causing any problem like toxicity.

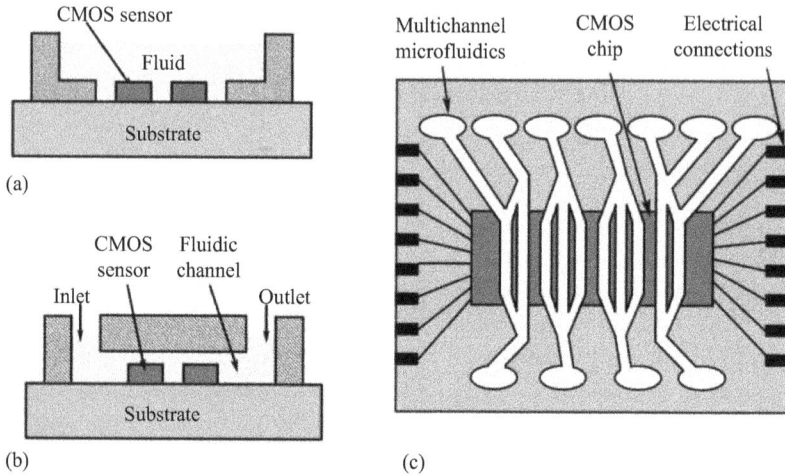

Figure 5.1   (a) Cross section of an open-top microfluidic packaging; (b) cross section of a single-channel microfluidic packaging; (c) top view of multichannel microfluidic packaging

2. Susceptibility of IC to the coupling of interfering signals by changing the medium surrounding the chip from air to liquid with different dielectric constant [166].
3. Uncertain and high-priced implementation processes: Heterogeneous integration of fluidics and electronics cannot rely on industry-standard wire bonding or flip-chip mounting. Since fluidic packaging has not been standardized by industry, two options are said to be useful to do so: adaptation of pre-existing standard packages or development of custom packages [166] which also might increase the prices of production.
4. Finding simple and reliable approaches for sealing microfluidics with IC.
5. Reusability of the package: The fluidic device should be reusable and allow for the remnants of the sample being washed out and removed after each experiment.
6. Lack of simulation and modeling software suitable for each application: The tools for designing hybrid CMOS-microfluidic systems are not as mature and reliable as IC design software.
7. Topographical mismatches between the footprints of electrical connections and microfluidic channels [295,296]: The sizes and locations of the channels and floor planning of different parts of electrical device and connection should be precise and matched to each other.

8. The utilized packaging method should isolate the electrical part from fluids while there is no contradiction between the conditions (like temperature) required for the implementation of each one of these two parts.
9. In applications in which the hybrid system should be placed into an incubator (like the incubator used for cell culture) or in specific conditions, the package should be usable in that atmosphere.

In this regard, some methods are proposed in the literature to step toward overcoming the mentioned challenges which are reviewed in this chapter.

5.2 IC-microfluidic packaging techniques

IC-microfluidic packaging techniques reviewed in this section are divided into four categories: (1) rapid prototyping; (2) soft-lithography-based packaging; (3) direct-write fabrication process (DWFP); (4) other microfabrication methods.

5.2.1 Rapid prototyping methods

Rapid prototyping methods have been widely used in literature. In this technique, open-top standard cell-culture well is usually used to package the CMOS chip in such a way that an insulating layer covers nonsensitive areas of the chip [13,171,189,215,297].

Parylene is one of the most common materials used for such an insulating process due to its bio-friendly property as well as adherence properties to Si, silicon oxides, and even epoxy [298]. Prodromakis *et al.* [168] utilized this material to encapsulate their CMOS chip and an automated laser micromachining system was used to ablate on parylene and create some individual opening for sensing areas. Liftoff process (see Appendix B, Section B.2) of the parylene membrane was performed by using ultrasonics. In another effort, Prakash *et al.* [189] Incorporated a CMOS capacitive sensor with a cell-culture well glued on top of the packaged chip for tracking the proliferation of cancer cells. Similarly, Lopez *et al.* [248] used a glass ring glued on top of the carrier printed circuit board (PCB) as the container of the cell medium and the cells. Ballini *et al.* [217] and Viswam *et al.* [218] also employed a polycarbonate ring and a biocompatible plastic epoxy well, respectively, to contain the medium and encapsulated the wire bonds by epoxy.

Datta-Chaudhuri *et al.* [299] embedded their chip in an epoxy handle wafer allowing photolithographic post-processing and microfluidic integration (Figure 5.2(a) and (b)). Thin-film metal traces passivated with parylene-C provide an off-chip electrical connection. Figure 5.3(a)–(e) shows the packaging procedure. The chip was located facedown into the center of a glass petri dish coated with PDMS (as a mold). Then, epoxy was poured around the chip and cured to form the handle wafer. After that, Cr/Au was sputter-deposited and patterned onto the front side of the surface. Subsequently, poly(3,4-ethylenedioxythiophene) doped with poly(styrenesulfonate) (PEDOT:PSS) was electrodeposited on the Au electrodes. Parylene, then, was deposited and patterned to cover the input/output (I/O) region. A two-step process of dry etching and liftoff was used for patterning the metal and

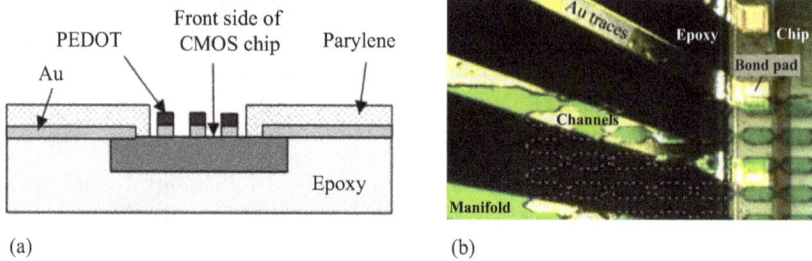

Figure 5.2   (a) Cross section and (b) the image of the microfluidic structure bonded to the surface of a packaged chip [299]

the parylene. This group reported a similar approach with well and microfluidic channels for the development of a bioelectronics nose [298].

In another effort, Musayev *et al.* [26] used epoxy to both insulate the CMOS chip wire-bonded to a ceramic package and make a reservoir for holding the solution above the chip (see Figure 5.4(a) and (b)).

Halonen *et al.* [300] combined a second generation of low-temperature co-fired ceramic (LTCC) technology with flip-chip bonding to provide a durable package compatible with cell culture. The conductor lines were screen printed on the used commercial Dupont 951 Green Tape^{TM} LTCC. While a standard LTCC process was used to manufacture the package, a special process was adapted for the microfluidic channels. The sensor chips were glued to the LTCC packages with isotropic conductive adhesive (ICA). As seen in Figure 5.5, ICA, which includes conductive silver particles embedded in adhesive polymer resins, was applied on the LTCC contact pads as *pumps* through a stamping process with laser-processed alumina stamp. The chip was aligned by using a flip-chip bonder and glued onto the bumps by epoxy and cured on a hot plate. An epoxy underfill was applied around the bumps and cured for sealing the chip and the LTCC against the liquid and providing additional attachment strength. The same underfill material was employed to glue the well on top of the package. Fluidic channels were created utilizing special lamination technique involving sacrificial carbon tape and lower pressure.

Another material useful for making a reservoir on top of the chip is PDMS [167]. Nabovati *et al.* [28] reported a rapid microfluidic packaging process using two layers of PDMS (PDMS2, PDMS1) and one layer of glass as shown in Figure 5.6. The packaging procedure is illustrated in Figure 5.7(a)–(f). Figure 5.7(a) illustrates the combined and individual layouts of the different layers of the package made of (1) chip, (2) PDMS2, (3) PDMS1, and (4) glass. The vertical diagram shows the epoxy (depicted in red color). The fabrication steps are as follows:

1.   The PDMS was prepared and spin coated on a petri dish to form a uniform layer with suitable thickness. Then, the PDMS samples were cured on a hot plate.

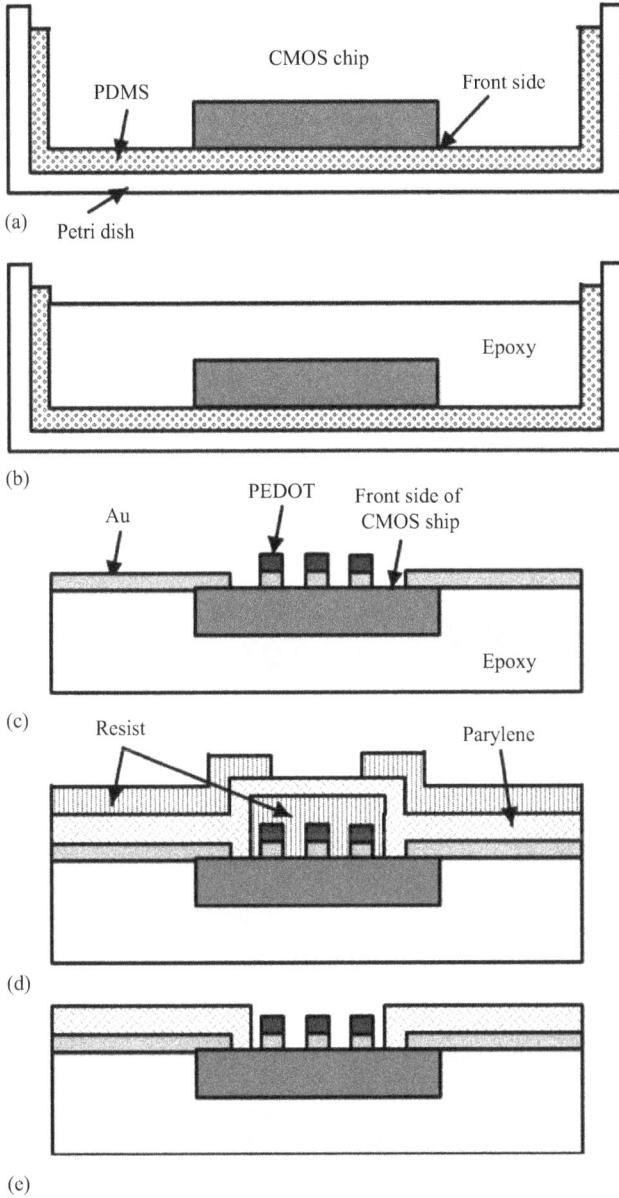

Figure 5.3   (a) Placing the chip facedown into the center of a glass petri dish
coated with PDMS; (b) pouring and curing epoxy around the chip to
form a handle wafer; (c) removing the wafer from the petri dish,
leaving the chip faceup, sputter deposition of Cr/Au onto the surface,
and electrodeposition of PEDOT:PSS on the Au electrodes in the
center of the chip; (d) and (e) deposition and patterning of the
parylene passivation layer

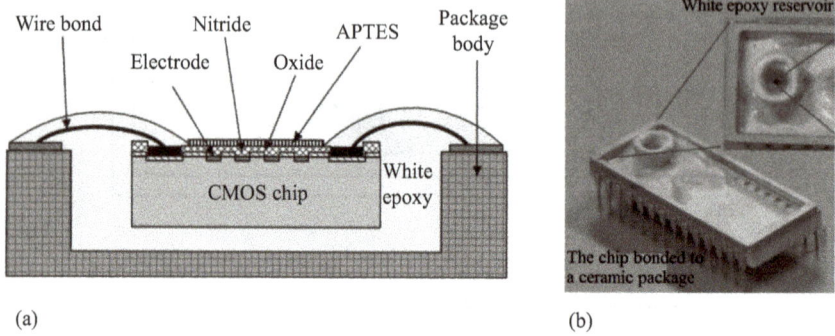

Figure 5.4   (a) Cross section and (b) the image of the packaging approach used in [26] where the CMOS chip is wire-bonded to a ceramic package and white epoxy is used to isolate the wires and build a reservoir for sample solution

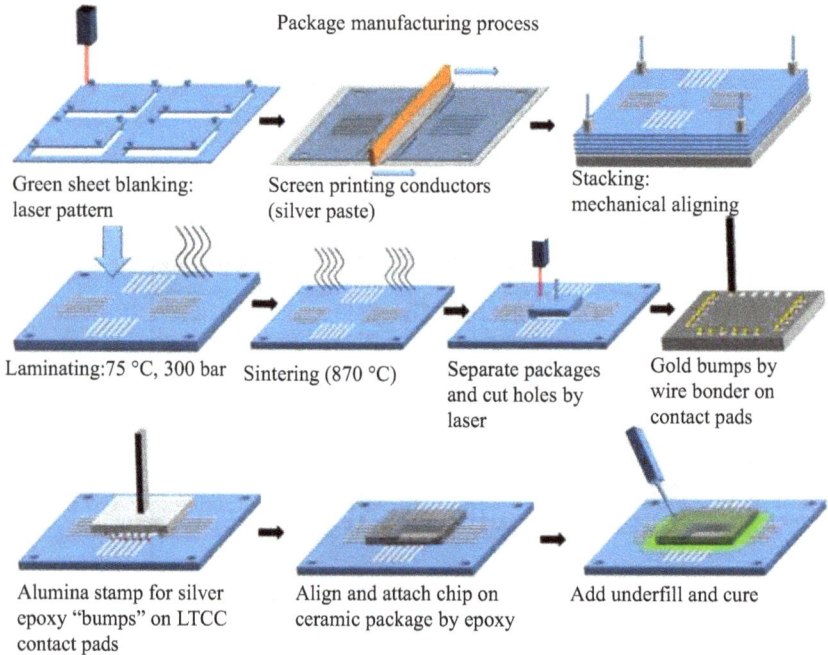

Figure 5.5   ICA stamping process, sensor chip mounting, and underfill application [300]

2. The PDMS sheets were patterned and cut by employing a charge pump laser machine. The top glass layer was also patterned and cut. A mask-less laser method was used to form micro-wells on PDMS sheets and the top glass layer.
3. The layers were thoroughly rinsed, cleaned, and dried.

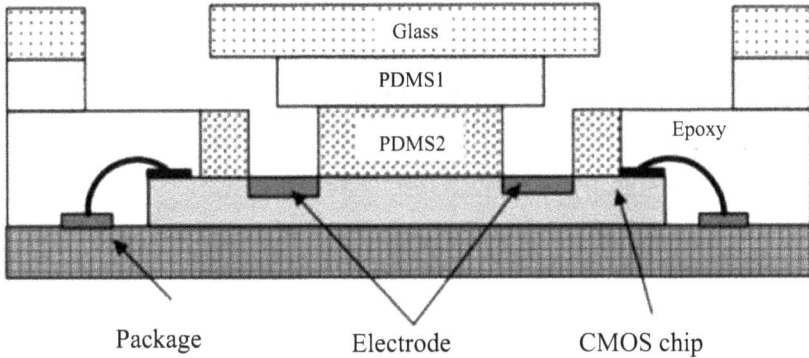

Figure 5.6 Cross section of the rapid microfluidic packaging method where the micro-wells are created on three stacked layers of PDMS2, PDMS1, and aligned on the open-top packaged die using a flip-chip bonder

4. The dies were packaged on open-top packages and located and fixed by tape on a vacuum plate.
5. A flip-chip bonder stacked and aligned the three layers of PDMS2, PDMS1, and glass on the open-top packaged die, respectively. Using two cameras mounted on the device helped to see the position of the die and the PDMS2 layer and achieve a very high-precision alignment.
6. By employing a pick-and-place nozzle, these layers were pressed on the die to remove any excess air between them. While pressing, the hot plate was turned on and the epoxy resin was poured around the layers. The epoxy resin was used to attach the layers, fill the distance between the PDMS1 and the chip, and cover the wire bonds. The leakage of epoxy resin underneath the PDMS1 layer is prevented by the temporary hermetic bond created by the soft PDMS1 layer under pressing force with the underlying substrate.
7. While the epoxy was cured, the layers were constantly pressed by the nozzle.
8. After curing the epoxy, the chip was removed from the hot plate.

Li *et al.* [171] wire-bonded the chip to a standard ceramic package and coated all surfaces within the package, consisting of package contact pads, wire bonds, and the electrode array chip with a layer of parylene. Then, parylene was removed from only the electrode array area (while all other surfaces were insulated) by reactive ion etch (RIE) utilizing oxygen gas. Since typical RIE masking materials like metal and photoresist could not be used for their complex 3D structure, a nontraditional process was employed. A PDMS cylinder was prepared whose size was matched with the area of the uncovered area of the chip's surface and attached to a silicon chip of slightly smaller size using oxygen plasma. The other side of the cylinder was attached to a glass slide. Thereafter, the slide/PDMS/silicon chip assembly were clamped to the parylene-coated package to cover the center of the chip as depicted in Figure 5.8(b). Crystal adhesive was melted where the PDMS/silicon cylinder was held to fill the cavity inside the package. Then, the slide/PDMS/silicon chip assembly was

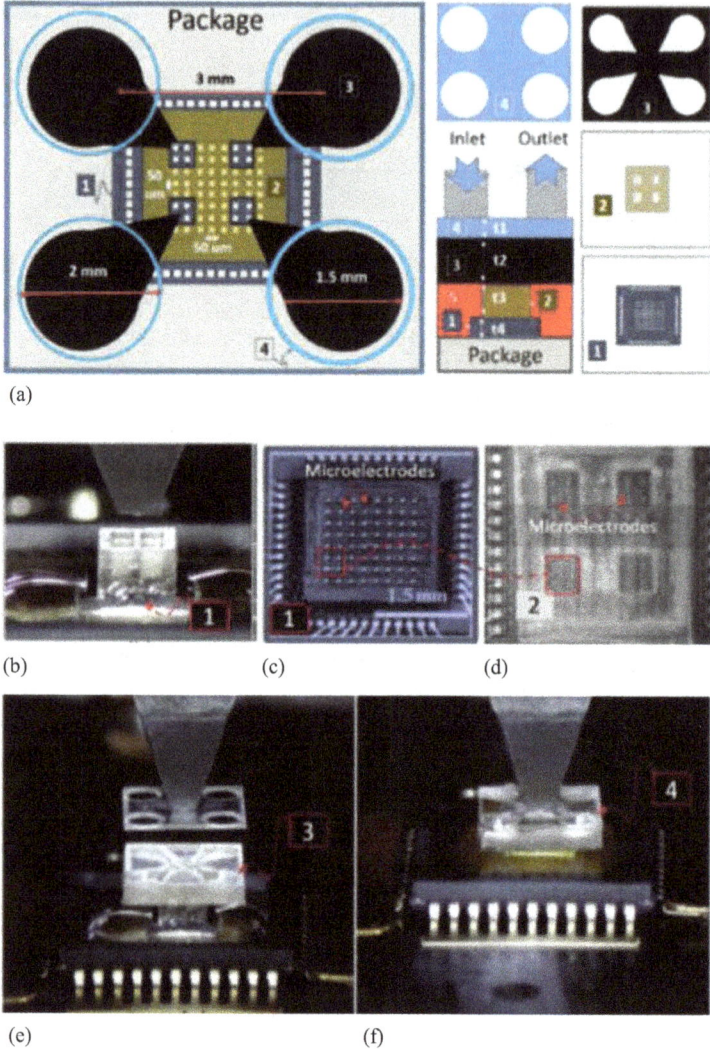

(a)

(b)   (c)   (d)

(e)   (f)

Figure 5.7   *(a) Combined and individual layouts of the different layers of the*
package; (b) PDMS2 layer alignment and bonding; (c) CMOS chip;
(d) four PDMS chambers; (e) placement and pressing of glass layer
above PDMS2; (f) pressing of glass layer above PDMS2 [28]

disassembled. The parylene over the on-chip electrodes was then etched by employing
RIE. The packaging was completed by removing the crystal adhesive using
acetone (ACE).

Although rapid prototyping techniques [13,26,28,167,171,189,301–303] are
low cost, they are not appropriate for high-precision applications with a large
number of wells and microchannels.

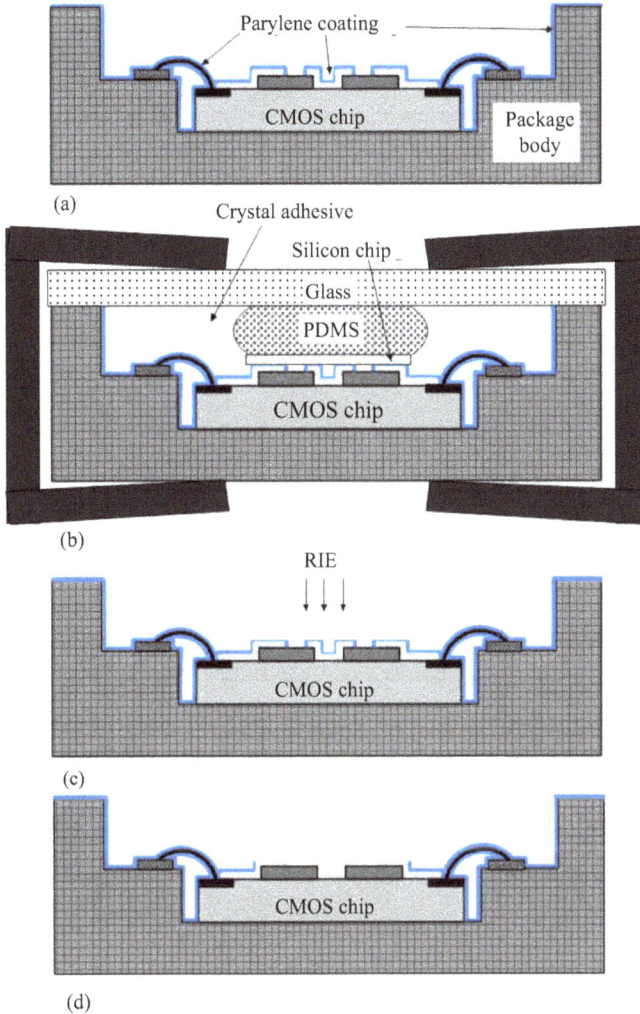

Figure 5.8 *Process flow for chip-in-package sealing: (a) chip is wire-bonded to package and coated by a layer of parylene; (b) a PDMS cylinder and silicon chip are attached to a glass slide and clamped to the parylene-coated package; (c) glass slide is detached and parylene is etched; (d) crystal adhesive removed to form final package*

5.2.2 Soft lithography

Soft lithography is a valuable standard technique for the development of integrated microfluidic systems and PDMS is among the most widely used materials to create and bond microfluidics on a CMOS chip for different applications [304,305]. For example, a PDMS microfluidic device is proposed by Welch *et al.* [296] which is bonded to a flexible PCB using oxygen plasma (see Figure 5.9). To enhance bonding, the surface of the PCB is coated with polyimide (PI). A CMOS chip is also a flip-chip bonded to the same PCB.

Another attempt was also made by Li and Yin *et al.* [173,305] to develop a CMOS-compatible epoxy chip-in-carrier process in which the chip was placed on the glass and fixed by epoxy as shown in Figure 5.10(a) and (b). The metallic conductors such as the screening silver (Ag) electrode was deposited and patterned by other microfabrication processes. To prepare a device for testing, the electrodes were encapsulated with parylene. Also, a PDMS microfluidic structure fabricated by using SU8 mold was bonded onto the chip-in-carrier assembly [173].

In a similar effort, Zhang *et al.* [27] developed a flexible microfluidic-CMOS packaging technique by bonding a PDMS substrate, a CMOS chip, and a PDMS

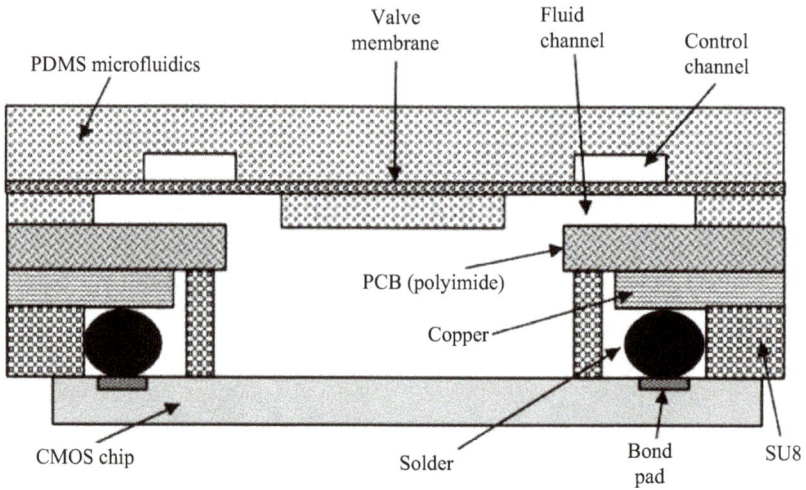

Figure 5.9   Cross section of the packaging technique proposed by Welch et al.

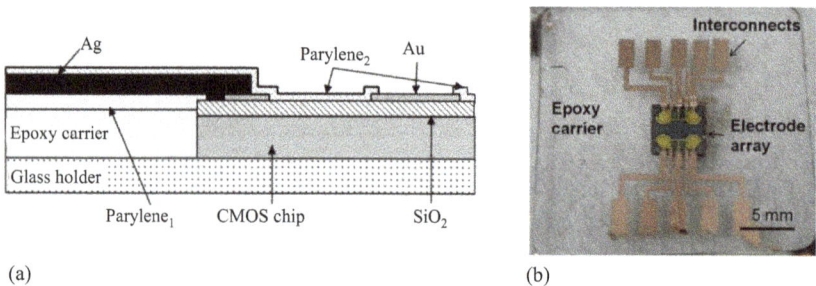

(a)   (b)

Figure 5.10   (a) Cross section and (b) the image of the chip-in-carrier device along with screen-printed planar interconnects used in [173]

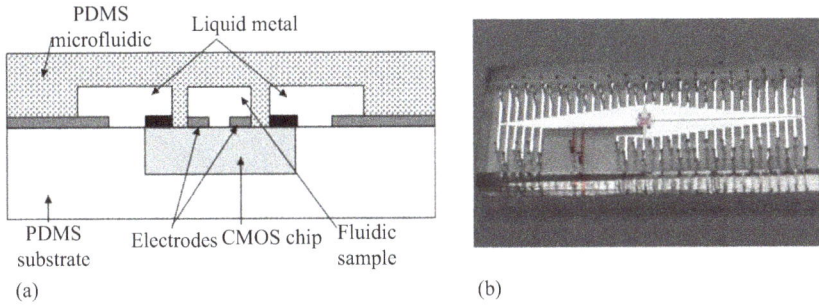

(a)   (b)

Figure 5.11   *(a) Cross section and (b) the image of a packaging technique where liquid metal is used for electrical connections of IC to the substrate [27]*

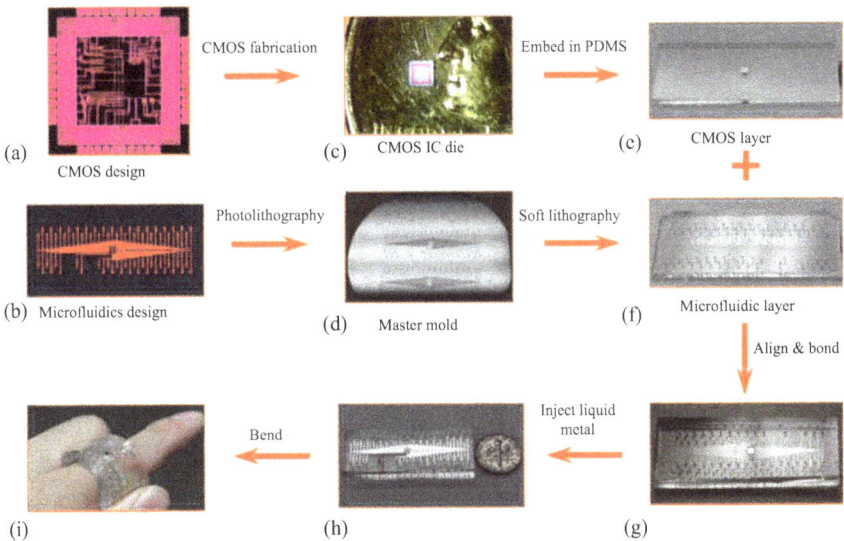

Figure 5.12   *The fabrication and packaging procedures to produce the flexible CMOS/microfluidic hybrid microsystem proposed in [27]*

microfluidic device as depicted in Figure 5.11(a) and (b) and used the microfluidic channels filled with liquid metals as conductive interconnections. In this work, the microfluidic channels and electrical connections are effectively created by soft lithography. Figure 5.12 illustrates the fabrication and packaging procedures. Two PDMS layers including a top microfluidic layer and a bottom CMOS layer form the packaged system. The bare CMOS die is located upside down on a flat silicon wafer and held in contact with wafer by applying a small pressure to its back side. Thereafter, degassed PDMS prepolymer is poured onto the CMOS die and baked. Then, the CMOS die is peeled off from the silicon wafer and embedded in a piece of flat PDMS while its active area is still exposed. A conventional soft lithography was used for the fabrication of microfluidic layer with a patterned SU8 photoresist master mold fabricated on a silicon wafer utilizing photolithography. After pouring PDMS prepolymer onto the mold and degassing in a vacuum chamber and baking it, the partially cured PDMS is peeled off from the mold and formed the micro-fluidic layer. The liquid inlet/outlet ports are formed by punching through the whole layer. A strong bond between microfluidic and CMOS layers were formed by the treatment of both layers with air or oxygen plasma. After alignment of the microfluidic layer on top of the CMOS layer, the package is baked. Finally, liquid metal Galinstan was used to fill into some of the microfluidic channels as the electrical interconnects to the CMOS chip.

In another effort, Wu *et al.* [170] integrated a hybrid system including a three-layer microfluidic structure on a flexible PCB using a flip-chip bonding technique as illustrated in Figure 5.13(a) and (b). The microfluidic layers consist of a pneumatic valve actuation layer made of glass, a valve membrane made of PDMS, as well as a glass fluidic layer including channels. After being etched by photolithography and wet etching, the glass wafers were drilled to form the holes of inlets, outlets, and pneu-matic ports. Then, they were bonded together using a PDMS membrane. Similarly, Burdallo *et al.* [174] used a PCB as the carrier with a hole in which the chip is bonded. A PDMS microfluidic structure was also bonded by the adhesive materials coated on the PCB. Berdat *et al.* [91] designed a molded PDMS microfluidic device clamped with a biochip in between two micromachined slides of PMMA.

In [306], a lab-on-CMOS integration process is reported using a chip-in-carrier assembly. First, a silicon substrate carrier (die carrier) was employed to expand the surface area beyond a CMOS chip. A photoresist was spin-coated on a silicon wafer and patterned with a mask. The through-wafer cavities within each carrier on the silicon wafer were opened by using deep reactive ion etching (DRIE) in such a way that they were wider than CMOS chip in order to provide some placement freedom. Then, the silicon carrier wafer was insulated by thermal oxide. After patterning and etching the dicing streets (at the same time with the cavities), the carriers were singulated along with them. Alternatively, the combination of all carriers and

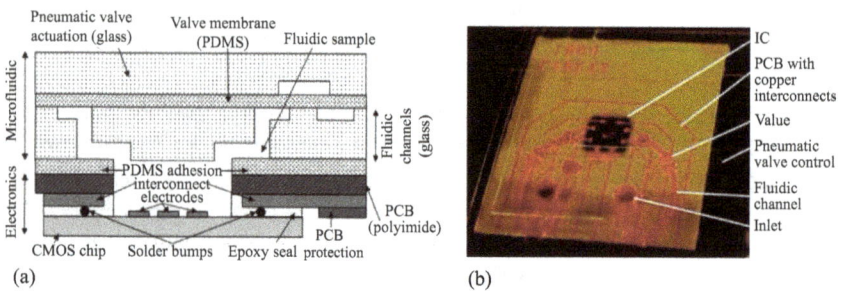

Figure 5.13   *(a) Cross section and (b) the image of the hybrid system proposed by Wu et al. [170]*

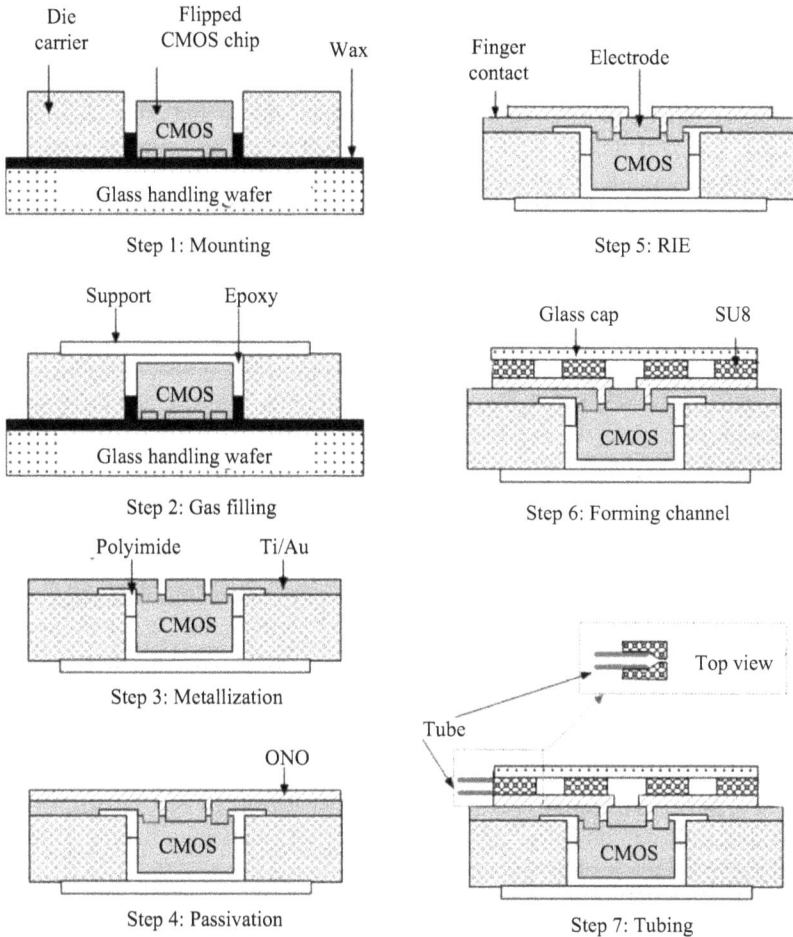

Figure 5.14   The integration process flow of the lab-on-CMOS proposed by Huang et al.

CMOS chips at the wafer scale could be singulated. Then, the lab-on-CMOS integration process flow was performed as depicted in Figure 5.14:

1. A Si substrate carrier was used to implement a chip-in-carrier assembly mounted onto a wax-coated glass handling wafer. The chip-carrier assembly is created by mounting down the die carrier onto a wax-coated glass handling wafer and then placing a CMOS chip into the cavity with its active surface down. After heating, a wax reflow process and cooling, the carrier and the chip were attached to the wax and the trapped air bubbles were expelled.

2. The gap between the chip and the carrier was filled with epoxy and a thin glass was attached to the top of the chip-carrier assembly to provide a vacuum seal and additional mechanical strength. Then, the glass is separated by softening

the wax. Oxygen plasma was also applied briefly to remove any epoxy reached to unwanted areas due to imperfect wax sealing.

3. A layer of PI is applied and patterned by photolithography to insulate the exposed silicon on chip and smooth the trench slopes and other sharp edges. Then, a Ti/Au thin film is deposited and patterned by using thermal vapor deposition and liftoff process, respectively.

4. After metallization, the interconnects were insulated by an oxide/nitride/oxide (ONO) passivation layer using plasma-enhanced chemical vapor deposition system at a low temperature that the PI and the epoxy within the assembly remain stable.

5. Then, the ONO layer over the sensor electrodes and the contact fingers at the carrier periphery was etched using RIE without exposing the Ti/Au metal film edges. The CMOS chip surface is protected by the added ONO layer and the inherent overglass layer. The ONO layer and Au metallization sealed all CMOS bonding pads.

6. Open microfluidic channels were formed by spin-coating SU8 on chip-carrier assembly, soft baking, exposure, post exposure baking, and development and then pressed by a glass slide to flatten the top surface. Thereafter, oxygen plasma was used to remove all SU8 residues from the Au electrodes, which also converts the hydrophobic SU8 surface to hydrophilic surface. Next, a thin glue layer of SU8 is applied to the top surface of the open channels and glass cap is attached to cover the open microfluidic channels.

The final device is baked and inlets and outlet tubes, which were routed to taper joints, were inserted laterally into the sidewall edge of the SU8 layer.

In another effort, Norian *et al.* [307] used SU8 to define the active microfluidic area and encapsulate the wire bonds and support the overlaying indium-tin-oxide-coated polyethylene naphthalate cover slip.

Huang *et al.* [308] spin-coated (see Appendix B, Section B.6) and developed SU8 on their device to fabricate the insulation layer and cell trapping micro-wells. They used a shifted Pt electrode strategy as shown in Figure 5.15. The chip was treated with a strong stream of oxygen plasma to make a more hydrophilic surface and achieve a high aspect ratio by removing any undeveloped SU8 residual on the

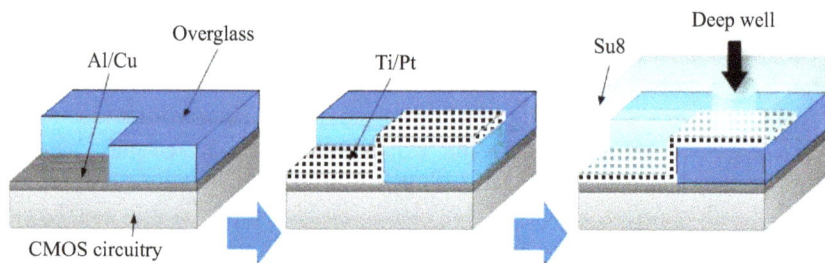

Figure 5.15   Post-fabrication flow of the shifted electrode design with SU8 micro-wells

electrode. The WEs cover only a portion of the bottom of the SU8 micro-wells in order to avoid the probable conducting connections between two adjacent electrodes after Pt liftoff process.

Soft lithography methods have been among the most popular ones for the fabrication of polymeric microfluidics, specifically made of PDMS [27,169,173,174,296,309]. However, silica and glass [170] fluidic devices are preferable for applications that require thermal and chemical stability.

5.2.3   Direct-write fabrication process

In recent years, there has been growing interest in 3D microfluidic fabrication processes like jetting technique, stereolithography (SLA or resin printing), laser melting, laser sintering, and DWFP [310–312]. Among these techniques, a DWFP was proposed by Ghafar-Zadeh *et al.* [68,172,280,313] for hybrid CMOS-microfluidic packaging. In this approach, the encapsulation of a paste-like sacrificial layer on the surface of the CMOS chip creates the microfluidic as shown in Figures 5.16 and 1.9(d).

In the DWFP reported in [172,223], the deposited sacrificial layer is a mixture of petroleum jelly included microcrystalline wax ink.

Figure 5.17(a)–(f) shows the packaging procedure which includes the following:

1. Encapsulation of bonding wires and pads: In this step, a partially cured epoxy resin is dispensed on the packaged chip. The high viscosity of the semi-cured epoxy and surface tension cause that the epoxy flows around the bonding wires but stops near the pads. The trajectory path of ink deposition over the sensing electrodes is precisely measured by a high-precision optical profiler and programmed into a robot-driven dispensing system.
2. Ink deposition: A paste-like ink is extruded through a micronozzle and deposited on the substrate. The velocity of the moving micronozzle over the trajectory, the air-pressure applied to extrude the ink through the micronozzle, the microcrystalline fraction of the organic ink mixture, and the relative height between the nozzle and substrate are the parameters that should be controlled in this process.

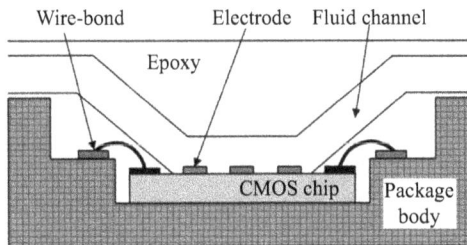

Figure 5.16   Cross section of the hybrid microfluidic-CMOS capacitive sensor whose microfluidic is fabricated by DWFP

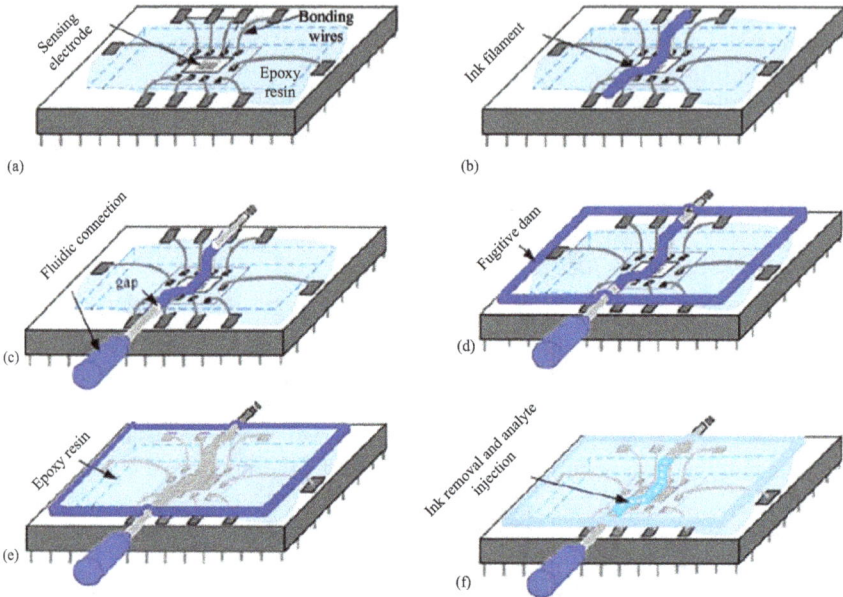

*Figure 5.17   The DWFP procedure reported in [223]: (a) encapsulation of
bonding wires and pads; (b) ink deposition; (c) fluidic connection;
(d) fugitive dam; (e) fugitive ink encapsulation; (f) ink removal and
analyte injection*

3. Fluidic connection: The microscale fluidic fittings are fixed close to the
    deposited ink using hot glue. In order to fill the space between fluidic con-
    nection and the ink filament, and consequently prevent the infiltration of epoxy
    into the fitting during the encapsulation process, extra fugitive ink is deposited.
4. Fugitive dam: Another ink deposition is done in the boundary of epoxy
    encapsulation.
5. Fugitive ink encapsulation: A low viscosity epoxy resin is dispensed and cured
    on the deposited ink within the encapsulation boundary.
6. Ink removal and analyte injection: The fugitive ink is melted and expelled
    under air pressure or light vacuum. Then, the ink remnants can be removed by
    injected hot water through the channel.

Compared to soft lithography, the DWFP [68,172,280] offers the advantage of
direct fabrication on IC. Additionally, there is no need for molds or masks in this
technique making it appropriate for complex polymer-based 3D microfluidics.

5.2.4   Other microfabrication methods

There are also some other techniques to fabricate IC-microfluidic packaging. For
instance, a microfluidic structure is proposed by Lee *et al.* [9] which is post-
fabricated on top of a SiGe IC as shown in Figure 5.18(a) and (b). This system is

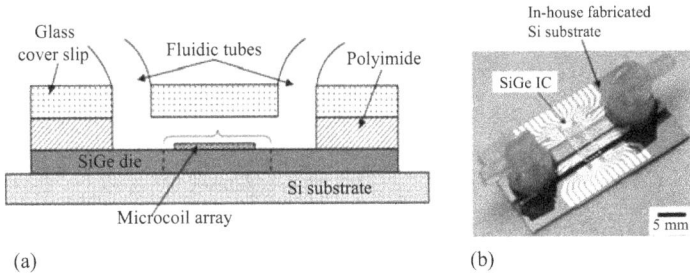

Figure 5.18   (a) Cross section and (b) the image of a microfluidic structure directly post-fabricated on top of a SiGe IC for magnetic manipulation [9]

reported for cell manipulation using magnetic micro-coils fabricated on a CMOS chip. However, the proposed packaging technique might be useful for packaging microfluidics along with other ICs like CMOS capacitive sensors. In this system, the IC is diced from the wafer with an area larger than the required active area to be used as the carrier for microfluidics. After the definition of the sidewalls of microfluidic channels by spin-coating (see Appendix B, Section B.6) and patterning a layer of PI, a glass cover slip is sealed on top of the structure. Then, plastic fluidic tubes are glued to its inlet and outlet. The SiGe IC-microfluidic system is located above a Si substrate with electrical leads as electrical connections.

Lindsay *et al.* [314] used a fan-out wafer-level packaging approach for scalable heterogeneous integration of Si ICs and microfluidics. They developed a stacked laser-cut fluidic assembly with a simple fluid Y-channel which was laser-cut into a two-sided pressure-sensitive adhesive sheet sandwiched between glass/acrylic layers.

Laminated photosensitive dry-film resist (DFR) can also be used both for the realization of passive microfluidic channels and chambers as well as adhesive wafer bonding and provide heterogeneous integration of different substrates on wafer level. The packaging proposed by Karl *et al.* [315] includes a three-layer bonded sandwich of CMOS substrate, patterned DFR, and a glass cap.

In another work, Mujika *et al.* [53,316] fabricated 3D multilayer microstructures by adhesive bonding of two photolithographically patterned SU8 layers (SU8 50 and SU8 5) on separate wafers. The process was based on the successive CMOS-compatible adhesive bonding and releasing steps proposed by Agirregabiria *et al.* [317]. Pyrex wafers were selected as substrates and propylene glycol monoether acetate (PGMEA) was used as a developer. The auxiliary polyimide film includes Kapton film. A positive photoresist (S1818) was used as adhesive layer between Pyrex and Kapton. This approach, which is illustrated in Figure 5.19, includes the following four main steps:

1. Substrate wafer preparation: Two substrates as the bottom and the temporary top substrate must be prepared:
    Bottom substrate: A highly polymerized thin SU8 2 layer was used along with an optimized photolithographic process of SU8 5 to enhance the

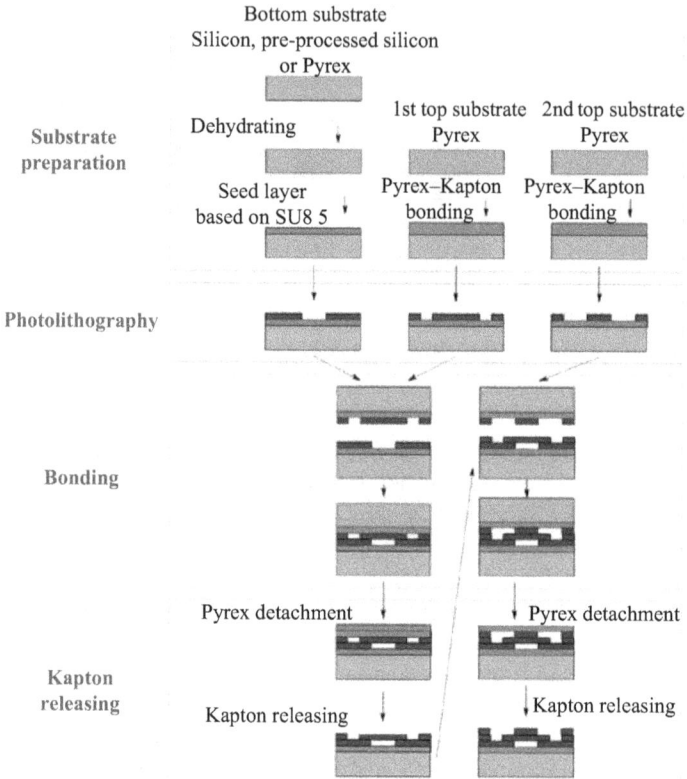

Figure 5.19   Fabrication process of SU8 multilayer microstructures using successive CMOS compatible adhesive bonding and releasing steps

adhesion between Pyrex glass and the thick SU8. After cleaning and rinsing the substrate wafer, the SU8 5 is spun and soft baked. Then, the photoresist is polymerized and baked. Finally, the SU8 5 is developed in a PGMEA bath and hard-baked.

Top substrate: Kapton films bonded to a Pyrex wafer were employed as the top SU8 substrate. The positive photoresist (S1818) was spun on the Pyrex wafer. Immediately, both substrates were brought into contact under a vacuum in a substrate bonder, pressed together, and heated.

2. Photolithography of SU8 50 layers: After preparing the two substrates, SU8 50 was photo-patterned on both by carrying out two photolithographic processes. To achieve uniform SU8 films with good adhesive properties, an approach including five main steps were employed: (1) dynamic spin coating, (2) soft baking with progressive temperature ramps, (3) ultraviolet (UV) exposure,

(4) post-baking using progressive temperature ramps, and (5) developing in a PGMEA bath and rinsing in isopropanol (IPA).

3. Bonding the two photo-patterned layers: both photo-patterned wafers were put into contact utilizing a substrate bonder, pressed together and heated to be bonded together.

4. Release of the temporary film: The SU8 stack and bottom substrate are separated from the top Pyrex wafer and the Kapton film. After rinsing the wafers in an IPA ultrasonic bath, the top Pyrex wafer and the Kapton film were detached by the insertion of a razor blade. Then, the Kapton film was peeled off from the SU8 stack.

After these four steps, multilayered SU8 stacks can be obtained by preparing, bonding, and releasing more Kapton substrates.

Inac *et al.* [318] reported a three-wafer stack packaging process as seen in Figure 5.20. In this process, the first wafer includes sensors and active circuitry and the inlet and outlet of microfluidics were etched from the backside of this wafer by using localize backside etching. The channels were patterned and etched from the second Si wafer. Then, the front sides of these wafers were bonded together using plasma-activated oxide–oxide fusion bonding. The desired height of the channel was formed by grinding the backside of the microfluidic channel. Finally, a glass wafer (the third wafer) was adhesively bonded to the top of the microfluidic channel to seal it.

Similarly, Matbaechi Ettehad *et al.* [319] integrated CMOS electronics, Si channels, and a glass wafer using this three-wafer-stack approach and reported a microfluidic LoC for the manipulation and characterization of yeast cells using DEP forces. They utilized an external macrofluidic technology to control the fluid flow and sample injection through the microfluidic channel. A PMMA fluidic manifold with a cavity with the same size of the chip was fabricated by using a commercial 3D printer.

Rasmussen *et al.*[320] reviewed some other techniques like surface micromachining where the microchannels can be formed after etching a sacrificial layer beneath a conformably deposited material as shown in Figure 5.21(a) and (b). In this figure, a shallow microchannel is formed by employing the standard metal layer inside the CMOS chip which was patterned by using traditional CAD tools. The conductors and vias which were etched play the role of the sacrificial layer.

Man *et al.* [309] employed a batch lithographic process to fabricate large volume plastic capillaries on planar substrates and integrated microfluidic and circuit elements for DNA analysis. They proposed two types of capillaries fabricated utilizing the simplified three-polymer, three-mask, low-temperature process depicted in Figure 5.22. A thin layer of low-tensile stress SiN or an equivalent thicker SiO_2 layer is deposited in p-type silicon substrates. SiN (or SiO_2) layer is then etched on the backside at the outlets and inlets of the capillaries. Then, the pyramidal outlet and inlet access holes are formed by anisotropical etching of the sample in ethylenediamine pyrocatechol (EDP) (or tetramethyl ammonium hydroxide and water (TMAHW)). After vapor deposition of a thin layer of the polymer poly-p-xylylene (parylene C) on the samples, a sacrificial

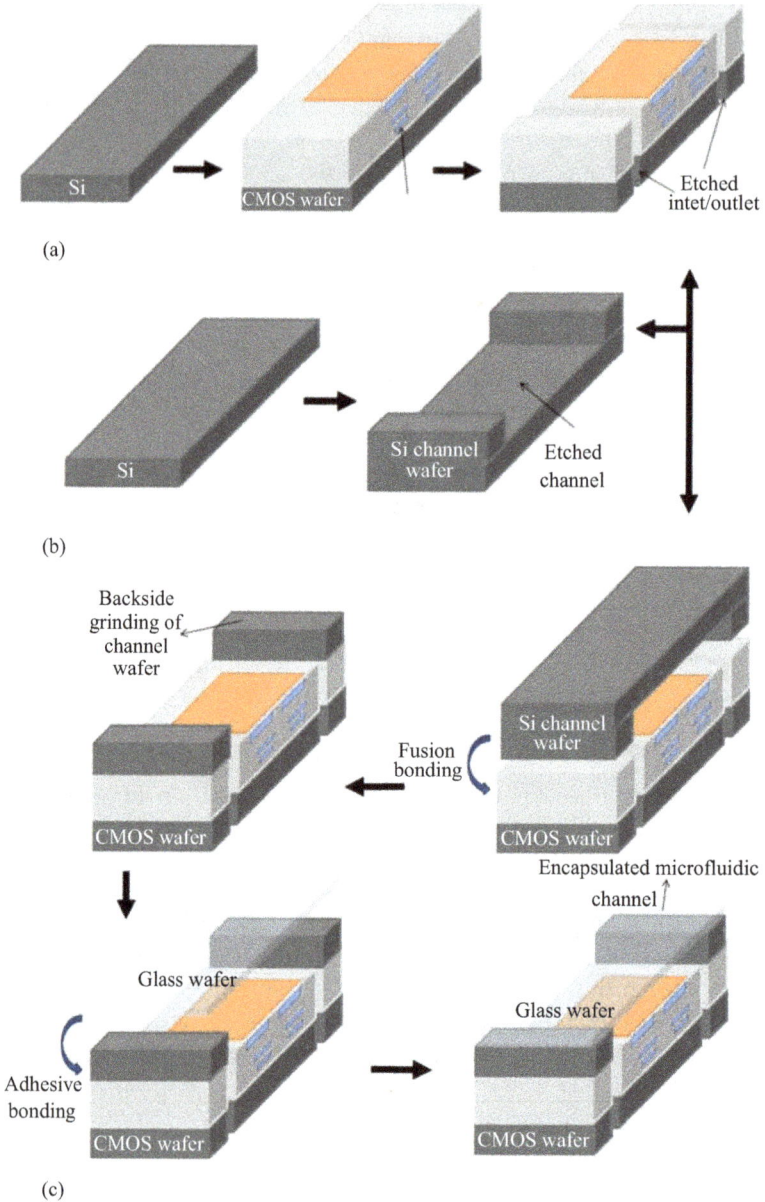

(a)

(b)

(c)

*Figure 5.20   (a) CMOS fabrication; (b) creation of the microfluidic channel;
(c) three-wafer-stack bonding process*

layer of photoresist is spin-coated and lithographically patterned on the front of the wafers to define the height of the capillaries and then hard-baked. Thereafter, a second layer of poly-p-xylylene is deposited and sealed the resist. After a short plasma

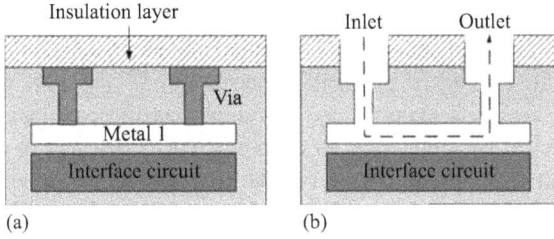

Figure 5.21 Microchannel realization through surface micromachining: (a) before etching and (b) after etching

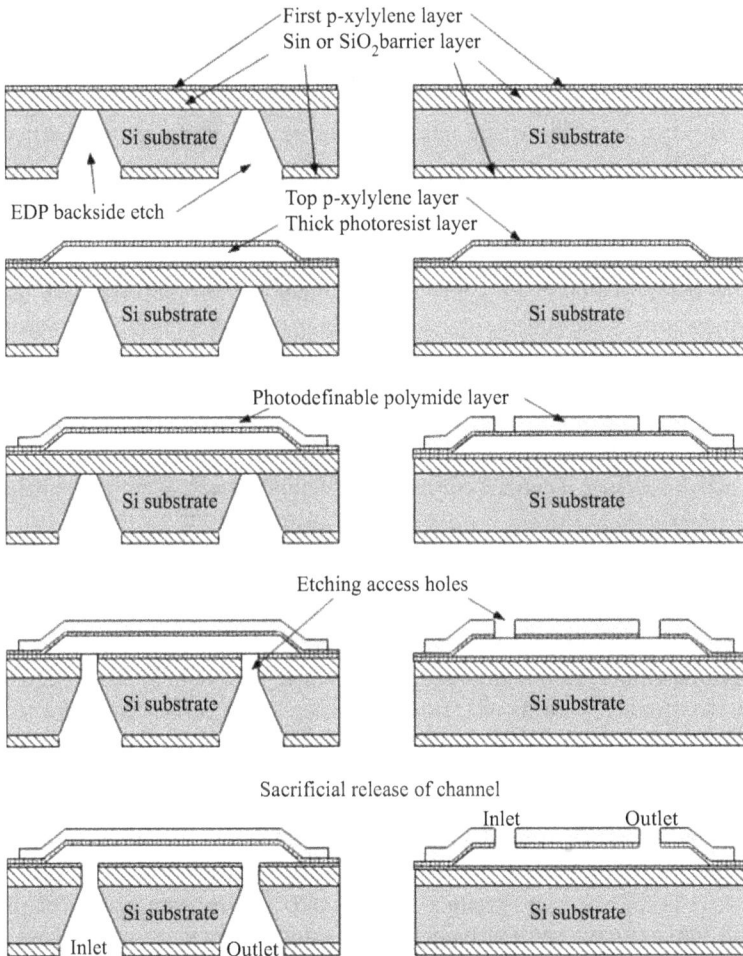

Figure 5.22 The fabrication procedure of front and back access capillaries

treatment to roughen the p-xylylene surface, a layer of photosensitive PI is spin-coated and soft-baked. The p-xylylene layer plays the role of a barrier to prevent the soft sacrificial resist being attacked by the PI solvent.

Thereafter, the PI is exposed and developed as the capillary wall. The top p-xylylene is etched by employing an evaporated Al mask and in oxygen plasma down to the sacrificial resist on the capillaries prepared for having front inlets. Then, the Al mask is stripped in wet Al etchant and a thick layer of a resist protects the wafer front side. Thereafter, a combination of oxygen, CF$_4$, and oxygen plasma was used to form the backside outlets and inlets. The protective resist and the sacrificial layer are eliminated by warm ACE. After rinsing, the samples are dried and diced.

5.3   Discussion on IC-microfluidic packaging

Different fluidic packaging methods are compared in Table 5.1, which is interesting in several ways. First, the whole microsystems need to be fixed on a carrier to be handled easily.

According to Table 5.1, a standard chip package [171], a Si wafer [173,305,323], or PCB [174] is a suitable option as a career. For some applications such as wearable health monitoring, the system needs to be flexible and skin-mountable. So, flexible substrates made of PDMS [27], PI [170], or other flexible materials can be used. Furthermore, one of the suitable structures for high throughput analysis is the chip-in-carrier structure [169,173] in which the extended surface area of the carrier is useful for microfluidics. Such carriers are usually made out of Si or epoxy. It is worth mentioning that one of the techniques used in chip-in-carrier systems is the chip-in-hole process, like [174], which is an appropriate option to integrate different chips into an active substrate heterogeneously [298]. In this process, a preformed hole in the carrier is allocated to the IC. But the mismatch between the cavity and the size of the IC might cause an uneven surface that should be even and the gap should be filled [306].

Table 5.1 also highlights the prevalent techniques used for interconnection between carriers and ICs including bump bonding (like flip-chip [170,296,324], tape-automated bonding (TAB) [326]), and wire bonding [168,174,309,321,322,325]. It is difficult to protect wire bonds from corrosion and utilize them for CMOS-microfluidic integration. In the systems where the chip is bonded to a substrate with a window for fluid containment, the flip-chip technique seems more suitable [170]. However, this technique is complicated and expensive due to the required preprocessing steps and various extra materials. TAB is a similar approach to flip-chip but it has temperature restrictions and is not suitable for CMOS-microfluidic integration [173]. Screen printing [173] and planar thin-film metallization [169,298,306] are also useful techniques for making electrical connections which are usually reported for chip-in-carrier systems to rout signals off the CMOS IC and onto the carrier. Additionally, these connections are appropriate for forming complex microfluidics on top of the chip and surrounding carrier and do not suffer from

Table 5.1 Summary of different IC-microfluidic packaging approaches

Application	Critical isolation material from aqueous environment	Carrier	Chip-in-carrier?	Metallic connections	Fluidic device	Size of CMOS chip (mm^2)	Integrated valves?	Ref.
Tracking cancer cell proliferation	Loctite 3340 (A polymer)	DIP40 ceramic carrier	No	Wire bonding	Well	2.25	No	[189]
Real-time measurements of cell proliferation	Epoxy	DIP40 ceramic carrier	Yes	Wire bonding	Well (cut Eppendorf flex tube, 1.5 mL)	3 × 3	No	[13]
CV of potassium ferrocyanide	Parylene	Ceramic package	No	Wire bonding	PDMS reservoir	3 × 3	No	[171]
Detection of cell- or tissue-specific responses	PDMS/plastic	Brass substrate	No	Wire bonding	A PDMS reservoir	2.07 × 1.44	No	[167]
Intracellular AP measurements and impedance spectroscopy in drug-screening applications	–	PCB	No	Wire bonding	A glass ring	19.2 × 10	No	[248]
Recording and stimulation of electrogenic cells *in vitro*	Epoxy	PCB	No	Wire bonding	A polycarbonate ring	3.85 × 2.10	No	[217]
Impedance spectroscopy and electrophysiological imaging of cells	Epoxy	PCB	No	Wire bonding	A plastic epoxy well	4.5 × 2.4	No	[218]
Cell viability monitoring	Epoxy	LTCC package	No	Wire bonding	Well	3 × 3	No	[300]
Real-time holistic cellular characterization and cell-based drug screening	Epoxy, PDMS	PCB	No	Wire bonding	Cell-culture plate with a laser-cut bottom	2 × 3	No	[297]
LoC devices	Pressure-sensitive adhesive, glass/acrylic	–	No	Planar thin-film metallization	Laser-cut pressure-sensitive adhesive sheet sandwiched between glass/acrylic layers	2.4 × 2.4	No	[314]
DNA detection	White epoxy	Ceramic package	No	Wire bonding	A reservoir made out of white epoxy	1.5 × 1.5	No	[26]
Cell culture monitoring and drug test	Epoxy	TQFP-44 package	No	Wire bonding	Micro-wells created on three layers of PDMS and glass	9	No	[28]

(Continues)

Table 5.1 (Continued)

Application	Critical isolation material from aqueous environment	Carrier	Chip-in-carrier?	Metallic connections	Fluidic device	Size of CMOS chip (mm²)	Integrated valves?	Ref.
Single-cell recording	–	Ceramic substrate	No	Flip-chip	Glass ring	8 × 8	No	[303]
pH sensing	Parylene	PCB	No	Wire bonding	Parylene encapsulated gate	–	No	[168]
Neuronal recordings and as an electric cell-substrate impedance sensor (ECIS)	Silastic 9161 RTV	Ceramic package	No	Wire bonding	Glass cylinder culture chamber	3.2 × 3.2	No	[302]
Neuronal recordings and as an ECIS	Hysol CB064	Ceramic package	No	Wire bonding	Epoxy window frames	3.2 × 3.2	No	
Monitoring of electrogenic cells	Epoxy and PDMS	Package	No	Wire bonding	A glass O-ring	4.4 × 4.4	No	[301]
Bacteria growth monitoring	Epoxy	Package	No	Wire bonding	Epoxy microfluidics (DWFP)	2	No	[68]
LoC devices	PI/SiO₂	Si	Yes (chip-in-hole)	Planar thin-film metallization	PDMS microfluidics	–	–	[169]
LoC devices	Parylene	Epoxy	Yes	Screen-printed metallization	PDMS microfluidics	5 × 5	No	[173]
pH sensing	DABA (Ebecryl 600)	PCB	Yes (chip-in-hole)	Wire bonding (protected by Si rubber)	PDMS microfluidics	3 × 3	No	[174]
Detection of DNA	PDMS, PMMA	PCB	No	–	PDMS microfluidics	3.34 × 4	No	[91]
Amperometry measurements	SU8	PCB	No	Wire bonding	SU8 micro-well	–	No	[308]
Quantitative-polymerase chain-reaction LoC	SU8, parylene-C	Package	No	Wire bonding	SU8 open microchannels covered with a glass cap	–	No	[307]
Bioelectronic nose	Parylene	Epoxy	Yes	Planar thin-film metallization	PDMS microfluidics/well (cut 15 mL centrifuge tube)	3 × 3	No	[298]
Wearable health monitoring and environmental sensing	PDMS	PDMS substrate	Yes	Galinstan (liquid metal interconnection)	PDMS microfluidics	1.5 × 1.5	No	[27]

Application	Material	Substrate		Interconnect	Cover / microfluidics	Dimensions		Ref.
Automated detection of magnetically labeled serum protein–PAH adducts	Epoxy	PI	No	Flip-chip	A glass microfluidic layer	–	Yes	[170]
LoC devices	SU8	PI flexible PCB	No	Flip-chip	PDMS microfluidics	1.5 × 3	Yes	[296]
Magnetic manipulation of biological cells	PI	Si substrate	No	Planar thin-film metallization	PI channel sidewalls covered with a glass cover slip	4 × 1	No	[9]
	SU8	Si/SiO$_2$ substrate	No	Planar thin-film metallization	SU8 open microfluidics sealed with a cured PDMS layer	4 × 1	No	
DEP manipulation of cells	Polycarbonate and soft glue	PCB	No	Wire bonding	Polycarbonate fluidic cap	7.7 × 8.1	No	[321]
DNA analysis systems	PI/p–xylylene	Si substrate	No	Wire bonding (protected by a lid)	Polymeric capillaries with etched vias as inlet and outlet	–	No	[309]
Multi-wavelength chemical detection systems-on-a-chip	SiO$_2$	Si substrate	No	Wire bonding	PDMS microfluidics	D = 4 mm	Yes	[322]
Amperometry experiments	Oxide/nitride	Si substrate	Yes (chip-in-hole)	Planar thin-film metallization	SU8 open microchannels covered with a glass cap	1.6 × 1.8	No	[306]
Biomedical applications	Spin-on-glass	Si substrate	Yes	Planar thin-film metallization	Acrylic microfluidic	3 × 3	No	[323]
Cytometry on a chip	PDMS	Glass substrate	No	Flip-chip	Glass microfluidic	1 × 2.4	No	[324]
DNA analysis	SiO$_2$	Si substrate	No	Wire bonding	Glass microfluidic	47 × 5	No	[325]
pH sensing	PI	–	No	TAB	Precut gate windows encapsulated by an adhesive like B-staged epoxies, pressure-sensitive thermosetting Si or acrylics	1.28 × 2.16	No	[326]
Detection of *E. coli* O157:H7	SU8	–	No	–	Two photolithographically patterned SU8 layers bonded on separate wafers	–	No	[316]
DEP immobilization of yeast cells	Si	Si	No	–	Si microfluidic channel covered by glass cap	5 × 5	No	[319]

D: Diameter

interference of interconnects. Precise photolithography paves the way for well-defined planar thin-film deposition. Similarly, in screen printing, a predefined mask, or screen is used to deposit conductive inks. Although thicker interconnections in this technique are more reliable, they are not as precisely patterned as planar thin films [173]. In the system reported in [27], the carrier is formed around the die and Galinstan is used to make liquid-metal-filled microchannels as electrical connections. In opposite to the chip-in-hole process, this approach does not need a very flat surface.

The CMOS circuits of a biosensor should be covered by an insulating layer like polymer, nitride, and oxide [298]. When electrical connections (usually, connections between PCB and the chip) remain in contact with fluid, chemical-inert covers and sealants (of materials like epoxy, PDMS, and parylene) should also be added around them to protect them from corrosion [294,327]. Materials like epoxy can be selectively applied or exposed to UV radiation. Standard epoxy is solidified by exposure to high temperatures (100 °C) and the substrate should withstand such temperatures which might not be useful for some applications. As an alternative, UV-curable polymers like UV-curable epoxy can protect the wire bonds as a patternable passivation material by easier procedures [174,328]. Flip-chips are underfilled with a waterproof adhesive or a passivation material leading electrical connections to be protected [170]. Wires are usually placed in a thick polymeric well [329] or covered by passivation materials [174]. In some structures [309], the wire is located in a cutout area in front of the chip separated from the aqueous environment and etched vias are used to make fluidic connections from the back of the chip.

In some systems [27,169,170,173,174,321,323], microfluidics are fabricated separately which are usually polymeric microchannels replicated by molding [169,173,174] or hot embossing [321] (see Box 5.1) or glass microfluidics [170], and then glue or plasma-activated bonding are utilized to adhesively bond them on top of the CMOS chip. PDMS, SU8, PI, laminated photosensitive DFR, diacrylate bisphenol A (DABA) photocurable polymer, and benzocyclobutene (BCB) are some of the prevalent materials used as adhesive. It is noticeable that the adhesive might not provide a hermetic bonding for different fluid samples [76] and can also occupy a large area.

Box 5.1:   Hot embossing

A typical micro-hot embossing process includes five steps:

- Loading and alignment of the substrate and the mold.
- Heating the substrate and the mold to molding temperature.
- Pressing the mold against the substrate and embossing the microstructure patterns at embossing temperature.
- Cooling the substrate and the mold to demolding temperature.
- Demolding the component.

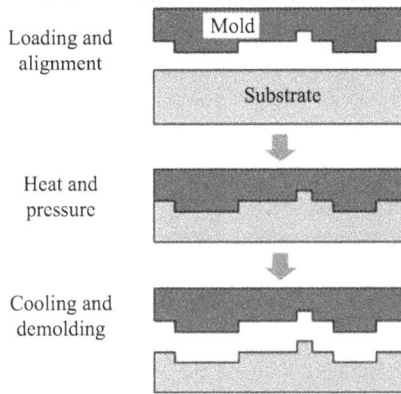

Figure B.5.1   The hot embossing process

5.4   Summary

We reviewed different IC-microfluidic packaging for CMOS-based LoC applications including rapid prototyping technique, soft lithography, DWFP, and other microfabrication techniques such as micromachining and alike. Rapid prototyping is a low-cost technique but is not suitable for large arrays of microchannels and wells. Soft lithography is a popular technique but thermal and chemical stability should be taken into consideration. DWFP can be used to fabricate complex 3D microfluidics directly on top of the IC without the need for masks or mold. Despite reviewing these techniques, different used materials and their challenges were discussed in this chapter.

There are standard techniques for the electrical packaging of CMOS chips (like wire bonding and flip-chip); however, the microfluidic packaging of CMOS chips are performed through nonconventional techniques and further studies are required in order to standardize low-cost and reliable packaging techniques for such hybrid systems. Although it is well-established that CMOS technology is efficient for affordable mass production of microelectronic devices, there is a growing interest to find generic and affordable microfluidics packaging methods incorporated with reusable CMOS biosensors for mass-production purposes. So, novel contact-free hybrid microfluidic/CMOS techniques are of the main objectives for developing affordable and portable biosensing platforms, especially for point-of-care applications.

Chapter 6
Biological/chemical applications

Various applications of capacitive biosensors have been developed for different kinds of target biospecies like proteins [96,100,102,330–334], nucleic acids [335,336], cells [94,337,338], saccharides [88], and small organic molecules [102,339,340]. The bio-interface layer coated on the sensing transducer might be formed by its own nature (like the biofilm created in cell growth), or be functionalized with biological recognition elements (BREs) capable of selective reaction with specific cells or molecules [76]. In this chapter, various types of biological applications of capacitive sensors are reviewed. Although all of them are not implemented in complementary metal-oxide semiconductor (CMOS) technology, they show the potential of this type of electrochemical sensors for life science applications. These applications can be categorized into three general groups: (1) organic solvent sensing, (2) nonselective cell monitoring and toxicity test, and (3) selective sensing.

6.1 Chemical sensing

Chemical sensing, especially gas sensing and organic solvent sensing, is among the most favorite applications of capacitive sensors playing an important role in monitoring the toxicity of drugs, food, soil, and water. The dielectric constants of organic solvents can be considered as the sensitivity and/or selectivity factor for their detection purposes. Additionally, the electrical conductivity of most polar solvents is low and suitable for capacitive sensing. It is worth mentioning that passivation layers above the sensing electrodes, as insulation layers, decrease the leakage current and consequently the capacitive sensing property will be improved [76].

For gaseous organic solvents, sensing electrodes are usually coated with polymer films such as polyether urethane (PEUT), poly(cyanopropylmethylsiloxane), ethylcellulose, and poly(epichlorohydrin) as an interface between the solvent and the electrodes (see Figure 6.1(a)) [228]. Physical properties of the polymeric film (like dielectric or mass) can be changed by analyte absorption [226]. The density of diffused molecules and consequently the dielectric constants of the polymer will be changed proportionally to the concentration of solvents. So, the dielectric properties of the analyte can be assessed by monitoring the variations of the polymer dielectric constant. Similarly, the presence of a liquid-phase chemical

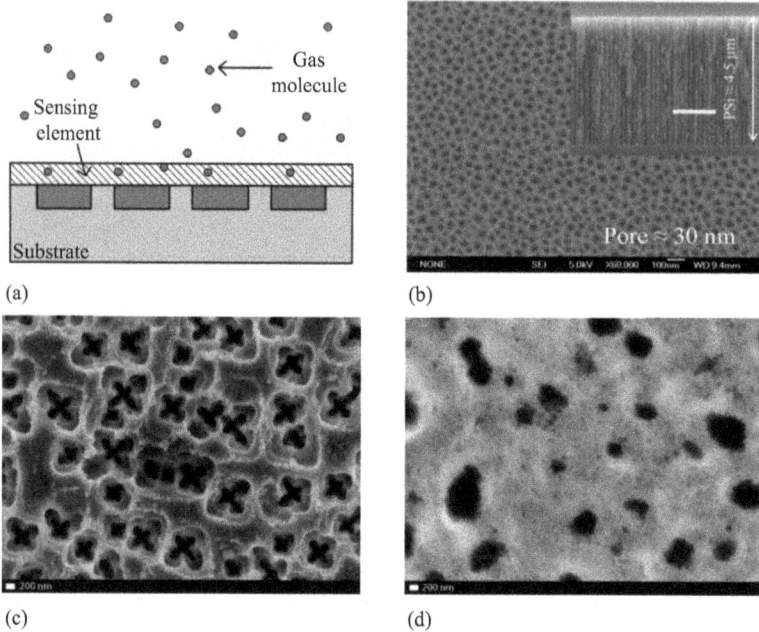

Figure 6.1   (a) Organic solvent (gas phase) sensor; (b) top-view and cross-sectional view of meso-PSi layer produced from p-type Si [341]; (c) n-type PSi sensors used in [342]; (d) p-type PSi sensors used in [342]

solvent above the electrodes, which can be injected into a microfluidic channel above the electrode, changes the sensing capacitance proportional to its dielectric constant [75,183,280].

Porous Si (PSi) has attracted widespread interest in sensing-based devices due to the compatibility with existing Si processing technology. They are often fabricated by an electrochemical etching procedure. For example, Harraz *et al.* [341] used a galvanostatic electrochemical etching of crystalline Si in a hydrogen fluoride (HF)-based solution (see Figure 6.1(b) and Table 6.1) to produce meso-PSi layers for capacitive chemical sensing of different polar (methanol (MeOH), ethanol (EtOH), acetonitrile (MeCN), acetone (ACE), chloroform (CHL), etc.) and nonpolar (toluene (TOL), *n*-hexane, etc.) organic solvents. Harraz *et al.* [341] fabricated meso-PSi layers by using a galvanostatic electrochemical etching of a heavily doped p-type Si wafer (see Appendix B, Section B.1.3). After cutting the wafers, the pieces were washed by deionized water (DI water) and sonicated in ACE. Then, they were dipped in HF solution to remove the native oxides. The electrochemical etching was performed in a cell made of Teflon. A Cu plate behind the silicon electrode played the role of a current collector. Anodization was performed using an anodizing solution composed of $HF/H_2O/EtOH$ and surface passivation was done by anodic oxidation in H_2SO_4 electrolyte under a galvanostatic condition for stability and proper operation of the

Table 6.1 Comparison of different CMOS capacitive sensors used for organic solvent sensing

Organic solvent	Intermediate layer	CMOS/ non-CMOS	Capacitive transducer	Substrate material	Ref.
EtOH, MeOH, ACE, MeCN, CHL, TOL, *n*-hexane	Meso-PSi	Non-CMOS	Floating electrode	Si	[341]
ACE, MeOH, EtOH, IPA, water, CHL, TOL, isoprene	PSi	Non-CMOS	Floating electrode	Si	[342]
TOL, EtOH	PEUT	CMOS	Q-IDE[1]	Al	[226]
ACE, MeOH, EtOH, DCM	SiO$_2$	CMOS	IDE	Al	[183]
DI water, ACE, MeOH, IPA	CMOS passiva- tion layers	CMOS	IDE	Al	[13]

[1]Quasi IDE.

device. Afterward, the porous layers were rinsed in DI water and EtOH and dried naturally. The electrical contacts were created on the front porous surface by colloidal Ag paint and Cu wires and baking.

In another work, Baker *et al.* [342] used metal-oxide nanostructure decorated n- and p-type PSi interfaces (Figure 6.1(c) and (d)) for the detection of liquid organic solvents including ACE, EtOH, MeOH, isopropanol (IPA), water, CHL, TOL, and isoprene. N-type and p-type PSi sensors reported in [342] were fabricated by a process including SiC deposition (etching chemical mask), photolithographic patterning, reactive ion etch (RIE) of Si, electrochemical anodization of the exposed Si, and finally deposition of metal contacts by e-beam evaporation. Metal-oxide nanostructures, Au$_x$O, and SnO$_x$, were deposited to the PSi interface. The PSi interfaces are exposed to electroless solutions and are then put in DI water and MeOH.

In the CMOS capacitive sensor reported in [226], PEUT is used as the polymeric sensing layer for the detection of TOL and EtOH. In another effort, Nabovati *et al.* [183] removed the whole passivation layer over the topmost metal of CMOS technology which was used for the fabrication of electrodes to reach the maximum sensitivity of chemical solvent detection and exposed their capacitive sensor to polar chemical solvents with different dielectric constants including EtOH, MeOH, ACE, and dichloromethane (DCM). Senevirathna *et al.* [13] also tested their CMOS capacitive sensor by sensing water, ACE, MeOH, and IPA.

6.2 Cell monitoring and toxicity test

In capacitive cell-based biosensors [20,21,68,189] (Figure 6.2(a)), living organisms such as bacteria and other types of cells deposited on the electrodes can form a bio-interface layer. The activities and metabolic status of cells such as viability, growth inhibition, adherence, proliferation, and migration can change the electrical

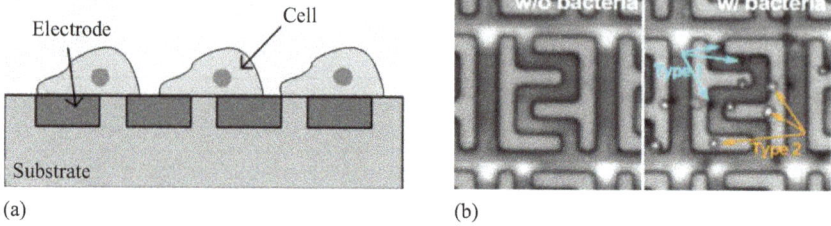

Figure 6.2   *(a) Cell biosensor; (b) before and after binding the bacteria to the electrodes used in [20] with two types of adherent bacteria (type 1: bacteria between electrodes, type 2: bacteria atop electrodes)*

properties of the surrounding micro-environment and affect the measured signal. Since cell culture is one of the required steps for these applications, some fundamentals of this context are briefly mentioned in Appendix E. Table 6.2 compares some capacitive sensors reported for cell monitoring applications.

For example, Couniot *et al.* [109] proposed a CMOS capacitive sensor for the detection of *Staphylococcus epidermidis* (*S. epidermidis*) suspended in phosphate-buffered saline (PBS). In this work, *S. epidermidis* was cultivated on agar plates and incubated at 37 °C. Afterward, a single colony was suspended in Tryptic Soy Broth (TSB) and incubated to reach a stationary phase bacterial culture. Then, the culture was centrifuged and cells were resuspended in PBS 1×. This step was repeated two times. They used Al/Al_2O_3-electrodes and could achieve a limit of detection (LoD) of 10^7 cfu mL^{-1}. The same group [20] developed another CMOS capacitive biosensor for detecting the suspension of the same bacteria in PBS 1:1000 and achieved an LoD of about seven bacteria. Figure 6.2(b) shows the binding of bacteria to the electrodes which might occur between the electrode (type 1), or atop the electrodes (type 2).

Ghafar-Zadeh *et al.* [68] proposed a CMOS capacitive biosensor for bacteria growth monitoring, which was examined for *Escherichia coli* (*E. coli*) suspended in Luria–Bertani (LB) medium. In another effort, Prakash *et al.* [189] developed a CMOS capacitive biosensor for tracking the adhesion and proliferation of MDA-MB-231 breast cancer cells (see Table 6.2). In this work, MDA-MB-231 cells were cultured in a flask in a growth medium containing improved minimum essential medium, fetal bovine serum, and penicillinstreptomycin (100×). Around 48 hours were required for the MDA-MB-231 cells to grow in the flask. Afterward, the cells were detached by using trypsin/ethylenediaminetetraacetic acid (EDTA) and the suspension in the growth medium with the desired density was prepared. Cells were loaded to the chip without additional surface functionalization or modification.

A CMOS-based capacitive biosensor was reported by Senevirathna *et al.* [13] for real-time measurements of the proliferation of ovarian cancer cells. The experiments were performed by adding a prepared cell suspension to the device and growing adherent cells onto the chip array. Figure 6.3 illustrates the adherence and proliferation of CP70 cells across the sensor array after 21 and 44 h of incubation.

Table 6.2 Comparison of different CMOS capacitive sensors used for cell monitoring

Cell	Bio-friendly layer	CMOS/ non-CMOS	Capacitive transducer	Transducer material	Application	Ref.
E. coli	CMOS passivation layers	CMOS	IDE (Op-fin)	Al	Bacteria growth monitoring	[68]
S. epidermidis	Al_2O_3	CMOS	IDE	Al	Bacteria detection	[109]
S. epidermidis	Al_2O_3	CMOS	IDE	Al	Detection of single bacterial cell	[20]
Breast cancer cells	One of the CMOS passivation layers	CMOS	Floating electrode	Al	Tracking cancer cell proliferation	[189]
Human lung carcinoma cell	PDMS	CMOS	IDE	Al	Cell growth screening	[21]
H1299 and Hek293 cells	PEM films	CMOS	IDE	Al	Cell culture monitoring and drug test	[28]
Ovarian cancer cell lines	CMOS passivation layers	CMOS	IDE	Al	Measurements of cell proliferation	[13]
Neutrophil granulocyte	SiO_2/Si_3N_4 passivation layer	CMOS	IDE	Al	Monitoring activation processes of cells	[343]
E. coli	SAM of MPA	Non-CMOS	IDE	Au	Toxicity screening of various NPs (different sizes of Fe_3O_4 NPs as models)	[94]
S. aureus	MUA	Non-CMOS	Surface stress-based (membrane)	RiE: Au/Cr/Si FlE: PDMS	Detection of *S. aureus*	[344]

Op-fin, passivated IDE with a window in between the fingers; RiE, rigid electrode; FlE, flexible electrode.

After 21 h   After 44 h

Figure 6.3   CP70 cells adhere and proliferate across the sensor array after 21 and 44 h of incubation [13]

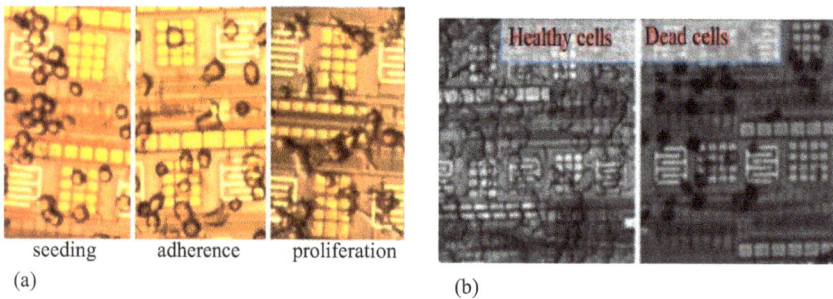

seeding   adherence   proliferation
(a)   (b)

Figure 6.4   (a) The response of the CMOS chip reported in [21] to 100k Cells mL^{-1} for 12 h; (b) cell viability before and after introducing antibiotics to the cell growth medium on the smart petri dish reported in [28]

Nabovati *et al.* [21] used CMOS-based capacitive biosensors for high-throughput cell growth monitoring by using H1299 (human lung carcinoma) cell line. The biocompatibility of the electrodes implemented on the topmost metal of CMOS technology was increased by a very thin layer of the PDMS spin-coated on top of them. After preparing the cell culture in a medium, appropriate aliquots of the cell suspension were diluted based on the well size on the CMOS chip. Figure 6.4(a) illustrates the response to 200 μL of cell suspension with the concentration of 100k Cells mL^{-1} in the media over 12 h including cell seeding, adherence, and proliferation phases, respectively. In another work, the same group [28] examined their chip for the growth monitoring of nonresistant H1299 and resistant Hek293 cell lines as well as drug cytotoxicity (see Table 6.2 and Figure 6.4(b)). In this work, the biocompatibility was increased by building up polyelectrolyte multilayer (PEM) films on CMOS chips using a layer-by-layer (LBL) polyelectrolyte deposition technique creating strong electrostatic interaction between PEMs layer. These layers include a positively charged poly(ethylenei-mine) (PEI) layer as a precursor base layer to begin the sequential adsorption of the PEMs as well as the polyanion (poly(sodium styrene sulfonate) (PSS)) and polycation (poly(allylamine) hydrochloride (PAH)) layer to form five polyelectrolyte bilayers [PSS/PAH]$_5$.

Bunnfors *et al.* [343] used LoC-based capacitive biosensors for monitoring the activity of neutrophil granulocytes and their oxidative stress when triggered by cerium-oxide-based nanoparticles (NPs). The neutrophils were isolated from venous whole blood which was anti-coagulated with heparin, equilibrated to room temperature, layered on an equal volume of a density gradient and centrifuged at room temperature. The separated neutrophils were resuspended in NaCl and PBS and centrifuged at room temperature. Twice brief hypotonic treatment in ice-cold distilled water and washing the cells in PBS including NaCl and 4-(2-hydroxyethyl)-1-piperazineethanesulfonic acid (HEPES) solution leads to remove the remaining erythrocytes. After resuspension of isolated neutrophils in HEPES solution and adjusting the cell concentration to the desired value, the isolated neutrophils were kept on ice until use. During experiment, they were incubated at 37 °C without CO_2. The NPs, Ce_2O_3, and gadolinium (Gd) alloyed Ce_2O_3 were prepared by employing the room temperature wet-chemical synthesis procedure reported by Eriksson *et al.* [345] and the concentrations of cerium (Ce) and Gd in the prepared NPs were measured. These NPs were employed as triggers for the initiation of cell activation. Figure 6.5 illustrates the neutrophils on insulator surface in initial activation stage and after that. The experimental results showed that the sensor was able to detect the neutrophil response, but not the stimuli.

Qureshi *et al.* developed a whole-cell-based capacitive biosensor in [94] to investigate the biological toxicity of NPs. In this sensor, *E. coli* cells were immobilized on Au IDEs and interacted with different sizes of iron oxide (Fe_3O_4) NPs (5, 20, and 100 nm) as models resulting in morphological changes of the surface of *E. coli* cells. Figure 6.6(a)–(c) illustrates healthy normal cells and their morphological changes when they are exposed to 5 and 100 nm Fe_3O_4 NPs. The results illustrated that the smallest NP interacted with *E. coli* more efficiently and caused maximum capacitance change. In another work, Jian *et al.* [344] reported a surface stress-based capacitive biosensor for the detection of *S. aureus*. As mentioned in Table 6.2, they used a PDMS micromembrane coated with a layer of 11-mercaptoundecanonic acid (MUA: SH-(CH2)10-COOH) as an indicator.

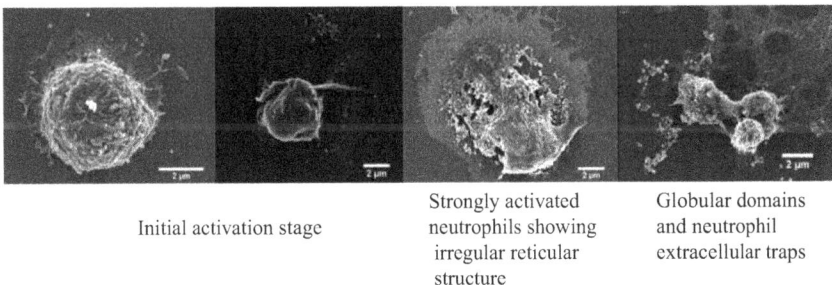

| Initial activation stage | Strongly activated neutrophils showing irregular reticular structure | Globular domains and neutrophil extracellular traps |

Figure 6.5 Activation process of neutrophil granulocytes on the chip reported in [343]

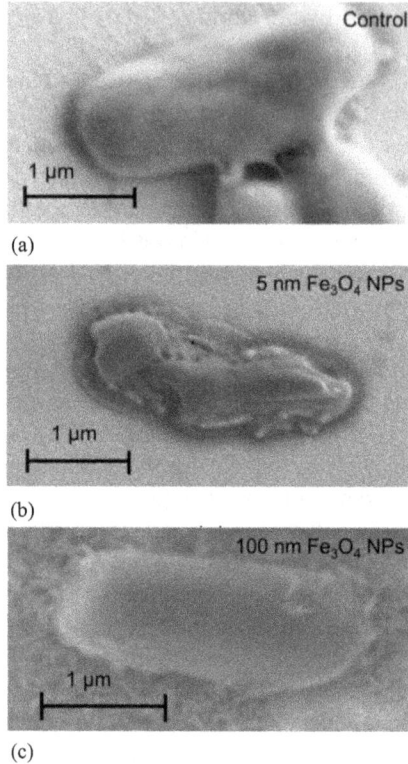

*Figure 6.6   (a) Healthy/control cells; cellular morphological changes when
exposed to (b) 5 nm and (c) 100 nm Fe_3O_4 NPs [94]*

6.3   Selective sensing

For many years, classical BREs or target receptors such as nucleic acids, anti-bodies, enzymes, and whole cells have been used in electrochemical biosensors (Figure 6.7(a)). Recent developments in biochemical sensors and nanosensors have led to artificial BREs such as phages, aptamers, affibodies, and molecularly imprinted polymers (MIPs) [82]. Furthermore, nano-sized materials such as NPs [346], carbon nanotubes (CNTs) [347], and nanorods (NRs) [84] are generating considerable interests in terms of the improvement of some analytical assay by combining them with the BREs like enzymes and antibodies [82].

The electrical material and the required method for forming the intermediate layer to capture target biospecies play important roles in capacitive biosensors. Several methods are reported for immobilization of receptors [348], such as covalent attach-ment with cross-linking agents like silanes [91,93,338,349], physical adsorption (noncovalent) [86,349], and embedding in membranes [83,350] or polymers [351,352]. In [85,353–357], receptors are immobilized on a Si/SiO_2 surface with a covalent attachment. In another effort [358], a receptor is incorporated in a Langmuir–Blodgett

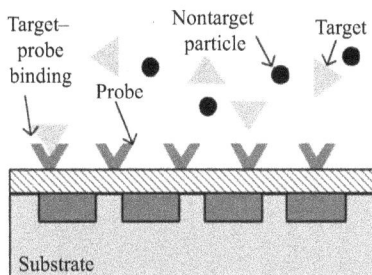

Figure 6.7   A selective sensor for detection of different biospecies such as DNA, antibody, antigen, and enzyme

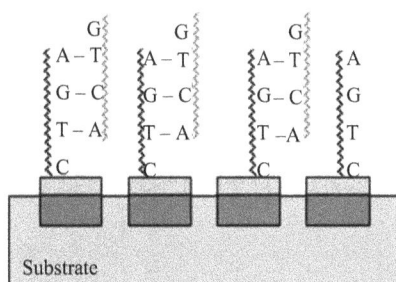

Figure 6.8   Hybridization-based DNA detection

film, on a Si/SiO$_2$ structure. A polymeric membrane can also be used to immobilize receptors on a Si/SiO$_2$ structure [357,359,360]. In a work in [361], receptors are immobilized on Si/SiO$_2$ surface, and poly(ethylene glycol) (PEG) is utilized to overlay the receptor layer and exclude ions from it. Immobilization of protein receptors on an Au electrode is typically based on covalent methods using functional moieties of amino acids or providing Au-thiol bonds to create SAMs [346,348,362–365]. Tantalum oxide is one of the materials for the surface of the electrodes that can be used for the immobilization of a receptor [366]. Pt electrode with a polymeric membrane is also useful for this issue [367,368].

Herein, according to the used BREs, selective sensors are divided into three general groups: (1) the assays based on nucleic acids (ribonucleic acid (RNA) and DNA), (2) the assays based on antibody, and (3) other techniques.

6.3.1   Nucleic acid-based methods

In genosensors [117,237], nucleic acids are immobilized on the surface of electrodes as the BRE or RNA/DNA probe (see Figure 6.8). The target RNA or ssDNA molecules can bind to their complementary RNA or ssDNA of recognition layer and subsequently affect the measured capacitance signal which is useful for RNA/DNA detection purposes. Table 6.3 compares some of such biosensors.

Table 6.3 Comparison of different CMOS capacitive sensors used for nucleic acid-based techniques

Transducer material	Capacitive transducer	CMOS/ non-CMOS	Linker	BRE	Application	Ref.
Au	IDE	Non-CMOS	Thiol	Oligomer ssDNA probe sequences	Early detection of viral infectious diseases	[336]
Au	IDE	CMOS	Thiol	DNA probe	DNA detection	[117]
RiE: Cr/Au FlE: Al- and Au-coated PDMS micromembrane	Surface stress-based (membrane)	Non-CMOS	Thiol	Oligomer ssDNA probe sequences	Detection of hybridization and single nucleotide polymorphism discrimination	[369]
RiE:- FlE: LBL-assembled polymers and SWNTs	Surface stress-based (membrane)	Non-CMOS	Phosphoramide linkage with amine group	Oligomer ssDNA probe sequences	Detection of anthrax DNA	[179]
RiE: phosphor-doped Si FlE: highly boron-doped Si membrane passivated by LTO	Surface stress-based (membrane)	Non-CMOS	Aminosilane activated by Glu, PDITC, and GOPTS	β-Thalassemia oligonucleotides	Detection of DNA mutations	[370]
RiE: Au FlE: Si membrane passivated by LTO	Surface stress-based (membrane)	Non-CMOS	Thiol and Au/ Thiol and GOPTS	Probe oligonucleotides	DNA hybridization	[93]

RiE, rigid electrode; FlE, flexible electrode.

As seen in Table 6.3, a nucleic acid-based capacitive biosensor was developed in [336] using Au IDEs functionalized with 24-nucleotide DNA probes. After pretreatment of the Au electrodes surface, thiolated ssDNA oligomers were pretreated. For thiol bonding formation between the oligo and the Au surface, the 5' thiol-modified oligomers need a reduction of the disulfide bonds. After washing immobilized tris(2-carboxyethyl)phosphine (TCEP) disulfide reducing gel with $1 \times$ TE-MgSO4 buffer, oligo was added and the mixture was incubated. After incubation, the supernatant contains reduced thiol-modified oligomers. To fill vacant Au sites and improve hybridization efficiency, electrodes were incubated with 11-mercapto-1-undecanol (MCU). This biosensor can detect as few as 20 complementary DNA target molecules (1.5 aM) without the need for amplification methods like using nanomaterials or nanostructures. In this work, covalent immobilization of purified ssDNA probe oligomers improved the reproducibility of sensor detection. Stagni *et al.* [117] also reported a CMOS capacitive biosensor for DNA detection whose Au electrodes are functionalized with DNA probes.

In another work, Cha *et al.* [369] proposed a dome-like membrane transducer in which probes were immobilized on an Au-coated PDMS micromembrane through thiol–Au bonding (Table 6.3). This sensor was examined for DNA hybridization detection as well as protein recognition. While the liquid side of the membrane was coated with Au, an Al layer plays the role of a conductive flexible electrode in the dry cavity side of the membrane. After cleaning the surface of Au-coated PDMS membranes, single-stranded thiolated probe DNA was immobilized on them. Then, 6-mercapto-1-hexanol (MCH) was used to avoid physisorption during hybridization and improve the accessibility of the DNA probe to targets as a spacer. Thereafter, DNA solution in PBS buffer containing various sequences was applied for hybridization experiments.

Another surface stress-based capacitive sensor was presented by Kang *et al.* [179] which was designed for detecting the hybridization of anthrax DNA and used multilayered nanocomposite membrane composed of single-walled carbon nanotubes (SWNTs) and polymers (see Table 6.3). The multilayered nanocomposite membrane comprised LBL-assembled cationic and anionic polymers as well as SWNTs ($[PAH/PSS]_5/PAH/SWNT_{10}/[PAH/PSS]_5$). The phosphorylated DNA probe was immobilized on an aminated substrate as shown in Figure 6.9. The nanomembrane was deprotonated by triethylamine solutions and then washed with distilled water. A solution mixed with 1-Ethyl-3-(3-dimethylaminopropyl)-carbodiimide (EDC) and 1-methylimidazole was used to immobilize the probe DNA. In this work, SWNT constitutes the conductor and makes it possible to control the temperature easily. In another effort, a surface stress-based sensor is proposed by Tsouti *et al.* [370] for the detection of DNA hybridization in which probe DNA molecules are immobilized on the surface of ultra-thin Si membranes passivated by low-temperature oxide (LTO) film. They compared five different immobilization procedures of biomolecules conjugated with amino reactive groups including 3-[2-(2-aminoethylamino) ethylamino] propyltrimethoxysilane (AEEPTMS), (3-aminopropyl) tris (trimethylsiloxy)

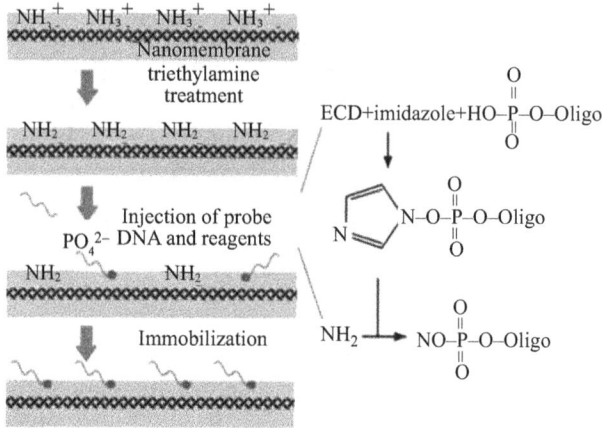

Figure 6.9   The immobilization procedure for the phosphorylated DNA to the aminated substrate

silane (APTTMS), 3-(aminopropyl) trimethoxysilane (APTMS), poly(amidoamine) (PAMAM) dendrimer (4 generation), and 3-aminopropyltriethoxysilane (APTES). The surface functionalization of Si samples coated with five different types of aminosilane was followed by activating with 1,4-phenylene diisothiocyanate (PDITC) and glutaraldehyde (Glu), and coating with (3-glycidoxypropyl) trimethoxysilane (GOPTS). Tsekenis *et al.* used micromembranes in [93] to compare the grafting density and hybridization efficiency of two functionalization methods for DNA detection. In both methods, thiol-modified ssDNA was immobilized on LTO on Si surfaces. In the first functionalization technique, Au was deposited onto the LTO surfaces. In the second technique, a GOPTS layer was self-assembled on the surface of the LTO substrate. This study showed that the hybridization efficiency is enhanced using the second functionalization method. They also used the direct laser printing of oligonucleotides onto planar surfaces with the use of laser-induced forward transfer (LIFT) technique (see Box 6.1) for selective probe immobilization (see Table 6.3).

Box 6.1:   LIFT

LIFT is a direct-write laser technique for the deposition of small volumes of material from a donor to a receiving substrate. A pulsed laser beam is irradiated as the driving force from a donor substrate to a receiving substrate which is placed in parallel to it as shown in Figure B.6.1. This technique can be used for deposition of cells, proteins, and DNA and offers the advantage of a contactless approach, high spatial resolution, and minimization of the waste of the biomaterials.

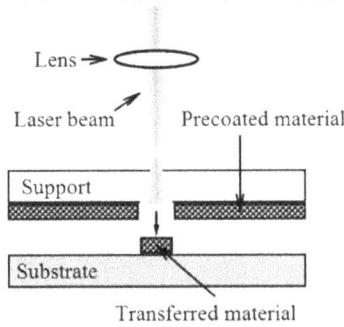

Figure B.6.1   LIFT experimental setup

6.3.2   Antibody-based assays

Antibodies are among the most widely used BREs [83–87] and can bind to specific piece (or pieces) of antigens, named epitope (see Table 6.4 and Figure 6.10). Antibodies might be monoclonal or polyclonal if they are generated from the clones of a single cell or several different immune cells, respectively. Polyclonal antibodies may bind to multiple epitopes on the same antigen whereas monoclonal antibodies bind to a unique epitope on the antigens [82]. The selectivity of monoclonal antibodies is more than polyclonal ones, but their production is more expensive. Many electrochemical biosensors work based on antibody–antigen reaction [359,362,363,368].

Yang *et al.* [348] proposed a strategy, as mentioned in Table 6.4, for antibody immobilization based on electrodeposition of nanometer-sized bioactive hydroxyapatite (HAP) on a self-assembled β-mercaptoethanol-modified Au electrodes (see Figure 6.11). The process includes (1) self-assembled sublayer of β-mercaptoethanol, (2) electrodeposition of nanometer-sized HAP on the modified Au surface, (3) blocking the film defects by 1-dodecanethiol (DDT), and (4) covalently coupling transferrin antiserum with divinyl sulfone (DS) on the HAP-coating film.

In another effort by Malvano *et al.* [338], Pt electrode was modified with perovskite-structured materials (see Figure 6.12). Monoclonal *E. coli* antibody was immobilized on a strontium titanate perovskite layer ($SrTiO_3$) on Pt electrode to detect *E. coli* cells. After immersion of $SrTiO_3$-modified electrode in an EtOH solution containing APTES, it was sonicated. Then, Glu solution was dropped onto the modified electrode. Next, it was rinsed with water and covered with anti-*E. coli* solution. After blocking the unreacted active sites with ethanolamine, the electrode was rinsed in PBS to remove unbound antibodies.

Lebogang *et al.* [371] proposed an immunosensor for broad-spectrum detection of microcystins as a harmful toxin in surface waters. As shown in Figure 6.13, in this biosensor, tyramine was electropolymerized onto an Au electrode surface to make an anchor layer for monoclonal antibodies and GNPs were used for dense attachment of them. In this work, the surface of Au electrode was modified with GNPs attached to a Glu-activated polytyramine layer. GNPs amplify the signal by

Table 6.4 *Comparison of different CMOS capacitive sensors used for protein-based assays*

Transducer material	Capacitive transducer	CMOS/ non-CMOS	Linker	BRE	Application	Ref.
SrTiO$_3$-modified Pt electrode	Floating electrode	Non-CMOS	APTES, Glu	Monoclonal anti-E. coli	Detection of E. coli O157:H7	[338]
Au	Floating electrode	Non-CMOS	A nano-HAP film on a SAM of β-mercaptoethanol monolayer, DS	Anti-human transferrin	Detection of human transferrin	[348]
Au	Floating electrode	Non-CMOS	Glu, GNP	Anti-Adda mAB	Detection of cyanotoxins in waters	[371]
Au	Floating electrode	Non-CMOS	SAM of MUA, GNP, EDC, NHS	Ab2, Ab1	Detection of HBsAg	[346]
Au	IDE	Non-CMOS	SAM of 2-mercaptoethylamine, GNP, EDC, NHS	Ab2, Ab1	Detection of HBsAg	[331]
PVDF Immobilon-P membranes attached Au electrode	3D IDE	Non-CMOS	-	Anti-HRP	Detection of HRP antigen	[83]
ZnO NR structures deposited on Au electrodes	3D IDE	Non-CMOS	sulfo-MBS	Anti-HRP	Detection of HRP antigen	[84]
RiE: Si/Au FiE: Au with PDMS micromembrane	Surface stress-based (membrane)	Non-CMOS	SAM of thiol, EDC, NHS	Anti-CSFV	Detection of CSFV antigens	[330]
Au electrode coate with Si$_3$N$_4$/parylene C	IDE	CMOS circuit/non-CMOS electrode	Glu	Anti-SSAT	Detection of cancer biomarker (SSAT)	[4]
Au	IDE	Non-CMOS	SAM of MPA, EDC, NHS	Anti-CRP, anti-TNFα, and anti-IL6	Detection of cardiovascular risk biomarkers (CRP, TNFα, and IL6)	[372]
Au with planar nanogap (Si/SiO$_2$)	Interdigitated nanogap electrode	Non-CMOS	APDMMS, Glu and 3-aminobenzoic acid, EDC, NHS	Anti-cTnT	Detection of cTnT	[373]

RiE, rigid electrode; FiE, flexible electrode.

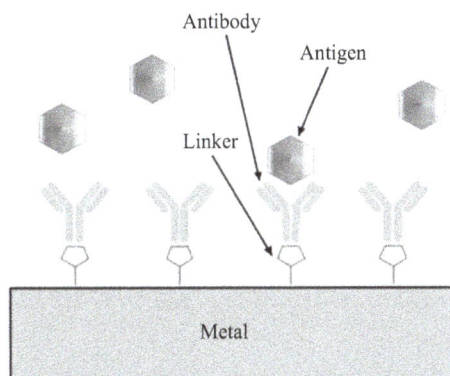

Figure 6.10   A typical configuration of an immunosensor to show antibody–antigen reaction

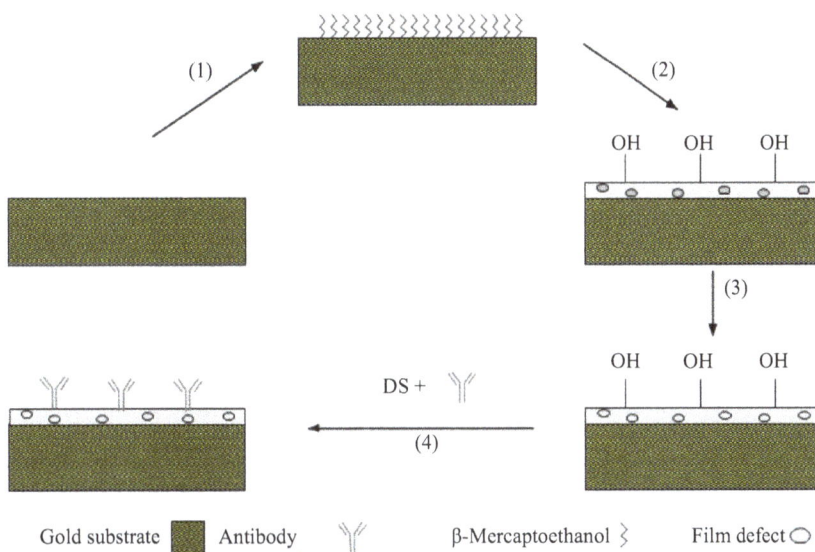

Gold substrate   Antibody   β-Mercaptoethanol   Film defect

Figure 6.11   The stepwise process involved in the immobilization of antibody onto the gold electrode

increasing the surface area for the immobilization of the antibody. After cleaning Au electrodes, a thin insulation layer of polytyramine was electrodeposited on its surface which also plays the role of an anchor for antibodies. Electropolymerization was performed in tyramine dissolved in MeOH containing NaOH solution. CV was used for electropolymerization. The unpolymerized tyramine was washed with pure

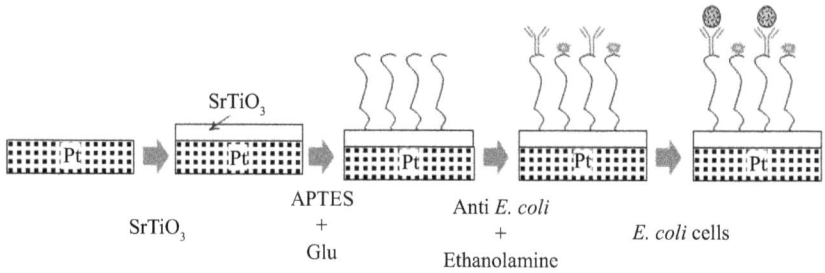

Figure 6.12   The fabrication process of an E. coli O157:H7 label-free immunosensor

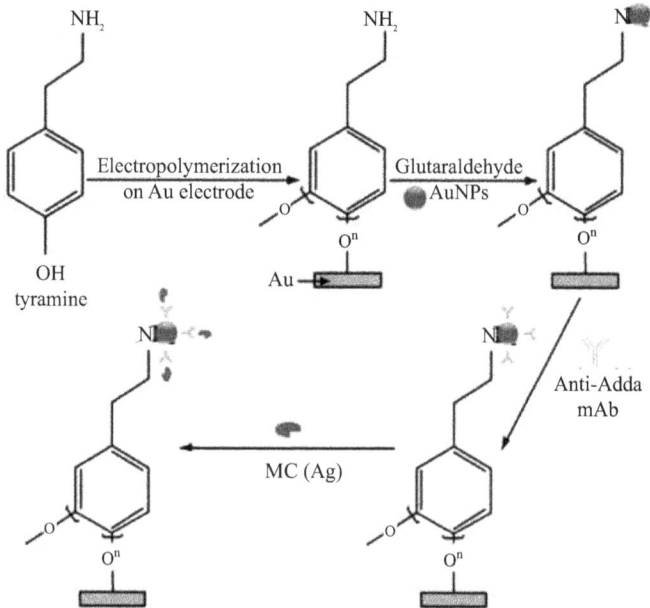

Figure 6.13   Process of the modification of Au electrode by electropolymerization of tyramine, immobilization of antibodies, and antigen binding

water. After activating the polytyramine layer by using Glu in potassium phosphate buffer, the electrodes were again washed with the buffer. Afterward, drop-coating the electrodes with a colloidal solution of GNPs and incubating allow the citrate-stabilized GNPs to adsorb via the aldehyde group and possible unreacted amino groups of tyramine. Thereafter, non-adsorbed GNPs were washed with pure water and phosphate buffer and dried with a stream of nitrogen gas. Then, anti-Adda monoclonal antibodies were deposited on the electrodes and left to adsorb. Afterwards, the electrodes were treated with DDT in EtOH to block any bare spaces on its surface. Finally, the electrodes were treated with ethanolamine to block any possible non-reacted aldehyde groups stemmed from activation with Glu.

Sandwich assay is a commonly approach used for protein detection which needs two probes binding to different regions of the target. In this approach, after immobilization of the first probe on the surface of the sensor, the analyte is introduced. Then, the secondary probe will be introduced after washing steps. The secondary probe might be labeled or detectable by another labeled probe binding to all the secondary probes. Although this technique improves selectivity, it increases development cost and limits the use in research settings [25].

As an example, Alipour *et al.* [346] developed a capacitive biosensor for the detection of hepatitis B surface antigen (HBsAg). They used two planar Au electrodes covered by an insulating SAM. Then, the primary antibody (Ab_1), HBsAg, and the secondary antibody (Ab_2) attached to gold NPs (GNPs) formed a sandwich-type structure (Ab_1-HBsAg-(Ab_2+GNP)) as mentioned in Table 6.4 and shown in Figure 6.14(a) and (b). To form insulating monolayer, after cleaning the Au electrode surface, they were polished with Al_2O_3. Thereafter, a SAM is formed by adding alkanethiols solution in EtOH to each well containing the electrodes and incubation overnight. Then, the wells sere rinsed with EtOH and PBS buffer, respectively. Next, each electrode was exposed to a mixed solution of 1-ethyl-3-(3-dimethylaminopropyl)-carbodiimide/*N*-hydroxysuccinimide (EDC/NHS) to

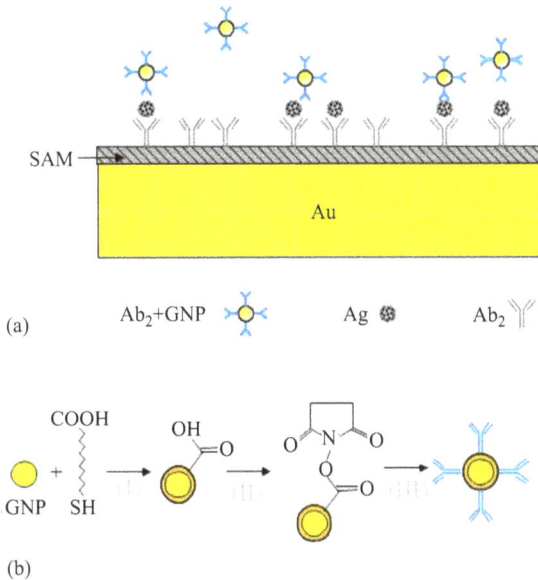

Figure 6.14 *(a) The sandwiches of Ab_1–HBsAg–(Ab_2+GNP) formed on the Au electrodes modified by a SAM of MUA. (b) The process for attachment of the secondary antibodies (Ab_2) to GNPs including (I) insulating monolayer formation on the GNPs via a SAM; (II) activation of the terminal carboxylic acid groups of the SAM via NHS and EDC binding; (III) connection of Ab_2 to the activated GNPs*

activate the terminal carboxylic acid groups of the SAM. To immobilize Ab_1, the modified electrodes were incubated in PBS containing Ab_1. Unreacted terminal carboxylic acid groups were blocked by further incubation using BSA in PBS as a blocking buffer. Then, any loosely non-bound Ab_1 or BSA were removed by washing each well with PBS. Thereafter, the colloidal GNPs were dialyzed against a NaOH, and mixed with PBS Tween-20 to stabilize the GNPs against aggregation. Then, MUA in absolute EtOH was added to the colloidal GNPs and the solution was stirred leading to the formation of a SAM of MUA on the GNPs (Step (I) in Figure 6.14(b)). After centrifuging the modified GNP to remove the unbound MUA, they were washed with PBS and these centrifugation and washing processes were repeated three times. To attach Ab_2 to the GNPs, MUA-coated GNPs was mixed and stirred with EDC/NHS solution to activate the terminal carboxylic acid groups of the SAM (Step (II) in Figure 6.14(b)). After centrifuging to remove the unbound molecules, activated GNPs was redispersed in PBS containing Ab_2, incubated, shaked, and kept in a refrigerator (Step (III) in Figure 6.14(b)). After centrifuging the mixture again, the attached Ab_2 on GNP (Ab_2 + GNP) was redispersed in PBS buffer. In another work, Zeinabad *et al.* [331] presented an ultrasensitive immunosensor using Au IDEs and GNP in a sandwich-type structure (Ab_1+HBsAg+(Ab_2+GNP)) for the detection of HBsAg (see Figure 6.15 and Table 6.4). A SAM of 2-mercaptoethylamine film coated on IDEs improved the efficiency of antibody immobilization.

Sanguino *et al.* presented two capacitive immunosensors in [83] and [84] where, respectively, poly(vinylidene difluoride) (PVDF) immobilon-P membranes and ZnO NR structures were used as 3D matrices for the immobilization of probe molecules across the region where the fringing electric field of microelectrodes can penetrate which is larger than the sum of the thickness of fingers of IDE and the space in between two fingers ($t_{IDE}+d_{IDE}$ in Figure 3.3(a) of Section 3.2.2). They used a horseradish peroxidase (HRP) antibody–antigen reaction to show the functionality of their proposed immunosensors.

A stress-based capacitive sensor was developed in [330] for the detection of classical swine fever virus (CSFV) antigens (see Figure 6.16). Two Au electrodes form the capacitor. The bottom Au electrode is placed on the Si substrate and the top Au electrode is located on the surface of a PDMS membrane which is functioned with antibodies. A naked Au membrane was immersed in thiol solution to form the Au–S bonds in the reaction between Au atoms and thiol molecules and

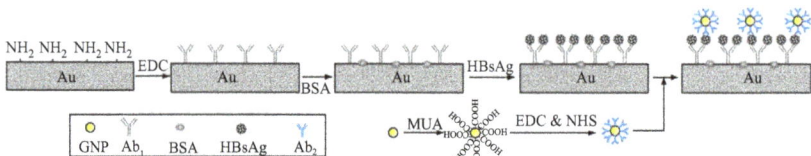

Figure 6.15   A sandwich-type capacitive immunosensor proposed by Zeinabad et al.

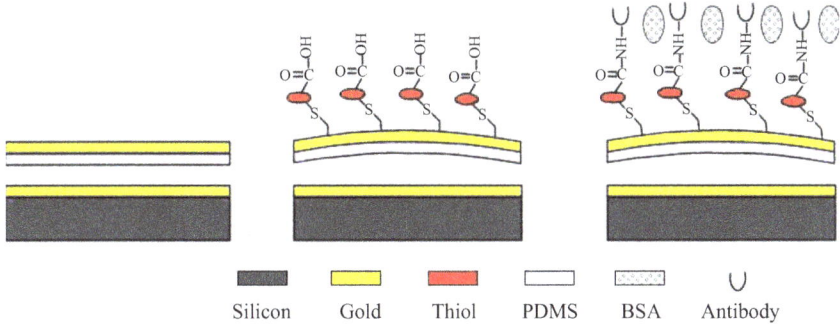

Figure 6.16 Modification process of a capacitive surface stress biosensor for the detection of CSFV

immobilization of thiol molecules on the surface of the membrane. Since there is no solution in the capacitive cavity, SAMs can be formed on PDMS-Au layer instead of Si–Au layer. Then, EDC/NHS solution as catalyst improved coupling efficiency between thiol molecules and antibodies. In this case, EDC as a coupling agent activated carboxyl groups, and NHS provided a stable active ester. So, the amide bond (–CONH–) provides the cross-linking of the antibodies to the SAM, and the electrode was modified with classical anti-CSFV. To avoid the impact of other substances, the nonspecific and unreacted sites were filled with BSA. Binding CSFV antigens to antibodies leads to convex deformation and changes the value of the measured capacitance.

Alhoshany *et al.* [4] developed a CMOS capacitive biosensor for label-free detection of cancer biomarkers whose sensing elements (Au IDE) were fabricated on a glass substrate and coated with a thin film of parylene C. The surface was treated to immobilize antibodies to screen and quantify spermidine/spermine N1 acetyl-transferase (SSAT) enzyme. The surfaces of the electrodes coated by parylene C were incubated in Glu solution as a cross-linking agent in $1 \times$ PBS solution and kept in an inert argon atmosphere for absorbing active agents on the surface and covalently coupling the active agent to the surface of the substrate. After incubation, the surfaces were washed with PBS Tween-20/DI water. The antibody was applied on the surfaces by using a drop-cast method with a fine micropipette and then the electrode was incubated at room temperature. After washing the surfaces of the electrodes similarly, they were immersed in a BSA solution as a blocking reaction to avoid any non-specific binding and the surfaces were again washed with the PBS buffer, removing most of the unbound antibodies from the surfaces. Thereafter, the samples were incubated in the dark with a solution of the SSAT enzyme.

Qureshi *et al.* developed a capacitive immunosensor in [372] to detect multiple disease biomarkers which is useful for the prediction of inflammation and cardio-vascular diseases. In [372], the Au electrodes were subjected to plasma cleaning and immersed in an ethanolic solution of β-mercaptopropionic acid (MPA). After being rinsed with distilled water, the SAM layer of MPA on the Au electrodes was dried over pure N_2 gas. Incubation with a mixture of EDC and NHS in distilled water activated the free carboxyl groups of MPA on the surface of the electrodes. The surface-activated capacitors were then divided into two groups for two formats. In the first format, each activated electrode was covalently coupled with a specific pure form of three types of antibodies (anti-CRP, -IL6, and -TNF) in PBS (see Table 6.4). In the second format, the antibodies were mixed and co-immobilized covalently on activated electrodes by incubating the antibody mixture. The LoDs of both formats were in the range of 25 pg mL^{-1} to 25 ng mL^{-1}.

Hsueh *et al.* proposed a planar interdigitated nanogap electrode in [373] capable of the measurement of cardiac-troponin T (cTnT). Aminopropyldimethylmethoxysilane (APDMMS), Glu, and 3-aminobenzoic acid are the three layers cross-linked on the Si/SiO$_2$ surface in this sensor. If the coating layer is on top of electrode surfaces, the nonuniform distribution of the coating molecules can deteriorate the detection SNR. To mitigate the uniformity problem, the cross-linker was immobilized on the dielectric surface between nanogap electrodes rather than on the top of the electrode surfaces.

6.3.3   Other selective techniques and artificial BREs

In many cases, antibodies may not have the required characteristics like specificity, affinity performance, or availability. So, several groups of non-antibody BREs have been developed to overcome the problems of antibodies such as aptamers [96,97], affibodies [374,375], and affimers [100,101]. Aptamers are short oligonucleotide (ssDNA or RNA molecules) or peptide (artificial proteins) molecules able to bind to a specific target molecule. Reusability, easy modification, versatility, and thermal stability are some of the advantages of aptamers (RNA, DNA, or peptides) [82]. Affibodies are new engineered proteins with high specificity and affinity binding to target proteins or peptides. Affimer is a revolution of peptide aptamers to display peptide loops. The size of affimer is twice of affibodies. Small size, high specificity, and flexible functionalization are some of the features of affimers [376].

Qureshi *et al.* reported an aptamer–antibody-based sandwich assay in [96] for detection of vascular endothelial growth factor (VEGF) in human serum which is useful for early cancer diagnosis. In this approach, the Au IDE was functionalized with a highly specific and selective anti-VEGF aptamer. After binding VEGF protein to the aptasensor surface, a sandwich complex was formed with antibody-conjugated magnetic beads (MB-Abs) ((anti-VEGF aptamer)-VEGF-(anti-VEGF antibody)) as shown in Figure 6.17.

Namhil *et al.* [377] proposed a horizontal thin-film nanogap capacitive sensor with no electrode polarization effects when used with water and ionic buffer solutions. The device was fabricated using standard photolithography structure and composed of top and bottom Au electrodes separated by an etched SiO$_2$ support layer. They modified the electrodes with thiol-functionalized ssDNA aptamers to detect human alpha thrombin. A capacitive aptasensor is also reported by Chen *et al.* [378] for thrombin. This non-CMOS sensor was designed based on face-to-face electrode pairs comprising a gold electrode and an indium tin oxide (ITO) film as the opposing electrode with a double-side polyethylene terephthalate tape as a

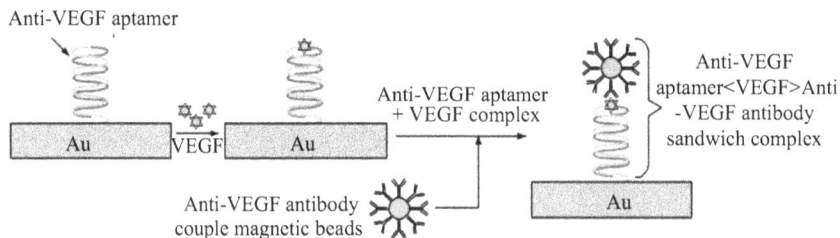

Figure 6.17   The sandwiching process of the apta-immunosensor reported in [96]

thin insulating spacer. Aptamers and DDT formed a SAM on the gold electrode and played the role of the BRE.

In another effort, Zhurauski *et al.* developed an affimer-functionalized Au IDE in [100] for the detection of a protein tumor biomarker, Her4, in undiluted serum. In this work, the electrodes were precleaned with isopropyl alcohol, ACE, and DI milli-Q water, and then treated using UV ozone. For affimer SAM formation, cysteine-modified Her4 affimer solution containing tris(2-carboxyehyl) phosphine hydrochloride (TCEP) was prepared in PBS buffer using optimized conditions and incubated to reduce any disulfide bonds. After reduction, this mixture was poured onto the electrodes and incubated. Afterward, the chips were washed with PBS Tween-20 and PBS to remove any unbound affimer. The chips were finally blocked with PBS Tween-20-based starting block.

Protein-based biosensors can be compatible with CMOS technology. Ghafar-Zadeh *et al.* [65] reported an impedimetric protein-based biosensor for glucose biosensing by employing genetically engineered GLK as receptor proteins. In this work, a SAM of octanedithiol on Au coupled with nitrilotriacetic acid (NTA) Ni^{2+} linker was used to link histidine-tagged recombinant human GLK enzyme (see Figure 6.18). Surface preparation procedure was as follows:

1. The Au electrodes were washed and rinsed using H_2SO_4/H_2O_2 and DI water and absolute EtOH. Then, the electrodes were dried under nitrogen gas.
2. By immersion in EtOH solution, SAMs of 1,8-octanedithiol were formed on the electrodes.
3. To remove any physioabsorbed material, the modified electrodes were rinsed with EtOH.
4. The modified electrodes were immersed into a solution of maleimide-C3-NTA to accomplish the coupling reaction of maleimide-C3-NTA and a sulf-hydryl surface.
5. To remove any physio-absorbed material, the electrodes were rinsed with DI water and MeOH.
6. Electrodes with NTA ligand were immersed in $NiCl_2$ to load Ni cations (Ni^{++}).
7. Electrodes were rinsed with DI water to remove the residual metal and ion solution.
8. His-tagged GLK proteins were coordinated to Ni^{++}.

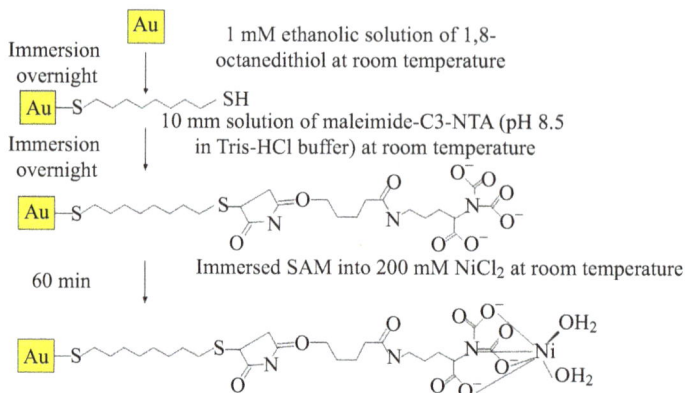

Figure 6.18   Surface reaction scheme of a protein-based biosensor illustrating attachment of NTA functional groups on to the free sulfhydryl groups of 1,8 octanedithiol SAM on an Au electrode

9. The electrodes were cleaned with DI water and buffer.
10. The sensor was freshly preserved in the buffer.

This group could achieve a high sensitivity of 0.5 mM in the range of 0.5 to 7.5 mM.

Bio-imprinting is a technique for the preparation of extremely selective recognition layers [102–104], which allows developing high-performance biomimetic sensors. Strategies like soft lithography [381], template immobilization [382], emulsion polymerization [339], surface grafting [383], and nano-imprint lithography (NIL) [384] are utilized to produce recognition cavities fitted with target (template) compounds from the point of their shape, position, size, and orientation [385]. MIP, also known as artificial antibodies, can be used instead of relatively unstable antibodies and enzymes, because they can bind to their target (template) molecules selectively [102]. Furthermore, MIPs offer other advantages such as high thermostability, short time of synthesis, and cost-effectiveness [82]. For example, Ertürk *et al.* used microcontact imprinting procedure in [332] for the detection of PSA which is an important biomarker for early diagnosis of prostate cancer (see Table 6.5). They prepared PSA-MIP electrodes in the presence of a functional monomer (methacrylic acid (MAA)) and a cross-linker (ethylene glycol dimethacrylate (EGDMA)) via UV polymerization.

The PSA-MIP capacitive electrodes used in [332] were prepared in three steps (see Figure 6.19):

1. Glass cover slips or protein stamps were prepared in the first step. After cleaning them in HCl, DI water, NaOH, DI water, and EtOH, respectively, they were dried with nitrogen gas and then immersed in APTES in EtOH to introduce the amino groups on the surface. After rinsing the surface with EtOH, for the activation of amino groups, the cover slips were immersed in Glu solution in phosphate buffer. After rinsing the electrodes with phosphate buffer and

Table 6.5   Comparison of different CMOS capacitive sensors used for other selective sensing techniques

Transducer material	Capacitive transducer	CMOS/ non-CMOS	Linker	BRE	Application	Ref.
Au	IDE	CMOS	A SAM of octanedithiol+melemide-NTA-Ni^{2+}	GLK	Glucose biosensing	[65]
Au	IDE	Non-CMOS	SAM of MPA, EDC, NHS, MB	Abs, anti-VEGF aptamer	Detection of VEGF cancer biomarker in serum	[96]
Au with SiO$_2$ nanogap	Nanogap face-to-face electrode pairs	Non-CMOS	Thiol	ssDNA aptamers	Detection of human alpha thrombin	[377]
Au and ITO	Nanogap face-to-face electrode pairs	Non-CMOS	A SAM of DDT	Aptamer	Thrombin sensing	[378]
Au	IDE	Non-CMOS	C-terminal cysteine	Anti-Her4 affimer	Detection of a protein tumor biomarker (Her4)	[100]
Au	Floating electrode	Non-CMOS	APTES and Glu* N-hydroxymethyl acrylamide and PEG-400 DMA**	*E. coli* phage MIP	Detection of *E. coli*	[379]
Au	Floating electrode	Non-CMOS	APTES and Glu* MAA and EGDMA**	PSA-MIP	Detection of PSA for early detection of prostate cancer	[332]
Au	Floating electrode	Non-CMOS	APTMS and Glu* MBAm, EGDMA, TRIM, NAPMA, AAm, PETMP and INIFERTER**	Trypsin-nano-MIP	Detection of trypsin	[102]
			GOPTS and EIPA* MBAm, EGDMA, TRIM, NAPMA, AAm, PETMP and INIFERTER**	THC MIP	Detection of THC	
Au	Floating electrode	Non-CMOS	Polytyramine, Glu	*Salmonella*-specific M13 bacteriophage	Detection of *salmonella* spp.	[98]
Au	Floating electrode	Non-CMOS	Polytyramine, Na-CMC, EDC, NHS, APBA	Tetrahedral boronate anion	Saccharide detection	[380]
Al/SiO$_2$	IDE	CMOS	APTES, EDC	CPBA	Detection of neurotransmitter DA	[115]

*For immobilization of template on glass.

**Monomer and cross-linker on electrode surface; PEG-400 DMA, poly(ethylene glycol) 400-dimethacrylate; MBAm, N,N'-methylene-bis-acrylamide; TRIM, trimethylolpropane trimethacrylate; NAPMA, N-(3-aminopropyl)methacrylamide hydrochloride; AAm, acrylamide; PETMP, pentaerythritol tetrakis(3-mercaptopropionate); INIFERTER: N,N-diethyl-dithiocarbamic acid benzyl ester.

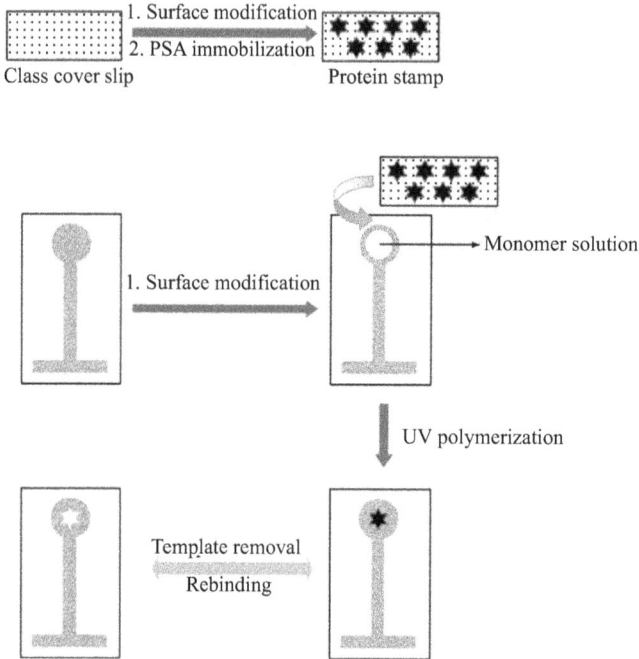

Figure 6.19   The microcontact imprinting of PSA onto the capacitive biosensor

drying them with nitrogen gas, the cover slips were immersed in PSA solution to immobilize PSA onto the surface. Next, they were washed with phosphate buffer and dried with pure nitrogen gas.

2. In the second step, Au electrodes were cleaned with EtOH, DI water, ACE, DI water, and acidic Piranha solution, respectively. After plasma cleaning, electropolymerization of tyramine was performed by CV in ethanolic solution of tyramine to introduce free primary amino groups on the surface of the electrode. Afterward, the electrodes were immersed in a solution containing acryloyl chloride and triethylamine (in TOL). Therefore, the amide groups were created on the surface which could interact with the protein stamp.

3. In the third step, α-α'-azobisisobutyronitrile (AIBN) as the initiator was added into a monomer solution containing MAA as the functional monomer and EGDMA as the cross-linker. The solution was pipetted onto the electrode surface and the protein stamp was put into contact with the electrode. After the polymerization under UV light, the protein stamp was removed from the surface. To block any pinholes in the coating polymer layer, the electrodes were immersed in DDT in EtOH.

In another work, Canfarotta *et al.* [102] utilized MIP NPs as a receptor for capacitive sensing of a protein, trypsin, and a small molecule, tetrahydrocannabinol (THC). Nano-MIPs were produced by the solid-phase method and their high specificity helped to improve LoD and the selectivity of the sensor.

A phage, also known as a bacteriophage, is a virus that infects host-bacteria and can capture and digest the target bacteria selectively which results in releasing ion in the outer medium. So, it can be used as a BRE to detect specific bacteria [98,99]. Robustness and low cost are some of their advantages and it is easy to produce them [82]. For example, Ertürk *et al.* used molecular imprinting in [379] for bacterial detection (see Table 6.5). Their bacteriophage-imprinted biosensor was capable of detecting *E. coli* with an LoD value of 100 cfu mL^{-1}. Niyomdecha *et al.* [98] presented a capacitive flow injection system for the detection of *Salmonella* species (spp.) in which bacteriophage was immobilized on a poly-tyramine/Au electrode utilizing a Glu as a cross-linker.

The detection of saccharide is another application of capacitive sensors. For example, Bergdahl *et al.* [380] used aminophenylboronic acid (APBA) immobilized on an Au electrode to recognize different types of saccharides (see Figure 6.20(a)). Au electrodes were cleaned with various solutions in the ultrasonic cleaner. After plasma cleaning, electro-polymerization of tyramine introduced free primary amino groups on the surface via the deposition of poly-tyramine. Then,

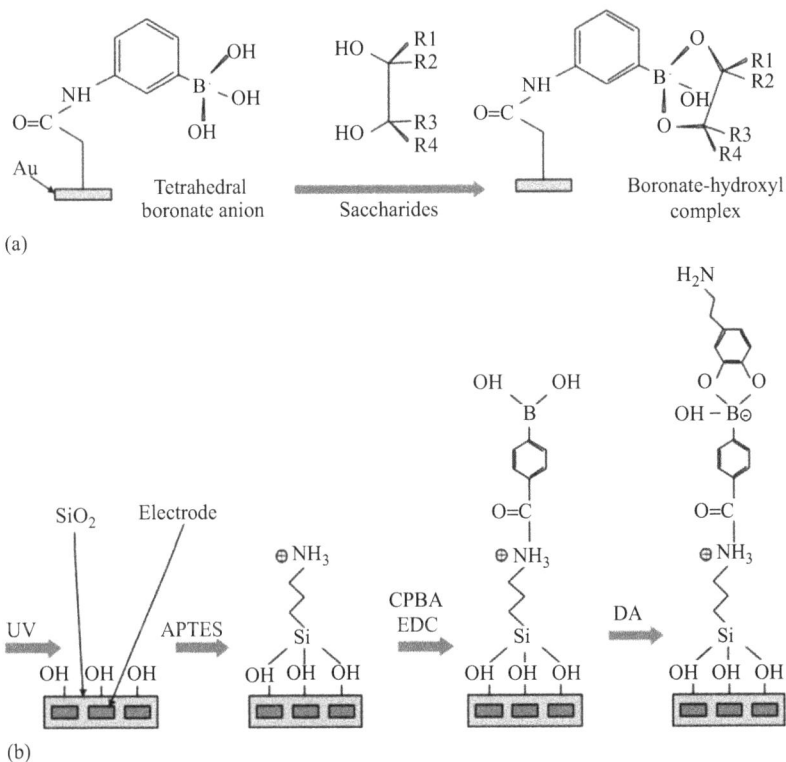

Figure 6.20 *Tetrahedral boronate anion formation on the Au electrode after APBA modification and interaction of them with saccharides; (b) capacitive detection of the neurotransmitter DA*

sodium carboxymethyl cellulose (Na-CMC) was dissolved in sodium phosphate buffer to the desired concentration. Poly-tyramine-coated electrodes were immersed in this solution to introduce carboxyl groups on the surface of the electrode. Then, the electrodes were immersed in EDC and NHS in phosphate buffer to activate carboxyl groups. NHS-activated carboxylic groups were then allowed to bind with the primary amino groups of APBA in phosphate buffer. By the deprotonation of the activated carboxyl groups after APBA treatment, tetrahedral boronate anion, interacting with the monosaccharides was introduced on the surface to form boronate-hydroxyl complexes. Then, the APBA-modified electrode was treated with DDT in EtOH to provide proper insulation of the Au electrodes.

In the CMOS capacitive biosensor presented by Lu *et al.* [115], after immobilization of 4-carboxyphenylboronic acid (CPBA), the positive charges produced from the protonation of boronate and amino group increase the electrode–solution capacitance and the negative charges produced by binding of the neurotransmitter DA and CPBA molecules reduce this capacitance (see Table 6.5 and Figure 6.20 (b)). The surface of the IDEs covered by the CMOS SiO_2 thin film was first treated by a UV-ozone dry stripper and immersed in APTES. Under the UV light, the ozone was decomposed to oxygen molecules and atoms which can clean and activate the surface by introducing –OH to the oxide surface. The amino group introduced by APTES was protonated to $-NH_3^+$, which was coupled to CPBA under the activation of EDC in the 2-(N-morpholino)ethanesulfonic acid (MES) buffer.

6.4   Summary

This chapter gave an overview of the applications of both CMOS and non-CMOS capacitive sensors for life science applications including chemical sensing (like gas sensors), cell monitoring and toxicity test (like bacteria growth monitoring and drug test), and selective sensing based on various BREs such as nucleic acids (like ssDNA), antibodies, and other BREs (like artificial proteins and aptamers).

According to the discussions throughout this chapter, capacitive biosensors have been generating considerable interest in terms of medical, biological, and chemical applications by detecting thousand types of organic solvents, bacteria, viruses, DNA fragments, antigens, and alike. These high-demand applications explain the need for further research for the development of advanced diagnostic devices with specific BREs for each application. Especially, the systems designed for high-throughput screening of multiple targets require reliable bio-interface layer to avoid cross-reactivity and provide multiplexing in microarray chips. There are also some capacitive biosensors which are designed by using some processes and technologies that are not compatible with CMOS processes. More investigations on such sensors and adapting them to CMOS technology can open a new avenue to develop more various CMOS-based capacitive biosensors with novel structures and functionalities.

Chapter 7

Current technology and future work

Electrochemical impedance spectroscopy (EIS) as a well-established approach is the most conventional technique used for capacitive measurement of sensing electrodes [386,387]. An array of sensing electrodes is connected to such a measurement device. In handheld EIS systems, an array of sensing sites can be employed for several point-of-care applications like bacteria detection [388] or blood analysis [389]. However, such portable systems are undergoing a revolution and there is a widespread interest to embed the whole system in a single chip. In this regard, a complementary metal-oxide semiconductor (CMOS)-based capacitive sensing laboratory-on-chip (LoC) is a candidate that can be implemented in a syringe style washable package to direct the analyte toward the sensing sites [76].

7.1 Conventional impedimetric and capacitive measurement systems

Frequency response analysis (FRA) is based on applying a frequency-dependent voltage or current test signal and measurement of the points on the frequency response of an impedance function or transfer function [390]. Usually, a sinusoidal voltage, $v(t)$, with different frequencies, is employed to perturb the signal between a working and a reference electrode and the resulting current is measured. Assumed the electrochemical cell is a linear system, the output current, $i(t)$, will be a sinusoid with the same frequency but shifted in phase. Resistance and capacitance can be determined by the measured impedance, $Z(t)$:

$$Z(t) = \frac{v(t)}{i(t)} = \frac{V_0 \sin(2\pi f t)}{I_0 \sin(2\pi f t + \varphi)} \tag{7.1}$$

where V_0 and I_0 stand for the maximum amplitude of voltage and current signals, respectively. f, t, and φ identify the applied frequency, the time, and the phase shift between $v(t)$ and $i(t)$. Impedance is expressed as a complex number, but the most popular formats for evaluation of the data are Nyquist and Bode plots [184]. This technique has been used for many applications like detecting viruses, DNA, lactate, and glucose [391,392].

In many applications [192,363,393,394], the potential pulse and current pulse techniques are used which are comparable with the principle introduced in Section 4.1.7. In comparison to EIS, these two methods do not need to measure

amplitude and phase shift of the impedance under measurement and both components of sensing impedance can be directly measured from the output current or
voltage of the sensor [118]. After applying a potential pulse with a known amplitude (e.g. 50 mV), the current response is sampled at an identified frequency (e.g.
50 kHz). The current decay will follow (7.2):

$$i(t) = \frac{V}{R} \exp\left(-\frac{t}{R.C_S}\right) \tag{7.2}$$

where $i(t)$ is the current response as a function of time, V identifies the amplitude of
the potential pulse, R is the dynamic resistance of the recognition layer, and C_S
stands for the measured capacitance at the working electrode–solution interface
[192]. This technique works based on an assumption that the interface could be
modeled as a resistor–capacitor circuit in series. R and C_S can be obtained by linear
least-squares fitting for the curve of $i(t)$ versus time. This technique has poor
baseline stability and is sensitive to external disturbances. Furthermore, it might be
required to replace the working electrode with a new one due to the damages
arising from sharp potential pulse [395].

There are also some nonconventional impedimetric measurement systems that
take advantage of lock-in amplifiers to synthesize a sinusoidal signal on the sensing
electrodes. Such a setup is reported for bacteria growth monitoring in [279]. The
voltage amplitude, the frequency, and the measurement time period are selected by
a user interface software (UIS). The electrodes are selected by using a multiplexer
and the UIS. The electrode array is provided from Applied Physics Inc.

As an alternative, a pulse current can be used for capacitance biosensing
[102,332]. Figure 7.1(a) shows the schematic of a measurement system useful for
current pulse method. This system is composed of a current source, an electrochemical flow cell connected to three electrodes (working, auxiliary, and reference

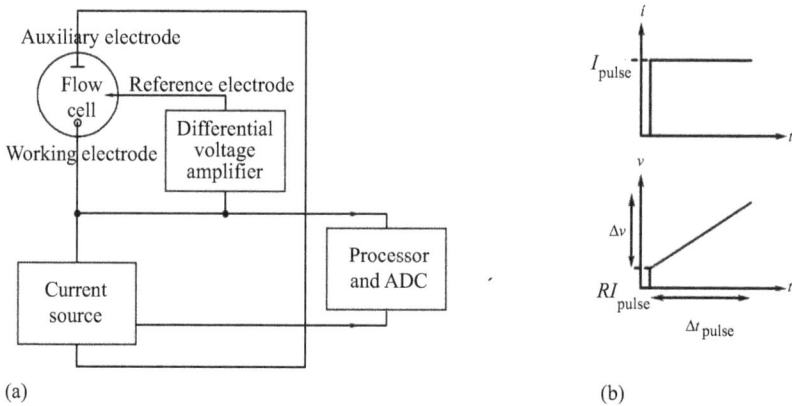

(a)   (b)

Figure 7.1   (a) A measurement system useful for current pulse method; (b) the
supplied current to the RC circuit model and the response voltage

electrodes), a differential voltage amplifier, a processor, and an analog-to-digital converter (ADC). An automated flow injection system is also utilized to perform current pulse measurement [395]. When the resistor-capacitor (RC) model sample is charged by a constant current, I_{pulse}, the potential response increases linearly with time (t) (see Figure 7.1(b)). The corresponding resistance, R, is obtained by Ohm law, $v(t = 0) = RI_{pulse}$, where $v(t = 0)$ is the voltage at $t = 0$. Then, the equivalent capacitance can be measured using the slope of the potential response ($\Delta v / \Delta t_{pulse}$) through the equation $C = (I_{pulse}\Delta t_{pulse})/\Delta v$, where Δt_{pulse} is the current pulse length in time (as shown in Figure 7.1 (b)) [396].

Potentiostat/galvanostat, I–V meter, and LCR meter are the most common apparatus useful for these measurements in laboratories. Table 7.1 lists some commercialized impedance measurement systems, most of which are frequency response analyzers. Figure 7.2(a), (b), and (c) shows an FRA made by Novocontrol Technologies, a potentiostat/galvanostat/zero-resistance ammeter (ZRA) made by Gamry Instruments, and an electric cell-substrate impedance sensor (ECIS) made by Applied Biophysics, respectively. The system developed by Applied Biophysics is capable of continuous-time multi-sample analysis.

Table 7.1 Commercialized impedance measurement systems

Company	Model
Novocontrol	Alpha-A
Gamry Instruments	Reference 600+
Clarke-Hess Communications Research	2505
Solartron Analytical	1255A
Core Technology Group	SA Series 01
Venable Instruments	6300
Applied Biophysics	ECIS ZΘ

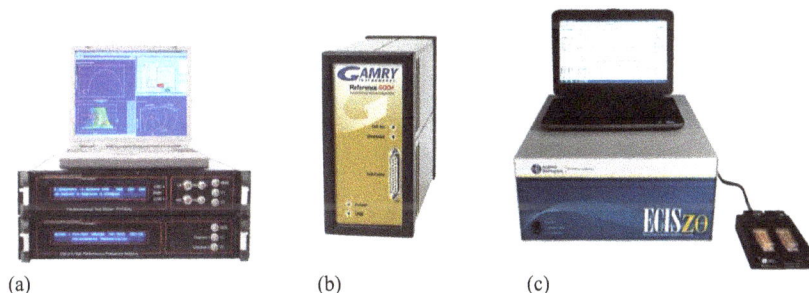

(a) (b) (c)

Figure 7.2 (a) Alpha-A analyzer made by Novocontrol Technologies [397]; (b) Reference 600+ potentiostat/galvanostat/ZRA made by Gamry Instruments [398]; (c) ECIS model ZΘ made by Applied Biophysics [399]

7.2   Handheld impedance or capacitance measurement systems

A handheld impedance or capacitance measurement system includes a disposable electrode (like carbon screen-printed electrode) fixed on the system as well as an impedance reader card such as AD5933 (Figure 7.3(a)), DIGIL 410-378 (Figure 7.3 (b)), impedance analyzer model 16777k/LCR meter (Figure 7.3(c)), and PalmSens3 (Figure 7.3(d)) made by Analog Devices, DIGILENT, SinePhase, and PalmSens, respectively. For example, AD5933 (shown in Figure 7.3 (a)) is a high-precision impedance converter system that combines an ADC and on-board frequency generator to excite external complex impedance with a known frequency. A large array of electrodes can be read out by the integration of the impedance reader card with other microelectronic circuitry. AD5933 has been used for various applications like blood glucose sensing [400], measurement of cholesterol concentration [401], wearable hydration monitoring [402], and kidney healthcare [403].

7.3   Toward fully integrated capacitive sensing LoC

An on-chip capacitive biosensing LoC is composed of biofunctionalized sensing electrodes, a microfluidic structure, and an ASIC [76]. Micofluidic structures are commercialized by several microfluidic manufacturers like Micralyn, Acamp, and FlowJEM companies and several microelectronic integrated circuit manufacturers

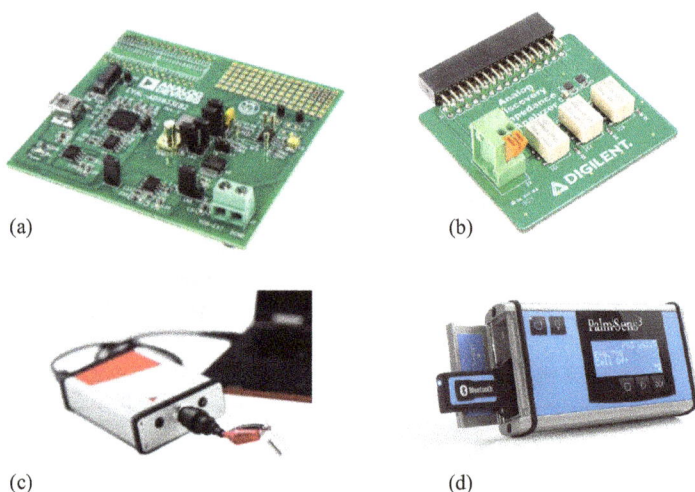

Figure 7.3   (a) *AD5933 impedance converter/analyzer [404]; (b) DIGIL 410-378 impedance analyzer for analog discovery [405]; (c) SinePhase impedance analyzer 16777k/LCR meter [406]; (d) PalmSens3 (a USB and battery-powered handheld potentiostat/galvanostat) [407]*

like Austria Mikro System (AMS) and Taiwan Semiconductor Manufacturing Company (TSMC) fabricate integrated circuits. Since the introduction of the idea of micro-total analysis system (μTAS) in the early 1990s to replace the laboratory liquid management tasks with microfluidics and miniaturize an entire laboratory to one chip, microelectromechanical systems (MEMS)/microelectronics and micro-fluidics communities have dedicated many efforts to integrate on-chip actuator, sensor, and signal processing circuits with the microfluidics in LoC devices [408]. However, many promising research prototypes of hybrid CMOS-microfluidic devices with active circuits for LoC applications could not be commercialized, except a few successful examples for specific applications, like Ion Torrent™ (Life Technologies, Thermo Fisher Scientific) DNA sequencer [1] and i-STAT 1 (Abbott Point of Care) blood analyzer system [409].

Although various ICs can be fabricated by the standard CMOS technology for a different applications such as capacitive and impedance sensors, several steps are required to achieve a hybrid technology for emerging LoC applications in the future. Current technologies allow for the integration of microfluidic structure and CMOS chip through nonconventional techniques like the IC-microfluidic packaging methods described in Chapter 5. Several modifications and further efforts are required to develop highly controllable microfluidic packaging process and reliable, repeatable, highly sensitive, and miniaturized capacitive sensors for future use. Some important issues are discussed in the following subsections [76].

7.3.1 Capacitance characterization

One of the important challenges in CMOS-based capacitive sensing LoCs is the characterization. The conventional computer-aided design (CAD) tools are appropriate for the capacitance characterization of the microelectronic devices implemented in deep CMOS chip like calculating the coupling capacitance between two conductors, not measuring the parasitic capacitance between the substrate and the topmost metal layer specifically with regard to complicated configurations. Estimating the parasitic capacitance due to the sensing electrode above the CMOS chip is the minimum requirement in designing capacitive biosensors. Moreover, the principle of the standard capacitance characterization is suitable for the capacitances between the grounded substrate and each conductor and between each two conductors. However, a grounded metal layer made from some materials like indium tin oxide might cover the structure of LoCs [76].

7.3.2 Multiphysics modeling of LoC-based capacitive biosensors

An important challenge is electrical modeling of biological/chemical samples which is a circuit network of some capacitors and resistors. This network topology and the dynamic range of each capacitor and resistor can be affected by the geometry and dimensions of electrodes, the concentration of nonspecific molecules or cells and the target molecules or cells, the topmost metal or passivation layers. Although the CAD tools implemented for designing microelectronic circuitries can provide very accurate models for CMOS transistors and other microelectronic elements, designing CMOS-based capacitive biosensor needs to exact models of biological/chemical samples and/or biological recognition elements (BREs) formed on top of the chip [76].

Generally, the multidisciplinary concept and complexity of LoC makes its modeling more difficult. Currently, there is no software that can simulate the LoC behavior completely for all configurations and applications. In spite of the development of very accurate multiphysics simulation software like COMSOL and ANSYS, these software are still incapable of efficient combination of electrical, fluidic, and biological/chemical effects. A vast amount of knowledge about physics, electricity, fluidic, injected liquid characteristics, studied bio-particles, and BRE properties is required to implement a multiphysics model. Moreover, the model should be reconfigurable for different applications. So, new modified and extended CAD tools are required to design LoC-based capacitive biosensors [181].

7.3.3 Generic LoC-based capacitive biosensor

Although several papers have been published on the design of generic MEMS-based capacitive sensors like accelerometers, designing LoC-based capacitive biosensors requires different design strategy as aforementioned in Chapter 4. Different features like calibration and offset cancellation module, high precision, and a large array of electrodes and their signal processing modules should be taken into consideration for designing an optimized generic system [76].

Since CMOS-based LoC is a small-scale industry, the biochemical processes in LoCs might be more complex than in conventional lab instruments due to some effects like surface roughness, capillary forces, and chemical interactions of construction materials on the processes which might be more dominant than the parameters under measurement. Miniaturization may lead to low signal-to-noise ratios. Furthermore, a long-lasting and reliable calibration integrated within LoC for biological/chemical analysis is a great challenge that causes the need to stable standardization solutions and consequently high production cost. Moreover, in medical applications, strict regulations are required to provide a reliable device [181].

7.3.4 Cleaning procedure

In standard biological laboratories, using disposable electrodes prevents transmission of infectious agents which are sterilized and ready-to-use without the need for further cleaning. But an integrated and non-disposable integrated capacitive sensor that should be employed for several measurements should be cleaned before and after each use. A cleaning procedure is required to be designed for the interaction of biological/chemical analyte with the sensor and might not be appropriate for every experiment [76].

7.3.5 Packaging

As aforementioned in Section 5.1, there are several challenges in the packaging of microelectronic devices and microfluidics. A microelectronic device only needs some I/O electrical signals, while in CMOS-based LoCs, some fluidic inlets and outlets are required in addition to electrical signals. For example, power supply and the ground are electrical input and output, respectively, while the connections to analyte and waste reservoir play the role of fluidic inlet and outlet, respectively. Standard techniques like wire bonding and flip-chip are utilized for electrical packaging of CMOS chips while the nonconventional techniques like direct-write fabrication process are required for microfluidic packaging. A standard micro-fluidic packaging is required for CMOS-based LoCs that would be repeatable, CMOS-compatible, and reliable with hermetic bonding [76].

7.4 Summary

This chapter gave an overview of the past, present, and future technologies for capacitance sensing in biological and chemical analytes. After a brief review of the conventional and handheld impedimetric and capacitive measurement systems like potentiostat/galvanostat, I–V meter, and LCR meter in this chapter, different challenges toward full integration of LoC-based capacitive biosensors including capacitance characterization, cleaning process, packaging as well as multiphysics modeling, and generic designing of these biosensors were discussed.

According to the issues outlined throughout this book and specifically this chapter, although various techniques are being developed for CMOS capacitive biosensors, LoC-based capacitive biosensors are still in the early stage of developments and further studies are needed to achieve standard technologies for low-cost mass production of generic capacitive biosensors useful for a wide range of biological/chemical applications.

Appendix A
Simulation of electrodes

Herein, a polymer (PDMS)-coated IDE covered with fibroblast cells is simulated by Sonnet software as an example. In this regard, the specifications of the passivation layers and the sample of bio-particles in contact with the electrodes should be determined.

Figure A.1 shows an IDE and its parasitic capacitors. This IDE consists of two combs called E_1 and E_2. The parasitic capacitances between E_1 and the substrate, between E_2 and the substrate, and between E_1 and E_2 are depicted with C_{1s}, C_{2s}, and C_{12}, respectively. Since E_1 is grounded, the total capacitance seen at the node of the reference electrode is equal to $C_R = C_{1s} + C_{12}$. If an aqueous sample is applied to the electrode, the corresponding capacitance of the sample should be added to C_R.

Here, fibroblast cells with a relative permittivity equal to 1 and an electrical conductivity of 5 S/m are used to examine the effect of the cell confluence percentage on the measured capacitance [410]. The simulation results measured at 100 kHz for two different sizes are reported in Table A.1. As seen in the graphs depicted in Table A.1, the more the cell confluence percentage, the larger the sensed capacitance.

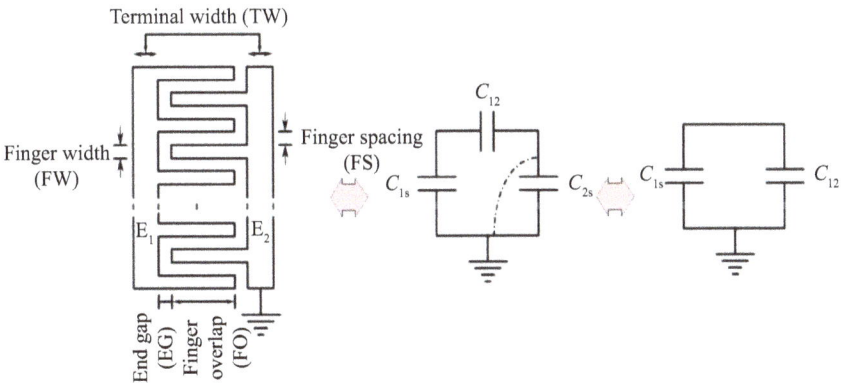

Figure A.1   IDEs and their parasitic capacitance

Table A.1   The simulation results of IDE

Characteristics		Sensed capacitance versus fibroblast cell confluence
FW	5 μm	
S	5 μm	
FO	5 μm	
EG	5 m	
TW	5 μm	
Finger pairs	2	

FW	10 μm	
FS	10 μm	
FO	10 μm	
EG	10 μm	
TW	10 μm	
Finger pairs	2	

Appendix B

Fabrication techniques

B.1 Etching

Etching is a technique to chemically remove layers from the surface of a substrate in microfabrication. Selectivity is one of the important criteria in this process which allows for etching one material without etching another [411]. There are two types of etching: (1) wet etching and (2) dry etching.

B.1.1 Wet etching

In wet etching, liquid-phase, usually acidic etchants, are used. Most wet-etching operations are isotropic and etch in all directions of the workpiece at the same rate. Isotropy causes undercuts beneath the mask material and erosion of the substrate which restricts the resolution of geometric features in the substrate. Furthermore, the solutions have to be refreshed periodically. The following conditions are required for effective etching [411]:

1. transport of the etchant to the surface;
2. a chemical reaction;
3. transport of the chemical reaction products away from the surface;
4. ability to rapid etch-stop in order to achieve superior pattern transfer, commonly by employing an underlying layer with high selectivity.

B.1.2 Dry etching

In dry etching, chemical reactants are used in a low-pressure system, and a plasma or discharge in areas of high magnetic and electric fields is involved. Typically, a masked pattern of material is removed by exposing it to the bombardment of ions, electrons, photons, or highly reactive molecules formed by the dissociation of suitable gases.

On the contrary to wet-etching process, this type of etching can provide a high degree of directionality and consequently highly anisotropic etching profiles. Furthermore, only small amounts of the reactant gases are required for dry etching.

Wet etching provides a high etching rate and selectivity, but it is generally isotropic and the etchant chemicals might remove substrate material. In the cases where high anisotropy is necessary, dry etching techniques (using only physical removal or both chemical reactions and physical removal) are preferred. There are several dry-etching techniques including sputter etching, reactive plasma etching, physical–chemical etching, and cryogenic dry etching.

B.1.2.1   Plasma etching (RIE)

Reactive ion etch (RIE) is a type of dry etching to remove deposited material on wafers by chemically reactive plasma that is generated in a vacuum by an electromagnetic field in between two electrodes. By applying RF power to the lower electrode, an electric field is created, which accelerates the ions from the plasma toward the surface of the sample (see Figure B.1). Then, the surface is etched due to the bombardment and reaction with ions [412].

This technique has the advantages of high aspect ratio and fine resolution. Many process parameters like RF power, gas flows, and pressure affect the conditions of an RIE system. An extension and modified version of RIE called deep reactive ion etching (DRIE) can be used to excavate deep features [412].

B.1.3   Electrochemical etching

In addition to chemical techniques, an external electrical current can be used for etching the sample substrate made of semiconductors and even other materials like metals. Electrochemical etching is an anodic dissolution process employing chemical and electrical reactions that happens when the current flows between the conductors immersed in an electrolyte.

In this technique, a potentiostat or galvanostat is used to generate the electrical charge required for etching and controls the etching current or potential as desired and consequently the profiles of the features to be etched and the etching process. In electrochemical etching, the sample substrate to be etched is located in an electrochemical cell, like Figure B.2, in such a way that its surface is exposed to an electrolyte containing a liquid and the ions conducting electrical charge and dissolving the substrate upon applying the current. The back of the substrate is connected to the anode of the power supply. Type of semiconductor, the type of doping (n or p), and doping density of the semiconductor can affect the etching process.

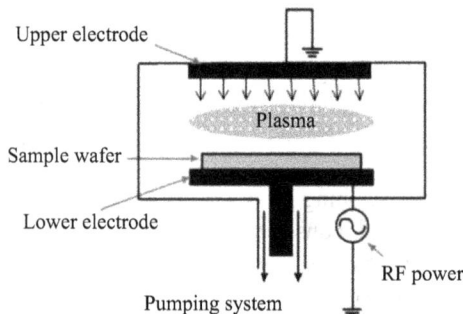

Figure B.1   An RIE system

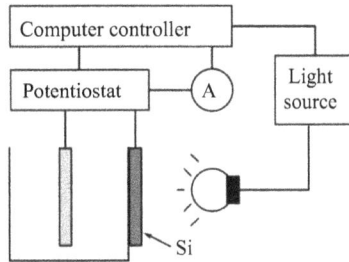

Figure B.2   The setup for photoelectrochemical of n-type semiconductors (like Si)

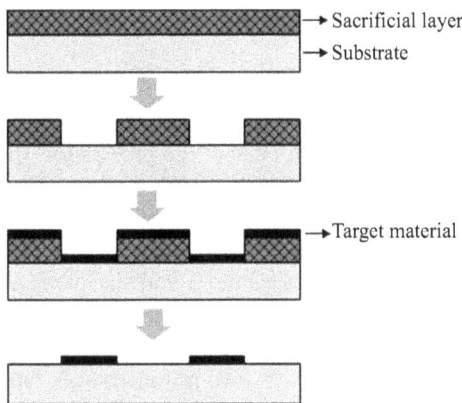

Figure B.3   Steps of liftoff process (from top to bottom)

To anodically dissolve the n-doped semiconductors (that does not have holes) in an acidic electrolyte containing negative ions, the illumination of the sample from the back can be used to produce holes travel to the opposite surface of the substrate. Figure B.2 shows a setup for photoelectrochemical of n-type semiconductors. The light source is not required for etching p-type semiconductors.

B.2   Liftoff process

This technique is metallization with sacrificial resist and a cheap alternative to etching in the applications where the effects of a direct etching of a material on the layer below are undesirable or etching tool with the suitable gases is not accessible, but a slower turnaround time is permitted.

As shown in Figure B.3, in this technique, after deposition of a sacrificial stencil layer (or resist), an inverse pattern is created on it. After lithography, the target material is deposited on the resist pattern. In the next step, the sacrificial layer is washed out together with the deposited target material on its surface which is not in contact with substrate. So, the final patterns of the target material are remained [412].

B.3   Chemical vapor deposition

Chemical vapor deposition (CVD) techniques are various vacuum deposition methods by chemical reactions occurred at the surface of a substrate exposed to one or more volatile reactants. Source gas molecules adsorb and react on surface to deposit a film. The frequently produced by-products are desorbed and pumped away as depicted in Figure B.4 [412].

B.3.1   Thermal CVD

Thermal CVD is the process of gas-phase heating and reducing or decomposition of a chemical vapor precursor species containing the material to be deposited by the high temperature to generate radical species that can be deposited on a substrate.

B.3.2   Plasma-enhanced CVD

Plasma-enhanced CVD (PECVD) is a low-temperature CVD process employing plasma to enhance chemical reaction rates of the precursors. The plasma of gaseous reactants is generally generated by AC (RF) or DC discharge between two electrodes, the space between which is filled with gaseous reactants. A simple parallel-plate diode reactor like Figure B.5 can be used for PECVD system. The sample wafer is located on the heater plate (heated bottom electrode) and the gases are introduced to the sample from the top and pumped away around the bottom electrode [412].

Temperature, flow rate, flow rate ratio, pressure, and the RF power are the main variables. In advanced PECVD reactors, two power sources formed by RF power to both electrodes provide different frequencies, power levels, and duty cycles [412].

B.3.3   Atomic layer deposition

Atomic layer deposition (ALD) techniques are subclasses of CVD techniques and a surface-controlled layer-by-layer process in which the thickness of the thin film deposited on the substrate surface is controllable up to the order of one atomic monolayer [412].

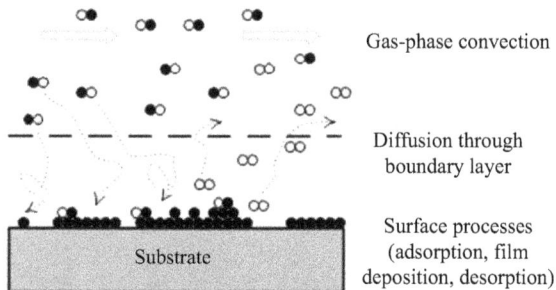

Figure B.4   Chemical vapor deposition

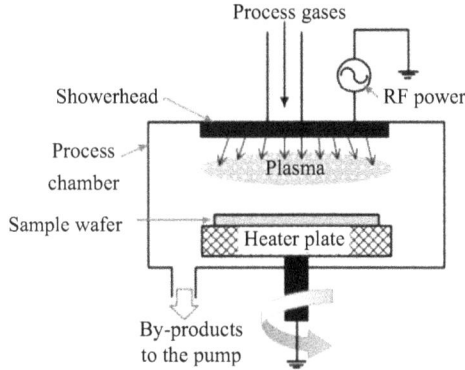

Figure B.5   PECVD system

In ALD, the film is deposited through a repeated exposure to separate chemical reactants. Sticking a single layer of the gas to the surface after a chemical reaction results in surface passivation. But, before deposition of the next layer, the passivating atoms must be removed in different ways such as thermal spikes and chemical reactions. ALD process is generally as follows (see Figure B.6) [412]:

1. Reactant A is applied and reacts with the surface. As a result, chemical bonds are created between the surface atoms and precursor gas molecules until all possible reaction sites are occupied and no more reactions can occur.
2. Excess reactant and all unreacted precursor molecules or reaction by-products are purged with an inert carrier gas (like nitrogen).
3. A second precursor or reactant B is applied and reacts with the first reacted layer until the surface saturates.
4. Excess reactant and reaction by-products purged with inert carrier gas.
5. Steps (1) to (4) are repeated until the appropriate thickness is achieved.

Similar coating of all surfaces paves the way for excellent coating over steps. Thermal ALD, plasma-enhanced ALD, metal ALD, photo-assisted ALD, and catalytic SiO_2 ALD are different types of this technique.

B.3.3.1   Plasma-enhanced atomic layer deposition

This type of ALD technique takes advantage of plasma species in chemical reactions whose high reactivity helps to reduce the deposition temperature. In this technique, passivating atoms are removed without a thermal spike.

B.4   Electrochemical deposition

B.4.1   Electroplating/galvanic deposition

In electroplating process, a wafer or conductive surface as a cathode is coated by another metal through electrodeposition process in metal ion-containing electrolyte

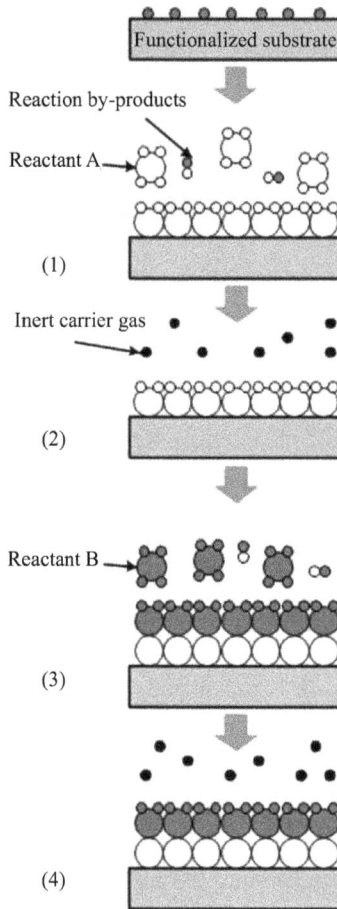

Figure B.6   A general ALD process

solution. The counter electrode is either made of the metal to be deposited, or passive [412].

Three transport processes involve in electrochemical deposition (ECD): (1) diffusion at the electrodes because of local depletion of the reactant; (2) migration in the electrolyte; and (3) convective transport in the plating bath affected by factors like heating, recirculation, and stirring [412].

In order to fabricate elaborate microstructures, electroplating can be conducted onto a photoresist pattern. A lithographic galvanic process includes, in turn, seed layer deposition and lithography, plating, resist stripping, as well as seed layer removal [412].

B.4.2   Electroless deposition

Electroless deposition is an autocatalytic process which depends on a reduction reaction in a plating bath containing a reducing agent and metal salts. In this technique, a suitable substrate surface is exposed to a catalyst as a reducing agent to start the reduction reaction which then proceeds locally. This paves the way for selective deposition. Gold (see page 47), copper, and nickel are the usual metals to be deposited by this technique. In comparison to electroplating, electroless deposition does not require to make electrical contacts to the substrate [412].

B.5   Physical vapor deposition

Physical vapor deposition (PVD) techniques are various vacuum deposition methods in which a target material is converted from a condensed phase to a vapor phase in a vacuum chamber and then layers of molecules or atoms from the vapor phase are deposited onto a solid substrate in the form of a condensed phase thin film.

Atoms can be ejected from the target material by different approaches such as ion bombardment, laser beam bombardment (laser ablation), electron beam (E-beam) heating, or resistive heating. Sputtering and evaporation are the most common PVD techniques [412].

B.5.1   Sputtering

In this technique, the target material is bombarded by ionized gas which is usually an inert gas like argon (Ar^+). In this technique, the substrate temperature is low and even materials with very high melting points can be sputtered (see Figure B.7).

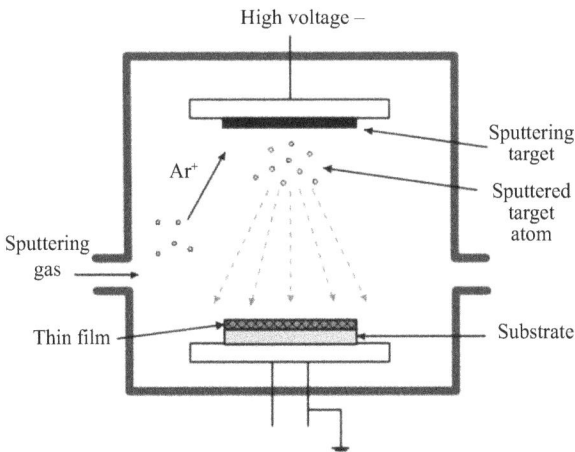

Figure B.7   Sputtering apparatus

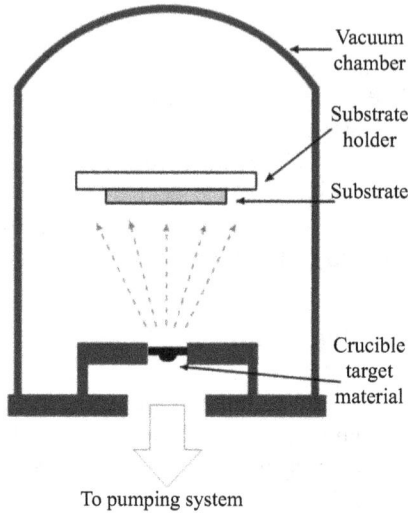

Figure B.8   A typical thermal or e-beam evaporation system

B.5.2   Evaporation

In this technique, a heated metal vaporizes and the evaporated atoms of the hot metal with high vapor pressures will be transported to the substrate in a high vacuum (see Figure B.8).

B.5.2.1   Thermal evaporation

In thermal evaporation, the target material is vaporized by using thermal energy. The material is heated to a temperature that molecules or atoms of the material are lost from the surface in the ultra-high vacuum. This technique is not suitable for materials with high melting points and is susceptible to contamination.

B.5.2.2   E-beam evaporation

In e-beam evaporation technique, a beam of electrons is used to heat the target material. This technique can provide localized heating and is suitable for evaporation of many materials with very little contamination and high efficiency and has the advantage of morphological and structural control of films.

B.6   Spin coating

This technique generally involves the deposition of uniform thin films (up to nanoscale thicknesses) onto the surface of flat substrates by using centrifugal force.

In this technique, a small amount of a solution of the desired material in a volatile solvent is applied to the center of the substrate which is either not spinning or spinning at low speed. By rotation of the substrate at high speed, the coating

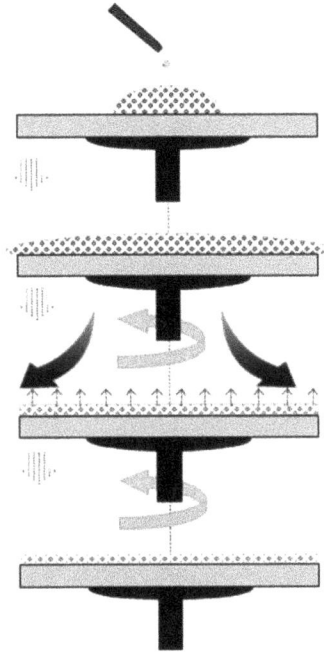

Figure B.9   The steps of spin coating (from top to bottom)

material is spread due to centrifugal force until the fluid spins off the edges of the substrate. Simultaneously, the applied solvent evaporates (Figure B.9).

The thickness of the films is controllable by the angular speed of spinning. The higher the speed, the thinner the film. Spin-coated films can fill recesses and cavities, but spinning is not appropriate if a uniform thickness is required over the topography. This technique is widely utilized for resist spinning and other materials like polymers (spin-on-dielectrics (SODs)) and spin-on-glasses (SOGs) [412].

Appendix C

Simulation of single-ended and fully differential core-CBCM CVCs

The core-CBCM CVC proposed by Ghafar-Zadeh *et al.* [81] shown in Figure 4.29(a) is simulated by using 0.18-μm CMOS technology and power supply voltage of 1.8 V. Figure C.1 illustrates the transient response of the output voltage for different values of capacitance changes (ΔC). Figure C.2 depicts the final output voltage versus the capacitance changes.

The output voltage of the sensor, across the integrating capacitor, changes linearly by the variations of ΔC from 0 to about 4 fF. The sensitivity of this sensor is about 225 mV/fF. According to (4.32), its sensitivity can be raised by increasing the gain of current mirrors (A_I). But, obviously, the limited output voltage swing restricted the dynamic range of this sensor (about 4 fF in this example). Like other voltage-mode circuits, there is a trade-off between the resolution and the IDR of the sensors—the more the sensitivity, the less the IDR.

As aforementioned, fully differential core-CBCM CVC was proposed as an improvement of the single-ended version. Figure 4.37 is one of the solutions which

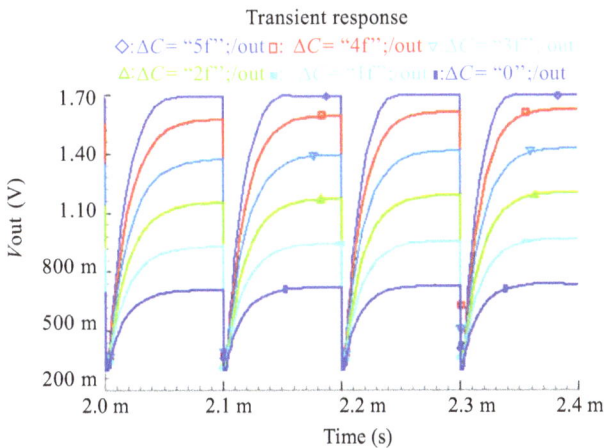

Figure C.1   The transient response of the output voltage of the single-ended core-CBCM CVC

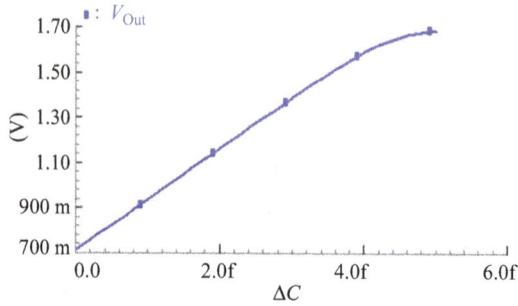

Figure C.2   Output voltage versus input capacitance variations (ΔC) for the single-ended core-CBCM CVC

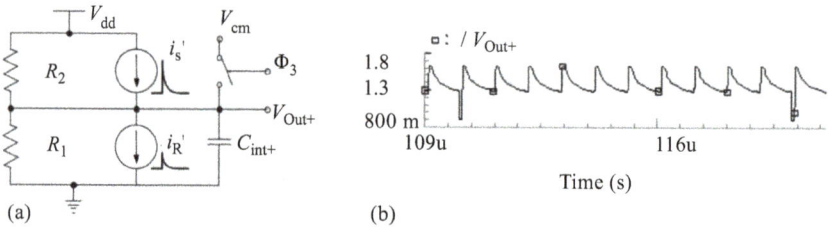

Figure C.3   (a) A simple model for the positive output of the first stage of the CVC; (b) the voltage across C_{int+} for the first stage of the CVC without CMFB circuit for $C_R = 50\ fF$ and $N = 10$ cycles

is simulated here. In Figure 4.37, the sharp exponential currents i_S and i_R of the CBCM core are amplified (i'_S and i'_R, respectively) using current mirrors composed of the transistors M_5–M_{14} and the current differences are injected into C_{int+} and C_{int-} and converted to the voltages V_{Out+} and V_{Out-}.

In the ideal case, after N cycles of sampling time ($f_{int} = f_S/N$), $\Delta i = i_S - i_R$ gradually charges the integrating capacitor C_{int+} to a voltage which is identified by V_f here and it is expected that until the reset clock (Φ_3) is inactive, V_{Out+} remain constant at V_f which is proportional to capacitance changes ΔC. But, in core-CBCM circuits, the currents are not DC. So, the transistors of the current mirrors do not always operate in saturation region and because of the exponential waveform of the current, after a while, the transistors are close to cutoff region. If R_1 and R_2 stand for the output resistances of M_9 and M_7, respectively, in cutoff region, the output voltage V_{Out+} tends to the voltage $V_x = R_1/(R_1+R_2)V_{dd}$. If the values of R_1 and R_2 are in such a way that $V_x << V_f$, while the current mirror transistors are off, the capacitor C_{int+} starts to be discharged and V_{Out+} tends to voltage less that V_f by a time constant equal to R_{total} and C_{total} where R_{total} and C_{total} denote the total resistor and capacitor at this node. As seen in Figure C.3, V_{Out+} will have overshoots in this situation leading to an error in the measurement. In addition, if the value of the final floating CM voltage (V_x) is farther than $V_{dd}/2$, the output voltages of the fully differential circuit might not be symmetric with respect to $V_{dd}/2$.

Thus, the fully differential circuit requires a CMFB like the one shown in Figure 4.37 to maintain the output CM voltage constant at $V_{cm} = V_{dd}/2$. So, the problem of output overshoot would also be relaxed. In Figure 4.37, the transistors M_{15}–M_{19} mirror the currents and inject them into the output nodes of the first stage, and V_{Out+} and V_{Out-}, to tune the output CM voltage of the first stage. A negative feedback helps to modify the output CM voltage to the desired V_{cm}. For example, if the output CM voltage of this stage is less than V_{cm}, the drain current of M_{19} and consequently the drain currents of M_{15}–M_{19} rise and lead to an increase in the output CM voltage of V_{Out+} and V_{Out-}.

In Figure 4.37, an amplifier was used to amplify the differential voltages V_{Out+} and V_{Out-} and to enhance the sensitivity as well as to suppress the CM noises presented in its input nodes.

Figure C.4 illustrates the transient response of the fully differential core-CBCM circuit shown in Figure 4.37 for four cycles of Φ_1 and Φ_2 using 0.18-μm

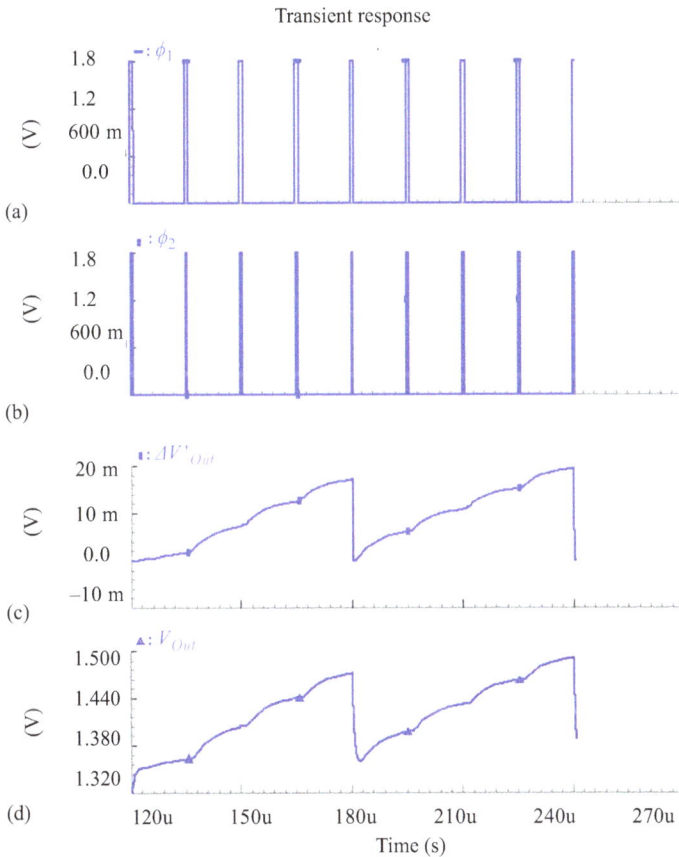

Figure C.4 (a) Φ_1; (b) Φ_2. The transient responses of (c) $\Delta V'_{Out} = (V_{Out+}) - (V_{Out-})$ of the first stage and (d) V_{Out} of the second stage for $N = 4$ cycles

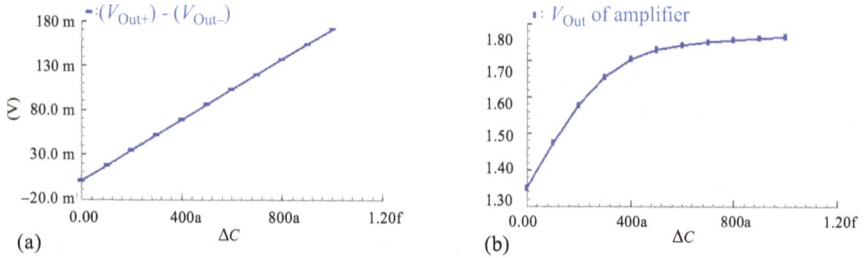

Figure C.5   (a) The differential output voltage of the first stage $((V_{Out+}) - (V_{Out-}))$ of the CVC versus sensing capacitance changes (ΔC_s); (b) the output voltage of the second stage $(V_{Out}$ of amplifier)

CMOS technology. There is still a trade-off between the sensitivity and IDR due to the limited output swing of the amplifier. So, a simple amplifier with a low gain is used here whose slew rate should be high enough to be able to follow the variations of its inputs. Although tuning the floating output voltage of the first stage with the CMFB circuit does not solve the problem of overshoots completely, the residual drop in the output voltage can be minimized by raising the sampling frequency, f_S. An increase in the number of integration cycles, N, can enhance the sensitivity, but the slight downturn of the output voltages may result in some errors. Moreover, although raising the gain of the current mirrors (A_I) can improve the sensitivity, it can decrease the output resistance of the current mirrors and intensify the problem of the output overshoots. Thus, a suitable sensitivity can be achieved by a compromise between the values of these two parameters.

Figure C.4 shows the transient responses of $\Delta V_{Out} = (V_{Out+}) - (V_{Out-})$ of the first stage and V_{Out} of the second stage for $N = 4$ cycles as well as the controlling signals Φ_1 and Φ_2. Figure C.5 illustrates the linear operating ranges of the first and the second stages. As seen in these figures, although the IDR is limited to about 300 aF due to the restricted output swing of the amplifier, the sensitivity of the circuit is increased by the gain of the amplifier. According to Figure C.5(a), the sensitivity of the first stage is about 198 mV/fF while the sensitivity of the sensor after amplification is roughly equal to 1.42 V/fF.

Appendix D
Simulation of a core-CBCM CFC

Current-mode core-CBCM capacitive sensors emerged as alternatives for voltage-mode core-CBCM sensors that suffer from limited IDR (as shown in Appendix C). Figure 4.40(a) shows such a circuit. This core-CBCM CFC is simulated by using 0.18-μm CMOS technology. Figure D.1(a)–(c) illustrates the transient response of this sensor for $\Delta C = 60$ fF including the input current (i_{CCO}), the output frequency

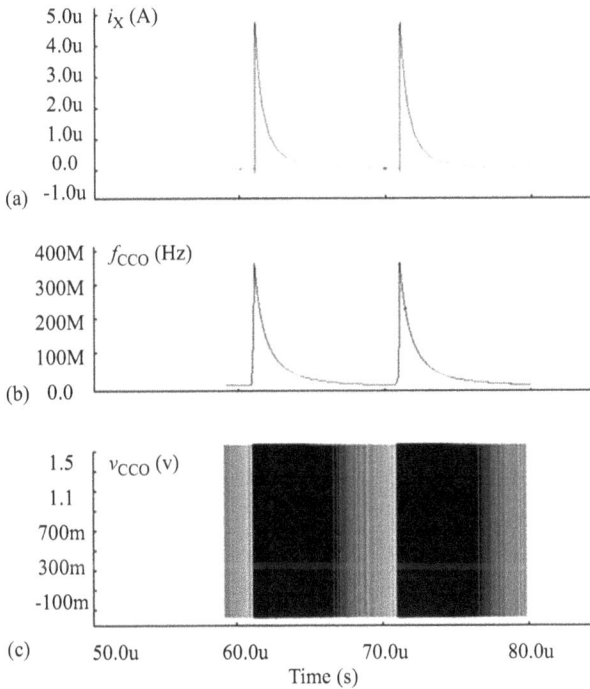

Figure D.1   Transient response of the core-CBCM CFC (a) $i_X = A_I(i_S - i_R)$; (b) the output frequency of the CCO (f_{CCO}); (c) the output voltage of the CCO (v_{CCO})

(f_{CCO}), and the output voltage of the CCO (v_{CCO}), respectively. The high-frequency output pulses are so compact that cannot be shown distinctly during one cycle of Φ_1 or Φ_2. As seen in Figure D.1(a)–(c), the CCO used in this circuit has enough speed to follow the variations of its input current which is the differential current of the core-CBCM ($i_X = A_1(i_S - i_R)$) amplified by the current mirror. It is worth mentioning that the jitter of the CCO averages out at zero and does not affect the sensor response considerably due to the long integration time ($T_{int} = NT_S$).

To bias the used CCO [413] in its linear region, a bias current is also added to the input current of CCO ($i_{CCO} = i_X + I_{bias}$). Figures D.2 and D.3 illustrate the input current (i_{CCO}) and the output frequency (f_{CCO}) of the CCO, respectively, for five different ΔCs versus time. As seen in Figure D.3, for $\Delta C = 70$ fF, the maximum output frequency of the CCO is less than 600 MHz. So, the required counter should be capable of counting the clock pulses with a frequency higher than this (about 1 GHz).

Figures D.4 and D.5 depict the integration of i_{CCO} and f_{CCO} over T_{int} versus ΔC_s, respectively. As seen in Figure D.4, the first stage of the sensor including the CBCM core, current mirrors, and bias current have a linear response versus the variations of ΔC. According to Figure D.5, the integration of f_{CCO} over T_{int} showed

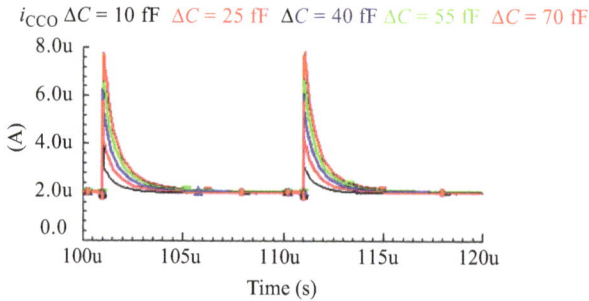

Figure D.2   Input current of the CCO versus time for five different ΔC_s

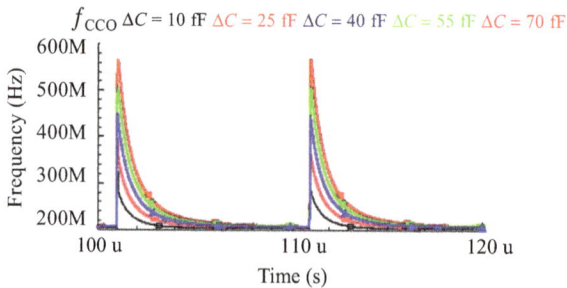

Figure D.3   CCO output frequency versus time for five different ΔC_s

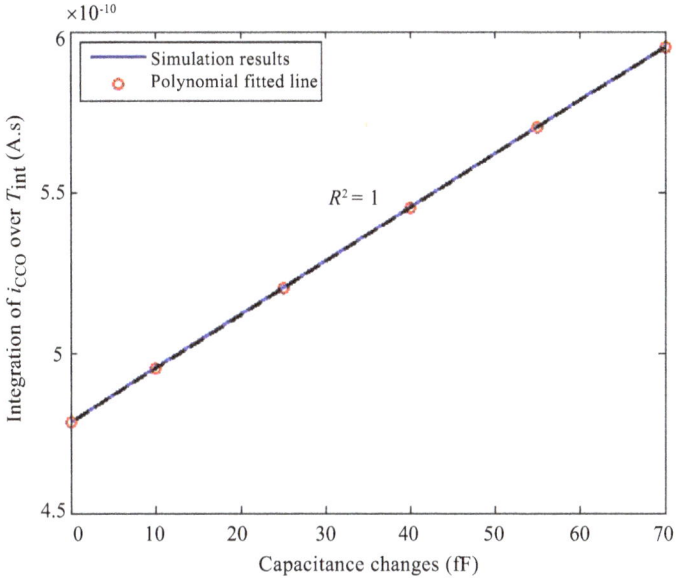

Figure D.4   Integration of i_{CCO} over T_{int} versus ΔC_s

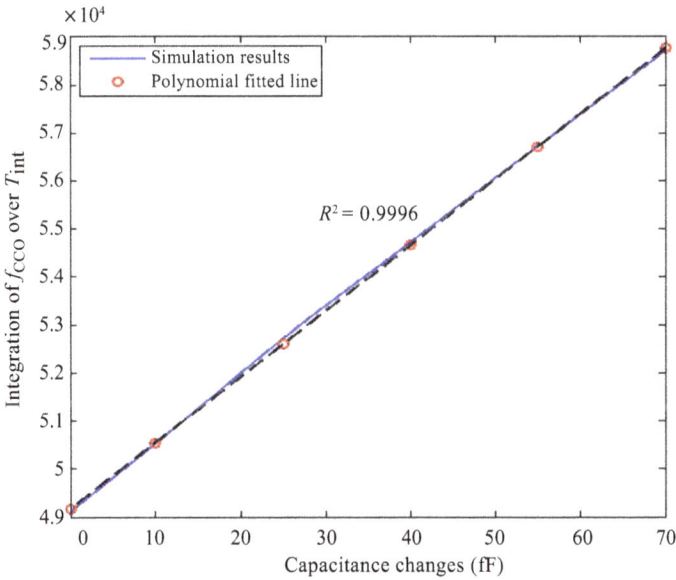

Figure D.5   Integration of f_{CCO} over T_{int} versus ΔC_s

a roughly linear variation with respect to the variations of ΔC. Based on the above-mentioned descriptions, it is expected that the number of pulses counted by the counter would also change linearly in response to various ΔC_s.

Figure D.6 illustrates the counted number of pulses with respect to the variations of ΔC up to 70 fF and the simulations are repeated for three different temperatures of 15 °C, 27 °C, and 45 °C by the assumption of the temperature being constant after calibration. The polynomial fitted line (using linear least squares) whose slope is used for the measurement of the sensor sensitivity showed that this sensor has a sensitivity about 138 pulses/fF at room temperature (27 °C). In the ideal case with a completely linear response, this sensitivity indicates a resolution of about 7.2 aF. But, in fact, the response has a small nonlinearity.

R-squared (R^2) is a statistical measure useful for measuring the nonlinearity that demonstrates how close the response data is to the fitted straight regression line. R^2 indicator of the lines in Figure D.6 is about 0.9996 ($R^2 = 0.9996$) that

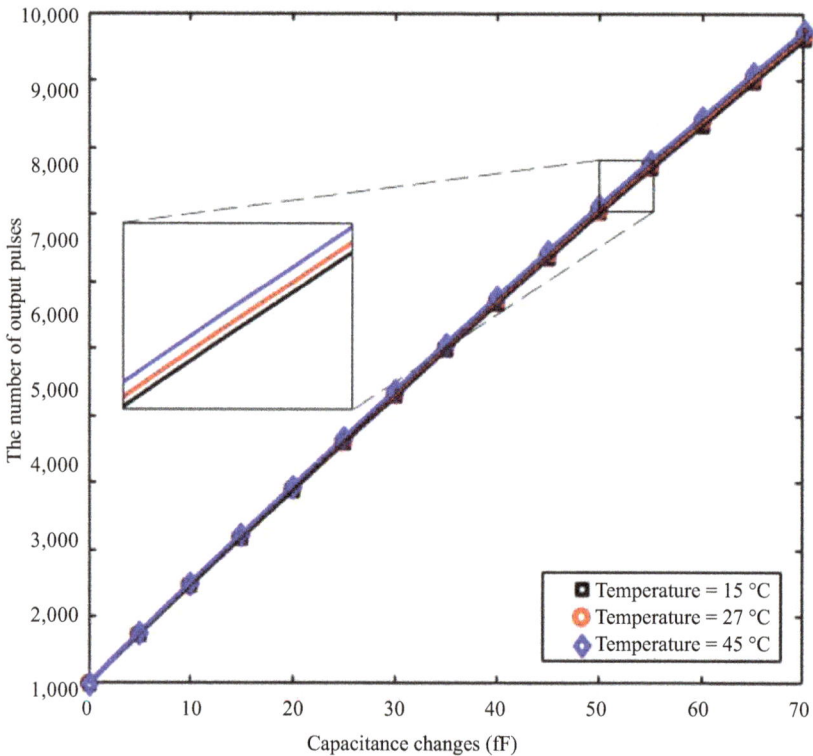

Figure D.6   Number of pulses versus capacitance changes ΔC (fF) (at three different temperatures of 15 °C, 27 °C, and 45 °C)

shows 99.96% of the counter number variations can be modeled linearly for capacitance changes in the range of 0 to 70 fF.

The error between the simulated curve obtained from Cadence and the polynomial fitted line with this curve is illustrated in Figure D.7 for the desired IDR. This figure shows that the maximum absolute error due to the nonlinearity is about 873 aF, which is a limiting factor for the sensing resolution.

As seen in Figures D.4 and D.5, R^2 is 1 and 0.9996 for these curves meaning that the first stage has a great linearity and the original source of the nonlinearity of the whole sensor is the CCO.

Figure D.6 shows that raising the temperature increases the slope of the fitted line slightly. As seen in Figures D.6 and D.7, the temperature dependency of the sensor is negligible.

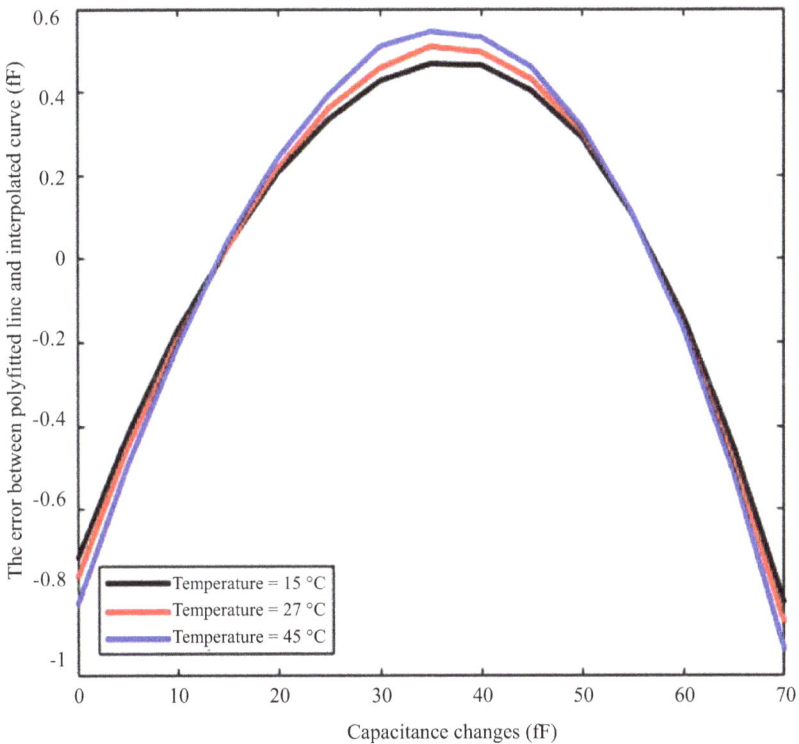

Figure D.7   Error between the simulation results for 15 different values of ΔC and the polynomial fitted line (at three different temperatures of 15 °C, 27 °C, and 45 °C)

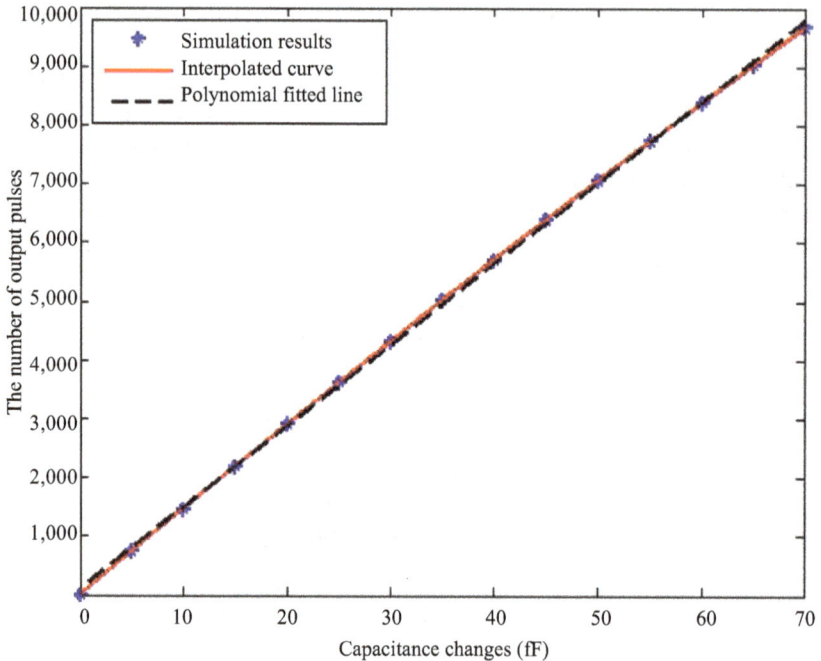

Figure D.8   Fifteen measured number of pulses versus capacitance changes ΔC (fF) at 27 °C, and the interpolated curve and the polynomial fitted line to these 15 points

To achieve better resolution, pre-distortion can be used. In this technique, a lookup table including 15 points is created by measuring the number of output pulses for different values from 0 to 70 fF at 5 fF steps, which can be done with a capacitor bank. The trend of the curve for other values of ΔC can be obtained by the interpolation of these 15 points. Then, the effect of the nonlinearity can be decreased by the pre-distortion based on the interpolated curve. The number of pulses for these 15 points as well as the interpolated curve and the polynomial-fitted line are illustrated in Figure D.8. For other counted number of output pulses, the nearest point of the interpolated curve to that value is considered. The corresponding ΔC can be estimated by using the interpolated curve. As seen in

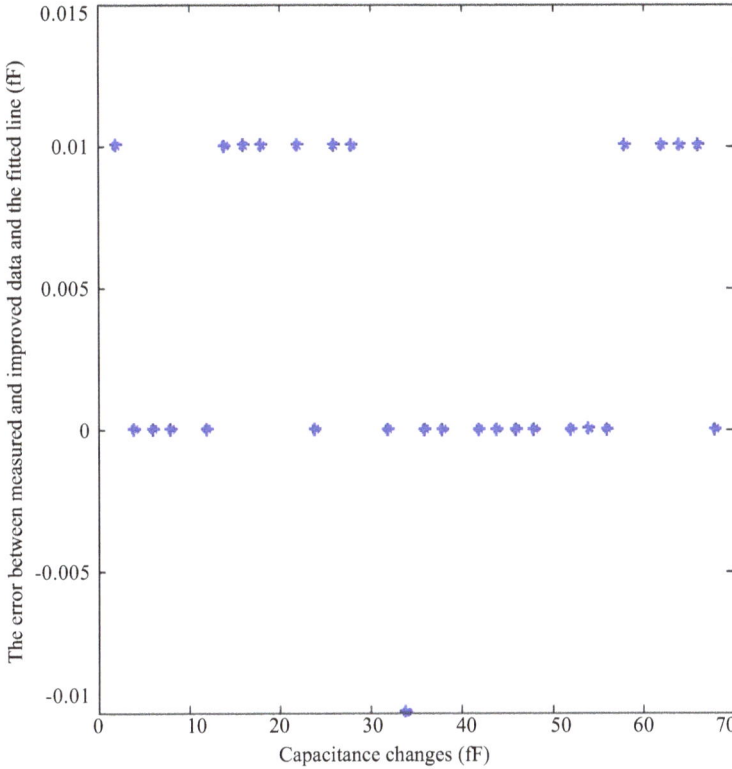

Figure D.9   *The error between the measured numbers of pulses related to the other values of* ΔC *after pre-distortion, and the polynomial fitted line*

Figure D.9, by subtracting the difference between the interpolated curve and polynomial fitted line for the associated ΔC from the counted number of output pulses and using the pre-distorted point and the fitted line, the corresponding ΔC can be estimated by an error less than 10 aF. In other word, a resolution of about 10 aF can be achieved by using interpolation and pre-distortion.

Appendix E
Cell culture

Cell culture is an *in vitro* tool for cellular assays isolated from their natural biological environment. Cells are removed either from a cell strain or cell line that has previously been prepared or from the organism directly and disaggregated before cultivation [414].

Most of the experiments are based on 2D cell culture *in vitro* because of its simplicity and low cost, reproducibility, long-term culture, and easy interpretability. In this method, cells grow as a monolayer on a flat surface such as the bottom of a culture flask or a flat petri dish [415]. Cellular functions like adhesion, growth, migration, and secretion are influenced by the dissolved molecules in the adjacent cells and/or the extracellular matrix (ECM) [416].

In 2D culture method, the interactions between the cellular and extracellular environments are disturbed, the morphology of the cells and the way of their division are changed, and the cells growing adherently lose their polarity. This method is usually monoculture and suitable for the study of only one cell type. Additionally, in contrast to *in vivo* assays where the availability of the ingredients of the medium (like nutrients, oxygen, metabolites, and signaling molecules) is variable, the access to these compounds in 2D culture method is unlimited [415].

On the other hand, 3D culture method offers the advantage of mimicry of the natural structure of tumor mass or tissue alike. Furthermore, it provides proper interactions of cell–extracellular and cell–cell environment. Cell-membrane-bound molecules and ECM complex molecules mediate the cell adhesion. Although 3D method overcame the disadvantages of 2D method, it is difficult to interpret and more time-consuming and expensive [415]. Figure E.1 compares 2D and 3D cell culture.

There are two types of cells. They might be grown either in suspension or adherent cultures. Suspension cells do not require attachment to a surface for growth such as the cells existing in the bloodstream that naturally live in suspension. On the contrary, adherent cells (anchorage-dependent cells) need to attach to a surface to grow. The cells that come from tissue are commonly of this type.

Culture medium and other factors affecting the culture environment such as pH, CO_2, and temperature should be controlled during the process. These conditions vary for each cell type.

To expand culture propagation, new cell culture is made by subculturing or passaging that refers to the removal of the medium and the diluting of cells from a previous culture with fresh growth medium. The protocols for subculturing adherent and suspension mammalian cells are as follows [414,417].

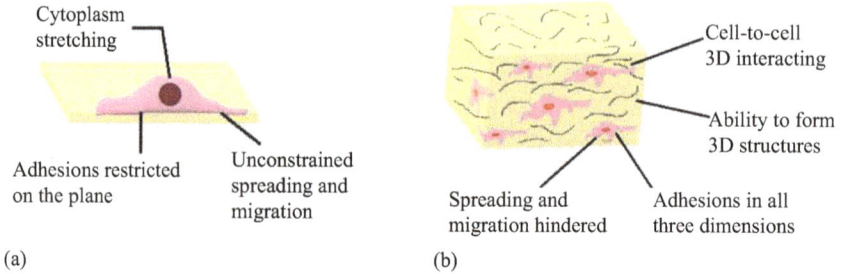

Figure E.1   Comparison between (a) 2D cell culture and (b) 3D cell culture

E.1   Subculturing adherent cells

1. Remove and discard the cell culture media from the culture vessel.
2. Wash cells using a balanced salt solution without magnesium and calcium like PBS, Hanks' balanced salt solution (HBSS), etc. (approximately 2 mL per 10 cm^2 culture surface area). To avoid disturbing the cell layer, gently add wash solution to the side of the culture vessel opposite the attached cell layer. Rock the culture vessel back and forth several times. Any traces of calcium, serum, and magnesium inhibiting the action of the detaching reagent would be removed by washing.
3. Remove and discard the wash solution from the culture vessel.
4. Add pre-warmed detaching reagent such as TrypLE™ or trypsin to the side of the flask. Employ enough reagent (approximately 0.5 mL per 10 cm^2) to cover the cell layer and gently rock the container to achieve complete coverage of the cell layer.
5. Incubate the culture vessel at room temperature for enough time depending on the cell line (e.g. about 2 min) until cells are fully detached from the vessel.
6. Observe the dissociation of the cells under the microscope. In case less than 90% of the cells are detached, increase the incubation time a few more minutes and check every 30 s. Tapping the vessel can also help to increase the speed of the cell detachment.
7. When more than or equal to 90% of the cells are detached, tilt the vessel for a short time to allow the cells to drain. Add pre-warmed complete growth medium with the volume twice the dissociation reagent. Pipet over the cell layer surface several times to disperse the medium.
8. Transfer the cells to a 15-mL conical tube. Next, centrifuge at 200×g for 5–10 min with a suitable time and speed depending on the cell type.
9. Resuspend cells in fresh pre-warmed complete growth medium and disperse the medium by pipetting over the cell layer surface several times. Take a sample for counting.
10. Measure the total number of cells and the percentage viability. If necessary, add growth media to obtain the desired cell concentration and recount them.
11. Dilute cell suspension to the desired seeding cell density for the cell line, and plate cells onto a new cell culture dish and return the cells to the incubator.

E.2   Subculturing suspension cells

Since suspension cells are suspended in growth medium, passaging them is less complicated and traumatic than subculturing adherent cells. There is no need to treat them enzymatically to dissociate them from the surface of the culture vessel. Instead of replacing growth medium, the cells are fed every 2 to 3 days until reach confluency. There are two approaches to do this. The former is direct dilution of the cells in the culture flask and proceed to expand them. The latter is to remove a portion of the cells from the culture flask and to dilute the remaining ones down to a seeding density recommended for the cell line. In comparison to adherent cultures, the lag period following the subculturing suspension cells is shorter.

E.2.1   Cell growth using shaker flasks

A general procedure for subculturing mammalian cells grown in suspension culture by employing shaker flasks in a shaking incubator is as follows:

1. When they are in log-phase growth before reaching confluency, remove the culture flask from the incubator and take a sample from the flask utilizing a sterile pipette. In case they settle down before sampling, distribute the cells evenly in the medium by swirling the flask.
2. Measure the total number of cells and the percentage viability.
3. Calculate the required growth media to add to dilute the culture down to the appropriate seeding density.
4. Add pre-warmed growth medium with the appropriate volume into the culture flask. If necessary, split the culture to several flasks.
5. Loosen the caps of the culture flasks in one full turn or utilize a gas-permeable cap to allow for proper gas exchange and return the culture flasks to the shaking with a shaking speed appropriate for the cell line.

Note: Gently centrifuge the cell suspension at $100 \times g$ for 5–10 min, and resuspend the cell pellet in fresh medium once every 3 weeks, if needed.

E.2.2   Cell growth using spinner flasks

A general procedure for subculturing mammalian cells grown in suspension culture by employing spinner flasks is as follows:

1. When they are in log-phase growth before reaching confluency, remove the culture flask from the incubator and take a sample from the flask utilizing a sterile pipette. In case they settle down before sampling, distribute the cells evenly in the medium by swirling the flask.
2. Measure the total number of cells and the percentage viability.
3. Calculate the required growth media to add to dilute the culture down to the appropriate seeding density.
4. Add pre-warmed growth medium with the appropriate volume into the culture flask. If necessary, split the culture to several flasks.

5. Loosen the side arm caps of the spinner flasks to allow for proper gas exchange and return the culture flasks to the incubator with a spinner speed appropriate for the cell line and the impeller type.

Note: Gently centrifuge the cell suspension at $100 \times g$ for 5–10 min, and resuspend the cell pellet in fresh medium once every 3 weeks, if needed.

References

[1] *Ion Torrent*. 2010. Available: http://tools.lifetechnologies.com/content/sfs/brochures/cms_090253.pdf

[2] J. M. Rothberg, W. Hinz, T. M. Rearick *et al.*, "An integrated semiconductor device enabling non-optical genome sequencing," *Nature*, vol. 475, no. 7356, pp. 348–352, 2011.

[3] J. Musayev, C. Altiner, Y. Adiguzel, H. Kulah, S. Eminoglu, and T. Akin, "Capturing and detection of MCF-7 breast cancer cells with a CMOS image sensor," *Sensors and Actuators A: Physical*, vol. 215, pp. 105–114, 2014.

[4] A. Alhoshany, S. Sivashankar, Y. Mashraei, H. Omran, and K. N. Salama, "A biosensor-CMOS platform and integrated readout circuit in 0.18-µm CMOS technology for cancer biomarker detection," *Sensors*, vol. 17, no. 9, pp. 1942 (1–12), 2017.

[5] G. Massicotte, S. Carrara, G. Di Micheli, and M. Sawan, "A CMOS amperometric system for multi-neurotransmitter detection," *IEEE Transactions on Biomedical Circuits and Systems*, vol. 10, no. 3, pp. 731–741, 2016.

[6] G. Xu, J. Abbott, and D. Ham, "Optimization of CMOS-ISFET-based biomolecular sensing: analysis and demonstration in DNA detection," *IEEE Transactions on Electron Devices*, vol. 63, no. 8, pp. 3249–3256, 2016.

[7] A. DeHennis, S. Getzlaff, D. Grice, and M. Mailand, "An NFC-enabled CMOS IC for a wireless fully implantable glucose sensor," *IEEE Journal of Biomedical and Health Informatics*, vol. 20, no. 1, pp. 18–28, 2016.

[8] R. Fior, J. Kwok, F. Malfatti, O. Sbaizero, and R. Lal, "Biocompatible optically transparent MEMS for micromechanical stimulation and multi-modal imaging of living cells," *Annals of Biomedical Engineering*, vol. 43, no. 8, pp. 1841–1850, 2015.

[9] H. Lee, Y. Liu, R. M. Westervelt, and D. Ham, "IC/microfluidic hybrid system for magnetic manipulation of biological cells," *IEEE Journal of Solid-State Circuits*, vol. 41, no. 6, pp. 1471–1480, 2006.

[10] M. Behnam, G. Kaigala, M. Khorasani, S. Martel, D. Elliott, and C. Backhouse, "Integrated circuit-based instrumentation for microchip capillary electrophoresis," *IET Nanobiotechnology*, vol. 4, no. 3, pp. 91–101, 2010.

[11] M. A. Al-Rawhani, B. C. Cheah, A. I. Macdonald *et al.*, "A colorimetric CMOS-based platform for rapid total serum cholesterol quantification," *IEEE Sensors Journal*, vol. 17, no. 2, pp. 240–247, 2017.

[12]   P. Murali, A. M. Niknejad, and B. E. Boser, "CMOS microflow cytometer for magnetic label detection and classification," *IEEE Journal of Solid-State Circuits*, vol. 52, no. 2, pp. 543–555, 2017.

[13]   B. P. Senevirathna, S. Lu, M. P. Dandin, J. Basile, E. Smela, and P. A. Abshire, "Real-time measurements of cell proliferation using a lab-on-CMOS capacitance sensor array," *IEEE Transactions on Biomedical Circuits and Systems*, vol. 12, no. 3, pp. 510–520, 2018.

[14]   E. P. Papageorgiou, H. Zhang, S. Giverts, C. Park, B. E. Boser, and M. Anwar, "Real-time cancer detection with an integrated lensless fluorescence contact imager," *Biomedical Optics Express*, vol. 9, no. 8, pp. 3607–3623, 2018.

[15]   Y. Zheng, N. Shang, P. S. Haddad, and M. Sawan, "A microsystem for magnetic immunoassay based on planar microcoil array," *IEEE Transactions on Biomedical Circuits and Systems*, vol. 10, no. 2, pp. 477–486, 2016.

[16]   Y. Jiang, X. Liu, T. C. Dang *et al.*, "A high-sensitivity potentiometric 65-nm CMOS ISFET sensor for rapid *E. coli* screening," *IEEE Transactions on Biomedical Circuits and Systems*, vol. 12, no. 2, pp. 402–415, 2018.

[17]   S.-J. Wu, Y.-C. Wu, H.-H. Tsai, H.-H. Liao, Y.-Z. Juang, and C.-H. Lin, "ISFET-based pH sensor composed of a high transconductance CMOS chip and a disposable touch panel film as the sensing layer," in *2015 IEEE Sensors*, 2015, pp. 1–4: IEEE.

[18]   V. Vijay, B. Raziyeh, S. Amir *et al.*, "High-density CMOS microelectrode array system for impedance spectroscopy and imaging of biological cells," in *2016 IEEE Sensors*, 2016, pp. 1–3: IEEE.

[19]   C.-L. Hsu, A. Sun, Y. Zhao, E. Aronoff-Spencer, and D. A. Hall, "A 16 × 20 electrochemical CMOS biosensor array with in-pixel averaging using polar modulation," in *Custom Integrated Circuits Conference (CICC), 2018 IEEE*, 2018, pp. 1–4: IEEE.

[20]   N. Couniot, L. A. Francis, and D. Flandre, "A 16 × 16 CMOS capacitive biosensor array towards detection of single bacterial cell," *IEEE Transactions on Biomedical Circuits and Systems*, vol. 10, no. 2, pp. 364–374, 2016.

[21]   G. Nabovati, E. Ghafar-Zadeh, A. Letourneau, and M. Sawan, "Towards high throughput cell growth screening: a new CMOS 8×8 biosensor array for life science applications," *IEEE Transactions on Biomedical Circuits and Systems*, 2016.

[22]   C. Stagni, C. Guiducci, L. Benini *et al.*, "A fully electronic label-free DNA sensor chip," *IEEE Sensors Journal*, vol. 7, no. 4, pp. 577–585, 2007.

[23]   S. Carrara, V. Bhalla, C. Stagni *et al.*, "Label-free cancer markers detection by capacitance biochip," *Sensors and Actuators B: Chemical*, vol. 136, no. 1, pp. 163–172, 2009.

[24]   L. Yao, M. Hajj-Hassan, E. Ghafar-Zadeh, A. Shabani, V. Chodavarapu, and M. Zourob, "CMOS capacitive sensor system for bacteria detection using phage organisms," in *Canadian Conference on Electrical and Computer Engineering. CCECE 2008*, 2008, pp. 000877–000880: IEEE.

[25]   J. S. Daniels and N. Pourmand, "Label-free impedance biosensors: opportunities and challenges," *Electroanalysis: An International Journal Devoted*

to Fundamental and Practical Aspects of Electroanalysis, vol. 19, no. 12, pp. 1239–1257, 2007.

[26]   J. Musayev, Y. Adlguzel, H. Kulah, S. Eminoglu, and T. Akln, "Label-free DNA detection using a charge sensitive CMOS microarray sensor chip," *IEEE Sensors Journal*, vol. 14, no. 5, pp. 1608–1616, 2014.

[27]   B. Zhang, Q. Dong, C. E. Korman, Z. Li, and M. E. Zaghloul, "Flexible packaging of solid-state integrated circuit chips with elastomeric microfluidics," *Scientific reports*, vol. 3, pp. 1098 (1–8), 2013.

[28]   G. Nabovati, E. Ghafar-Zadeh, A. Letourneau, and M. Sawan, "Smart cell culture monitoring and drug test platform using CMOS capacitive sensor array," *IEEE Transactions on Biomedical Engineering*, vol. 66, no. 4, pp. 1094–1104, 2018.

[29]   Y.-J. Huang, C.-W. Huang, T.-H. Lin *et al.*, "A CMOS cantilever-based label-free DNA SoC with improved sensitivity for hepatitis B virus detection," *IEEE Transactions on Biomedical Circuits and systems*, vol. 7, no. 6, pp. 820–831, 2013.

[30]   O. Tokel, U. H. Yildiz, F. Inci *et al.*, "Portable microfluidic integrated plasmonic platform for pathogen detection," *Scientific Reports*, vol. 5, no. 1, pp. 1–9, 2015.

[31]   W. Bishara, U. Sikora, O. Mudanyali *et al.*, "Holographic pixel superresolution in portable lensless on-chip microscopy using a fiber-optic array," *Lab on a Chip*, vol. 11, no. 7, pp. 1276–1279, 2011.

[32]   A. Ozcan and U. Demirci, "Ultra wide-field lens-free monitoring of cells on-chip," *Lab on a Chip*, vol. 8, no. 1, pp. 98–106, 2008.

[33]   T. W. Su, S. Seo, A. Erlinger, and A. Ozcan, "High-throughput lensfree imaging and characterization of a heterogeneous cell solution on a chip," *Biotechnology and Bioengineering*, vol. 102, no. 3, pp. 856–868, 2009.

[34]   S. Moon, H. O. Keles, A. Ozcan *et al.*, "Integrating microfluidics and lensless imaging for point-of-care testing," *Biosensors and Bioelectronics*, vol. 24, no. 11, pp. 3208–3214, 2009.

[35]   S. Seo, T.-W. Su, D. K. Tseng, A. Erlinger, and A. Ozcan, "Lensfree holographic imaging for on-chip cytometry and diagnostics," *Lab on a Chip*, vol. 9, no. 6, pp. 777–787, 2009.

[36]   L. Hong, H. Li, H. Yang, and K. Sengupta, "Fully integrated fluorescence biosensors on-chip employing multi-functional nanoplasmonic optical structures in CMOS," *IEEE Journal of Solid-State Circuits*, vol. 52, no. 9, pp. 2388–2406, 2017.

[37]   E. P. Papageorgiou, B. E. Boser, and M. Anwar, "An angle-selective CMOS imager with on-chip micro-collimators for blur reduction in near-field cell imaging," in 2016 *IEEE 29th International Conference on Micro Electro Mechanical Systems (MEMS)*, 2016, pp. 337–340: IEEE.

[38]   M. Velasco-Garcia, "Optical biosensors for probing at the cellular level: A review of recent progress and future prospects," in *Seminars in Cell & Developmental Biology*, 2009, vol. 20, no. 1, pp. 27–33: Elsevier.

[39]   K.-M. Lei, P.-I. Mak, M.-K. Law, and R. P. Martins, "CMOS biosensors for in vitro diagnosis—transducing mechanisms and applications," *Lab on a Chip*, vol. 16, no. 19, pp. 3664–3681, 2016.

[40]   B. Jang, P. Cao, A. Chevalier, A. Ellington, and A. Hassibi, "A CMOS fluorescent-based biosensor microarray," in *2009 IEEE International Solid-State Circuits Conference-Digest of Technical Papers*, 2009, pp. 436–437: IEEE.

[41]   D. H. Ta-chien, S. Sorgenfrei, P. Gong, R. Levicky, and K. L. Shepard, "A 0.18-μm CMOS array sensor for integrated time-resolved fluorescence detection," *IEEE Journal of Solid-State Circuits*, vol. 44, no. 5, pp. 1644–1654, 2009.

[42]   L. Hong, S. McManus, H. Yang, and K. Sengupta, "A fully integrated CMOS fluorescence biosensor with on-chip nanophotonic filter," in *2015 Symposium on VLSI Circuits (VLSI Circuits)*, 2015, pp. C206–C207: IEEE.

[43]   L. Sandeau, C. Vuillaume, S. Contié *et al.*, "Large area CMOS bio-pixel array for compact high sensitive multiplex biosensing," *Lab on a Chip*, vol. 15, no. 3, pp. 877–881, 2015.

[44]   A. Arandian, Z. Bagheri, H. Ehtesabi *et al.*, "Optical imaging approaches to monitor static and dynamic cell-on-chip platforms: a tutorial review," *Small*, vol. 15, no. 28, p. 1900737, 2019.

[45]   J. Lee, Y. H. Kwak, S.-H. Paek, S. Han, and S. Seo, "CMOS image sensor-based ELISA detector using lens-free shadow imaging platform," *Sensors and Actuators B: Chemical*, vol. 196, pp. 511–517, 2014.

[46]   Z. Göröcs and A. Ozcan, "On-chip biomedical imaging," *IEEE Reviews in Biomedical Engineering*, vol. 6, pp. 29–46, 2012.

[47]   S. B. Kim, H. Bae, J. M. Cha *et al.*, "A cell-based biosensor for real-time detection of cardiotoxicity using lensfree imaging," *Lab on a Chip*, vol. 11, no. 10, pp. 1801–1807, 2011.

[48]   S. B. Kim, K.-i. Koo, H. Bae *et al.*, "A mini-microscope for *in situ* monitoring of cells," *Lab on a Chip*, vol. 12, no. 20, pp. 3976–3982, 2012.

[49]   Y. Adiguzel and H. Kulah, "CMOS cell sensors for point-of-care diagnostics," *Sensors*, vol. 12, no. 8, pp. 10042–10066, 2012.

[50]   F. J. Blanco, M. Agirregabiria, J. Berganzo *et al.*, "Microfluidic-optical integrated CMOS compatible devices for label-free biochemical sensing," *Journal of Micromechanics and Microengineering*, vol. 16, no. 5, p. 1006, 2006.

[51]   A. Pai, A. Khachaturian, S. Chapman, A. Hu, H. Wang, and A. Hajimiri, "A handheld magnetic sensing platform for antigen and nucleic acid detection," *Analyst*, vol. 139, no. 6, pp. 1403–1411, 2014.

[52]   T. Aytur, J. Foley, M. Anwar, B. Boser, E. Harris, and P. R. Beatty, "A novel magnetic bead bioassay platform using a microchip-based sensor for infectious disease diagnosis," *Journal of Immunological Methods*, vol. 314, no. 1-2, pp. 21–29, 2006.

[53]   M. Mujika, S. Arana, E. Castano *et al.*, "Magnetoresistive immunosensor for the detection of *Escherichia coli* O157: H7 including a microfluidic network," *Biosensors and Bioelectronics*, vol. 24, no. 5, pp. 1253–1258, 2009.

[54]   N. Sun, T.-J. Yoon, H. Lee, W. Andress, R. Weissleder, and D. Ham, "Palm NMR and 1-chip NMR," *IEEE Journal of Solid-State Circuits*, vol. 46, no. 1, pp. 342–352, 2011.

[55] D. Ha, J. Paulsen, N. Sun, Y.-Q. Song, and D. Ham, "Scalable NMR spectroscopy with semiconductor chips," *Proceedings of the National Academy of Sciences*, vol. 111, no. 33, pp. 11955–11960, 2014.

[56] K.-M. Lei, P.-I. Mak, M.-K. Law, and R. P. Martins, "A μ NMR CMOS transceiver using a butterfly-coil input for integration with a digital microfluidic device inside a portable magnet," *IEEE Journal of Solid-State Circuits*, vol. 51, no. 10, pp. 2274–2286, 2016.

[57] V. Tsouti, C. Boutopoulos, I. Zergioti, and S. Chatzandroulis, "Capacitive microsystems for biological sensing," *Biosensors and Bioelectronics*, vol. 27, no. 1, pp. 1–11, 2011.

[58] M. M. Varma. Limitations of label-free sensors in serum based molecular diagnostics [Online]. Available: https://arxiv.org/ftp/arxiv/papers/1505/1505.01032

[59] S. S. Ghoreishizadeh, I. Taurino, G. De Micheli, S. Carrara, and P. Georgiou, "A differential electrochemical readout ASIC with heterogeneous integration of bio-nano sensors for amperometric sensing," *IEEE Transactions on Biomedical Circuits and Systems*, 2017.

[60] K. Niitsu, S. Ota, K. Gamo, H. Kondo, M. Hori, and K. Nakazato, "Development of microelectrode arrays using electroless plating for CMOS-based direct counting of bacterial and HeLa cells," *IEEE Transactions on Biomedical Circuits and Systems*, vol. 9, no. 5, pp. 607–619, 2015.

[61] K. Gamo, K. Nakazato, and K. Niitsu, "Design, theoretical analysis, and experimental verification of a CMOS current integrator with $1.2 \times 2.05 \ \mu m^2$ microelectrode array for high-sensitivity bacterial counting," *Japanese Journal of Applied Physics*, vol. 56, no. 1S, p. 01AH01, 2016.

[62] C. Giagkoulovits, M. A. Al-Rawhani, B. C. Cheah *et al.*, "Hybrid amperometric and potentiometrie sensing based on a CMOS ISFET array," in *Sensors, 2017 IEEE*, 2017, pp. 1–3: IEEE.

[63] S. Martinoia, N. Rosso, M. Grattarola, L. Lorenzelli, B. Margesin, and M. Zen, "Development of ISFET array-based microsystems for bioelectrochemical measurements of cell populations," *Biosensors and Bioelectronics*, vol. 16, no. 9-12, pp. 1043–1050, 2001.

[64] M. J. Milgrew, M. O. Riehle, and D. R. Cumming, "A 16×16 CMOS proton camera array for direct extracellular imaging of hydrogen-ion activity," in *2008 IEEE International Solid-State Circuits Conference-Digest of Technical Papers*, 2008, pp. 590–638: IEEE.

[65] E. Ghafar-Zadeh, S. F. Chowdhury, A. Aliakbar *et al.*, "Handheld impedance biosensor system using engineered proteinaceous receptors," *Biomedical microdevices*, vol. 12, no. 6, pp. 967–975, 2010.

[66] J. Dragas, V. Viswam, A. Shadmani *et al.*, "In vitro multi-functional microelectrode array featuring 59 760 electrodes, 2048 electrophysiology channels, stimulation, impedance measurement, and neurotransmitter detection channels," *IEEE Journal of Solid-State Circuits*, vol. 52, no. 6, pp. 1576–1590, 2017.

[67] L. Yao, P. Lamarche, N. Tawil *et al.*, "CMOS conductometric system for growth monitoring and sensing of bacteria," *IEEE Transactions on Biomedical Circuits and Systems*, vol. 5, no. 3, pp. 223–230, 2011.

[68]   E. Ghafar-Zadeh, M. Sawan, V. P. Chodavarapu, and T. Hosseini-Nia, "Bacteria growth monitoring through a differential CMOS capacitive sensor," *IEEE Transactions on Biomedical Circuits and Systems*, vol. 4, no. 4, pp. 232–238, 2010.

[69]   E. Ghafar-Zadeh, "Wireless integrated biosensors for point-of-care diagnostic applications," *Sensors*, vol. 15, no. 2, pp. 3236–3261, 2015.

[70]   A. Manickam, A. Chevalier, M. McDermott, A. D. Ellington, and A. Hassibi, "A CMOS electrochemical impedance spectroscopy (EIS) biosensor array," *IEEE Transactions on Biomedical Circuits and Systems*, vol. 4, no. 6, pp. 379–390, 2010.

[71]   H. Jafari, L. Soleymani, and R. Genov, "16-Channel CMOS impedance spectroscopy DNA analyzer with dual-slope multiplying ADCs," *IEEE Transactions on Biomedical Circuits and Systems*, vol. 6, no. 5, pp. 468–478, 2012.

[72]   R. Venugopalan and R. Ideker, "Bioelectrodes," in *Biomaterials Science*: Elsevier, 2013, pp. 957–966.

[73]   S. Forouhi, R. Dehghani, and E. Ghafar-Zadeh, "Toward high throughput core-CBCM CMOS capacitive sensors for life science applications: a novel current-mode for high dynamic range circuitry," *Sensors*, vol. 18, no. 10, pp. 3370 (1–29), 2018.

[74]   A. C. Fischer, F. Forsberg, M. Lapisa *et al.*, "Integrating mems and ics," *Microsystems & Nanoengineering*, vol. 1, no. 1, pp. 1–16, 2015.

[75]   E. Ghafar-Zadeh and M. Sawan, "Charge-based capacitive sensor array for CMOS-based laboratory-on-chip applications," *IEEE Sensors Journal*, vol. 8, no. 4, pp. 325–332, 2008.

[76]   E. Ghafar-Zadeh and M. Sawan, *CMOS Capacitive Sensors for Lab-on-chip Applications*. New York: Springer, 2010.

[77]   I. Evans and T. York, "Microelectronic capacitance transducer for particle detection," *IEEE Sensors Journal*, vol. 4, no. 3, pp. 364–372, 2004.

[78]   A. Tashtoush, "Nano-amplification strategy using charge based capacitance measurement for pathogenic bacteria detection," *Journal of Biotechnology and Strategic Health Research*, vol. 2, no. 2, pp. 87–100.

[79]   S. B. Prakash and P. Abshire, "On-chip capacitance sensing for cell monitoring applications," *IEEE Sensors Journal*, vol. 7, no. 3, pp. 440–447, 2007.

[80]   K. Asami, E. Gheorghiu, and T. Yonezawa, "Real-time monitoring of yeast cell division by dielectric spectroscopy," *Biophysical Journal*, vol. 76, no. 6, pp. 3345–3348, 1999.

[81]   E. Ghafar-Zadeh, M. Sawan, and D. Therriault, "A 0.18-μm CMOS capacitive sensor lab-on-chip," *Sensors and Actuators A: Physical*, vol. 141, no. 2, pp. 454–462, 2008.

[82]   C. I. Justino, A. C. Freitas, R. Pereira, A. C. Duarte, and T. A. R. Santos, "Recent developments in recognition elements for chemical sensors and biosensors," *TrAC Trends in Analytical Chemistry*, vol. 68, pp. 2–17, 2015.

[83] P. Sanguino, T. Monteiro, F. Marques, C. J. Dias, R. Igreja, and R. Franco, "Interdigitated capacitive immunosensors with PVDF immobilization layers," *IEEE Sensors Journal*, vol. 14, no. 4, pp. 1260–1265, 2014.

[84] P. Sanguino, T. Monteiro, S. Bhattacharyya, C. Dias, R. Igreja, and R. Franco, "ZnO nanorods as immobilization layers for interdigitated capacitive immunosensors," *Sensors and Actuators B: Chemical*, vol. 204, pp. 211–217, 2014.

[85] A. Balasubramanian, B. Bhuva, R. Mernaugh, and F. R. Haselton, "Si-based sensor for virus detection," *IEEE Sensors Journal*, vol. 5, no. 3, pp. 340–344, 2005.

[86] G.-Y. Lee, Y.-H. Choi, H.-W. Chung, H. Ko, S. Cho, and J.-C. Pyun, "Capacitive immunoaffinity biosensor based on vertically paired ring-electrodes," *Biosensors and Bioelectronics*, vol. 40, no. 1, pp. 227–232, 2013.

[87] G.-Y. Lee, J.-H. Park, Y. W. Chang, M.-J. Kang, S. Cho, and J.-C. Pyun, "Capacitive biosensor based on vertically paired electrode with controlled parasitic capacitance," *Sensors and Actuators B: Chemical*, 2018.

[88] A. Hartono, E. Sanjaya, and R. Ramli, "Glucose sensing using capacitive biosensor based on polyvinylidene fluoride thin film," *Biosensors*, vol. 8, no. 1, p. 12, 2018.

[89] X. Wang, X. Lu, and J. Chen, "Development of biosensor technologies for analysis of environmental contaminants," *Trends in Environmental Analytical Chemistry*, vol. 2, pp. 25–32, 2014.

[90] D. Berdat, A. Marin, F. Herrera, and M. A. Gijs, "DNA biosensor using fluorescence microscopy and impedance spectroscopy," *Sensors and Actuators B: Chemical*, vol. 118, no. 1-2, pp. 53–59, 2006.

[91] D. Berdat, A. C. M. Rodríguez, F. Herrera, and M. A. Gijs, "Label-free detection of DNA with interdigitated micro-electrodes in a fluidic cell," *Lab on a Chip*, vol. 8, no. 2, pp. 302–308, 2008.

[92] Y.-S. Liu, P. P. Banada, S. Bhattacharya, A. K. Bhunia, and R. Bashir, "Electrical characterization of DNA molecules in solution using impedance measurements," *Applied Physics Letters*, vol. 92, no. 14, p. 143902, 2008.

[93] G. Tsekenis, M. Chatzipetrou, J. Tanner *et al.*, "Surface functionalization studies and direct laser printing of oligonucleotides toward the fabrication of a micro-membrane DNA capacitive biosensor," *Sensors and Actuators B: Chemical*, vol. 175, pp. 123–131, 2012.

[94] A. Qureshi, A. Pandey, R. S. Chouhan, Y. Gurbuz, and J. H. Niazi, "Whole-cell based label-free capacitive biosensor for rapid nanosize-dependent toxicity detection," *Biosensors and Bioelectronics*, vol. 67, pp. 100–106, 2015.

[95] M. Varshney and Y. Li, "Double interdigitated array microelectrode-based impedance biosensor for detection of viable *Escherichia coli* O157: H7 in growth medium," *Talanta*, vol. 74, no. 4, pp. 518–525, 2008.

[96] A. Qureshi, Y. Gurbuz, and J. H. Niazi, "Capacitive aptamer–antibody based sandwich assay for the detection of VEGF cancer biomarker in serum," *Sensors and Actuators B: Chemical*, vol. 209, pp. 645–651, 2015.

[97] M. C. Rodriguez, A.-N. Kawde, and J. Wang, "Aptamer biosensor for label-free impedance spectroscopy detection of proteins based on recognition-induced switching of the surface charge," *Chemical Communications*, no. 34, pp. 4267–4269, 2005.

[98]   S. Niyomdecha, W. Limbut, A. Numnuam *et al.*, "Phage-based capacitive biosensor for Salmonella detection," *Talanta*, 2018.

[99]   M. Tolba, M. U. Ahmed, C. Tlili, F. Eichenseher, M. J. Loessner, and M. Zourob, "A bacteriophage endolysin-based electrochemical impedance biosensor for the rapid detection of *Listeria* cells," *Analyst*, vol. 137, no. 24, pp. 5749–5756, 2012.

[100] P. Zhurauski, S. K. Arya, P. Jolly *et al.*, "Sensitive and selective affimer-functionalised interdigitated electrode-based capacitive biosensor for Her4 protein tumour biomarker detection," *Biosensors and Bioelectronics*, vol. 108, pp. 1–8, 2018.

[101] A. Johnson, Q. Song, P. Ko Ferrigno, P. R. Bueno, and J. J. Davis, "Sensitive affimer and antibody based impedimetric label-free assays for C-reactive protein," *Analytical Chemistry*, vol. 84, no. 15, pp. 6553–6560, 2012.

[102] F. Canfarotta, J. Czulak, A. Guerreiro *et al.*, "A novel capacitive sensor based on molecularly imprinted nanoparticles as recognition elements," *Biosensors and Bioelectronics*, 2018.

[103] Z. Cheng, E. Wang, and X. Yang, "Capacitive detection of glucose using molecularly imprinted polymers," *Biosensors and Bioelectronics*, vol. 16, no. 3, pp. 179–185, 2001.

[104] A. Aghaei, M. R. M. Hosseini, and M. Najafi, "A novel capacitive bio-sensor for cholesterol assay that uses an electropolymerized molecularly imprinted polymer," *Electrochimica Acta*, vol. 55, no. 5, pp. 1503–1508, 2010.

[105] D. De Venuto, S. Carrara, and B. Riccò, "Design of an integrated low-noise read-out system for DNA capacitive sensors," *Microelectronics Journal*, vol. 40, no. 9, pp. 1358–1365, 2009.

[106] N. Petrellis, C. Spathis, K. Georgakopoulou, and A. Birbas, "Capacitive sensor estimation based on self-configurable reference capacitance," *Recent Patents on Signal Processing*, vol. 3, no. 1, pp. 12–21, 2013.

[107] A. Romani, N. Manaresi, L. Marzocchi *et al.*, "Capacitive sensor array for localization of bioparticles in CMOS lab-on-a-chip," in *Solid-State Circuits Conference, 2004. Digest of Technical Papers. ISSCC. 2004 IEEE International*, 2004, pp. 224–225: IEEE.

[108] K. Mohammad, D. A. Buchanan, and D. J. Thomson, "Integrated 0.35 pm CMOS capacitance sensor with atto-farad sensitivity for single cell analysis," in *Biomedical Circuits and Systems Conference (BioCAS), 2016 IEEE*, 2016, pp. 22–25: IEEE.

[109] N. Couniot, D. Bol, O. Poncelet, L. A. Francis, and D. Flandre, "A capacitance-to-frequency converter with on-chip passivated microelectrodes for bacteria detection in saline buffers up to 575 MHz," *IEEE Transactions on Circuits and Systems II: Express Briefs*, vol. 62, no. 2, pp. 159–163, 2014.

[110] K. Mohammad and D. J. Thomson, "Differential ring oscillator based capacitance sensor for microfluidic applications," *IEEE Transactions on Biomedical Circuits and Systems*, vol. 11, no. 2, pp. 392–399, 2017.

[111] B. Senevirathna, S. Lu, and P. Abshire, "Characterization of a high dynamic range lab-on-CMOS capacitance sensor array," in *2017 IEEE International Symposium on Circuits and Systems (ISCAS)*, 2017, pp. 1–4: IEEE.

[112] J. S. Gaggatur and G. Banerjee, "High gain capacitance sensor interface for the monitoring of cell volume growth," in *2017 30th International Conference on VLSI Design and 2017 16th International Conference on Embedded Systems (VLSID)*, 2017, pp. 201–206: IEEE.

[113] B. Senevirathna, A. Castro, M. Dandin, E. Smela, and P. Abshire, "Lab-on-CMOS capacitance sensor array for real-time cell viability measurements with I2C readout," in *Circuits and Systems (ISCAS), 2016 IEEE International Symposium on*, 2016, pp. 2863–2866: IEEE.

[114] E. Naviasky, T. Datta-Chaudhuri, and P. Abshire, "High resolution capacitance sensor array for real-time monitoring of cell viability," in *2014 IEEE International Symposium on Circuits and Systems (ISCAS)*, 2014, pp. 634–637: IEEE.

[115] M. S.-C. Lu, Y.-C. Chen, and P.-C. Huang, "5 × 5 CMOS capacitive sensor array for detection of the neurotransmitter dopamine," *Biosensors and Bioelectronics*, vol. 26, no. 3, pp. 1093–1097, 2010.

[116] A.-Y. Chang and M. S.-C. Lu, "A CMOS magnetic microbead-based capacitive biosensor array with on-chip electromagnetic manipulation," *Biosensors and Bioelectronics*, vol. 45, pp. 6–12, 2013.

[117] C. Stagni, C. Guiducci, L. Benini *et al.*, "CMOS DNA sensor array with integrated A/D conversion based on label-free capacitance measurement," *IEEE Journal of Solid-State Circuits*, vol. 41, no. 12, pp. 2956–2964, 2006.

[118] K.-H. Lee, J.-O. Lee, M.-J. Sohn *et al.*, "One-chip electronic detection of DNA hybridization using precision impedance-based CMOS array sensor," *Biosensors and Bioelectronics*, vol. 26, no. 4, pp. 1373–1379, 2010.

[119] S. Druart, D. Flandre, and L. A. Francis, "A self-oscillating system to measure the conductivity and the permittivity of liquids within a single triangular signal," *Journal of Sensors*, vol. 2014, 2014.

[120] P. Ciccarella, M. Carminati, M. Sampietro, and G. Ferrari, "Multichannel 65 zF rms resolution CMOS monolithic capacitive sensor for counting single micrometer-sized airborne particles on chip," *IEEE Journal of Solid-State Circuits*, vol. 51, no. 11, pp. 2545–2553, 2016.

[121] M. Bakhshiani, M. A. Suster, and P. Mohseni, "A 9 MHz–2.4 GHz fully integrated transceiver IC for a microfluidic-CMOS platform dedicated to miniaturized dielectric spectroscopy," *IEEE Transactions on Biomedical Circuits and Systems*, vol. 9, no. 6, pp. 849–861, 2015.

[122] M. Bakhshiani, M. A. Suster, and P. Mohseni, "21.4 A microfluidic-CMOS platform with 3D capacitive sensor and fully integrated transceiver IC for palmtop dielectric spectroscopy," in *Solid-State Circuits Conference-(ISSCC), 2015 IEEE International*, 2015, pp. 1–3: IEEE.

[123] M. Bakhshiani, M. A. Suster, and P. Mohseni, "A broadband sensor interface IC for miniaturized dielectric spectroscopy from MHz to GHz," *IEEE Journal of Solid-State Circuits*, vol. 49, no. 8, pp. 1669–1681, 2014.

[124] M. M. Bajestan, A. A. Helmy, H. Hedayati, and K. Entesari, "A 0.62–10GHz CMOS dielectric spectroscopy system for chemical/biological material characterization," in *2014 IEEE MTT-S International Microwave Symposium (IMS)*, 2014, pp. 1–4: IEEE.

[125] M. M. Bajestan, A. A. Helmy, H. Hedayati, and K. Entesari, "A 0.62–10 GHz complex dielectric spectroscopy system in CMOS," *IEEE Transactions on Microwave Theory and Techniques*, vol. 62, no. 12, pp. 3522–3537, 2014.

[126] M. Daphtary and S. Sonkusale, "Broadband capacitive sensor CMOS interface circuit for dielectric spectroscopy," in *Circuits and Systems, 2006. ISCAS 2006. Proceedings. 2006 IEEE International Symposium on*, 2006, pp. 4285–4288: IEEE.

[127] Y. Kim, A. Agarwal, and S. R. Sonkusale, "Broadband dielectric spectroscopy CMOS readout circuit for molecular sensing," in *Circuits and Systems, 2005. ISCAS 2005. IEEE International Symposium on*, 2005, pp. 5906–5909: IEEE.

[128] P. Ciccarella, M. Carminati, M. Sampietro, and G. Ferrari, "CMOS monolithic airborne-particulate-matter detector based on 32 capacitive sensors with a resolution of 65zF rms," in *2016 IEEE International Solid-State Circuits Conference (ISSCC)*, 2016, pp. 486–488: IEEE.

[129] M. Carminati, G. Ferrari, F. Guagliardo, and M. Sampietro, "ZeptoFarad capacitance detection with a miniaturized CMOS current front-end for nanoscale sensors," *Sensors and Actuators A: Physical*, vol. 172, no. 1, pp. 117–123, 2011.

[130] G. Vlachogiannakis, M. A. Pertijs, M. Spirito, and L. C. de Vreede, "A 40-nm CMOS complex permittivity sensing pixel for material characterization at microwave frequencies," *IEEE Transactions on Microwave Theory and Techniques*, vol. 66, no. 3, pp. 1619–1634, 2018.

[131] V. Sekar, W. J. Torke, S. Palermo, and K. Entesari, "A self-sustained microwave system for dielectric-constant measurement of lossy organic liquids," *IEEE Transactions on Microwave Theory and Techniques*, vol. 60, no. 5, pp. 1444–1455, 2012.

[132] A. A. Helmy, H.-J. Jeon, Y.-C. Lo *et al.*, "A self-sustained CMOS microwave chemical sensor using a frequency synthesizer," *IEEE Journal of Solid-State Circuits*, vol. 47, no. 10, pp. 2467–2483, 2012.

[133] J. Nehring, M. Bartels, R. Weigel, and D. Kissinger, "A permittivity sensitive PLL based on a silicon-integrated capacitive sensor for microwave biosensing applications," in *Biomedical Wireless Technologies, Networks, and Sensing Systems (BioWireleSS), 2015 IEEE Topical Conference on*, 2015, pp. 1–3: IEEE.

[134] S. Guha, K. Schmalz, C. Wenger, and F. Herzel, "Self-calibrating highly sensitive dynamic capacitance sensor: towards rapid sensing and counting

of particles in laminar flow systems," *Analyst*, vol. 140, no. 9, pp. 3262–3272, 2015.

[135]   O. Elhadidy, M. Elkholy, A. A. Helmy, S. Palermo, and K. Entesari, "A CMOS fractional-N PLL-based microwave chemical sensor with 1.5% permittivity accuracy," *IEEE Transactions on Microwave Theory and Techniques*, vol. 61, no. 9, pp. 3402–3416, 2013.

[136]   T. Mitsunaka, D. Sato, N. Ashida *et al.*, "CMOS biosensor IC focusing on dielectric relaxations of biological water with 120 and 60 GHz oscillator arrays," *IEEE Journal of Solid-State Circuits*, vol. 51, no. 11, pp. 2534–2544, 2016.

[137]   J.-C. Chien, M. Anwar, E.-C. Yeh, L. P. Lee, and A. M. Niknejad, "A 6.5/17.5-GHz dual-channel interferometer-based capacitive sensor in 65-nm CMOS for high-speed flow cytometry," in *2014 IEEE MTT-S International Microwave Symposium (IMS)*, 2014, pp. 1–4: IEEE.

[138]   J.-C. Chien, E.-C. Yeh, L. P. Lee, M. Anwar, and A. M. Niknejad, "A near-field modulation chopping stabilized injection-locked oscillator sensor for protein conformation detection at microwave frequency," in *2015 Symposium on VLSI Circuits (VLSI Circuits)*, 2015, pp. C332–C333: IEEE.

[139]   J.-C. Chien and A. M. Niknejad, "Oscillator-based reactance sensors with injection locking for high-throughput flow cytometry using microwave dielectric spectroscopy," *IEEE Journal of Solid-State Circuits*, vol. 51, no. 2, pp. 457–472, 2016.

[140]   O. Elhadidy, S. Shakib, K. Krenek, S. Palermo, and K. Entesari, "A wide-band fully-integrated CMOS ring-oscillator PLL-based complex dielectric spectroscopy system," *IEEE Transactions on Circuits and Systems I: Regular Papers*, vol. 62, no. 8, pp. 1940–1949, 2015.

[141]   T. York, T. Phua, L. Reichelt, A. Pawlowski, and R. Kneer, "A miniature electrical capacitance tomograph," *Measurement Science and Technology*, vol. 17, no. 8, p. 2119, 2006.

[142]   S. B. Prakash and P. Abshire, "A fully differential rail-to-rail CMOS capacitance sensor with floating-gate trimming for mismatch compensation," *IEEE Transactions on Circuits and Systems I: Regular Papers*, vol. 56, no. 5, pp. 975–986, 2009.

[143]   Y. Yusof, K. Sugimoto, H. Ozawa, S. Uno, and K. Nakazato, "On-chip microelectrode capacitance measurement for biosensing applications," *Japanese Journal of Applied Physics*, vol. 49, no. 1S, pp. 01AG05 (1–6), 2010.

[144]   F. Widdershoven, A. Cossettini, C. Laborde *et al.*, "A CMOS pixelated nanocapacitor biosensor platform for high-frequency impedance spectroscopy and imaging," *IEEE Transactions on Biomedical Circuits and Systems*, 2018.

[145]   E. Ghafar-Zadeh and M. Sawan, "A core-CBCM sigma delta capacitive sensor array dedicated to lab-on-chip applications," *Sensors and Actuators A: Physical*, vol. 144, no. 2, pp. 304–313, 2008.

[146]   E. Ghafar-Zadeh and M. Sawan, "High accuracy differential capacitive circuit for bioparticles sensing applications," in *48th Midwest Symposium on Circuits and Systems*, 2005, pp. 1362–1365: IEEE.

[147] Y.-W. Chang, H.-W. Chang, T.-C. Lu *et al.*, "Charge-based capacitance measurement for bias-dependent capacitance," *IEEE Electron Device Letters*, vol. 27, no. 5, pp. 390–392, 2006.

[148] B. Sell, A. Avellán, and W. H. Krautschneider, "Charge-based capacitance measurements (CBCM) on MOS devices," *IEEE Transactions on Device and Materials Reliability*, vol. 2, no. 1, pp. 9–12, 2002.

[149] H. Zhao, R. Kim, A. Paul *et al.*, "Characterization and modeling of sub-femtofarad nanowire capacitance using the CBCM technique," *IEEE Electron Device Letters*, vol. 30, no. 5, pp. 526–528, 2009.

[150] P. Zhang, Q. Wan, C. Feng, and H. Wang, "All regimes parasitic capacitances extraction using a multi-channel CBCM technique," *IEEE Transactions on Semiconductor Manufacturing*, vol. 30, no. 2, pp. 121–125, 2017.

[151] K. Sawada, G. Van der Plas, Y. Miyamori *et al.*, "Characterization of capacitance mismatch using simple difference charge-based capacitance measurement (DCBCM) test structure," in *Microelectronic Test Structures (ICMTS), 2013 IEEE International Conference on*, 2013, pp. 49–52: IEEE.

[152] K. Sawada, G. Van der Plas, S. Mori *et al.*, "In-line monitoring test structure for charge-based capacitance measurement (CBCM) with a start-stop self-pulsing circuit," in *Microelectronic Test Structures (ICMTS), 2015 International Conference on*, 2015, pp. 145–149: IEEE.

[153] C.-F. Huang and C.-C. Chen, "Capacitor arrays for minimizing gradient effects and methods of forming the same," ed: Google Patents, 2014.

[154] D. Sylvester, J. C. Chen, and C. Hu, "Investigation of interconnect capacitance characterization using charge-based capacitance measurement (CBCM) technique and three-dimensional simulation," *IEEE Journal of Solid-State Circuits*, vol. 33, no. 3, pp. 449–453, 1998.

[155] J. C. Chen, B. W. McGaughy, D. Sylvester, and C. Hu, "An on-chip, attofarad interconnect charge-based capacitance measurement (CBCM) technique," in *Electron Devices Meeting, 1996. IEDM'96, International*, 1996, pp. 69–72: IEEE.

[156] B. Ward, J. Bordelon, S. Prior, B. Tranchina, and J. Liu, "Test structure and method for capacitance extraction in multi-conductor systems," in *Microelectronic Test Structures, 2001. ICMTS 2001. Proceedings of the 2001 International Conference on*, 2001, pp. 189–193: IEEE.

[157] A. Bogliolo, L. Vendrame, L. Bortesi, and E. Barachetti, "Charge-based on-chip measurement technique for the selective extraction of cross-coupling capacitances," *Signal Propagation on Interconnects*, 2002.

[158] M.-C. Tang, C.-C. Cheng, M.-F. Wang *et al.*, "A nondestructive method of extracting the width and thickness of interconnects for a 40-nm technology," *IEEE Transactions on Electron Devices*, vol. 56, no. 9, pp. 1891–1896, 2009.

[159] P. Zhang, C. Feng, H. Wang, and W. Shan, "Analysis and characterization of capacitance variation using capacitance measurement array," *IEEE Transactions on Semiconductor Manufacturing*, vol. 27, no. 2, pp. 301–311, 2014.

[160] N. D. Arora and J. Wang, "Method of measuring interconnect coupling capacitance in an IC chip," ed: Google Patents, 1999.

[161]   C.-T. Fan and J.-H. Wang, "Chip capacitance measurement circuit," ed: Google Patents, 2002.

[162]   B. Froment, F. Paillardet, M. Bely *et al.*, "Ultra low capacitance measurements in multilevel metallisation CMOS by using a built-in electronmeter," in *Electron Devices Meeting, 1999. IEDM'99. Technical Digest International*, 1999, pp. 897–900: IEEE.

[163]   S. Saxena and Y. Yu, "Method for accurate measurement of leaky capacitors using charge based capacitance measurements," ed: Google Patents, 2018.

[164]   S. Mori, K. Ogawa, H. Oishi *et al.*, "Monitoring test structure for plasma process induced charging damage using charge-based capacitance measurement (PID-CBCM)," in *Microelectronic Test Structures (ICMTS), 2015 International Conference on*, 2015, pp. 132–137: IEEE.

[165]   Y.-W. Chang, H.-W. Chang, T.-C. Lu, Y.-C. King, K.-C. Chen, and C.-Y. Lu, "Combining a novel charge-based capacitance measurement (CBCM) technique and split C-V method to specifically characterize the STI stress effect along the width direction of MOSFET devices," *IEEE Electron Device Letters*, vol. 29, no. 6, pp. 641–644, 2008.

[166]   T. Datta-Chaudhuri, E. Smela, and P. A. Abshire, "System-on-chip considerations for heterogeneous integration of CMOS and fluidic bio-interfaces," *IEEE Transactions on Biomedical Circuits and Systems*, vol. 10, no. 6, pp. 1129–1142, 2016.

[167]   H. Wang, A. Mahdavi, D. A. Tirrell, and A. Hajimiri, "A magnetic cell-based sensor," *Lab on a Chip*, vol. 12, no. 21, pp. 4465–4471, 2012.

[168]   T. Prodromakis, K. Michelakis, T. Zoumpoulidis, R. Dekker, and C. Toumazou, "Biocompatible encapsulation of CMOS based chemical sensors," in *Sensors, 2009 IEEE*, 2009, pp. 791–794: IEEE.

[169]   Y. Huang and A. J. Mason, "Lab-on-CMOS: integrating microfluidics and electrochemical sensor on CMOS," in *2011 6th IEEE International Conference on Nano/Micro Engineered and Molecular Systems*, Kaohsiung, Taiwan, 2011, pp. 690–693: IEEE.

[170]   A. Wu, L. Wang, E. Jensen, R. Mathies, and B. Boser, "Modular integration of electronics and microfluidic systems using flexible printed circuit boards," *Lab on a Chip*, vol. 10, no. 4, pp. 519–521, 2010.

[171]   L. Li and A. J. Mason, "Post-CMOS parylene packaging for on-chip biosensor arrays," in *Sensors, 2010 IEEE*, 2010, pp. 1613–1616: IEEE.

[172]   E. Ghafar-Zadeh, M. Sawan, and D. Therriault, "Novel direct-write CMOS-based laboratory-on-chip: design, assembly and experimental results," *Sensors and Actuators A: Physical*, vol. 134, no. 1, pp. 27–36, 2007.

[173]   L. Li, H. Yin, and A. J. Mason, "Epoxy chip-in-carrier integration and screen-printed metalization for multichannel microfluidic lab-on-CMOS microsystems," *IEEE Transactions on Biomedical Circuits and Systems*, 2018.

[174]   I. Burdallo, C. Jimenez-Jorquera, C. Fernández-Sánchez, and A. Baldi, "Integration of microelectronic chips in microfluidic systems on printed circuit board," *Journal of Micromechanics and Microengineering*, vol. 22, no. 10, pp. 105022 (1–7), 2012.

[175] A. Hassibi, S. Zahedi, R. Navid, R. W. Dutton, and T. H. Lee, "Biological shot-noise and quantum-limited signal-to-noise ratio in affinity-based biosensors," *Journal of Applied Physics*, vol. 97, no. 8, p. 084701, 2005.

[176] N. Couniot, "Highly-sensitive CMOS capacitive biosensors towards detection of single bacterial cell in electrolyte solutions," PhD thesis, UCL, 2015.

[177] D. R. Thévenot, K. Toth, R. A. Durst, and G. S. Wilson, "Electrochemical biosensors: recommended definitions and classification," *Biosensors and Bioelectronics*, vol. 16, no. 1-2, pp. 121–131, 2001.

[178] A. Shabani, M. Zourob, B. Allain, C. A. Marquette, M. F. Lawrence, and R. Mandeville, "Bacteriophage-modified microarrays for the direct impedimetric detection of bacteria," *Analytical Chemistry*, vol. 80, no. 24, pp. 9475–9482, 2008.

[179] T. J. Kang, D.-K. Lim, J.-M. Nam, and Y. H. Kim, "Multifunctional nanocomposite membrane for chemomechanical transducer," *Sensors and Actuators B: Chemical*, vol. 147, no. 2, pp. 691–696, 2010.

[180] N. Van Overstraeten-Schlögel, O. Lefèvre, N. Couniot, and D. Flandre, "Assessment of different functionalization methods for grafting a protein to an alumina-covered biosensor," *Biofabrication*, vol. 6, no. 3, p. 035007, 2014.

[181] Y. H. Ghallab and Y. Ismail, "CMOS based lab-on-a-chip: applications, challenges and future trends," *IEEE Circuits and Systems Magazine*, vol. 14, no. 2, pp. 27–47, 2014.

[182] N.-T. Nguyen, M. Hejazian, C. H. Ooi, and N. Kashaninejad, "Recent advances and future perspectives on microfluidic liquid handling," *Micromachines*, vol. 8, no. 6, p. 186, 2017.

[183] G. Nabovati, E. Ghafar-Zadeh, M. Mirzaei, G. Ayala-Charca, F. Awwad, and M. Sawan, "A new fully differential CMOS capacitance to digital converter for lab-on-chip applications," *IEEE Transactions on Biomedical Circuits and Systems*, vol. 9, no. 3, pp. 353–361, 2015.

[184] C. Kokkinos, A. Economou, and M. I. Prodromidis, "Electrochemical immunosensors: critical survey of different architectures and transduction strategies," *TrAC Trends in Analytical Chemistry*, vol. 79, pp. 88–105, 2016.

[185] A. J. Bard and L. R. Faulkner, "Fundamentals and applications," *Electrochemical Methods*, vol. 2, p. 482, 2001.

[186] Q. Liu and P. Wang, *Cell-Based Biosensors: Principles and Applications.* Norwood, MA, USA: Artech House, 2009.

[187] M. Gouy, "Sur la constitution de la charge électrique à la surface d'un électrolyte," *Journal of Theoretical and Applied Physics*, vol. 9, no. 1, pp. 457–468, 1910.

[188] D. L. Chapman, "LI. A contribution to the theory of electrocapillarity," *The London, Edinburgh, and Dublin Philosophical Magazine and Journal of Science*, vol. 25, no. 148, pp. 475–481, 1913.

[189] S. B. Prakash and P. Abshire, "Tracking cancer cell proliferation on a CMOS capacitance sensor chip," *Biosensors and Bioelectronics*, vol. 23, no. 10, pp. 1449–1457, 2008.

[190] W. M. Siu and R. S. Cobbold, "Basic properties of the electrolyte–SiO_2–Si system: physical and theoretical aspects," *IEEE Transactions on Electron Devices*, vol. 26, no. 11, pp. 1805–1815, 1979.

[191] M. W. den Otter, "Approximate expressions for the capacitance and electrostatic potential of interdigitated electrodes," *Sensors and Actuators A: Physical*, vol. 96, no. 2-3, pp. 140–144, 2002.

[192] C. Berggren, B. Bjarnason, and G. Johansson, "Capacitive biosensors," *Electroanalysis: An International Journal Devoted to Fundamental and Practical Aspects of Electroanalysis*, vol. 13, no. 3, pp. 173–180, 2001.

[193] O. Laczka, E. Baldrich, F. X. Muñoz, and F. J. del Campo, "Detection of *Escherichia coli* and *Salmonella typhimurium* using interdigitated microelectrode capacitive immunosensors: the importance of transducer geometry," *Analytical Chemistry*, vol. 80, no. 19, pp. 7239–7247, 2008.

[194] P. Van Gerwen, W. Laureys, G. Huyberechts *et al.*, "Nanoscaled interdigitated electrode arrays for biochemical sensors," in *Solid State Sensors and Actuators, 1997. Transducers'97 Chicago, 1997 International Conference on*, 1997, vol. 2, pp. 907–910: IEEE.

[195] S. O. Blume, R. Ben-Mrad, and P. E. Sullivan, "Modelling the capacitance of multi-layer conductor-facing interdigitated electrode structures," *Sensors and Actuators B: Chemical*, vol. 213, pp. 423–433, 2015.

[196] R. Igreja and C. Dias, "Extension to the analytical model of the interdigital electrodes capacitance for a multi-layered structure," *Sensors and Actuators A: Physical*, vol. 172, no. 2, pp. 392–399, 2011.

[197] R. Igreja and C. Dias, "Analytical evaluation of the interdigital electrodes capacitance for a multi-layered structure," *Sensors and Actuators A: Physical*, vol. 112, no. 2-3, pp. 291–301, 2004.

[198] S. O. Blume, R. Ben-Mrad, and P. E. Sullivan, "Characterization of coplanar electrode structures for microfluidic-based impedance spectroscopy," *Sensors and Actuators B: Chemical*, vol. 218, pp. 261–270, 2015.

[199] J. Wei, "Distributed capacitance of planar electrodes in optic and acoustic surface wave devices," *IEEE Journal of Quantum Electronics*, vol. 13, no. 4, pp. 152–158, 1977.

[200] R. Igreja and C. Dias, "Dielectric response of interdigital chemocapacitors: The role of the sensitive layer thickness," *Sensors and Actuators B: Chemical*, vol. 115, no. 1, pp. 69–78, 2006.

[201] C. Jungreuthmayer, G. M. Birnbaumer, J. Zanghellini, and P. Ertl, "3D numerical simulation of a lab-on-a-chip—increasing measurement sensitivity of interdigitated capacitors by passivation optimization," *Lab on a Chip*, vol. 11, no. 7, pp. 1318–1325, 2011.

[202] A. Bratov, J. Ramón-Azcón, N. Abramova *et al.*, "Three-dimensional interdigitated electrode array as a transducer for label-free biosensors," *Biosensors and Bioelectronics*, vol. 24, no. 4, pp. 729–735, 2008.

[203] H.-W. Jung, Y. W. Chang, G.-Y. Lee, S. Cho, M.-J. Kang, and J.-C. Pyun, "A capacitive biosensor based on an interdigitated electrode with nanoislands," *Analytica Chimica Acta*, vol. 844, pp. 27–34, 2014.

[204] S. Rana, R. H. Page, and C. J. McNeil, "Comparison of planar and 3-D interdigitated electrodes as electrochemical impedance biosensors," *Electroanalysis*, vol. 23, no. 10, pp. 2485–2490, 2011.

[205] R. B. Schoch and P. Renaud, "Ion transport through nanoslits dominated by the effective surface charge," *Applied Physics Letters*, vol. 86, no. 25, p. 253111, 2005.

[206] D. Stein, M. Kruithof, and C. Dekker, "Surface-charge-governed ion transport in nanofluidic channels," *Physical Review Letters*, vol. 93, no. 3, p. 035901, 2004.

[207] R. B. Schoch, H. Van Lintel, and P. Renaud, "Effect of the surface charge on ion transport through nanoslits," *Physics of Fluids*, vol. 17, no. 10, p. 100604, 2005.

[208] M. Bäcker, F. Kramer, C. Huck *et al.*, "Planar and 3D interdigitated electrodes for biosensing applications: the impact of a dielectric barrier on the sensor properties," *Physica Status Solidi (A)*, vol. 211, no. 6, pp. 1357–1363, 2014.

[209] E. Ghafar-Zadeh, G. Ayala-Charca, B. Gholamzadeh, S. Ghasemi, and S. Magierowski, "Towards scalable capacitive cantilever arrays for emerging biomedical applications," *Sensors and Actuators A: Physical*, vol. 260, pp. 90–98, 2017.

[210] A. O. Manzanares and F. M. de Espinosa, "Air-coupled MUMPs capacitive micromachined ultrasonic transducers with resonant cavities," *Ultrasonics*, vol. 52, no. 4, pp. 482–489, 2012.

[211] E. T. Carlen, M. S. Weinberg, A. M. Zapata, and J. T. Borenstein, "A micromachined surface stress sensor with electronic readout," *Review of Scientific Instruments*, vol. 79, no. 1, pp. 015106 (1–7), 2008.

[212] E. T. Carlen, M. S. Weinberg, C. E. Dubé, A. M. Zapata, and J. T. Borenstein, "Micromachined silicon plates for sensing molecular interactions," *Applied Physics Letters*, vol. 89, no. 17, pp. 173123 (1–4), 2006.

[213] A. Hassibi and T. H. Lee, "A programmable 0.18-μm CMOS electrochemical sensor microarray for biomolecular detection," *IEEE Sensors Journal*, vol. 6, no. 6, pp. 1380–1388, 2006.

[214] N. Nikkhoo, C. Man, K. Maxwell, and P. G. Gulak, "A 0.18 μm CMOS integrated sensor for the rapid identification of bacteria," in *Solid-State Circuits Conference, 2008. ISSCC 2008. Digest of Technical Papers. IEEE International*, 2008, pp. 336–617: IEEE.

[215] J. Abbott, T. Ye, L. Qin *et al.*, "CMOS nanoelectrode array for all-electrical intracellular electrophysiological imaging," *Nature Nanotechnology*, vol. 12, no. 5, pp. 460–467, 2017.

[216] U. Frey, J. Sedivy, F. Heer *et al.*, "Switch-matrix-based high-density microelectrode array in CMOS technology," *IEEE Journal of Solid-State Circuits*, vol. 45, no. 2, pp. 467–482, 2010.

[217] M. Ballini, J. Müller, P. Livi *et al.*, "A 1024-channel CMOS microelectrode array with 26,400 electrodes for recording and stimulation of electrogenic

cells *in vitro*," *IEEE Journal of Solid-State Circuits*, vol. 49, no. 11, pp. 2705–2719, 2014.

[218] V. Viswam, R. Bounik, A. Shadmani *et al.*, "Impedance spectroscopy and electrophysiological imaging of cells with a high-density CMOS micro-electrode array system," *IEEE Transactions on Biomedical Circuits and Systems*, vol. 12, pp. 1356–1368, 2018.

[219] B. Miccoli, C. M. Lopez, E. Goikoetxea *et al.*, "High-density electrical recording and impedance imaging with a multi-modal CMOS multi-electrode array chip," *Frontiers in Neuroscience*, vol. 13, p. 641, 2019.

[220] S. B. Prakash, P. Abshire, M. Urdaneta, and E. Smela, "A CMOS capacitance sensor for cell adhesion characterization," in *2005 IEEE International Symposium on Circuits and Systems*, 2005, pp. 3495–3498: IEEE.

[221] S. B. Prakash and P. Abshire, "A CMOS capacitance sensor that monitors cell viability," in *IEEE Sensors, 2005.*, 2005, p. 4: IEEE.

[222] S.-W. Wang and M. S.-C. Lu, "CMOS capacitive sensors with sub-μm microelectrodes for biosensing applications," *IEEE Sensors Journal*, vol. 10, no. 5, pp. 991–996, 2010.

[223] E. Ghafar-Zadeh, M. Sawan, and D. Therriault, "A microfluidic packaging technique for lab-on-chip applications," *IEEE Transactions on Advanced Packaging*, vol. 32, no. 2, pp. 410–416, 2009.

[224] E. Ghafar-Zadeh, M. Sawan, and D. Therriault, "CMOS based capacitive sensor laboratory-on-chip: a multidisciplinary approach," *Analog Integrated Circuits and Signal Processing*, vol. 59, no. 1, pp. 1–12, 2009.

[225] A. M. Kummer, A. Hierlemann, and H. Baltes, "Tuning sensitivity and selectivity of complementary metal oxide semiconductor-based capacitive chemical microsensors," *Analytical Chemistry*, vol. 76, no. 9, pp. 2470–2477, 2004.

[226] C. Hagleitner, D. Lange, A. Hierlemann, O. Brand, and H. Baltes, "CMOS single-chip gas detection system comprising capacitive, calorimetric and mass-sensitive microsensors," *IEEE Journal of Solid-State Circuits*, vol. 37, no. 12, pp. 1867–1878, 2002.

[227] A. Hierlemann, "Integrated chemical microsensor systems in CMOS-technology," in *Solid-State Sensors, Actuators and Microsystems, 2005. Digest of Technical Papers. TRANSDUCERS'05. The 13th International Conference on*, 2005, vol. 2, pp. 1134–1137: IEEE.

[228] C. Cornila, A. Hierlemann, R. Lenggenhager *et al.*, "Capacitive sensors in CMOS technology with polymer coating," *Sensors and Actuators B: Chemical*, vol. 25, no. 1-3, pp. 357–361, 1995.

[229] A. Hierlemann, U. Weimar, and H. Baltes, "Hand-held and palm-top chemical microsensor systems for gas analysis," *in Handbook of Machine Olfaction: Electronic Nose Technology*, USA, 2002, pp. 201–229.

[230] L. Moreno-Hagelsieb, B. Foultier, G. Laurent *et al.*, "Electrical detection of DNA hybridization: three extraction techniques based on inter-digitated Al/Al2O3 capacitors," *Biosensors and Bioelectronics*, vol. 22, no. 9-10, pp. 2199–2207, 2007.

[231] X. Chen, X. Yan, K. A. Khor, and B. K. Tay, "Multilayer assembly of positively charged polyelectrolyte and negatively charged glucose oxidase on a 3D Nafion network for detecting glucose," *Biosensors and Bioelectronics*, vol. 22, no. 12, pp. 3256–3260, 2007.

[232] S. Zhang, W. Yang, Y. Niu, Y. Li, M. Zhang, and C. Sun, "Construction of glucose biosensor based on sorption of glucose oxidase onto multilayers of polyelectrolyte/nanoparticles," *Analytical and Bioanalytical Chemistry*, vol. 384, no. 3, pp. 736–741, 2006.

[233] B. Thierry, F. M. Winnik, Y. Merhi, J. Silver, and M. Tabrizian, "Bioactive coatings of endovascular stents based on polyelectrolyte multilayers," *Biomacromolecules*, vol. 4, no. 6, pp. 1564–1571, 2003.

[234] G. Nabovati, Y. W. Zhu, and M. Sawan, "Capacitive sensor arrays," *Wiley Encyclopedia of Electrical and Electronics Engineering*, pp. 1–12, 2015.

[235] A. Sun, E. Alvarez-Fontecilla, A. Venkatesh, E. Aronoff-Spencer, and D. A. Hall, "A 64 × 64 high-density redox amplified coulostatic discharge-based biosensor array in 180 nm CMOS," in *ESSCIRC 2017-43rd IEEE European Solid State Circuits Conference*, 2017, pp. 368–371: IEEE.

[236] F. Hofmann, A. Frey, B. Holzapfl *et al.*, "Passive DNA sensor with gold electrodes fabricated in a CMOS backend process," in *Solid-State Device Research Conference, 2002. Proceeding of the 32nd European*, 2002, pp. 487–490: IEEE.

[237] C. Guiducci, C. Stagni, A. Fischetti, U. Mastromatteo, L. Benini, and B. Riccoricco, "Microelectrodes on a silicon chip for label-free capacitive DNA sensing," *IEEE Sensors Journal*, vol. 6, no. 5, pp. 1084–1093, 2006.

[238] R. K. Shervedani, A. H. Mehrjardi, and N. Zamiri, "A novel method for glucose determination based on electrochemical impedance spectroscopy using glucose oxidase self-assembled biosensor," *Bioelectrochemistry*, vol. 69, no. 2, pp. 201–208, 2006.

[239] K. Kerman, D. Ozkan, P. Kara, B. Meric, J. J. Gooding, and M. Ozsoz, "Voltammetric determination of DNA hybridization using methylene blue and self-assembled alkanethiol monolayer on gold electrodes," *Analytica Chimica Acta*, vol. 462, no. 1, pp. 39–47, 2002.

[240] L. Berdondini, P. D. van der Wal, N. F. de Rooij, and M. Koudelka-Hep, "Development of an electroless post-processing technique for depositing gold as electrode material on CMOS devices," *Sensors and Actuators B: Chemical*, vol. 99, no. 2-3, pp. 505–510, 2004.

[241] G. O. Mallory and J. B. Hajdu, *Electroless Plating: Fundamentals and Applications*. Orlando: William Andrew, 1990.

[242] F. Widdershoven, D. Van Steenwinckel, J. Überfeld *et al.*, "CMOS biosensor platform," in *2010 IEEE International Electron Devices Meeting (IEDM)*, 2010, pp. 36.1. 1–36.1. 4: IEEE.

[243] J. W. Ko, H. C. Koo, D. W. Kim *et al.*, "Electroless gold plating on aluminum patterned chips for CMOS-based sensor applications," *Journal of the Electrochemical Society*, vol. 157, no. 1, pp. D46–D49, 2010.

[244] M. Helou, M. Reisbeck, S. F. Tedde *et al.*, "Time-of-flight magnetic flow cytometry in whole blood with integrated sample preparation," *Lab on a Chip*, vol. 13, no. 6, pp. 1035–1038, 2013.

[245] S. G. Lemay, C. Laborde, C. Renault, A. Cossettini, L. Selmi, and F. P. Widdershoven, "High-frequency nanocapacitor arrays: concept, recent developments, and outlook," *Accounts of Chemical Research*, vol. 49, no. 10, pp. 2355–2362, 2016.

[246] F. Greve, J. Lichtenberg, K. Kirstein, U. Frey, J. Perriard, and A. Hierlemann, "A perforated CMOS microchip for immobilization and activity monitoring of electrogenic cells," *Journal of Micromechanics and Microengineering*, vol. 17, no. 3, p. 462, 2007.

[247] F. Heer, S. Hafizovic, W. Franks, A. Blau, C. Ziegler, and A. Hierlemann, "CMOS microelectrode array for bidirectional interaction with neuronal networks," *IEEE Journal of Solid-State Circuits*, vol. 41, no. 7, pp. 1620–1629, 2006.

[248] C. M. Lopez, H. S. Chun, S. Wang *et al.*, "A multimodal CMOS MEA for high-throughput intracellular action potential measurements and impedance spectroscopy in drug-screening applications," *IEEE Journal of Solid-State Circuits*, vol. 53, no. 11, pp. 3076–3086, 2018.

[249] K. Georgakopoulou, C. Spathis, N. Petrellis, and A. Birbas, "A capacitive-to-digital converter with automatic range adaptation for readout instrumentation," *IEEE Transactions on Instrumentation and Measurement*, vol. 65, no. 2, pp. 336–345, 2016.

[250] H. Ha, D. Sylvester, D. Blaauw, and J.-Y. Sim, "12.6 A 160nW 63.9 fJ/conversion-step capacitance-to-digital converter for ultra-low-power wireless sensor nodes," in *Solid-State Circuits Conference Digest of Technical Papers (ISSCC), 2014 IEEE International*, 2014, pp. 220–221: IEEE.

[251] K. Tanaka, Y. Kuramochi, T. Kurashina, K. Okada, and A. Matsuzawa, "A 0.026 mm^2 capacitance-to-digital converter for biotelemetry applications using a charge redistribution technique," in *IEEE Asian Solid-State Circuits Conference, 2007. ASSCC'07*, 2007, pp. 244–247: IEEE.

[252] A. Alhoshany and K. N. Salama, "A precision, energy-efficient, oversampling, noise-shaping differential SAR capacitance-to-digital converter," *IEEE Transactions on Instrumentation and Measurement*, vol. 68, no. 2, pp. 392–401, 2018.

[253] R. Nagulapalli, K. Hayatleh, S. Barker, S. Raparthy, and F. J. Lidgey, "A CMOS blood cancer detection sensor based on frequency deviation detection," *Analog Integrated Circuits and Signal Processing*, vol. 92, no. 3, pp. 437–442, 2017.

[254] X. Liu, L. Li, and A. J. Mason, "High-throughput impedance spectroscopy biosensor array chip," *Philosophical Transactions of the Royal Society A*, vol. 372, no. 2012, pp. 20130107 (1–14), 2014.

[255] A. Manickam, C. A. Johnson, S. Kavusi, and A. Hassibi, "Interface design for CMOS-integrated electrochemical impedance spectroscopy (EIS) biosensors," *Sensors*, vol. 12, no. 11, pp. 14467–14488, 2012.

[256] C. Guiducci, C. Stagni, G. Zuccheri *et al.*, "DNA detection by integrable electronics," *Biosensors and Bioelectronics*, vol. 19, no. 8, pp. 781–787, 2004.

[257] M. A. Miled and M. Sawan, "Dielectrophoresis-based integrated lab-on-chip for nano and micro-particles manipulation and capacitive detection," *IEEE Transactions on Biomedical Circuits and Systems*, vol. 6, no. 2, pp. 120–132, 2012.

[258] R. Bach, B. Davis, and R. Laubhan, "Improvements to CBCM (charge-based capacitance measurement) for deep submicron CMOS technology," in *Proceedings of the 7th International Symposium on Quality Electronic Design*, 2006, pp. 324–329: IEEE Computer Society.

[259] Y.-W. Chang, H.-W. Chang, C.-H. Hsieh *et al.*, "A novel simple CBCM method free from charge injection-induced errors," *IEEE Electron Device Letters*, vol. 25, no. 5, pp. 262–264, 2004.

[260] Y. Chang, H. Chang, T. Lu *et al.*, "A novel CBCM method free from charge injection induced errors: investigation into the impact of floating dummy-fills on interconnect capacitance," in *Microelectronic Test Structures, 2005. ICMTS 2005. Proceedings of the 2005 International Conference on*, 2005, pp. 235–238: IEEE.

[261] L. Vendrame, L. Bortesi, and A. Bogliolo, "Accuracy assessment and improvement of on-chip charge-based capacitance measurements," *in Proc. 7th IEEE SPI Workshop*, 2003, pp. 117–120.

[262] K. H. Dia, W. Tsao, C. H. Chien, and Z. Zeng, "A novel compact CBCM method for high resolution measurement in 28 nm CMOS technology," in *Microelectronic Test Structures (ICMTS), 2012 IEEE International Conference on*, 2012, pp. 119–121: IEEE.

[263] Z. Ning, H.-X. Delecourt, L. De Schepper, R. Gillon, and M. Tack, "Precise analogue characterization of MIM capacitors using an improved charge-based capacitance measurement (CBCM) technique," in *Solid-State Device Research Conference, 2005. ESSDERC 2005. Proceedings of 35th European*, 2005, pp. 269–272: IEEE.

[264] S. B. Prakash and P. Abshire, "A fully differential rail-to-rail capacitance measurement circuit for integrated cell sensing," in *2007 IEEE Sensors*, 2007, pp. 1444–1447: IEEE.

[265] P. Zhang, Q. Wan, C. Feng, and H. Wang, "Gate capacitance measurement using a self-differential charge-based capacitance measurement method," *IEEE Electron Device Letters*, vol. 36, no. 12, pp. 1271–1273, 2015.

[266] X. P. Yu, R. Q. Tian, W. L. Xu, and Z. Shi, "A new on-chip signal generator for charge-based capacitance measurement circuit," *Journal of Electronic Testing*, vol. 31, no. 3, pp. 329–333, 2015.

[267] Y. Yusof and N. Kazuo, "Ultra sensitive CMOS biosensor array for label free DNA detection: circuit design consideration," in *Biomedical Engineering and Sciences (IECBES), 2014 IEEE Conference on*, 2014, pp. 365–368: IEEE.

[268] C. Guiducci, C. Stagni, G. Zuccheri *et al.*, "A biosensor for direct detection of DNA sequences based on capacitance measurements," in *Proceeding of the 32nd European Solid-State Device Research Conference*, 2002, pp. 24–26.

[269] C. Stagni, C. Guiducci, M. Lanzoni, L. Benini, and B. Ricco, "Hardware-software design of a smart sensor for fully-electronic DNA hybridization detection," in *Proceedings of the conference on Design, Automation and Test in Europe-Volume 3*, 2005, pp. 198–203: IEEE Computer Society.

[270] S. Carrara, L. Benini, V. Bhalla *et al.*, "New insights for using self-assembly materials to improve the detection stability in label-free DNA-chip and immuno-sensors," *Biosensors and Bioelectronics*, vol. 24, no. 12, pp. 3425–3429, 2009.

[271] S. Carrara, A. Cavallini, Y. Leblebici *et al.*, "Capacitance DNA bio-chips improved by new probe immobilization strategies," *Microelectronics Journal*, vol. 41, no. 11, pp. 711–717, 2010.

[272] A. M. Sampson, E. T. Peterson, and I. Papautsky, "Interdigitated array microelectrode capacitive sensor for detection of paraffinophilic myco-bacteria," in *Microfluidics, BioMEMS, and Medical Microsystems VI*, 2008, vol. 6886, p. 68860X: International Society for Optics and Photonics.

[273] A. Tanskanen, B. Bahreyni, and M. Syrzycki, "Charge-based femto-farad capacitance measurement technique for MEMS applications," in *2016 IEEE Canadian Conference on Electrical and Computer Engineering (CCECE)*, 2016, pp. 1–4: IEEE.

[274] T. York, I. Evans, Z. Pokusevski, and T. Dyakowski, "Particle detection using an integrated capacitance sensor," *Sensors and Actuators A: Physical*, vol. 92, no. 1-3, pp. 74–79, 2001.

[275] B. Kim, S. Uno, and K. Nakazato, "Wireless charge based capacitance measurement circuits with on-chip spiral inductor for radio frequency identification biosensor," *Japanese Journal of Applied Physics*, vol. 51, no. 4S, p. 04DE08, 2012.

[276] K. Nakazato, "BioCMOS LSIs for portable gene-based diagnostic inspection system," in *Circuits and Systems (ISCAS), 2012 IEEE International Symposium on*, 2012, pp. 2287–2290: IEEE.

[277] K. Nakazato, "Potentiometric, amperometric, and impedimetric CMOS biosensor array," in *State of the Art in Biosensors-General Aspects*: InTech, 2013.

[278] E. Ghafar-Zadeh and M. Sawan, "A high precision and linearity differential capacitive sensor circuit dedicated to bioparticles detection," in *The 3rd International IEEE-NEWCAS Conference, 2005*, 2005, pp. 299–302: IEEE.

[279] E. Ghafar-Zadeh, M. Sawan, and V. P. Chodavarapu, "Micro-organism-on-chip: emerging direct-write CMOS-based platform for biological applications," *IEEE Transactions on Biomedical Circuits and Systems*, vol. 3, no. 4, pp. 212–219, 2009.

[280] E. Ghafar-Zadeh and M. Sawan, "A hybrid microfluidic/CMOS capacitive sensor dedicated to lab-on-chip applications," *IEEE Transactions on Biomedical Circuits and Systems*, vol. 1, no. 4, pp. 270–277, 2007.

[281]   M. A. Miled, M. Sawan, and E. Ghafar-Zadeh, "A dynamic decoder for first-order $\Sigma\Delta$ modulators dedicated to lab-on-chip applications," *IEEE Transactions on Signal Processing*, vol. 57, no. 10, pp. 4076–4084, 2009.

[282]   M. A. Miled and M. Sawan, "Subthreshold transistor operation for a high sensitivity capacitive sensor," in *Canadian Conference on Electrical and Computer Engineering, 2008. CCECE 2008*, 2008, pp. 001671–001674: IEEE.

[283]   M. A. Miled and M. Sawan, "A new fully integrated CMOS interface for a dielectrophoretic lab-on-a-chip device," in *Circuits and Systems (ISCAS), 2011 IEEE International Symposium on*, 2011, pp. 2349–2352: IEEE.

[284]   M. A. Miled, G. Massicotte, and M. Sawan, "Low-voltage lab-on-chip for micro and nanoparticles manipulation and detection: experimental results," *Analog Integrated Circuits and Signal Processing*, vol. 73, no. 3, pp. 707–717, 2012.

[285]   T. Datta, E. Naviasky, and P. Abshire, "Floating-gate capacitance sensor array for cell viability monitoring," in *2013 IEEE Biomedical Circuits and Systems Conference (BioCAS)*, 2013, pp. 65–68: IEEE.

[286]   G. Nabovati, E. Ghafar-Zadeh, and M. Sawan, "A novel multifunctional integrated biosensor array for simultaneous monitoring of cell growth and acidification rate," in *2016 IEEE International Symposium on Circuits and Systems (ISCAS)*, 2016, pp. 2855–2858: IEEE.

[287]   Z. Jun-Rui, I. A. Nanolab, and M. Mazza, "Low-energy biomarker detection through charge-based impedance measurements," in *2016 IEEE SENSORS*, 2016, pp. 1–3: IEEE.

[288]   S. Forouhi, E. Ghafar-Zadeh, and R. Dehghani, "Multimodal CMOS based biosenors for lab-on-chip applications: a new multidisciplinary approach," in *Electrical and Computer Engineering (CCECE), 2017 IEEE 30th Canadian Conference on*, 2017, pp. 1–4: IEEE.

[289]   D. Zhu, J. Mo, S. Xu *et al.*, "A new capacitance-to-frequency converter for on-chip capacitance measurement and calibration in CMOS technology," *Journal of Electronic Testing*, pp. 1–5, 2016.

[290]   R. Yamane, H. Iwasaki, Y. Dei, J. Cui, and T. Matsuoka, "A capacitance detection circuit for on-chip microparticle manipulation," in *2014 IEEE International Meeting for Future of Electron Devices, Kansai (IMFEDK)*, 2014, pp. 1–2: IEEE.

[291]   A. M. Tashtoush, "Miniaturized IC biosensors on a single chip for biological/biomedical applications," *International Journal of Simulation—Systems, Science & Technology*, vol. 15, no. 6, pp. 8–18, 2014.

[292]   S. Forouhi, O. Farhanieh, R. Dehghani, and E. Ghafar-Zadeh, "A current based capacitance-to-frequency converter for lab-on-chip applications." in *2017 IEEE 60th International Midwest Symposium on Circuits and Systems (MWSCAS)*, Boston, MA, USA, 2017, pp. 92–95: IEEE.

[293]   C. Laborde, F. Pittino, H. Verhoeven *et al.*, "Real-time imaging of microparticles and living cells with CMOS nanocapacitor arrays," *Nature Nanotechnology*, vol. 10, no. 9, pp. 791 (1–6), 2015.

[294] S. M. Khan, A. Gumus, J. M. Nassar, and M. M. Hussain, "CMOS enabled microfluidic systems for healthcare based applications," *Advanced Materials*, pp. 1705759 (1–26), 2018.

[295] S.-S. Li and C.-M. Cheng, "Analogy among microfluidics, micromechanics, and microelectronics," *Lab on a Chip*, vol. 13, no. 19, pp. 3782–3788, 2013.

[296] D. Welch and J. B. Christen, "Seamless integration of CMOS and micro-fluidics using flip chip bonding," *Journal of Micromechanics and Microengineering*, vol. 23, no. 3, pp. 035009 (1–7), 2013.

[297] J. S. Park, M. K. Aziz, S. Li *et al.*, "1024-Pixel CMOS multimodality joint cellular sensor/stimulator array for real-time holistic cellular characterization and cell-based drug screening," *IEEE Transactions on Biomedical Circuits and Systems*, vol. 12, no. 1, pp. 80–94, 2017.

[298] T. B. Datta-Chaudhuri, "Integration of CMOS technology into lab-on-chip systems applied to the development of a bioelectronic nose," *Electrical and Computer Engineering*, University of Maryland, College Park, 2015.

[299] T. Datta-Chaudhuri, P. Abshire, and E. Smela, "Packaging commercial CMOS chips for lab on a chip integration," *Lab on a Chip*, vol. 14, no. 10, pp. 1753–1766, 2014.

[300] N. Halonen, J. Kilpijärvi, M. Sobocinski *et al.*, "Low temperature co-fired ceramic packaging of CMOS capacitive sensor chip towards cell viability monitoring," *Beilstein Journal of Nanotechnology*, vol. 7, p. 1871, 2016.

[301] F. Heer, W. Franks, A. Blau *et al.*, "CMOS microelectrode array for the monitoring of electrogenic cells," *Biosensors and Bioelectronics*, vol. 20, no. 2, pp. 358–366, 2004.

[302] A. H. Graham, C. R. Bowen, S. M. Surguy, J. Robbins, and J. Taylor, "New prototype assembly methods for biosensor integrated circuits," *Medical Engineering and Physics*, vol. 33, no. 8, pp. 973–979, 2011.

[303] R. Huys, D. Braeken, D. Jans *et al.*, "Single-cell recording and stimulation with a 16k micro-nail electrode array integrated on a 0.18 μm CMOS chip," *Lab on a Chip*, vol. 12, no. 7, pp. 1274–1280, 2012.

[304] P. K. Challa, T. Kartanas, J. Charmet, and T. P. Knowles, "Microfluidic devices fabricated using fast wafer-scale LED-lithography patterning," *Biomicrofluidics*, vol. 11, no. 1, pp. 014113 (1–8), 2017.

[305] H. Yin, L. Li, and A. J. Mason, "Screen-printed planar metallization for lab-on-CMOS with epoxy carrier," in *2016 IEEE International Symposium on Circuits and Systems (ISCAS)*, 2016, pp. 2887–2890: IEEE.

[306] Y. Huang and A. J. Mason, "Lab-on-CMOS integration of microfluidics and electrochemical sensors," *Lab on a Chip*, vol. 13, no. 19, pp. 3929–3934, 2013.

[307] H. Norian, R. M. Field, I. Kymissis, and K. L. Shepard, "An integrated CMOS quantitative-polymerase-chain-reaction lab-on-chip for point-of-care diagnostics," *Lab on a Chip*, vol. 14, no. 20, pp. 4076–4084, 2014.

[308] M. Huang, J. B. Delacruz, J. C. Ruelas, S. S. Rathore, and M. Lindau, "Surface-modified CMOS IC electrochemical sensor array targeting single

chromaffin cells for highly parallel amperometry measurements," *Pflügers Archiv: European Journal of Physiology*, vol. 470, no. 1, pp. 113–123, 2018.

[309] P. Man, D. Jones, and C. Mastrangelo, "Microfluidic plastic capillaries on silicon substrates: a new inexpensive technology for bioanalysis chips," in *Proceedings IEEE The Tenth Annual International Workshop on Micro Electro Mechanical Systems, MEMS'97*, 1997, pp. 311–316: IEEE.

[310] N. Bhattacharjee, A. Urrios, S. Kang, and A. Folch, "The upcoming 3D-printing revolution in microfluidics," *Lab on a Chip*, vol. 16, no. 10, pp. 1720–1742, 2016.

[311] R. Amin, S. Knowlton, A. Hart *et al.*, "3D-printed microfluidic devices," *Biofabrication*, vol. 8, no. 2, pp. 022001 (1–21), 2016.

[312] S. Waheed, J. M. Cabot, N. P. Macdonald *et al.*, "3D printed microfluidic devices: enablers and barriers," *Lab on a Chip*, vol. 16, no. 11, pp. 1993–2013, 2016.

[313] D. Therriault, S. R. White, and J. A. Lewis, "Chaotic mixing in three-dimensional microvascular networks fabricated by direct-write assembly," *Nature Materials*, vol. 2, no. 4, pp. 265 (1–7), 2003.

[314] M. Lindsay, K. Bishop, S. Sengupta *et al.*, "Heterogeneous integration of CMOS sensors and fluidic networks using wafer-level molding," *IEEE Transactions on Biomedical Circuits and Systems*, vol. 12, no. 5, pp. 1046–1055, 2018.

[315] W. J. Karl, M. Schikowski, J.-E. Thon, and R. Knechtel, "Adhesive wafer bonding for CMOS based lab-on-a-chip devices," *Japanese Journal of Applied Physics*, vol. 59, no. SB, p. SBBD04, 2020.

[316] M. Mujika, S. Arana, E. Castano *et al.*, "Microsystem for the immuno-magnetic detection of *Escherichia coli* O157: H7," *Physica Status Solidi (A)*, vol. 205, no. 6, pp. 1478–1483, 2008.

[317] M. Agirregabiria, F. Blanco, J. Berganzo *et al.*, "Fabrication of SU-8 multi-layer microstructures based on successive CMOS compatible adhesive bonding and releasing steps," *Lab on a Chip*, vol. 5, no. 5, pp. 545–552, 2005.

[318] M. Inac, M. Wietstruck, A. Göritz *et al.*, "BiCMOS integrated microfluidic packaging by wafer bonding for lab-on-chip applications," in *2017 IEEE 67th Electronic Components and Technology Conference (ECTC)*, 2017, pp. 786–791: IEEE.

[319] H. Matbaechi Ettehad, P. Soltani Zarrin, R. Hölzel, and C. Wenger, "Dielectrophoretic immobilization of yeast cells using CMOS integrated microfluidics," *Micromachines*, vol. 11, no. 5, p. 501, 2020.

[320] A. Rasmussen, M. Gaitan, L. E. Locascio, and M. E. Zaghloul, "Fabrication techniques to realize CMOS-compatible microfluidic microchannels," *Journal of Microelectromechanical Systems*, vol. 10, no. 2, pp. 286–297, 2001.

[321] I. Chartier, C. Bory, A. Fuchs *et al.*, "Fabrication of hybrid plastic-silicon microfluidic devices for individual cell manipulation by dielectrophoresis," in *Microfluidics, BioMEMS, and Medical Microsystems II*, 2003, vol. 5345, pp. 7–16: International Society for Optics and Photonics.

[322] J. A. Chediak, Z. Luo, J. Seo, N. Cheung, L. P. Lee, and T. D. Sands, "Heterogeneous integration of CdS filters with GaN LEDs for fluorescence detection microsystems," *Sensors and Actuators A: Physical*, vol. 111, no. 1, pp. 1–7, 2004.

[323] A. Uddin, K. Milaninia, C.-H. Chen, and L. Theogarajan, "Wafer scale integration of CMOS chips for biomedical applications via self-aligned masking," *IEEE Transactions on Components, Packaging and Manufacturing Technology*, vol. 1, no. 12, pp. 1996–2004, 2011.

[324] L. Hartley, K. V. Kaler, and O. Yadid-Pecht, "Hybrid integration of an active pixel sensor and microfluidics for cytometry on a chip," *IEEE Transactions on Circuits and Systems I: Regular Papers*, vol. 54, no. 1, pp. 99–110, 2007.

[325] M. A. Burns, B. N. Johnson, S. N. Brahmasandra *et al.*, "An integrated nanoliter DNA analysis device," *Science*, vol. 282, no. 5388, pp. 484–487, 1998.

[326] N. Ho, J. Kratochvil, G. Blackburn, and J. Janata, "Encapsulation of polymeric membrane-based ion-selective field effect transistors," *Sensors and Actuators*, vol. 4, pp. 413–421, 1983.

[327] H. Li, X. Liu, L. Li, X. Mu, R. Genov, and A. J. Mason, "CMOS electrochemical instrumentation for biosensor microsystems: a review," *Sensors*, vol. 17, no. 1, p. 74, 2016.

[328] A. Miled and M. Sawan, "High throughput microfluidic rapid and low cost prototyping packaging methods," *JoVE (Journal of Visualized Experiments)*, no. 82, p. e50735, 2013.

[329] M. Dandin, I. D. Jung, M. Piyasena *et al.*, "Post-CMOS packaging methods for integrated biosensors," in *SENSORS, 2009 IEEE*, 2009, pp. 795–798: IEEE.

[330] J. Tian, Y. Bai, X. Tang, S. Sang, and F. Wang, "A capacitive surface stress biosensor for CSFV detection," *Microelectronic Engineering*, vol. 159, pp. 55–59, 2016.

[331] H. A. Zeinabad, H. Ghourchian, M. Falahati, M. Fathipour, M. Azizi, and S. M. Boutorabi, "Ultrasensitive interdigitated capacitance immunosensor using gold nanoparticles," *Nanotechnology*, vol. 29, no. 26, p. 265102, 2018.

[332] G. Ertürk, M. Hedström, M. A. Tümer, A. Denizli, and B. Mattiasson, "Real-time prostate-specific antigen detection with prostate-specific antigen imprinted capacitive biosensors," *Analytica Chimica Acta*, vol. 891, pp. 120–129, 2015.

[333] G. Ertürk, M. Hedström, and B. Mattiasson, "A sensitive and real-time assay of trypsin by using molecular imprinting-based capacitive biosensor," *Biosensors and Bioelectronics*, vol. 86, pp. 557–565, 2016.

[334] A. Foubert, N. V. Beloglazova, M. Hedström, and S. De Saeger, "Antibody immobilization strategy for the development of a capacitive immunosensor detecting zearalenone," *Talanta*, 2018.

[335]   A. Mahadhy, *Development of an Ultrasensitive Capacitive DNA-sensor: A Promising Tool Towards Microbial Diagnostics*. Lund University (Media-Tryck), 2015.

[336]   L. Wang, M. Veselinovic, L. Yang, B. J. Geiss, D. S. Dandy, and T. Chen, "A sensitive DNA capacitive biosensor using interdigitated electrodes," *Biosensors and Bioelectronics*, vol. 87, pp. 646–653, 2017.

[337]   N. Idil, M. Hedström, A. Denizli, and B. Mattiasson, "Whole cell based microcontact imprinted capacitive biosensor for the detection of *Escherichia coli*," *Biosensors and Bioelectronics*, vol. 87, pp. 807–815, 2017.

[338]   F. Malvano, L. Maritato, G. Carapella *et al.*, "Fabrication of $SrTiO_3$ layer on Pt electrode for label-free capacitive biosensors," *Biosensors*, vol. 8, no. 1, p. 26, 2018.

[339]   P. Lenain, S. De Saeger, B. Mattiasson, and M. Hedström, "Affinity sensor based on immobilized molecular imprinted synthetic recognition elements," *Biosensors and Bioelectronics*, vol. 69, pp. 34–39, 2015.

[340]   M. Hedström and B. Mattiasson, "Bioimprinting as a tool for the detection of aflatoxin B1 using a capacitive biosensor," *Biotechnology Reports*, vol. 11, pp. 12–17, 2016.

[341]   F. A. Harraz, A. A. Ismail, H. Bouzid, S. Al-Sayari, A. Al-Hajry, and M. Al-Assiri, "A capacitive chemical sensor based on porous silicon for detection of polar and non-polar organic solvents," *Applied Surface Science*, vol. 307, pp. 704–711, 2014.

[342]   C. Baker and J. L. Gole, "Detection of liquid organic solvents on metal oxide nanostructure decorated porous silicon interfaces," *ACS Sensors*, vol. 1, no. 3, pp. 235–242, 2016.

[343]   K. Bunnfors, N. Abrikossova, J. Kilpijärvi *et al.*, "Nanoparticle activated neutrophils-on-a chip: a label-free capacitive sensor to monitor cells at work," *Sensors and Actuators B: Chemical*, p. 128020, 2020.

[344]   A. Jian, X. Tang, Q. Feng *et al.*, "A PDMS surface stress biosensor with optimized micro-membrane: fabrication and application," *Sensors and Actuators B: Chemical*, vol. 242, pp. 969–976, 2017.

[345]   P. Eriksson, A. A. Tal, A. Skallberg *et al.*, "Cerium oxide nanoparticles with antioxidant capabilities and gadolinium integration for MRI contrast enhancement," *Scientific Reports*, vol. 8, no. 1, pp. 1–12, 2018.

[346]   E. Alipour, H. Ghourchian, and S. M. Boutorabi, "Gold nanoparticle based capacitive immunosensor for detection of hepatitis B surface antigen," *Analytical Methods*, vol. 5, no. 17, pp. 4448–4453, 2013.

[347]   K. Besteman, J.-O. Lee, F. G. Wiertz, H. A. Heering, and C. Dekker, "Enzyme-coated carbon nanotubes as single-molecule biosensors," *Nano Letters*, vol. 3, no. 6, pp. 727–730, 2003.

[348]   L. Yang, W. Wei, X. Gao, J. Xia, and H. Tao, "A new antibody immobilization strategy based on electrodeposition of nanometer-sized hydroxyapatite for label-free capacitive immunosensor," *Talanta*, vol. 68, no. 1, pp. 40–46, 2005.

[349] L. C. Shriver-Lake, W. B. Gammeter, S. S. Bang, and M. Pazirandeh, "Covalent binding of genetically engineered microorganisms to porous glass beads," *Analytica Chimica Acta*, vol. 470, no. 1, pp. 71–78, 2002.

[350] N.-P. Huang, J. Vörös, S. M. De Paul, M. Textor, and N. D. Spencer, "Biotin-derivatized poly (L-lysine)-g-poly (ethylene glycol): a novel polymeric interface for bioaffinity sensing," *Langmuir*, vol. 18, no. 1, pp. 220–230, 2002.

[351] D. A. Butterfield, J. Colvin, J. Liu, J. Wang, L. Bachas, and D. Bhattacharrya, "Electron paramagnetic resonance spin label titration: a novel method to investigate random and site-specific immobilization of enzymes onto polymeric membranes with different properties," *Analytica Chimica Acta*, vol. 470, no. 1, pp. 29–36, 2002.

[352] S. Cosnier, "Biomolecule immobilization on electrode surfaces by entrapment or attachment to electrochemically polymerized films. A review," *Biosensors and Bioelectronics*, vol. 14, no. 5, pp. 443–456, 1999.

[353] F. Gardies, C. Martelet, B. Colin, and B. Mandrand, "Feasibility of an immunosensor based upon capacitance measurements," *Sensors and Actuators*, vol. 17, no. 3-4, pp. 461–464, 1989.

[354] P. Bataillard, F. Gardies, N. Jaffrezic-Renault, C. Martelet, B. Colin, and B. Mandrand, "Direct detection of immunospecies by capacitance measurements," *Analytical Chemistry*, vol. 60, no. 21, pp. 2374–2379, 1988.

[355] E. Souteyrand, J. Martin, and C. Martelet, "Direct detection of biomolecules by electrochemical impedance measurements," *Sensors and Actuators B: Chemical*, vol. 20, no. 1, pp. 63–69, 1994.

[356] C. Saby, N. Jaffrezic-Renault, C. Martelet *et al.*, "Immobilization of antibodies onto a capacitance silicon-based transducer," *Sensors and Actuators B: Chemical*, vol. 16, no. 1-3, pp. 458–462, 1993.

[357] H. Maupas, C. Saby, C. Martelet *et al.*, "Impedance analysis of Si/SiO_2 heterostructures grafted with antibodies: an approach for immunosensor development," *Journal of Electroanalytical Chemistry*, vol. 406, no. 1-2, pp. 53–58, 1996.

[358] A. Barraud, H. Perrot, V. Billard, C. Martelet, and J. Therasse, "Study of immunoglobulin G thin layers obtained by the Langmuir–Blodgett method: application to immunosensors," *Biosensors and Bioelectronics*, vol. 8, no. 1, pp. 39–48, 1993.

[359] V. Billard, C. Martelet, P. Binder, and J. Therasse, "Toxin detection using capacitance measurements on immunospecies grafted onto a semiconductor substrate," *Analytica Chimica Acta*, vol. 249, no. 2, pp. 367–372, 1991.

[360] A. Sibai, K. Elamri, D. Barbier, N. Jaffrezic-Renault, and E. Souteyrand, "Analysis of the polymer-antibody-antigen interaction in a capacitive immunosensor by FTIR difference spectroscopy," *Sensors and Actuators B: Chemical*, vol. 31, no. 1-2, pp. 125–130, 1996.

[361] H. Berney, J. Alderman, W. Lane, and J. K. Collins, "A differential capacitive biosensor using polyethylene glycol to overlay the biolayer," *Sensors and Actuators B: Chemical*, vol. 44, no. 1-3, pp. 578–584, 1997.

[362] C. Berggren and G. Johansson, "Capacitance measurements of anti-body–antigen interactions in a flow system," *Analytical Chemistry*, vol. 69, no. 18, pp. 3651–3657, 1997.

[363] C. Berggren, B. Bjarnason, and G. Johansson, "An immunological Interleukin-6 capacitive biosensor using perturbation with a potentiostatic step," *Biosensors and Bioelectronics*, vol. 13, no. 10, pp. 1061–1068, 1998.

[364] C. E. Nwankire, A. Venkatanarayanan, T. Glennon, T. E. Keyes, R. J. Forster, and J. Ducrée, "Label-free impedance detection of cancer cells from whole blood on an integrated centrifugal microfluidic platform," *Biosensors and Bioelectronics*, vol. 68, pp. 382–389, 2015.

[365] M. Knichel, P. Heiduschka, W. Beck, G. Jung, and W. Göpel, "Utilization of a self-assembled peptide monolayer for an impedimetric immunosensor," *Sensors and Actuators B: Chemical*, vol. 28, no. 2, pp. 85–94, 1995.

[366] A. Gebbert, M. Alvarez-Icaza, W. Stoecklein, and R. D. Schmid, "Real-time monitoring of immunochemical interactions with a tantalum capacitance flow-through cell," *Analytical Chemistry*, vol. 64, no. 9, pp. 997–1003, 1992.

[367] H. Maupas, A. Soldatkin, C. Martelet, N. Jaffrezic-Renault, and B. Mandrand, "Direct immunosensing using differential electrochemical measurements of impedimetric variations," *Journal of Electroanalytical Chemistry*, vol. 421, no. 1-2, pp. 165–171, 1997.

[368] S. Ameur, H. Maupas, C. Martelet *et al.*, "Impedimetric measurements on polarized functionalized platinum electrodes: application to direct immunosensing," *Materials Science and Engineering: C*, vol. 5, no. 2, pp. 111–119, 1997.

[369] M. Cha, J. Shin, J.-H. Kim *et al.*, "Biomolecular detection with a thin membrane transducer," *Lab on a Chip*, vol. 8, no. 6, pp. 932–937, 2008.

[370] V. Tsouti, C. Boutopoulos, P. Andreakou *et al.*, "Detection of DNA mutations using a capacitive micro-membrane array," *Biosensors and Bioelectronics*, vol. 26, no. 4, pp. 1588–1592, 2010.

[371] L. Lebogang, M. Hedström, and B. Mattiasson, "Development of a real-time capacitive biosensor for cyclic cyanotoxic peptides based on Adda-specific antibodies," *Analytica Chimica Acta*, vol. 826, pp. 69–76, 2014.

[372] A. Qureshi, J. H. Niazi, S. Kallempudi, and Y. Gurbuz, "Label-free capacitive biosensor for sensitive detection of multiple biomarkers using gold interdigitated capacitor arrays," *Biosensors and Bioelectronics*, vol. 25, no. 10, pp. 2318–2323, 2010.

[373] H.-T. Hsueh and C.-T. Lin, "An incremental double-layer capacitance of a planar nano gap and its application in cardiac-troponin T detection," *Biosensors and Bioelectronics*, vol. 79, pp. 636–643, 2016.

[374] C. I. Justino, A. C. Duarte, and T. A. Rocha-Santos, "Analytical applications of affibodies," *TrAC Trends in Analytical Chemistry*, vol. 65, pp. 73–82, 2015.

[375] B. Renberg, J. Nordin, A. Merca *et al.*, "Affibody molecules in protein capture microarrays: evaluation of multidomain ligands and different detection formats," *Journal of Proteome Research*, vol. 6, no. 1, pp. 171–179, 2007.

[376] P. K. Ferrigno, "Non-antibody protein-based biosensors," *Essays in Biochemistry*, vol. 60, no. 1, pp. 19–25, 2016.

[377] Z. G. Namhil, C. Kemp, E. Verrelli *et al.*, "A label-free aptamer-based nanogap capacitive biosensor with greatly diminished electrode polarization effects," *Physical Chemistry Chemical Physics*, vol. 21, no. 2, pp. 681–691, 2019.

[378] H.-J. Chen, R. L. Chen, B.-C. Hsieh *et al.*, "Label-free and reagentless capacitive aptasensor for thrombin," *Biosensors and Bioelectronics*, vol. 131, pp. 53–59, 2019.

[379] G. Ertürk and R. Lood, "Bacteriophages as biorecognition elements in capacitive biosensors: phage and host bacteria detection," *Sensors and Actuators B: Chemical*, vol. 258, pp. 535–543, 2018.

[380] G. E. Bergdahl, M. Hedström, and B. Mattiasson, "Capacitive saccharide sensor based on immobilized phenylboronic acid with diol specificity," *Applied Biochemistry and Biotechnology*, vol. 188, no. 1, pp. 124–137, 2019.

[381] T. Cohen, J. Starosvetsky, U. Cheruti, and R. Armon, "Whole cell imprinting in sol-gel thin films for bacterial recognition in liquids: macromolecular fingerprinting," *International Journal of Molecular Sciences*, vol. 11, no. 4, pp. 1236–1252, 2010.

[382] R. Ouyang, J. Lei, and H. Ju, "Surface molecularly imprinted nanowire for protein specific recognition," *Chemical Communications*, no. 44, pp. 5761–5763, 2008.

[383] E. Yildirim, E. Turan, and T. Caykara, "Construction of myoglobin imprinted polymer films by grafting from silicon surface," *Journal of Materials Chemistry*, vol. 22, no. 2, pp. 636–642, 2012.

[384] L. J. Guo, "Recent progress in nanoimprint technology and its applications," *Journal of Physics D: Applied Physics*, vol. 37, no. 11, p. R123, 2004.

[385] A. Mujahid, N. Iqbal, and A. Afzal, "Bioimprinting strategies: from soft lithography to biomimetic sensors and beyond," *Biotechnology Advances*, vol. 31, no. 8, pp. 1435–1447, 2013.

[386] M. Grossi and B. Riccò, "Electrical impedance spectroscopy (EIS) for biological analysis and food characterization: a review," *Sensors and Sensor Systems*, pp. 303–325, 2017.

[387] B.-Y. Chang and S.-M. Park, "Electrochemical impedance spectroscopy," *Annual Review of Analytical Chemistry*, vol. 3, pp. 207–229, 2010.

[388] K. Bourzac. (2009, August 4, 2009). Rapid TB Detector. Available: www.technologyreview.com/2009/08/04/211191/rapid-tb-detector/

[389] (2020). *Piccolo Xpress*. Available: www.abaxis.com/index.php/piccolo-xpress?language_content_entity=en

[390] H. Morgan, T. Sun, D. Holmes, S. Gawad, and N. G. Green, "Single cell dielectric spectroscopy," *Journal of Physics D: Applied Physics*, vol. 40, no. 1, p. 61, 2006.

[391] H. Huang, Z. Liu, and X. Yang, "Application of electrochemical impedance spectroscopy for monitoring allergen–antibody reactions using gold

nanoparticle-based biomolecular immobilization method," *Analytical Biochemistry*, vol. 356, no. 2, pp. 208–214, 2006.

[392] F. Lucarelli, G. Marrazza, A. P. Turner, and M. Mascini, "Carbon and gold electrodes as electrochemical transducers for DNA hybridisation sensors," *Biosensors and Bioelectronics*, vol. 19, no. 6, pp. 515–530, 2004.

[393] K. Teeparuksapun, M. Hedström, E. Y. Wong, S. Tang, I. K. Hewlett, and B. Mattiasson, "Ultrasensitive detection of HIV-1 p24 antigen using nanofunctionalized surfaces in a capacitive immunosensor," *Analytical Chemistry*, vol. 82, no. 20, pp. 8406–8411, 2010.

[394] M. Hedström, I. Y. Galaev, and B. Mattiasson, "Continuous measurements of a binding reaction using a capacitive biosensor," *Biosensors and Bioelectronics*, vol. 21, no. 1, pp. 41–48, 2005.

[395] D. Erlandsson, K. Teeparuksapun, B. Mattiasson, and M. Hedström, "Automated flow-injection immunosensor based on current pulse capacitive measurements," *Sensors and Actuators B: Chemical*, vol. 190, pp. 295–304, 2014.

[396] D. Erlandsson, M. Hedstrom, B. Mattiasson, and J. Larsson, "Method of measuring a capacitance," ed: Google Patents, 2016.

[397] (2021-04-09). *Alpha-A Analyzer*. Available: www.novocontrol.de/php/ana_alpha.php

[398] (2021-04-09). *Reference 600+ Potentiostat/Galvanostat 5 MHz EIS*. Available: www.gamry.com/potentiostats/reference-600-plus/

[399] (2021-04-09). *ECIS Z-Theta—Applied BioPhysics*. Available: www.bio-physics.com/ztheta.php

[400] P. S. H. Jose, K. Rajasekaran, P. Rajalakshmy, and B. Jebastina, "A non-invasive method for measurement of blood glucose using bio impedance technique," in *2019 2nd International Conference on Signal Processing and Communication (ICSPC)*, 2019, pp. 138–142: IEEE.

[401] K. Ain, R. A. Wibowo, S. Soelistiono, and L. Muniroh, "Measurement of cholesterol concentration based on bioimpedance with AD5933-EVAL," in *2017 5th International Conference on Instrumentation, Communications, Information Technology, and Biomedical Engineering (ICICI-BME)*, 2017, pp. 1–4: IEEE.

[402] T. Agcayazi, G. J. Hong, B. Maione, and E. Woodard, "Wearable infant hydration monitor," in *2016 IEEE Virtual Conference on Applications of Commercial Sensors (VCACS)*, 2017, pp. 1–10: IEEE.

[403] Y.-P. Lu, J.-W. Huang, I.-N. Lee *et al.*, "A portable system to monitor saliva conductivity for dehydration diagnosis and kidney healthcare," *Scientific Reports*, vol. 9, no. 1, pp. 1–9, 2019.

[404] (2021-04-09). *Analog Devices EVAL-AD5933EBZ ADC Evaluation Board for AD5933 | RS Components*. Available: https://uk.rs-online.com/web/p/signal-conversion-development-tools/8031419/

[405] (2021-04-09). *DIGIL 410-378: Impedance analyzer for analog discovery at reichelt elektronik*. Available: www.reichelt.com/de/en/impedance-analyzer-for-analog-discovery-digil-410-378-p259325.html

[406] (2021-04-09). *AO-16777k Impedance Analyzer / LCR Meter (16.7 MHz) – Maranata-Madrid SL – NIF B-85746204*. Available: www.alphaomega-electronics.com/en/impedance/5252-ao-16777k-impedance-analyzer-lcr-meter-167-mhz.html

[407] (2021-04-09). *PalmSens3 potentiostat/ galvanostat/ Impedance analyser*. Available: www.palmsens.com/wp-content/uploads/2016/12/PalmSens3-description.pdf

[408] Y. Temiz, R. D. Lovchik, G. V. Kaigala, and E. Delamarche, "Lab-on-a-chip devices: how to close and plug the lab?" *Microelectronic Engineering*, vol. 132, pp. 156–175, 2015.

[409] (2021, 2021). *i-STAT 1*. Available: www.pointofcare.abbott/int/en/offerings/istat/istat-handheld

[410] E. Ghafar-Zadeh, B. Gholamzadeh, F. Awwad, and M. Sawan, "On-chip electroporation: characterization, modeling and experimental results," in *2012 Annual International Conference of the IEEE Engineering in Medicine and Biology Society*, San Diego, CA, USA, 2012, pp. 2583–2586: IEEE.

[411] S. Kalpakjian and S. Schmid, *Manufacturing Engineering and Technology*, SI edition, 6th ed. Singapore: Pearson Publications, 2009.

[412] S. Franssila, *Introduction to Microfabrication*. UK: John Wiley & Sons, 2010.

[413] K. Hassanli, S. M. Sayedi, R. Dehghani, A. Jalili, and J. J. Wikner, "A low-power wide tuning-range CMOS current-controlled oscillator," *Integration, the VLSI Journal*, vol. 55, pp. 57–66, 2016.

[414] *Cell culture protocol*. Available: www.ptglab.com/support/cell-culture-protocol/cell-culture-protocol/

[415] M. Kapałczyńska, T. Kolenda, W. Przybyła *et al.*, "2D and 3D cell cultures—a comparison of different types of cancer cell cultures," *Archives of Medical Science: AMS*, vol. 14, no. 4, p. 910, 2018.

[416] Q. Liu, C. Wu, H. Cai, N. Hu, J. Zhou, and P. Wang, "Cell-based biosensors and their application in biomedicine," *Chemical Reviews*, vol. 114, no. 12, pp. 6423–6461, 2014.

[417] *Cell culture basics handbook*. Available: www.vanderbilt.edu/viibre/CellCultureBasicsEU.pdf

Index

Nutrition in Traditional Therapeutic Foods
– Volume 2 –

THE AUTHORS

Dr. Subbulakshmi is an eminent nutritionist with over 45 years of teaching, research and administrative experience. She has a Ph.D in Food Science and Nutrition and has also worked as a post doctoral fellow in CFTRI, Mysore in early 1970s. She has worked as the First Principal of the S.M. Patel College of Home Science in Vallabh Vidyanagar, Gujarat and Indramani Mandelia College of Home Science in Pilani, Rajasthan. She has been instrumental in introducing internship for the students in Home Science colleges and Sports Nutrition and Food Science and Technology in the Nutrition curriculum.

She has published several research papers and her book on Food Processing and Preservation has been widely used in the colleges. She has guided more than 25 Ph.D. and M.Phil. students in Food Science and Nutrition apart from more than 60 M.Sc. students in their Dissertations. She has published more than 70 Research Papers in National and International Journals. Dr. Subbulakshmi has worked towards the improvement of community Health and Nutrition. Having worked on research projects funded by National and International Agencies, she has prepared a number of booklets and pamphlets on nutrition related topics for educating the common man.

She has received Fulbright award for studying the college administration in USA; PL480 Research Scholarship for doing Ph.D.; UGC's National Research Associateship award; as well as Hari Om Ashram Prizes for 3 different best Research papers in Nutrition and many more that speak of her abilities. She has been awarded M.L. Khurana Memorial Award –2002; T.K. Basu Award for excellent contributions to Nutritional Sciences from International College of Nutrition USA; Life Time Achievement Award from Alumni Association of the M.S. University of Baroda; Life Time Achievement Award from Nutrition Society of India, Mumbai.

She is an active life member of many Professional Organizations like Nutrition Society of India, Association of Food Scientists and Technologists of India, Indian Dietetic Association, Home Science Association of India, Association for Women Scientists, Indian Association for Preschool Education and many others.

Dr. Mandalika Subhadra is an expert in Food Science and Nutrition. She is currently working as an Associate Professor of Nutrtion in the college of Home Science, Nirmala Niketan, University of Mumbai. She has a post-graduate degree from Sri Satya Sai Institute of Higher Learning, Anantapur (A.P., India) and awardee of gold medal for securing first rank. She has received her Ph.D. in Food Science and Nutrition from SNDT Women's University, Mumbai, India. She has contributed her valuable expertise to various institutions during her 20 years of service as an academician and researcher. Her contribution was appreciated with a best teacher award by the Rajamahendri Degree College for Women, Andhra Pradesh. She is also a recognised Ph.D. guide of the University of Mumbai and selected as Ph.D. referee by various other reputed Universities. She has been actively involved in teaching and research guidance at under graduate and post graduate levels. She has guided around 25 postgraduate students and 6 Ph.D. scholars on research topics from various branches of nutrition including Clinical Nutrition, Community Nutrition, Nutritional Food Product Development, Food Science and Technology, Biochemistry, Micronutrients, Antioxidants, Sports and Exercise Nutrition. She has got several research publications to her credit in various peer reviewed journals and has been a co-editor of a book on Methodologies of Nutritional assessment. She has shared her research achievements in various national and international conferences. Her deep interest in research is evident through a Patency (No. 225689) she has obtained on 'Antiestrogenic activity of phytosterols from fenugreek seeds'. She has coordinated major and minor research projects sponsored by DAE/BRNS, UGC, University of Mumbai, and reputed food industries.

Dr. Subhadra has been invited as a resource person at various seminars and conferences. She was honoured with an Ambassadorial Scholarship of Good Will by the Rotary International to take up an academic assignment at the University of Mauritius for 3 and half months in Mauritius which was highly appreciated. Besides being an academician, she is also actively involved in community outreach programmes.

Nutrition in Traditional Therapeutic Foods
– *Volume 2* –

— *Authors* —
G. Subbulakshmi
Nutrition Consultant,
Former Director,
Department of PGSR in Home Science,
SNDT Women's University, Mumbai

and
M. Subhadra
Associate Professor,
Department of Food, Nutrition and Dietetics,
College of Home Science, Nirmala Niketan, Mumbai

2015
Daya Publishing House®
A Division of
Astral International Pvt. Ltd.
New Delhi – 110 002

Cataloging in Publication Data--DK
Courtesy: D.K. Agencies (P) Ltd. <docinfo@dkagencies.com>
Subbulakshmi, G. author.
Nutrition in traditional therapeutic foods / authors, G. Subbulakshmi and M. Subhadra.
Volume   2   cm
Includes bibliographical references and index.
ISBN 9789351306580 (International Edition)

1. Spices--Therapeutic use. 2. Spices--Physiological effect. 3. Medicinal plants--Analysis. 4. Diet therapy. 5. Nutrition. I. Subhadra, M. (Mandalika), author. II. Title.

DDC 615.32   23

Published by   : **Daya Publishing House®**
A Division of
Astral International Pvt. Ltd.
– ISO 9001:2008 Certified Company –
4760-61/23, Ansari Road, Darya Ganj
New Delhi-110 002
Ph. 011-43549197, 23278134
E-mail: info@astralint.com
Website: www.astralint.com

Laser Typesetting   : **Classic Computer Services**, Delhi - 110 035

Printed at   : **Replika Press Pvt. Ltd.**

PRINTED IN INDIA

Acknowledgements

We are grateful to Dr. B. Sesikeran, former Director of National Institute of Nutrition, Hyderabad, India, for kindly accepting our request to write the foreword for this volume which indeed has complemented the value of the book. Our heartfelt acknowledgements to all the research scholars, friends and others who have directly or indirectly inspired and helped us towards this achievement.

The globally renowned traditional therapeutic dietary practices are the strongest source of inspiration to us towards this endeavour. Our intense desire was to compile the available scientific evidence on these practices and present it in a nutshell for the benefit of current and future generation of nutritionists.

We sincerely thank all the scientists for sharing the outcome of their excellent research through publications. We are also immensely thankful to all the web search engines for providing access to the authentic scientific information without which this venture would have been impossible. The valuable contribution of all the scientists has been duly acknowledged both in the text as well as in the references given towards the end of each chapter. The copyright of the sources of information remains with the original authors. However, if reference to any author/s is missing in the book, it is purely unintentional and inadvertent, and deeply regretted.

Dr. G. Subbulakshmi

Dr. M. Subhadra

Foreword

It gives me great pleasure to write this foreword for the 2nd volume of an elegantly compiled set of facts and traditional knowledge. Each chapter is devoted to one food entity and a thorough in depth information is provided to the reader. This includes descriptive information, taxonomy, compositional analysis, basic biochemistry, clinical and pharmacological aspects, and every aspect of research done on each of these entities. Epidemiological studies and community practices also form part of the information. While most people know a bit about each of them, a good amount of information provided in this volume is new knowledge for most of us. The ultimate aim of this kind of compilation is to empower students, teachers, nutritionists, dieticians, biochemists, analytical chemists, basic scientists and physicians with knowledge that would help in their day to research activities and patient care. This would also provide lay persons who are of late yearning for health benefits through food and natural sources, the required authentic knowledge.

I complement the authors for the time and energies that has gone into this volume going by the elaborate bibliography given at the end of each chapter. I am sure there is scope for a few more such volumes that could cover lesser known traditional therapeutic foods.

Dr. B. Sesikeran

Former Director,
NIN-Hyderabad

Preface

The second volume of our book on *Nutrition in Traditional Therapeutic Foods* includes the therapeutic role of few more foods as evidenced by the available clinical studies. Our aim of infusing newer approach in scientific research on the traditional foods taking into consideration the therapeutic value of different parts of the plant including roots, leaves, flowers, fruits and seeds has been upheld in this volume too.

Although several books are available in the market on Clinical nutrition and/or Therapeutic diets, the uniqueness of the book is to bring out the role of chemical and nutritional constituents in foods as anti-oxidants, anti-inflammatory and anti-microbial agents responsible for ameliorating various metabolic, degenerative and other disorders. During the compilation, we have found that some of the scientists while exploring the influence of foods on the bioavailability and pharmaceutical effects of drugs, have also compared the therapeutic efficacy of foods and drugs. It is encouraging to note that the foods/active principles enhance the bioavailability of the drugs; and certain foods are clinically more efficient than even drugs. Thus research emphasizes the need for a synergistic approach of food and medicine in treating diseases. Hence, this book is of great value in the present day context of Medical Nutrition Therapy wherein the medical professionals, nutritionists and dieticians jointly work towards the treatment of patients.

Our book also gives due emphasis to the concept of preventive nutrition which insists on regular consumption of variety of foods including fruits, vegetables, spices and condiments in order to enjoy disease free long life. It is indeed heartwhelming to observe that, based on the information given in the first volume many students opted to research on the various aspects of the therapeutic value of foods aiming at promotion of their regular consumption as a preventive measure.

The book comprises of 13 chapters. Each chapter covers information on details of the food/plant in terms of its traditional usage, chemical and nutritional composition, overview of therapeutic benefits and safety issues (wherever applicable) supported by the available scientific evidence.

Dr. G. Subbulakshmi

Dr. M. Subhadra

Contents

Chapter 1

Almonds
(*Amygdalus communis* L./ *Prunus amygdalus/Prunus dulcis/ Amygdalus dulcis*)

Almond is a species of Prunus belonging to the family Rosaceae. The global production of almonds is around 1.7 million metric tons, with California producing 80 per cent of the world's almonds. The principal almond varieties are: Butte, Carmel, Fritz, Mission, Monterey, Nonpareil, Padre, and Price (Milbury *et al.,* 2006).

Nutritional and Phytochemical Composition

In almond seeds lipid is the main storage component, which comprises ~50 per cent of the total weight of the seed and is located as intracellular oil bodies of diameter size ranging from ~1 to 5 µm (Ren *et al.,* 2001). Almonds are low in saturated fatty acids and rich in unsaturated fatty acids (Milbury *et al.,* 2006; Haider *et al.,* 2012). Almonds are good source of protein, arginine, dietary fiber, minerals (magnesium, copper, manganese, calcium, and potassium), vitamins (α-tocopherol) and numerous bioactive substances, such as phytosterols and flavonoids. They contain phytates and phenolics that confer anti-oxidant, anti-inflammatory and lipid-lowering properties while inhibiting trypsin and amylase activity.

Total phenols in almonds range from 127 to 241 mg gallic acid equivalents/100 g of fresh weight depending on the variety. Defatted almond whole seed, brown skin, and green shell cover extracts exhibit the presence of quercetin, isorhamnetin, quercitrin, kaempferol 3-O-rutinoside, isorhamnetin 3-O-glucoside, catechin, epicatechin and morin as the major flavonoids (Wijeratne *et al.,* 2006; Milbury *et al.,* 2006).

Almond skins, removed from the nut by hot water blanching during preparation of almond meat, constitute 4 per cent of the total almond weight and are generally treated as a waste product. However, the polyphenols in almonds are mainly located in the skin and include both nonflavonoid (*i.e.* hydroxybenzoic acids and aldehydes, and hydroxycinnamic acids) and flavonoid compounds (*i.e.* flavan-3-ols, flavonols, dihydroflavonols, and flavanones) (Sang *et al.,* 2002; Frison and Sporns 2002a; b; Milbury *et al.,* 2006; Wijeratne *et al.,* 2006; Monagas *et al.,* 2007; Bolling *et al.,* 2009). Flavan-3-ols in almond skins occur as monomers of (+)-catechin and (-)-epicatechin, and as a complex series of A and B type procyanidin, propelargonidin, and prodelphinidin polymers up to decamers (Amarowicz *et al.,* 2005; Brieskorn and Betz, 1988; Monagas *et al.,* 2007). Flavanols and flavonol glycosides, the most abundant phenolic compounds in almond skins, represent 38-57 per cent and 14-35 per cent of the total quantified phenolics, respectively. Rutinosides, in particular isorhamnetin-3-O-rutinoside, are the predominant forms of flavonol glycosides (Monagas *et al.,* 2007).

Nutritional Value of Almonds (per 100 g)

Nutrient	Amount	Nutrient	Amount
Energy (kcal)	576	Glutamic acid (g)	6.81
Sugars (g)	3.89	Glycine (g)	1.469
Dietary fiber (g)	12.2	Proline (g)	1.032
Fat (g)	49.42	Serine (g)	0.948
Saturated (g)	3.73	Vitamin A (IU)	4.70
Polyunsaturated (g)	12.07	Riboflavin (vit. B_2) (mg)	0.211
Monounsaturated (g)	30.89	Niacin (vit. B_3) (mg)	1.014
Protein (g)	21.22	Pantothenic acid (B_5) (mg)	3.385
Lysine (g)	0.58	Vitamin B_6 (mg)	0.469
Tryptophan (g)	0.21	Folate (vit. B_9) (µg)	0.143
Methionine (g)	0.15	Choline (mg)	50.0
Leucine (g)	1.49	Vitamin E (mg)	52.1
Phenylalanine (g)	1.12	Vitamin K (µg)	26.2
Cystine (g)	0.19	Iron (mg)	264.0
Threonine (g)	0.60	Magnesium (mg)	3.72
Tyrosine (g)	0.45	Manganese (mg)	268.0
Valine (g)	0.82	Phosphorus (mg)	2.285
Arginine (g)	2.45	Potassium (mg)	484.0
Histidine (g)	0.56	Sodium (mg)	705.0
Aspartic acid (g)	2.92	Zinc (mg)	1.0
Alanine (g)	1.03		

Source. USDA Nutrient Database.

Therapeutic Benefits

Almond consumption is associated with amelioration of obesity, hyper-lipidemia, hyper-tension, hyper-glycemia and sharpening of memory. Rats fed almond paste orally for 28 days exhibited enhanced tryptophan levels and serotonergic turnover in the brain showing that almonds possess significant hypo-phagic and nootropic effects. Significant improvement in learning and memory was also observed (Haider *et al.,* 2012).

Long-term supplementation of almonds (52 g/d) significantly increased the intakes of MUFA, PUFA, fibre, protein, alpha-tocopherol, Cu and Mg in healthy men and women by 42, 24, 12, 19, 66, 15 and 23 per cent respectively and significantly decreased the intakes of trans fatty acids, animal protein, Na, cholesterol and sugars by 14, 9, 21, 17 and 13 per cent respectively. These spontaneous nutrient changes closely matched the dietary recommendations. Hence, a daily supplement of almonds can induce favourable nutrient modifications to an individual's habitual diet and thereby prevent cardiovascular and other chronic diseases (Jaceldo-Siegl *et al.,* 2004).

Almond oil had been used in ancient Chinese, Ayurvedic and Greco-Persian schools of Medicine to treat dry skin conditions such as psoriasis and eczema. Further, there are anecdotal evidence and clinical experiences that almond oil seemingly reduces hyper-trophic scarring post-operatively, smoothes and rejuvenates skin. Almond oil has emollient and sclerosant properties and, therefore, has been used to improve complexion and skin tone (Ahmad, 2010).

Anti-microbial Agent

Almond skin fractions displayed anti-microbial activity against *L. monocytogenes* and *S. aureus* in the range of 250-500 µg/ml. Natural skins showed potential against the Gram-negative *S. enterica.* Pairwise combinations of proto-catechuic acid, naringenin and epicatechin showed both synergistic and indifferent interactions against *S. enterica and S. aureus.* Antagonism was observed against *L. monocytogenes* with all combinations tested (Mandalari *et al.,* 2010a).

H. pylori is known to be a gastric pathogen of humans and the treatment regimen posing problems with side effects, poor compliance, relapses, and antibiotic resistance. Polyphenols from almond skins are active against a wide range of food-borne pathogens. Bisignano *et al.* (2013) found the natural almond skin to be most effective against *H. pylori* under simulated gastrointestinal conditions.

Supplementation of 56 g of roasted almonds or 10 g almond skins as a single daily dose significantly increased in the populations of *Bifidobacterium* spp. and *Lactobacillus* spp. in the fecal samples of the subjects. Thus, ingestion of almond and almond skin improves the intestinal microbiota profile and modifies the intestinal bacterial activities due to their fiber and other constituents (Liu *et al.,* 2014; Ukhanova *et al.,* 2014).

Anti-inflammatory Agent

Almond protein hydrolysates find application against inflammatory conditions. The high molecular size fraction modulated the levels of pro-inflammatory cytokines, interleukin (IL)-6, IL-1β, and tumour necrosis factor (TNF)-α in the activated cells whereas the hydrolysate fraction inhibited the relative expression of pro-inflammatory IL-6, IL-1β, TNF-α, iNOS and COX-2 genes (Udenigwe *et al.*, 2013). Liu *et al.* (2013) confirmed that almond diet for 4 weeks decreased IL-6 and CRP by a median 10.3 per cent (each), and TNF-α by a median 15.7 per cent in Chinese patients with type 2 diabetes mellitus and mild hyper-lipidemia.

Rajaram *et al.* (2010) reported that the consumption of almonds influenced a few but not all of the markers of inflammation and haemostasis. A clear dose response was not observed for any of the markers studied as serum E-selectin was significantly lower on the high-almond diet; C-reactive protein (CRP) was lower in both low and high almond diets and no effect of diet on IL-6 or fibrinogen.

Cardio-protective Agent

As early as 1994, Abbey *et al.*, demonstrated that the cholesterol-lowering effect of almonds was better as compared to walnuts in healthy and hyper-cholesterolemic subjects. They found a significant reduction in total and LDL cholesterol by 7 per cent and 10 per cent respectively, while the same was 5 per cent and 9 per cent respectively with walnuts. Experimental studies have shown improvements in the lipo-protein profiles of persons consuming diets high in almonds (Berry *et al.*, 1992; Spiller *et al.*, 1992; 1998). Almonds being rich in cardio-protective nutrients reduce LDL-C, decrease (re)absorption of cholesterol and bile acid; increase bile acid and cholesterol excretion; increase LDL-C receptor activity; and regulate enzymes involved in *de-novo* cholesterol synthesis and bile acid production (Berryman *et al.*, 2011). Almond supplementation for 6 months resulted in statistically significant changes in TC, LDL, TC:HDL and LDL:HDL in subjects with high LDL but not among normo-cholesterolemic individuals. Further, the TC:HDL, and LDL:HDL ratios were significantly reduced among those with BMI < 25 kg/m², but not in heavier individuals. Therefore, Jaceldo-Siegl *et al.* (2011) concluded that the cholesterol-lowering effect of almonds is responsive among hyper-cholesterolemic individuals.

Hyson *et al.* (2002) found that the whole almonds as well as the almond oil were equally effective in significantly reducing triglyceride, total and LDL cholesterol by 14, 4 and 6 per cent respectively, while increasing the HDL cholesterol by 6 per cent. In a randomized, crossover feeding trial participants were fed for 4 weeks each: a control diet, a low-almond diet, and a high-almond diet, wherein almonds contributed 0 per cent, 10 per cent, and 20 per cent of total energy, respectively. A significant dose-response effect was observed between percent energy in the diet from almonds and plasma ratio of alpha-tocopherol to total cholesterol (Jambazian *et al.*, 2005). Wien *et al.* (2010) reported that an American Dietetic Association diet which provides 20 per cent of calories from almonds over a 16-week period is effective in improving markers of insulin sensitivity and yields clinically significant improvements in LDL-C in adults with pre-diabetes.

Choudhury *et al.* (2014) found that consuming 50 g of almonds per day along with daily diet for 4 weeks increases plasma α-tocopherol and improves vascular function in asymptomatic healthy men aged between 20 and 70 years without any effect on plasma lipids or markers of oxidative stress. In a randomized crossover study, incorporation of 68 g of almonds (20 per cent of energy) into Step I diet markedly improved the serum lipid profile of healthy and mildly hyper-cholesterolemic adults. A dose response revealed a reduction in total and LDL cholesterol concentrations with progressively higher intakes (Sabate *et al.,* 2003). Upon reviewing the available data on the impact of almond and dietary SFA intake on reducing plasma total cholesterol and LDL cholesterol, Ortiz *et al.* (2012) found relative almond intake contributes to the reduction of plasma total cholesterol and LDL-C to a greater extent than a reduction in dietary SFA. Thus they suggest that along with almond intake simultaneous SFA reduction should further improve lipid profile.

A common background diet enriched with either virgin olive oil, walnuts or almonds for 4 weeks each in a crossover study on hyper-cholesterolemic patients was found to reduce LDL-cholesterol from baseline by 7.3 per cent, 10.8 per cent and 13.4 per cent respectively. Total cholesterol and LDL/HDL ratios also decreased on par confirming that nuts are better than olive oil, and almonds are better than walnuts (Damasceno *et al.,* 2011). Both walnuts and almonds exerted beneficial effects on plasma lipids and androgens in PCOS (Kalgaonkar *et al.,* 2011).

The bioavailability of almond skin flavonoids (ASF) was found to enhance the *ex-vivo* resistance of hamster LDL by 18.0 per cent and the *in-vitro* addition of 5.5 μmol/L vitamin E synergistically extended the lag time of the 60-min sample by 52.5 per cent. The anti-oxidant capacity of ASF is proved to be bioavailable and to act in synergy with vitamins C and E to protect LDL against oxidation in hamsters (Chen *et al.,* 2005). Almond skin polyphenolics (ASP) and quercetin in combination reduced the oxidative modification of apo B-100 and stabilized LDL conformation in a dose-dependent manner, acting in an additive or synergistic fashion with Vitamins C and E (Chen *et al.,* 2007). *In-vitro,* ASP acts as anti-oxidants and induces quinone reductase activity depending on the dose, method of extraction and interaction with anti-oxidant vitamins (Chen *et al.,* 2008).

Jalali-Khanabadi *et al.* (2010) demonstrated that almond supplementation, in addition to lowering CHD lipid risk factors in serum, may contribute to a dramatic change in the relation of lipid risk factors and susceptibility of serum lipids to oxidative modification probably due to the distribution of different almond phenolic anti-oxidants in different components of serum including nonlipoprotein molecules such as serum albumin.

Multiple benefits with almond consumption have been found in patients with metabolic syndrome: 1. moderate weight loss, decreased adiposity, and lower blood pressure and 2. reduction in fasting insulin by 2.60 μU/mL and HOMA-insulin resistance by 0.72, and 3. changes in median plasma IL-6 of -1.1 ng/L. But adjustment for weight loss attenuated the significance of the association (Casas-Agustench *et al.,* 2011).

Almond intake has preventive effects on oxidative stress and DNA damage caused by smoking (Jia *et al.,* 2006). Consumption of almond enhanced the anti-oxidant defenses and diminished the biomarkers of oxidative stress in smokers. Almond intervention in smokers significantly increased serum alpha-tocopherol, SOD, and GPX by 10, 35, and 16 per cent, respectively and decreased 8-OHdG, MDA, and DNA strand breaks by 28, 34, and 23 per cent respectively (Li *et al.*, 2007). The emerging evidence supports that almond consumption beneficially influences chronic degenerative disease risk beyond cholesterol reduction, particularly in populations with metabolic syndrome and type 2 diabetes mellitus (Kamil and Chen, 2012).

A significant reduction in 24-hour urinary C-peptide output (a marker of 24-hour insulin secretion) on almond diet (22 per cent of the daily energy intake) explains the association of almond consumption with reduced CHD risk in future (Jenkins *et al.*, 2008b).

Combining a number of foods and food components such as low SFA, soya protein, plant sterols and viscous fibers in a single dietary portfolio lowered LDL-C similar to statins and thus increased the potential effectiveness of dietary therapy (Jenkins *et al.*, 2003; Lamarche *et al.*, 2004).

Anti-obesity Agent

Cell walls were found to be rich in nonstarch arabinose-rich polysaccharides, with a high concentration of phenolic compounds in the almond seed coat. The cell walls of almond seeds reduce lipid bioaccessibility by hindering the release of lipid in the small intestine (Ellis *et al.*, 2004). This effect is regulated by the structure and properties of cell walls and thus plays a primary role in determining postprandial lipemia (Berry *et al.*, 2008). Zemaitis and Sabaté, (2001) had also reported that subjects on diets rich in almonds excreted significant amounts of lipid. Indeed, acute and longer-term almond ingestion helps regulate body weight (Fraser *et al.*, 2002; Sabate, 2003).

As it is a challenge to consume almonds without compromising body weight, an interesting experiment used mastication as a measure for the availability of the nut lipids by the number of chews of 56 g of almonds for 5 days. Fecal fat excretion as well as fecal energy losses were significantly higher after less number of chews (Cassady *et al.*, 2009). Zaveri and Drummond (2009) reported that snacking on almonds promoted significantly higher eating frequency (P ≤0.05), but not higher energy intake, body weight or percentage body fat as compared to cereal bars. Similarly, Foster *et al.* (2012) observed that obese individuals who consumed almond enriched diet or hypo-caloric nut free diet experienced clinically significant and comparable weight loss at 18 months despite almond group showing smaller weight loss at 6 months. In addition, the almond group experienced greater improvements in lipid profiles.

An almond-enriched low calorie diet (LCD) improved a preponderance of the abnormalities associated with the metabolic syndrome. Both the almond diet with low carbohydrate and high fat content (almond-LCD *i.e.* 39 per cent total fat, 25 per cent MUFA and 32 per cent carbohydrate as percent of dietary energy) and self-selected high complex carbohydrate and low fat diet (CHO-LCD *i.e.* 18 per cent total fat, 5 per cent MUFA and 53 per cent carbohydrate as percent of dietary energy) decreased body weight beyond the weight loss observed during long-term pharmacological interventions. However, the almond-LCD group experienced a sustained and greater weight reduction for the duration of the 24-week intervention (Wien *et al.*, 2003).

Anti-diabetic Agent

Almonds lowered serum glucose responses post-prandially. Even when consumed as snacks, almonds reduced hunger and desire to eat during the acute-feeding session. It provided appetite post-ingestive metabolic benefits without increasing the risk for weight gain (Tan and Mattes, 2013). Although almond consumption did not change insulin sensitivity significantly, changes in serum lipids in healthy adults and diabetic patients did occur (Lovejoy *et al.*, 2002).

Almond consumption lowered fasting insulin levels, fasting glucose, and insulin resistance index. Thus incorporation of almonds into a healthy diet has beneficial effects on adiposity, glycemic control, and the lipid profile, thereby potentially decreasing the risk for cardiovascular disease in patients with type 2 diabetes mellitus (Li *et al.*, 2011; Jenkins *et al.*, 2002; 2006). Further, the anti-oxidant activity of almonds and the reduced risk of CVD were confirmed by Jenkins *et al.* (2008a) through reduction in serum MDA and urinary isoprostanes, biomarkers of lipid peroxidation.

According to Cohen *et al.* (2011) regular ingestion of almond for 12 weeks reduced hemoglobin A(1c) by 4 per cent but did not influence fasting glucose concentrations. In addition, a standard serving of almonds reduced postprandial glycemia significantly in participants with diabetes but did not influence glycemia in participants without diabetes.

Oral administration of the purified polyphenol fraction from almond seed skin to rats fed corn starch significantly suppressed the increase in blood glucose levels and area under the curve (AUC), in a dose-dependent manner. This indicates that almond seed skin contains highly polymerized polyphenols with strong α-amylase inhibitory activity which retards absorption of carbohydrate (Tsujita *et al.*, 2013).

Josse *et al.* (2007) demonstrated that adding almonds to white bread reduces progressively the glycemic index of the composite meal in a dose-dependent manner. Adding 1 to 3 oz. of almonds to a 50 g carbohydrate meal brought down the glycemic index from 105.8 ± 23.3 to 45.2 ± 5.8 thus making it suitable as a food for diabetics. Inclusion of almonds in the breakfast meal decreased blood glucose concentrations and increased satiety both immediately and after a second-meal in adults with impaired glucose tolerance (IGT). The lipid component of almonds is likely to be responsible for the immediate post-ingestive response (Mori *et al.*, 2011).

Other Benefits

In comparison with the wheat bran and cellulose diet, whole almond significantly lowered aberrant crypt foci (ACF) and cell turnover (labeling index, LI) suggesting that almond consumption may reduce colon cancer risk despite their high fat content (Davis and Iwahashi, 2001).

Administration of natural almond skin (NS) powder was found to be beneficial for treatment of inflammatory bowel disease. NS powder significantly reduced diarrhea and loss of body weight along with a significant reduction in colonic myeloperoxidase activity. NS powder also reduced NF-κB and p-JNK activation; the pro-inflammatory cytokines release; the appearance of i-NOS, nitro-tyrosine and PARP in the colon; the up-regulation of ICAM-1 and the expression of P-selectin (Mandalari *et al.*, 2011).

Sultana *et al.* (2007) also found the topical application of almond oil is capable of preventing the structural damage caused by UV irradiation and is useful in decelerating the photoaging process. UV exposure causes oxidative stress, inflammation, erythema, and skin cancer. α-Tocopherol and polyphenols in almonds applied to medium or topically provided some degree of photo-protection against UVA radiation (Evans-Johnson *et al.*, 2013).

Almond oil in different concentrations (0.5 per cent, 1 per cent, 1.5 per cent, 2 per cent, 2.5 per cent and 3 per cent) significantly enhanced the penetration of drug from transdermal gels and patch across synthetic membrane/rabbit skin with 3 per cent concentration being the most effective (Hussain *et al.*, 2012).

The bioactive almond skin constituents in the non-lipophilic polyphenol extract were the most effective at protecting hepatocytes against hydro-peroxide induced hepatocyte oxidative stress and in protecting against dicarbonyl induced cytotoxicity. Catechins were also effective at preventing glyoxal (GO) or methyl-glyoxal (MGO) cyto-toxicity probably by trapping GO and MGO and/or rescuing hepatocytes from protein carbonylation (Dong *et al.*, 2010).

Effect of Processing on the Therapeutic Potential

Processing as with any other food raises a question on the therapeutic effects of almonds. Raw or roasted salted almonds or almond butter fed for four weeks significantly lowered low-density lipoprotein-cholesterol in hyper-cholesterolemic men. But only raw and roasted almonds and not almond butter lowered TC significantly. Interestingly, almond butter increased the HDL slightly. None of the forms changed the blood pressure proving that unblanched almonds whether raw, dry roasted, or roasted in butter can play an effective role in lowering the blood cholesterol level (Spiller *et al.*, 2003).

Mandalari *et al.* (2008a) confirmed that the bioaccessibility of the almond constituents could be improved by increasing its transit time in the gut which is regulated by almond cell walls. Finely ground almonds were the most digestible with 39, 45, and 44 per cent of lipid, vitamin E, and protein released after duodenal digestion respectively, as compared to natural almonds, blanched almonds and defatted finely ground almonds. They have also shown that the addition of finely ground almonds showed higher prebiotic index by altering the composition of gut bacteria by stimulating the growth of *bifidobacteria* and *E. rectale* (Mandalari *et al.*, 2008b). Further, they (Mandalari *et al.*, 2010b) demonstrated that the polyphenols present in almond skins did not affect bacterial fermentation revealing that almond skins resulting from industrial blanching could be used as potential prebiotics.

The oxygen-radical absorbance capacity (ORAC) of almond skins is 0.398-0.500 mmol Trolox/g (Monagas *et al.*, 2007). Garrido *et al.* (2008) observed that the anti-oxidant activity (ORAC values) of the roasted almond skins (0.803 to 1.08 m mol Trolox/g) was higher followed by that of blanched + dried (0.398 to 0.575 m mol Trolox/g) and blanched and freeze-dried (0.331 to 0.451 mmol Trolox/g) samples.

Hence, roasting is found to be the most suitable industrial processing of almonds to obtain almond skin extracts with the greatest anti-oxidant capacity.

The blanch water (BW), a by-product of the almond processing industry, showed a pro-anthocyanidin content of 71.84 ± 5.21 cyanidin equivalents/g. which posseses good anti-radical as well as *in-vivo* photo-protective activities (Mandalari *et al.,* 2013).

Safety Concerns of Almond Skin

No mortality, body weight, ophthalmic abnormalities or treatment-related findings in clinical tests (hematology, coagulation, urinalysis parameters, macroscopic or microscopic examinations) were observed with almond consumption. Differences between treated and control groups in weight gain, food consumption, clinical chemistry, and organ weight were not considered treatment-related. The no-observed-adverse-effect-level (NOAEL) for almond skins was considered to be 10 per cent (w/w) for both genders (females, 9.7 g/kg body weight/day; males, 8.2 g/kg body weight/day) (Song *et al.,* 2010).

Conclusions

Whole almonds, skin, and oil are all rich sources of various phytochemicals and hence individually and collectively offer innumerable health benefits that are comparable or sometimes superior to other foods such as walnuts. Therefore, it is advisable to consume almonds along with the skin.

References

Abbey M, Noakes M, Belling GB, Nestel PJ (1994). Partial replacement of saturated fatty acids with almonds or walnuts lowers total serum cholesterol and low-density-lipoprotein cholesterol. *Am. J. Clin. Nutr.* 59: 995–99.

Ahmad Z (2010). The uses and properties of almond oil. *Complement. Ther. Clin. Pract.* 16 (1): 10–12.

Amarowicz R, Troszynska A, Shahidi F (2005). Anti-oxidant activity of almond seed extract and its fractions. *J. Food Lipids.* 12: 344–58.

Berry EM, Eisenberg S, Friedlander Y, Norman Y, Kaufmann NA, Stein Y. (1992) l. Effects of diets rich in monounsaturated fatty acids on plasma lipoproteins–the Jerusalem Nutrition Study. II. Monounsaturated fatty acids vs. carbohydrates. *Am. J. Clin. Nutr.* 56: 394–403.

Berry SE, Tydeman EA, Lewis HB, Phalora R, Rosborough J, Picout DR, Ellis PR (2008). Manipulation of lipid bioaccessibility of almond seeds influences postprandial lipemia in healthy human subjects. *Am. J. Clin. Nutr.* 88 (4): 922–29.

Berryman CE, Preston AG, Karmally W, Deckelbaum RJ, Kris-Etherton PM (2011). Effects of almond consumption on the reduction of LDL-cholesterol: a discussion of potential mechanisms and future research directions. *Nutr. Rev.* 69 (4): 171–85.

Bisignano C, Filocamo A, La Camera E, Zummo S, Fera MT, Mandalari G (2013). Anti-bacterial activities of almond skins on cagA-positive and-negative clinical isolates of *Helicobacter pylori*. *BMC. Microbiol.* 13 (1): 103.

Bolling BW, Dolnikowski G, Blumberg JB, Chen CYO (2009). Quantification of almond skin polyphenols by liquid chromatography-mass spectrometry. *J. Food Sci.* 74: C326–32.

Brieskorn CH, Betz R (1988). Procyanidin polymers, the crucial ingredients of the almond seedcoat. *Z. Lebensm Unters Forsch.* 187: 347–53.

Casas-Agustench P, López-Uriarte P, Bulló M, Ros E, Cabré-Vila JJ, Salas-Salvadó J (2011). Effects of one serving of mixed nuts on serum lipids, insulin resistance and inflammatory markers in patients with the metabolic syndrome. *Nutr. Metab. Cardiovasc. Dis.* 21 (2): 126–35.

Cassady BA, Hollis JH, Fulford AD, Considine RV, Mattes RD (2009). Mastication of almonds: effects of lipid bioaccessibility, appetite, and hormone response. *Am. J. Clin. Nutr.* 89: 794–800.

Chen CY, Blumberg JB (2008). In-vitro activity of almond skin polyphenols for scavenging free radicals and inducing quinone reductase. *J. Agric. Food Chem.* 56 (12): 4427–34.

Chen CY, Milbury PE, Chung SK, Blumberg J (2007). Effect of almond skin polyphenolics and quercetin on human LDL and apolipoprotein B-100 oxidation and conformation. *J. Nutr. Biochem.* 18 (12): 785–94.

Chen CY, Milbury PE, Lapsley K, Blumberg JB (2005). Flavonoids from almond skins are bioavailable and act synergistically with vitamins C and E to enhance hamster and human LDL resistance to oxidation. *J. Nutr.* 135 (6): 1366–73.

Choudhury K, Clark J, Griffiths HR (2014). An almond-enriched diet increases plasma α-tocopherol and improves vascular function but does not affect oxidative stress markers or lipid levels. *Free Radic. Res.* 48 (5): 599–606.

Cohen AE, Johnston CS (2011). Almond ingestion at mealtime reduces postprandial glycemia and chronic ingestion reduces hemoglobin A(1c) in individuals with well-controlled type 2 diabetes mellitus. *Metabolism.* 60 (9): 1312–17.

Damasceno NR, Pérez-Heras A, Serra M, Cofán M, Sala-Vila A, Salas-Salvadó J, Ros E (2011). Crossover study of diets enriched with virgin olive oil, walnuts or almonds. Effects on lipids and other cardiovascular risk markers. *Nutr. Metab. Cardiovasc. Dis.* 21 Suppl 1: S14–20.

Davis PA, Iwahashi CK (2001). Whole almonds and almond fractions reduce aberrant crypt foci in a rat model of colon carcinogenesis. *Cancer Lett.* 165 (1): 27–33.

Dong Q, Banaich MS, O'Brien PJ (2010). Cytoprotection by almond skin extracts or catechins of hepatocyte cytotoxicity induced by hydroperoxide (oxidative stress model) versus glyoxal or methylglyoxal (carbonylation model). *Chem. Biol. Interact.* 185 (2): 101–109.

Ellis PR, Kendall CW, Ren Y, Parker C, Pacy JF, Waldron KW, Jenkins DJ (2004). Role of cell walls in the bioaccessibility of lipids in almond seeds. *Am. J. Clin. Nutr.* 80 (3): 604–613.

Evans-Johnson JA, Garlick JA, Johnson EJ, Wang XD, Oliver Chen CY (2013). A pilot study of the photoprotective effect of almond phytochemicals in a 3D human skin equivalent. *J. Photochem. Photobiol*. B. 126: 17–25.

Foster GD, Shantz KL, Vander Veur SS, Oliver TL, Lent MR, Virus A, Szapary PO, Rader DJ, Zemel BS, Gilden-Tsai A (2012). A randomized trial of the effects of an almond-enriched, hypocaloric diet in the treatment of obesity. *Am. J. Clin. Nutr.* 96: 249–54.

Fraser GE, Bennett HW, Jaceldo KB, Sabate J (2002). Effect on body weight of a free 76 Kilojoule (320 calorie) daily supplement of almonds for six months. *J. Am. Coll. Nutr.* 21: 275–83.

Frison S, Sporns P (2002a). Variation in the flavonol glycoside composition of almond seedcoats as determined by MALDI-TOF mass spectrometry. *J. Agric. Food Chem.* 50: 6818–22.

Frison S, Sporns P (2002b). Identification and quantification of flavonol glycosides in almond seedcoats using MALDI-TOF MS. *J. Agric. Food Chem.* 50: 2782–87.

Garrido I, Monagas M, Gómez-Cordovés C, Bartolomé B (2008). Polyphenols and anti-oxidant properties of almond skins: influence of industrial processing. *J. Food Sci.* 73 (2): C106–115.

Garrido I, Urpi-Sarda M, Monagas M, Gómez-Cordovés C, Martín-Alvarez PJ, Llorach R, Bartolomé B, Andrés-Lacueva C (2010). Targeted analysis of conjugated and microbial-derived phenolic metabolites in human urine after consumption of an almond skin phenolic extract. *J. Nutr.* 140 (10): 1799–807.

Haider S, Batool Z, Haleem DJ (2012). Nootropic and hypo-phagic effects following long term intake of almonds (*Prunus amygdalus*) in rats. *Nutr. Hosp.* 27 (6): 2109–115.

Hussain A, Khan GM, Shah SU, Shah KU, Rahim N, Wahab A, Rehman AU (2012). Development of a novel ketoprofen transdermal patch: effect of almond oil as penetration enhancers on in-vitro and ex-vivo penetration of ketoprofen through rabbit skin. *Pak. J. Pharm. Sci.* 25 (1): 227–32.

Hyson DA, Schneeman BO, Davis PA (2002). Almonds and almond oil have similar effects on plasma lipids and LDL oxidation in healthy men and women. *J. Nutr.* 132 (4): 703–707.

Jaceldo-Siegl K, Sabaté J, Batech M, Fraser GE (2011). Influence of body mass index and serum lipids on the cholesterol-lowering effects of almonds in free-living individuals. *Nutr. Metab. Cardiovasc. Dis.* 21 (Suppl 1): S7–13.

Jaceldo-Siegl K, Sabaté J, Rajaram S, Fraser GE (2004). Long-term almond supplementation without advice on food replacement induces favourable nutrient modifications to the habitual diets of free-living individuals. *Br. J. Nutr.* 92 (3): 533–40.

Jalali-Khanabadi BA, Mozaffari-Khosravi H, Parsaeyan N (2010). Effects of almond dietary supplementation on coronary heart disease lipid risk factors and serum lipid oxidation parameters in men with mild hyperlipidemia. *J. Altern. Complement. Med.* 16 (12): 1279–83.

Jambazian PR, Haddad E, Rajaram S, Tanzman J, Sabaté J (2005). Almonds in the diet simultaneously improve plasma alpha-tocopherol concentrations and reduce plasma lipids. *J. Am. Diet. Assoc.* 105 (3): 449–54.

Jenkins DJA, Kendall CWC, Josse AR, Salvatore S, Brighenti F, Augustin LS, Ellis PR, Vidgen E, Rao AV (2006). Almonds decrease postprandial glycemia, insulinemia, and oxidative damage in healthy individuals. *J. Nutr.* 136: 2987–92.

Jenkins DJ, Kendall CW, Marchie A, Faulkner D, Vidgen E, Lapsley KG, Trautwein EA, Parker TL, Josse RG, Leiter LA, Connelly PW (2003). The effect of combining plant sterols, soy protein, viscous fibers, and almonds in treating hyper-cholesterolemia. *Metabolism.* 52 (11): 1478–83.

Jenkins DJ, Kendall CW, Marchie A, Josse AR, Nguyen TH, Faulkner DA, Lapsley KG, Blumberg J (2008a). Almonds reduce biomarkers of lipid peroxidation in older hyperlipidemic subjects. *J. Nutr.* 138 (5): 908–13.

Jenkins DJ, Kendall CW, Marchie A, Josse AR, Nguyen TH, Faulkner DA, Lapsley KG, Singer W (2008b). Effect of almonds on insulin secretion and insulin resistance in nondiabetic hyperlipidemic subjects: a randomized controlled crossover trial. *Metabolism.* 57 (7): 882–87.

Jenkins DJA, Kendall CWC, Marchie A, Parker TL, Connelly PW, Qian W, Haight JS, Faulkner D, Vidgen E, Lapsley KG, Spiller GA (2002). Dose response of almonds on coronary heart disease risk factors: blood lipids, oxidized low-density lipoproteins, lipoprotein (a), homocysteine, and pulmonary nitric oxide: a randomized, controlled, crossover trial. *Circulation.* 106: 1327–32.

Jia X, Li N, Zhang W, Zhang X, Lapsley K, Huang G, Blumberg J, Ma G, Chen J (2006). A pilot study on the effects of almond consumption on DNA damage and oxidative stress in smokers. *Nutr. Cancer.* 54 (2): 179–83.

Josse AR, Kendall CW, Augustin LS, Ellis PR, Jenkins DJ (2007). Almonds and postprandial glycemia–a dose-response study. *Metabolism.* 56 (3): 400–404.

Kalgaonkar S, Almario RU, Gurusinghe D, Garamendi EM, Buchan W, Kim K, Karakas SE (2011). Differential effects of walnuts vs almonds on improving metabolic and endocrine parameters in PCOS. *Eur. J. Clin. Nutr.* 65 (3): 386–93.

Kamil A, Chen CY (2012). Health benefits of almonds beyond cholesterol reduction. *J. Agric. Food Chem.* 60 (27): 6694–702.

Lamarche B, Desroches S, Jenkins DJ, Kendall CW, Marchie A, Faulkner D, Vidgen E, Lapsley KG,Trautwein EA, Parker TL, Josse RG, Leiter LA, Connelly PW (2004). Combined effects of a dietary portfolio of plant sterols, vegetable protein, viscous fibre and almonds on LDL particle size. *Br. J. Nutr.* 92 (4): 657–63.

Li N, Jia X, Chen CY, Blumberg JB, Song Y, Zhang W, Zhang X, Ma G, Chen J (2007). Almond consumption reduces oxidative DNA damage and lipid peroxidation in male smokers. *J. Nutr.* 137 (12): 2717–22.

Li SC, Liu YH, Liu JF, Chang WH, Chen CM, Chen CY (2011). Almond consumption improved glycemic control and lipid profiles in patients with type 2 diabetes mellitus. *Metabolism.* 60 (4): 474–79.

Liu JF, Liu YH, Chen CM, Chang WH, Chen CY (2013). The effect of almonds on inflammation and oxidative stress in Chinese patients with type 2 diabetes mellitus: a randomized crossover controlled feeding trial. *Eur. J. Nutr.* 52 (3): 927–35.

Liu Z, Lin X, Huang G, Zhang W, Rao P, Ni L (2014). Prebiotic effects of almonds and almond skins on intestinal microbiota in healthy adult humans. *Anaerobe.* 26: 1–6.

Lovejoy JC, Most MM, Lefevre M, Greenway FL, Rood JC (2002). Effect of diets enriched in almonds on insulin action and serum lipids in adults with normal glucose tolerance or type 2 diabetes. *Am. J. Clin. Nutr.* 76 (5): 1000–1006.

Mandalari G, Arcoraci T, Martorana M, Bisignano C, Rizza L, Bonina FP, Trombetta D, Tomaino A (2013). Anti-oxidant and photo-protective effects of blanch water, a byproduct of the almond processing industry. *Molecules.* 18 (10): 12426–40.

Mandalari G, Bisignano C, D'Arrigo M, Ginestra G, Arena A, Tomaino A, Wickham MS (2010a). Anti-microbial potential of polyphenols extracted from almond skins. *Lett. Appl. Microbiol.* 51 (1): 83–89

Mandalari G, Bisignano C, Genovese T, Mazzon E, Wickham MS, Paterniti I, Cuzzocrea S (2011). Natural almond skin reduced oxidative stress and inflammation in an experimental model of inflammatory bowel disease. *Int. Immunopharmacol.* 11 (8): 915–24.

Mandalari G, Faulks RM, Bisignano C, Waldron KW, Narbad A, Wickham MS (2010b). *In-vitro* evaluation of the prebiotic properties of almond skins (*Amygdalus communis* L.). *FEMS Microbiol Lett.* 304 (2): 116–22.

Mandalari G, Faulks RM, Rich GT, Lo Turco V, Picout DR, Lo Curto RB, Bisignano G, Dugo P, Dugo G, Waldron KW, Ellis PR, Wickham MS (2008a). Release of protein, lipid, and vitamin E from almond seeds during digestion. *J. Agric. Food. Chem.* 56 (9): 3409–416.

Mandalari G, Nueno-Palop C, Bisignano G, Wickham MS, Narbad A (2008b). Potential prebiotic properties of almond (*Amygdalus communis* L.) seeds. *Appl. Environ. Microbiol.* 74 (14): 4264–70.

Milbury PE, Chen CY, Dolnikowski GG, Blumberg JB (2006). Determination of flavonoids and phenolics and their distribution in almonds. *J. Agric. Food Chem.* 54 (14): 5027–33.

Monagas M, Garrido I, Lebrón-Aguilar R, Bartolome B, Gómez-Cordovés C (2007). Almond (*Prunus dulcis* (Mill.) D.A. Webb) skins as a potential source of bioactive polyphenols. *J. Agric. Food Chem.* 55: 8498–507.

Mori AM, Considine RV, Mattes RD (2011). Acute and second-meal effects of almond form in impaired glucose tolerant adults: a randomized crossover trial. *Nutr. Metab.* (Lond). 8 (1): 6.

Ortiz RM, Garcia S, Kim AD (2012). Is Almond Consumption More Effective Than Reduced Dietary Saturated Fat at Decreasing Plasma Total Cholesterol and LDL-c Levels? A Theoretical Approach. *J. Nutr. Metab.* 2012: 265712.

Rajaram S, Connell KM, Sabaté J (2010). Effect of almond-enriched high-monounsaturated fat diet on selected markers of inflammation: a randomised, controlled, crossover study. *Br. J. Nutr.* 103 (6): 907–912.

Ren, Y., K. W. Waldron, J. F. Pacy, A. Brain, and P. R. Ellis (2001). Chemical and histo-chemical characterisation of cell wall polysaccharides in almond seeds in relation to lipid bioavailability, pp. 448–452. In W. Pfannhauser, G. R. Fenwick, and S. Khokhar (ed.), *Biologically-active phytochemicals in food*. The Royal Society of Chemistry, Cambridge, United Kingdom.

Sabate J (2003). Nut consumption and body weight. *Am. J. Clin. Nutr.* 78: 647S–650S.

Sabate J, Haddad E, Tanzman J, Jambazian PR, Rajaram S (2003). Serum lipid response to the graduated enrichment of a Step I diet with almonds: a randomized feeding trial. *Am. J. Clin. Nutr.* 77: 1379–84.

Sang S, Lapsley K, Jeong WS, Lachance PA, Ho CT, Rosen RT (2002). Anti-oxidative phenolic compounds isolated from almond skins (*Prunus amygdalus* Batsch). *J. Agric. Food Chem.* 50: 2459–63.

Song Y, Wang W, Cui W, Zhang X, Zhang W, Xiang Q, Liu Z, Li N, Jia X (2010). A subchronic oral toxicity study of almond skins in rats. *Food Chem. Toxicol.* 48 (1): 373–76.

Spiller GA, Jenkins DA, Bosello O, Gates JE, Cragen LN, Bruce B (1998). Nuts and plasma lipids: an almond-based diet lowers LDL-C while preserving HDL-C. *J. Am. Coll. Nutr.* 17 (3): 285–90.

Spiller GA, Jenkins DJ, Cragen LN, Gates JE, Bosello O, Berra K, Rudd C, Stevenson M, Superko R (1992). Effect of a diet high in mono-unsaturated fat from almonds on plasma cholesterol and lipoproteins. *J. Am. Coll. Nutr.* 11 (2): 126–30.

Spiller GA, Miller A, Olivera K, Reynolds J, Miller B, Morse SJ, Dewell A, Farquhar JW (2003). Effects of plant-based diets high in raw or roasted almonds, or roasted almond butter on serum lipoproteins in humans. *J. Am. Coll. Nutr.* 22 (3): 195–200.

Sultana Y, Kohli K, Athar M, Khar RK, Aqil M (2007). Effect of pre-treatment of almond oil on ultraviolet B-induced cutaneous photoaging in mice. *J. Cosmet. Dermatol.* 6 (1): 14–19.

Tan SY, Mattes RD (2013). Appetitive, dietary and health effects of almonds consumed with meals or as snacks: a randomized, controlled trial. *Eur. J. Clin. Nutr.* 67 (11): 1205–214.

Tsujita T, Shintani T, Sato H (2013). α-Amylase inhibitory activity from nut seed skin polyphenols. 1. Purification and characterization of almond seed skin polyphenols. *J. Agric. Food Chem.* 61 (19): 4570–4576.

Udenigwe CC, Je JY, Cho YS, Yada RY (2013). Almond protein hydrolysate fraction modulates the expression of pro-inflammatory cytokines and enzymes in activated macrophages. *Food. Funct.* 4 (5): 777–83.

Ukhanova M, Wang X, Baer DJ, Novotny JA, Fredborg M, Mai V (2014). Effects of almond and pistachio consumption on gut microbiota composition in a randomised cross-over human feeding study. *Br. J. Nutr.* 18: 1–7.

USDA, National Nutrition Data Base for standard reference, Release-26, NDB No-12061.

Wien MA, Bleich D, Raghuwanshi M, Gould-Forgerite S, Gomes J, Monahan-Couch L, Oda K (2010). Almond consumption and cardiovascular risk factors in adults with prediabetes. *J. Am.Coll. Nutr.* 29 (3): 189–97.

Wien MA, Sabaté JM, Iklé DN, Cole SE, Kandeel FR (2003). Almonds vs complex carbohydrates in a weight reduction program. *Int. J. Obes. Relat. Metab. Disord.* 27 (11): 1365–72.

Wijeratne SS, Abou-Zaid MM, Shahidi F (2006). Anti-oxidant polyphenols in almond and its coproducts. *J. Agric. Food Chem.* 54 (2): 312–18.

Zaveri S, Drummond S (2009). The effect of including a conventional snack (cereal bar) and a nonconventional snack (almonds) on hunger, eating frequency, dietary intake and body weight. *J. Hum. Nutr. Diet.* 22 (5): 461–68.

Zemaitis J, Sabaté J (2001). Effect of almond consumption on stool weight and stool fat. *Fed. Am. Soc. Exp. Biol. J.* 15 (4): A602.

Chapter 2
Banana/Plantain
(*Musa acuminata, Musa balbisiana, Musa paradisiaca*)

Banana is the most popular fruit especially of the poor, available and consumed throughout the year. It is thought to have originated in Malaysia around 4,000 years ago from where it spread throughout the Philippines and India. Bananas, also called

Four Varieties of Banana Fruit.

plantains, are staple food in many Pacific Island countries. Bananas belong to the genus Musa from the family *Musaceae.* Some of the species include the scarlet banana (*Musa coccinea*), pink banana (*Musa velutina*) and the Fe'i bananas (Wikipedia.org). Unripe banana is more popular in the southern parts of India as a vegetable, consumed in fried or baked form in Western Africa and the Caribbean countries as well. Cream-fleshed cultivar, Williams, of the Cavendish group, is the most commonly marketed banana worldwide.

Nutritional and Phytochemical Composition of Banana

Banana flour obtained from unripe banana (*Musa acuminata,* var. Nanicão) contains a high amount of total dietary fiber (56.24 g/100 g), including resistant starch (48.99 g/100 g) and fructans (0.05 g/100 g). The contents of available starch (27.78 g/100 g) and soluble sugars (1.81 g/100 g) were low. Oxalate content (1.1-1.6 per cent) of banana flour was nutritionally insignificant. The main phytosterols found were campesterol (4.1 mg/100 g), stigmasterol (2.5 mg/100 g) and β-sitosterol (6.2 mg/100 g). The total polyphenol content was 50.65 mg GAE/100 g with a moderate anti-oxidant activity of 358.67 and 261.00 µmol of Trolox equivalent/100 g by the FRAP and ORAC methods respectively. The content and bioavailability of Zn, Ca and Fe were low (Menezes *et al.,* 2011).

Nutritive Value of Banana

Nutrient	Amount	Nutrient	Amount
Energy (kcal)	89.0	Folate (µg)	20.0
Carbohydrates (g)	22.8	Choline (mg)	9.8
Sugars (g)	12.2	Vitamin C (mg)	8.7
Dietary fiber (g)	2.6	Iron (mg)	0.26
Fat (g)	0.3	Magnesium (mg)	27.00
Protein (g)	1.1	Manganese (mg)	0.27
Thiamine (mg)	0.03	Phosphorus (mg)	22.00
Riboflavin (mg)	0.07	Potassium (mg)	358.0
Niacin (mg)	0.67	Sodium (mg)	1.0
Pantothenic acid (mg)	0.33	Zinc mg)	0.15
Vitamin B$_6$ (mg)	0.4	Fluoride (mg)	2.2

Source: USDA Nutrient Database.

Iron- and zinc-fortified bananas were developed by transforming embryogenic cells of banana cv. Rasthali (AAB) with soybean ferritin cDNA using two different expression cassettes pSF and pEFE-SF to express ferritin by Kumar *et al.* (2011) which could serve as a functional food to overcome the malnutrition-related iron deficiency. A 6.32-fold increase in iron and a 4.58-fold increase in the zinc levels were noted in the leaves of transgenic plants indicating the increased content of these two minerals in the fruits as well.

Bananas are a very good source of vitamin B_6, and a good source of vitamin C and manganese. An average sized banana contains 358 mg of potassium and just 1 mg of sodium due to which it helps maintain normal blood pressure and heart function. The Karat cultivars of bananas are exceptionally rich in Riboflavin, Niacin, α-tocopherol as well as flavonoids (Englberger *et al.,* 2006a).

Ripened and unripened bontha, poovan, nendran, cavendish and rasthali bananas inhibited the proteolysis of casein by trypsin, chymotrypsin and papain. The inhibitory factors of trypsin and papain were found to be dissimilar. The inhibition was much more by ripened banana than by unripened banana cultivars. Unripened banana cultivars inhibit papain and cure stomach ulcers and the ripened banana cultivars inhibit the anti-nutritional factors (Rao, 1991).

Banana Flower.

Banana inflorescences are popularly known as 'navels,' and they are used in Brazil as nutritional complements both in raw and dried form. The moisture, protein, fat and ash contents of the inflorescence the *Musa acuminata*, cultivar "ouro", from Brazil were found to be 8.21, 14.50, 4.04 and 14.43 g per cent respectively. The high potassium (5008.26 mg/100 g) and fiber 49.83 per cent (lignin, cellulose and hemicelluloses) contents reveal the important functional and nutritional properties of this cultivar (Fingolo *et al.,* 2012).

Banana fruit peels and banana trunk are also used as food in India. The peels are rich in trace elements compared to the trunks (except in the case of Zn) (Selema and Farago, 1996).

Varietal Differences

Cultivation method (greenhouse and outdoors), farming style (conventional and organic) and region of production influence the sugars, ash and protein content of bananas. The chemical compositions in the bananas (Gran Enana and Pequeña

Enana) harvested in Tenerife and in those (Gran Enana) from Ecuador did not differ except for insoluble fiber content (Forster *et al.,* 2002). Interestingly length of the banana was also found to influence the protein, ash, and ascorbic acid content (higher are the values shorter the banana) (Forster *et al.,* 2002).

Green bananas/unripe bananas also called plantains contain starch, which gets converted into sugars only on ripening. They are also known to contain substantial amount of resistant starch. Differences in the resistant starch content was reported in Micronesian cultivars *i.e.* Utin Kerenis and Utin Ruk were found to contain the highest amounts of resistant starch as compared to Daiwang, Inahsio and Karat (Thakorlal *et al.,* 2010).

Several varieties of bananas were analyzed for their provitamin A content. Total carotene and retinol activity equivalents (RAE) ranged from 150 to 2176 µg/100 g and 8 RAE/100 g to 136 RAE/100 g, for Cavendish Williams, and Kirkirnan respectively (Blades *et al.,* 2003). Carotenoid-rich banana cultivars have also been identified in other Micronesian bananas (Englberger *et al.,* 2003). But, the yellow/orange-fleshed Asupina (a Fe'i banana) was found to contain the highest level (1,412 µg/100 g) of trans beta-carotene, which is >20 times higher than that of Williams. Hence banana is found to be potentially useful for the prevention of vitamin A deficiency (Englberger *et al.,* 2006b). Four banana varieties (Horn Plantain, Kirkirnan, Asupina and Pisang Raja) were also identified as good sources (>75 RAE/100 g) of provitamin A (Blades *et al.,* 2003).

The carotenoid content of banana was reported to vary according to its colour. All 10 yellow or yellow/orange-fleshed cultivars (Asupina, Kirkirnan, Pisang Raja, Horn Plantain, Pacific Plantain, Kluai Khai Bonng, Wain, Red Dacca, Lakatan, and Sucrier) had significant carotenoid levels, potentially meeting half or all of the estimated vitamin A requirements for a non-pregnant, non-lactating adult woman within normal consumption patterns. Karat is also an exceptionally carotenoid-rich variety (Englberger *et al.,* 2006b).

Banana flour prepared from two cooking banana varieties, namely 'Alukehel' (yield–31.3 per cent) and 'Monthan' (yield–25.5 per cent) showed 3.2 per cent crude protein, 1.3 per cent crude fat, 3.7 per cent ash, 8.9 per cent neutral detergent fiber, 3.8 per cent acid detergent fiber, 3.1 per cent cellulose, 1.0 per cent lignin and 5.0 per cent hemicelluloses of the dry matter. Potassium was found to be the predominant mineral in banana flour (Suntharalingam and Ravindran, 1993).

Effect of Cooking/Processing on Nutrients

Deep-fat frying had no significant effect on carotenoid and potassium content of banana at any frying conditions *i.e.* 120 -180° C and from 24 to 4 min, except that, at 120° C for 24 minutes, a loss of (≤11 per cent) potassium was noticed. However, there was a significant loss of (≤45 per cent) ascorbic acid. Thus, low temperature and long time (120°C/24 min) frying was detrimental to the micro-nutrients in bananas (Rojas-Gonzalez *et al.,* 2006). But, Avallone *et al.* (2009) reported that preparation of plantain chips called "tostones" by deep-fat frying resulted in retention of potassium and decrease in carotenoid contents with beta-carotene appearing to be more heat-resistant than alpha-carotene and L-ascorbic acid. According to Bresnahan *et al.* (2012) retinol bio-efficacy of bananas is improved by cooking as compared to ripening. A loss of 65 per cent of vitamin C was observed during the preparation of banana flour (Suntharalingam and Ravindran, 1993).

Anti-oxidant Property

Six different cultivars of banana flowers–Kathali, Bichi, Shingapuri, Kacha, Champa, and Kalabou were found to be a potential source of natural anti-oxidants. The Kacha variety contains the maximum amount of polyphenol and flavonoid; and thereby the highest total anti-oxidant capacity, DPPH radical scavenging activity and ABTS•(+) radical scavenging activity (China *et al.*, 2011).

A positive correlation has been noticed between free radical scavenging capacity and the content of phenolic compound of unripe, ripe, and leaky ripe stages of banana peel. The peel from unripe banana (*Musa paradisica*) showed higher anti-oxidant potency than that of ripe and leaky ripe ones. The ethyl acetate and water soluble fractions of unripe peel showed higher anti-oxidant activity than CHCl and hexane fractions (Sundaram *et al.*, 2011).

The reducing capacity of the methanolic extract of Musa ABB cv Pisang Awak increased with increasing concentration (31.5-1000 mg/ml) of the fruit extract and the activity was comparable with the standard BHT. Strong and positive correlations were obtained between total phenol and flavonoid contents ($R^2 = 0.693$-1.0). The free radical scavenging ability was attributed to the major anti-oxidants–polyphenols (Darsini *et al.*, 2012).

All parts of banana plant, including the flowers, fruits, peels and trunk have been identified with health benefits in various clinical conditions as presented below.

Ulcer

Peptic ulcer disease encompassing gastric and duodenal ulcers is the most prevalent gastro-intestinal disorder. People from the South-Western Nigeria use dried *Musa sapientum* peels blended with yam flour to ameliorate the gastric pain and ulcer. Siddha system of medicine, one of the ancient systems of medicine in India, recommends Bhasma (calcinated metals and minerals) of *Musa paradisiaca* as anti-ulcer agent which was confirmed by Vadivelan *et al.* (2006) in albino rats.

Unripe Banana is consumed in South India as a bland recipe for treating peptic ulcer. The flour made of plantain is quite often prescribed in dyspepsia. Several research reports are available on the anti-ulcerogenic properties of unripe banana (Hanszen, 1934; Elliot and Heward, 1976; Sanyal *et al.*, 1961). Green banana protects the gastric mucosa against NSAIDs and other ulcerogens in animals. The powder prepared from banana pulp showed anti-ulcerogenic activity in aspirin, indomethacin, phenylbutazone and prednisolone induced gastric ulcers in rats and cysteamine- and histamine-induced duodenal ulcers in guinea-pigs (Goel and co-workers., 1986). The anti-ulcerogenic action of banana preparations is believed to be due to growth stimulation of the gastric mucosa (Best *et al.*, 1984). Interestingly, even after the damage has occurred green banana has been said to stimulate growth of the gastric lining (Goel *et al.*, 1986; 2001).

Moreover, the banana-treated rats (0.5 g/kg orally, twice daily for 3 days) showed: (i) a significant increase in the [3H] thymidine incorporation into mucosal cell DNA; (ii) a significant increase in the total carbohydrate (sum of total hexoses, hexosamine, fucose and sialic acid) content of gastric mucosa; (iii) a significant decrease in gastric juice DNA and protein; and (iv) a significant increase in the total carbohydrates and carbohydrate/protein ratio of gastric juice (Mukhopadhyaya *et al.*, 1987).

The suspensions of bananas have been found to be highly surface active at both liquid-air and solid-liquid interfaces which they render hydrophobic by adsorption of phospholipid. Hills and Kirwood (1989) found banana to impart appreciable (75 per cent) protection against acid insult (1 ml of 0.8 N HCl) in a dose-dependent manner not attributable to "bulking" or buffering as it was equally effective at a pH of 2. The best protection (89 per cent by ulcer length) was obtained with banana vortexed with milk (1:1). Thus, banana gastric mucosal barrier is provided by an adsorbed layer of surface-active phospholipid (surfactant).

Dunjiæ and co-workers (1993) reported that pectin and phosphatidyl-choline in green banana strengthens the mucous-phospholipid layer that protects the gastric mucosa. Lewis *et al.* (1999); and Lewis and Shaw (2001) proved that Leucocyanidin a natural flavonoid from unripe banana pulp prevented the erosion of gastric mucosa. The synthetic analogues, hydroxyl-ethylated leucocyanidin and tetra-allyl leucocyanidin increase the thickness of gastric mucus in aspirin-induced erosions in rats.

Pannangpetch *et al.* (2001) indicated that bananas of different varieties have varying anti-peptic ulcer effects. The extracts of Palo and Hom bananas showed a prominent gastro-protective effect, whereas only the extract of Hom banana demonstrated an ulcer-healing effect. Methanolic extract of *Musa sapientum* (100 mg/kg) showed better ulcer protective effect in NIDDM rats compared with anti-diabetic drug glibenclamide (GLC, 0.6 mg/kg); and sucralfate (SFT, 500 mg/kg) in gastric ulcers induced by cold-restraint stress. From these observations Mohan Kumar *et al.* (2006) reported that the ulcer protective effect could be due to its predominant effect on mucosal glycoprotein, cell proliferation, free radicals and anti-oxidant systems.

Agarwal *et al.* (2009) reported that incision and dead space wounds in rats treated with 100 mg/kg/day of aqueous and methanolic extracts of *M. sapientum var. paradisiaca*, for a period of 10-21 days increased wound breaking strength and levels of hydroxyl-proline, hexuronic acid, hexosamine, superoxide dismutase, reduced glutathione in the granulation tissue and decreased percentage of wound area, scar area and lipid peroxidation when compared with the control group.

Prabha *et al.* (2011) identified an active compound a monomeric flavonoid - leucocyanidin with anti-ulcerogenic activity in *Musa sapientum*. Onasanwo *et al.* (2013) found that the anti-ulcer effect of the methanolic extract of the peels on *Musa sapientum* may be due to its anti-secretory and cyto-protective activity and that the healing of the ulcer might also be connected with basic fibroblast growth factors responsible for epithelial regeneration. Dadoo *et al.* (1995) reported that ripe banana should not be recommended for peptic ulcer as it increases acid production in the stomach.

Colitis

Banana plant is rich in resistant starch, which is used by colonic micro-biota for the anaerobic production of the short-chain fatty acids that serve as a major fuel source for colono-cytes. Scarminio *et al.* (2012) found green dwarf banana flour to show protective effects on the intestinal inflammation acting as a prebiotic and combination of this flour with prednisolone presented synergistic effects. Thus the authors recommended green dwarf banana flour to be an important dietary supplement and complementary medicine in the prevention and treatment of human inflammatory bowel disease.

Langkilde *et al.* (2002) reported that the addition of raw green banana flour containing a type of resistant starch RS-2 to the diet of ileostomy subjects did not interfere with small-bowel absorption of nutrients or total sterols, except for a small increase in iron excretion.

Diarrhea

Treatment of Bangladeshi children suffering from diarrhea and dehydration with banana (just like pectin) significantly reduced lactulose recovery, increased mannitol recovery, and decreased the lactulose-mannitol (LM) ratio, indicating improvement of permeability which was associated with a reduction in stool weight. Increased colonic as well as small bowel absorption mediated by the short chain fatty acids, produced in the colon upon fermentation is responsible for the anti-diarrheal effects of green banana (Rabbani *et al.*, 2004).

A. hydrophilia and *S. flexneri* were the most common causative organisms of diarrhea in children. The experimental group fed on a green plantain diet had significantly diminished stool output, consistency and stool weight with simultaneous increase in body weight gain than the yogurt-based control group proving the higher probiotic potential of green banana; and the average duration of diarrhea in the plantain-based diet group was 18 hours shorter (Alvarez-Acosta *et al.*, 2009). Further, Kosek *et al.* (2010) confirmed that the adjunctive nutritional therapy in endemic areas supported the use of green bananas in shortening the duration and frequency of Shigella dysentery as well as improving weight gain in early convalescence. Addition of cooked green banana to the diets of diarrhoeal children in rural Bangladesh in a community based trial quickened the recovery rates of children with acute diarrhoea by the 3rd day while the same in children with prolonged diarrhea was by the 10th day (Rabbani *et al.*, 2010).

Kidney Diseases

Diet is a major factor in idiopathic calcium oxalate urolithiasis. Musa stem juice (3 mL/rat/day orally) was found to be effective in reducing the formation and also in dissolving the pre-formed stones mainly of magnesium ammonium phosphate with traces of calcium oxalate in rats (Prasad *et al.*, 1993).

Poonguzhali and Chegu (1994) suggested that banana stem extract from the Musaceae family may be a useful agent in the treatment of patients with hyper-oxaluric urolithiasis as the extract reduced urinary oxalate, glycollic and glyoxylic acid and phosphorus excretion in the hyper-oxaluric rats. Kailash and Varalakshmi (1992) also found banana stem extract to significantly lower the activity of glycolic acid oxidase in rats compared to that of the glycolate-fed rats and decrease lactate dehydrogenase enzymes. During hyper-oxaluric state the levels of calcium, phosphorus, oxalate and glycolic acid showed marked alterations in liver tissue.

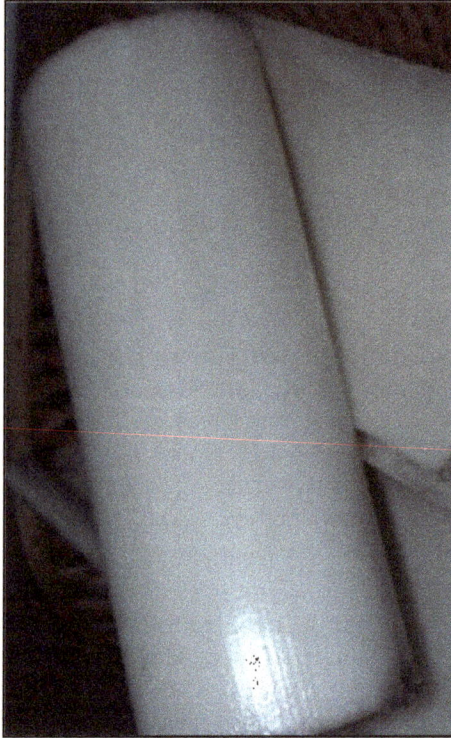

Banana Stem.

Using the measure of morphometrical area density and numerical density of crystallization for hyper-oxaluric urolithiasis, Li *et al.* (1998) found the vitamin B_6 group and the banana-stem extract group to be significantly less as compared to the crystallization group and encouragingly, the banana-stem extract group showing the least. Both vitamin B_6 and banana stem extract had no effect on renal tissue calcium. The extract appeared to have no effect on urinary calcium excretion. According to Penniston and Nakada (2009) gastro-intestinal binding of oxalate by calcium is an effective clinical strategy for hyper-oxaluria, whether mediated by calcium citrate with meals or by inclusion of calcium-containing foods with meals.

Diabetes

The starch in green (raw) banana is rich in resistant starch and hence low in glycemic index. Therefore, raw banana is a good choice for diabetics too. Ojewale and Adewunmi (2003) fed methanolic extract of mature, green fruits of *Musa paradisiaca* (100-800 mg/kg p.o.) in normal and streptozotocin induced diabetic mice using chlorpropamide as the reference anti-diabetic agent. The fruit extract significantly reduced the blood glucose concentrations of both normal and diabetic mice probably due, at least in part, to stimulation of insulin production and subsequent glucose utilization thus, lending credence to the folkloric use of the plant in the management and/or control of type-2 diabetic mellitus among the Yoruba-speaking people of South-Western Nigeria.

Banana flowers are also used in Indian folklore medicine for the treatment of diabetes mellitus. Oral administration of the ethanolic extract showed significant blood glucose lowering effect at 200 mg/kg in alloxan induced diabetic rats (120 mg/kg, i.p.) and the extract was also found to significantly scavenge oxygen free radicals, *viz*.., superoxide dismutase (SOD), catalase (CAT) plus protein, malondialdehyde and ascorbic acid *in-vivo* (Dhanabal *et al.,* 2005; Pari and Umamaheswari, 2000). The chloroform extract of *Musa sapientum* ('Ney Poovan') flowers displayed a significant reduction in blood glucose and glycosylated haemoglobin; and an increase in total haemoglobin concentration. The animals fed 0.25 g/kg body weight of banana flower was more effective than even glibenclamide (Pari and Umamaheswari, 1999; 2000).

The extracts of *Musa paradisiaca* and *Coccinia indica* protected the protein metabolic disorders significantly in streptozotocin-induced diabetes as assessed by serum levels of urea, uric acid, albumin, and creatinine along with urinary urea and albumin levels (Mallick *et al.,* 2009).

Banana flower and pseudo-stem extracts of *Musa sp.* "elakki bale" were found to promote glucose uptake into the cells through glucose transporters 1 and 3 in diabetes. Methanol and aqueous extracts of banana flower and pseudo stem were more potent in promoting glucose uptake as compared to the acetone and ethanol extracts. Thus, Bhaskar *et al.* (2011) hypothesized that consumption of nutraceutical-rich banana flower and pseudo-stem extract could reduce the dose of insulin being taken by diabetics.

Dikshit *et al.* (2012) also demonstrated the anti-diabetic and anti-hyperlipidemic effects of banana stem in streptozotocin induced-diabetic rats. Treatment with lyophilized stem juice of *M. sapientum* (50 mg/kg.) for four weeks significantly decreased fasting and post-prandial blood glucose and HbA1c. and increased serum insulin. Lipid profile, muscle and liver glycogen, and enzyme activity (*i.e.* glucokinase, glucose-6-phosphatase, and HMG-CoA reductase) were all restored to near normal levels. Kaimal *et al.* (2010) also reported that oral administration of the ethanol extract of fruits of Musa AAA (Chenkadali) significantly decreased the levels of serum triacylglycerol, cholesterol and alanine amino transferase (ALT) activity as well as lipid peroxides while glutathione content increased substantially in liver and pancreas similar to that of normal control and glibenclamide treated groups.

The anti-diabetic activity of banana stem and the flower may be attributed to the presence of flavonoids, alkaloids, steroid and glycosides.

Other Benefits

Individuals prone to exercise-associated muscle cramps (EAMCs) are instructed to eat ripe banana because of their high potassium (K(+)) concentration and carbohydrate content and the perception that K(+) imbalances and fatigue contribute to the genesis of EAMCs. But according to Miller (2012) the changes in [K(+)](p), plasma K(+) content, and [glucose](p) do not occur quickly enough to treat acute EAMCs, especially if they develop near the end of competition.

Due to its high concentration of anti-oxidants, banana flower extract reduced hepatic cell damage caused by free radicals (China *et al.*, 2011). When L-ascorbic acid or *M. suerier* were administered to UVB-irradiated mice, the reduction in skin elasticity was significantly inhibited. A significant increase in total glutathione was also found suggesting the potential effect of daily consumption of *M. suerier* on prevention of skin damage from repeated UVB exposure (Viyoch *et al.*, 2012).

Green bananas are considered a sub-product of low commercial value with little industrial use. Zandonadi *et al.* (2012) showed the possibility of developing gluten-free products (low lipid pastas) with green banana flour for people with celiac disease.

A methanol extract of banana peel (BPEx, 200 mg/kg, p.o.) significantly suppressed the re-growth of ventral prostates and seminal vesicles induced by testosterone in castrated mice. BPEx inhibited dose-dependently testosterone-induced cell growth, while the inhibitory activities of BPEx did not appear against dehydro-testosterone-induced cell growth indicating that methanol extract of banana peel can inhibit 5-alpha-reductase and might be useful in the treatment of benign prostate hyperplasia (Akamine *et al.*, 2009).

Interestingly, animals receiving 1000 mg/kg/day of the plantain fruit showed marked and very significant reduction in sperm cell concentration and percentage of morphologically normal spermatozoa thus suggesting an anti-fertility effect. But those receiving low dose of 500 mg/kg/day of plantain fruit showed significant increment in the semen parameters (Alabi *et al.*, 2013).

Conclusions

The traditional use of banana in various clinical conditions has been supported with substantial scientific evidence. However, there is still a wide scope to commercially exploit the therapeutic applications of economically viable banana peels, flowers and stem.

References

Agarwal PK, Singh A, Gaurav K, Goel S, Khanna HD, Goel RK (2009). Evaluation of wound healing activity of extracts of plantain banana (*Musa sapientum var. paradisiaca*) in rats. *Indian J. Exp. Biol.* 47 (1): 32–40.

Akamine K, Koyama T, Yazawa K (2009). Banana peel extract suppressed prostate gland enlargement in testosterone-treated mice. *Biosci. Biotechnol. Biochem.* 73 (9): 1911–14.

Alabi AS, Omotoso GO, Enaibe BU, Akinola OB, Tagoe CN (2013). Beneficial effects of low dose *Musa paradisiaca* on the semen quality of male Wistar rats. *Niger. Med. J.* 54 (2): 92–95.

Alvarez-Acosta T, León C, Acosta-González S, Parra-Soto H, Cluet-Rodriguez I, Rossell MR, Colina-Chourio JA (2009). Beneficial role of green plantain [*Musa paradisiaca*] in the management of persistent diarrhea: a prospective randomized trial. *J. Am. Coll. Nutr.* 28 (2): 169–76.

Avallone S, Rojas-Gonzalez JA, Trystram G, Bohuon P (2009). Thermal sensitivity of some plantain micronutrients during deep-fat frying. *J. Food Sci.* 74 (5): C339–47.

Best R, Lewis DA, Nasser N (1984). The anti-ulcerogenic activity of the unripe plantain banana (*Musa species*). *Br. J. Pharmacol.* 82 (1): 107–116.

Bhaskar JJ, Salimath PV, Nandini CD (2011). Stimulation of glucose uptake by *Musa sp.* (cv. elakki bale) flower and pseudo-stem extracts in Ehrlich ascites tumor cells. *J. Sci. Food Agric.* 91 (8): 1482–87.

Blades BL, Dufficy L, Englberger L, Daniells JW, Coyne T, Hamill S, Wills RB (2003). Bananas and plantains as a source of provitamin A. *Asia Pac. J. Clin. Nutr.* 12 (Suppl): S36.

Bresnahan KA, Arscott SA, Khanna H, Arinaitwe G, Dale J, Tushemereirwe W, Mondloch S, Tanumihardjo JP, De Moura FF, Tanumihardjo SA (2012). Cooking enhances but the degree of ripeness does not affect provitamin A carotenoid bioavailability from bananas in Mongolian gerbils. *J. Nutr.* 142 (12): 2097–2104.

China R, Dutta S, Sen S, Chakrabarti R, Bhowmik D, Ghosh S, Dhar P (2011). *In-vitro* anti-oxidant activity of different cultivars of banana flower (*Musa paradicicus* L.) extracts available in India. *J. Food Sci.* 76 (9): C 1292–99.

Dadoo RC, Khatri HL, Singla S (1995). Comparative evaluation of gastric secretory response to banana and porridge. *Indian J. Med. Sci.* 49 (1): 5–8.

Darsini DT, Maheshu V, Vishnupriya M, Sasikumar JM (2012). *In-vitro* anti-oxidant activity of banana (Musa spp. ABB cv. Pisang Awak). *Indian J. Biochem. Biophys.* 49 (2): 124–29.

Dhanabal SP, Sureshkumar M, Ramanathan M, Suresh B (2005). Hypoglycemic effect of ethanolic extract of *Musa sapientum* on alloxan induced diabetes mellitus in rats and its relation with antioxidant potential. *J. Herb. Pharmacother.* 5 (2): 7–19.

Dikshit P, Shukla K, Tyagi MK, Garg P, Gambhir JK, Shukla R (2012). Anti-diabetic and anti-hyperlipidemic effects of the stem of *Musa sapientum* Linn. in streptozotocin-induced diabetic rats. *J. Diabetes.* 4 (4): 378–85.

Dunjiæ BS, Svensson J, Axelson J, Adler Creutz P, Bengmark S (1993). Green banana protection of gastric mucosa against experimentally induced injuries in rats a multi component mechanism? Scand. *J. Gastroenterol.* 28: 894–98.

Elliot RC, Heward EJF (1976). The influence of a banana supplemented diet on gastric ulcers in mice. *Pharmac. Res. Commun.* 8: 167–71.

Englberger L, Darnton-Hill I, Coyne T, Fitzgerald MH, Marks GC (2003). Carotenoid-rich bananas: a potential food source for alleviating vitamin A deficiency. *Food Nutr. Bull.* 24 (4): 303–18.

Englberger L, Schierle J, Aalbersberg W, Hofmann P, Humphries J, Huang A, Lorens A, Levendusky A, Daniells J, Marks GC, Fitzgerald MH (2006a). Carotenoid and vitamin content of Karat and other Micronesian banana cultivars. *Int. J. Food Sci. Nutr.* 57 (5-6): 399–418.

Englberger L, Wills RB, Blades B, Dufficy L, Daniells JW, Coyne T (2006b). Carotenoid content and flesh color of selected banana cultivars growing in Australia. *Food Nutr. Bull.* 27 (4): 281–91.

Fingolo CE, Braga JM, Vieira AC, Moura MR, Kaplan MA (2012). The natural impact of banana inflorescences (*Musa acuminata*) on human nutrition. *An Acad Bras Cienc.* 84 (4): 891–98.

Forster MP, Rodríguez Rodríguez E, Díaz Romero C (2002). Differential characteristics in the chemical composition of bananas from Tenerife (Canary Islands) and Ecuador. *J. Agric. Food Chem.* 50 (26): 7586–92.

Goel RK, Gupta S, Shankar R, Sanyal AK (1986). Anti-ulcerogenic effect of banana powder (*Musa sapientum var. paradisiaca*) and its effect on mucosal resistance. *J. Ethnopharmacol.* 18: 33–44.

Goel RK, Sairam K, Rao CV (2001). Role of gastric antioxidant and anti-Helicobactor pylori activities in anti-ulcerogenic activity of plantain banana (*Musa sapientum var.paradisiaca*). *Indian J. Exp. Biol.* 39: 719–22.

Hanszen A (1934). The bactericidal power of the stomach and some factors which influence it. *Am. J. Dig. Dis.* 1: 725–28.

Hills BA, Kirwood CA (1989). Surfactant approach to the gastric mucosal barrier: protection of rats by banana even when acidified. *Gastro-enterology.* 97 (2): 294–303.

Kailash P, Varalakshmi P (1992). Effect of banana stem juice on biochemical changes in liver of normal and hyper-oxaluric rats. *Indian J. Exp. Biol.* 30 (5): 440–42.

Kaimal S, Sujatha KS, George S (2010). Hypo-lipidaemic and antioxidant effects of fruits of *Musa AAA* (Chenkadali) in alloxan induced diabetic rats. *Indian J. Exp. Biol.* 48 (2): 165–73.

Kosek M, Yori PP, Olortegui MP (2010). Shigellosis update: advancing antibiotic resistance, investment empowered vaccine development, and green bananas. *Curr. Opin. Infect. Dis.* 23 (5): 475–80.

Kumar GB, Srinivas L, Ganapathi TR (2011). Iron fortification of banana by the expression of soybean ferritin. *Biol. Trace Elem. Res.* 142 (2): 232–41.

Langkilde AM, Champ M, Andersson H (2002). Effects of high-resistant-starch banana flour (RS(2)) on *in-vitro* fermentation and the small-bowel excretion of energy, nutrients, and sterols: an ileostomy study. *Am. J. Clin. Nutr.* 75 (1): 104–111.

Lewis DA, Fields WN, Shaw GP (1999). A natural flavonoid present in unripe plantain banana pulp (*Musa sapientum L. var. paradisiaca*) protects the gastric mucosa from aspirin-induced erosions. *J. Ethnopharmacol*. 65 (3): 283–88.

Lewis DA, Shaw GP (2001). A natural flavonoid and synthetic analogues protect the gastric mucosa from aspirin-induced erosions. *J. Nutr. Biochem*. 12: 95–100.

Li S, Wu C, Nong H, Deng Y (1998). Morphometrical study on inhibitory effect of vitamin B6 and banana-stem extract on calcium crystallization. *Zhonghua. Wai. Ke. Za. Zhi*. 36 (12): 763–65. Article in Chinese.

Mallick C, De D, Ghosh D (2009). Correction of protein metabolic disorders by composite extract of *Musa paradisiaca* and *Coccinia indica* in streptozotocin-induced diabetic albino rat: an approach through the pancreas. *Pancreas*. 38 (3): 322–29.

Menezes EW, Tadini CC, Tribess TB, Zuleta A, Binaghi J, Pak N, Vera G, Dan MC, Bertolini AC, Cordenunsi BR, Lajolo FM (2011). Chemical composition and nutritional value of unripe banana flour (*Musa acuminata*, var. Nanicão). *Plant Foods Hum. Nutr*. 66 (3): 231–7.

Miller KC (2012). Plasma potassium concentration and content changes after banana ingestion in exercised men. *J. Athl. Train*. 47 (6): 648–54.

Mohan Kumar M, Joshi MC, Prabha T, Dorababu M, Goel RK (2006). Effect of plantain banana on gastric ulceration in NIDDM rats: role of gastric mucosal glycoproteins, cell proliferation, anti-oxidants and free radicals. *Indian J. Exp. Biol*. 44 (4): 292–99.

Mukhopadhyaya K, Bhattacharya D, Chakraborty A, Goel RK, Sanyal AK (1987). Effect of banana powder (*Musa sapientum* var. paradisiaca) on gastric mucosal shedding. *J. Ethnopharmacol*. 21 (1): 11–19.

Ojewole JA, Adewunmi CO (2003). Hypo-glycemic effect of methanolic extract of *Musa paradisiaca* (Musaceae) green fruits in normal and diabetic mice. *Methods. Find. Exp. Clin. Pharmacol*. 25 (6): 453–56.

Onasanwo SA, Emikpe BO, Ajah AA, Elufioye TO (2013). Anti-ulcer and ulcer healing potentials of *Musa sapientum* peel extract in the laboratory rodents. *Pharmacognosy. Res*. 5 (3): 173–78.

Pannangpetch P, Vuttivirojana A, Kularbkaew C, Tesana S, Kongyingyoes B, Kukongviriyapan V (2001). The anti-ulcerative effect of Thai Musa species in rats. *Phytother. Res*. 15 (5): 407–410.

Pari L, Umamaheswari J (1999). Hypo-glycaemic effect of *Musa sapientum* L. in alloxan-induced diabetic rats. *J. Ethnopharmacol*. 68 (1-3): 321–25.

Pari L, Umamaheswari J (2000). Anti-hyperglycaemic activity of *Musa sapientum* flowers: effect on lipid peroxidation in alloxan diabetic rats. *Phytother. Res*. 14 (2): 136–38.

Penniston KL, Nakada SY. (2009). Effect of dietary changes on urinary oxalate excretion and calcium oxalate supersaturation in patients with hyperoxaluric stone formation. *Urology*. 73 (3): 484–89.

Poonguzhali PK, Chegu H. (1994). The influence of banana stem extract on urinary risk factors for stones in normal and hyperoxaluric rats. *Br. J. Urol.* 74 (1): 23–25.

Prabha P, Karpagam T, Varalakshmi B, Packiavathy AS (2011). Indigenous anti-ulcer activity of *Musa sapientum* on peptic ulcer. *Pharmacognosy. Res.* 3 (4): 232–38.

Prasad KV, Bharathi K, Srinivasan KK (1993). Evaluation of (*Musa Paradisiaca* Linn. cultivar)–"Puttubale" stem juice for antilithiatic activity in albino rats. *Indian J. Physiol.Pharmacol.* 37 (4): 337–41.

Rabbani GH, Larson CP, Islam R, Saha UR, Kabir A (2010). Green banana-supplemented diet in the home management of acute and prolonged diarrhoea in children: a community-based trial in rural Bangladesh. *Trop. Med. Int. Health.* 15 (10): 1132–39.

Rabbani GH, Teka T, Saha SK, Zaman B, Majid N, Khatun M, Wahed MA, Fuchs GJ (2004). Green banana and pectin improve small intestinal permeability and reduce fluid loss in Bangladeshi children with persistent diarrhea. *Dig. Dis. Sci.* 49 (3): 475–84.

Rao NM (1991). Protease inhibitors from ripened and unripened bananas. *Biochem. Int.* 24 (1): 13–22.

Rojas-Gonzalez JA, Avallone S, Brat P, Trystram G, Bohuon P (2006). Effect of deep-fat frying on ascorbic acid, carotenoids and potassium contents of plantain cylinders. *Int. J. Food Sci. Nutr.* 57 (1–2): 123–36.

Sanyal AK, Das PK, Sinha S, Sinha YK (1961). Banana and gastric secretion. *J. Pharm. Pharmac.* 13: 318–19.

Scarminio V, Fruet AC, Witaicenis A, Rall VL, Di Stasi LC (2012). Dietary intervention with green dwarf banana flour (Musa sp AAA) prevents intestinal inflammation in a trinitro-benzenesulfonic acid model of rat colitis. *Nutr. Res.* 32 (3): 202–209.

Selema MD, Farago ME (1996). Trace element concentrations in the fruit peels and trunks of *Musa paradisiaca*. *Phytochemistry.* 42 (6): 1523–25.

Sundaram S, Anjum S, Dwivedi P, Rai GK (2011). Anti-oxidant activity and protective effect of banana peel against oxidative hemolysis of human erythrocyte at different stages of ripening. *Appl. Biochem. Biotechnol.* 164 (7): 1192–206.

Suntharalingam S, Ravindran G (1993). Physical and biochemical properties of green banana flour. *Plant Foods Hum. Nutr.* 43 (1): 19–27.

Thakorlal J, Perera CO, Smith B, Englberger L, Lorens A (2010). Resistant starch in Micronesian banana cultivars offers health benefits. *Pac. Health Dialog.* 16 (1): 49–59.

USDA, National Nutrition Data Base for standard reference, Release-26, NDB No-09040.

Vadivelan R, Elango K, Suresh B, Ramesh BR (2006). Pharmacological validation of *Musa paradisiaca* bhasma for anti-ulcer activity in albino rats–a preliminary study. *Anc. Sci. Life.* 25 (3–4): 67–70.

Viyoch J, Mahingsa K, Ingkaninan K (2012). Effects of Thai Musa species on prevention of UVB-induced skin damage in mice. *Food Chem. Toxicol.* 50 (12): 4292–301.

Wikipedia.org. (2014). Wikipedia Foundation, inc. wikipedia.org/wiki/Banana.

Zandonadi RP, Botelho RB, Gandolfi L, Ginani JS, Montenegro FM, Pratesi R (2012). Green banana pasta: an alternative for gluten-free diets. *J. Acad. Nutr. Diet.* 112 (7): 1068–72.

Chapter 3

Betel Leaf
(*Piper betle*)

Betel vine belongs to the family of Piperaceae. It is an aromatic slender creeper probably native of Malaysia, also cultivated in India, Srilanka, Bangladesh, Burma and Nepal (Kumar *et al.,* 2010; Guha *et al.,* 2006). The leaves are large, 15-20 cms long, broadly ovate, often unequal at the base, and have alternate heart shaped, smooth, shining and long stalked with pointed apex. They are yellowish or bright green in color and shining on both sides. The petiole is stout.

Varieties

There are about 100 varieties of betel vine in the world, of which about 40 are found in India and 30 in West Bengal (Guha, 1997; Maity, 1989; Samanta, 1994). The leaves are classified into different varieties according to various characteristics such as the place of cultivation, color, pungency, size, shape etc. For example, in Philippines, three varieties are described based on leaf shape and pubescence. In Indonesia and Malaysia, several varieties are known some with clove like flavor and others with red colored vein petioles. In India, more than 100 types/ cultivars of betel vine are recognized with the Bangla, Madras and Mysore varieties as the three main popular ones. The Madras variety leaves, also referred to as 'Kapoori' are green in color, spreading and not very pungent. The small and fragile 'chiguruyale' of Karnataka and the thicker leaves of 'ambadi' and 'kariyale' are few other prominent leaves from the southern states. The Calcutta variety, also referred to as Bangla has leaves which are dark green in color, thicker and have short tips.

Consumption Pattern

In India betel leaves are consumed in the form of pan which is a combination of the leaves with arecanut, spices, condiments etc., known by different names such as 'pan ki gilori' in Lucknow, and 'beeda' in the South. A wide variety of pan preparations are available in India, with the difference mainly in the type of leaf and the ingredients used. The most popular Indian varieties are the Magadhi, Venmony, Mysore, Salem, Calcutta, Banarasi, Kauri, Ghanagete and Bagerhati (Satyavati *et al.*, 1987; Warrier *et al.*, 1995).

Phytochemical Constituents in Betel Leaf

The phytochemical investigations of betel leaves showed that it had high amount of tannins. Total phenols amount to 27.6 mg/g of dry wt. However, the tannin and nitrate contents do not seem to influence the quality. The enzymes present in the leaves are diastase and catalase. However, alkaloids and glycosides are reported to be absent.

The various phytochemicals found in the betel plants are chavibetol, chavicol, hydroxychavicol, estragole, eugenol, methyl eugenol, hydroxycatechol, caryophyllene, eugenol methyl ether, cadinene, γ-lactone, allyl catechol, p-cymene, cepharadione A, dotriacontanoic acid,tritriacontane, p-cymene, terpinene, eucalyptol, carvacrol, sesquiterpenes, cadinene, caryophyllene, dotriacontanoic acid, hentriacontane, pentatriacontane, stearic acid, n-triacontanol, triotnacontane, piperlonguminine, allylpyrocatechol diacetate, isoeugenol, 1, 8-cineol, α-pinene, β-pinene, sitosterol, β-sitosteryl palmitate, γ-sitosterol, stigmasterol, ursolic acid, ursolic acid 3β-acetate (Rastogi and Mehrotra, 1993; Kumar *et al.*, 2010).

The essential oil present in betel leaves determines the aromatic nature of the leaf which was reported to be 0.55 (mg/g dry wt) in Palghat and 0.59 (mg/g dry wt) in Kapoori varieties. The ethanol extracts of three varieties (Bangla, sweet, and Mysore) of *Piper betel* revealed that the concentrations of chevibetol, allylpyrocatechol were significantly less in the sweet and Mysore varieties than the Bangla one (Rathee *et al.*, 2006). Chevibetol (3-hydroxy-4-methoxyallybenzene) is believed to be responsible for the characteristic smoky aroma of betel leaf. Other aroma active compounds include linalool, (Z)-3-hexenal, a-pinene and methanol.

Hydroxychavicol is the major phenolic component, isolated from the aqueous extract of *P. betle* leaf has been reported to possess anti-nitrosation, anti-mutagenic, anti-carcinogenic activities. It also has a tendency to act as an antioxidant, and a chemo-preventive agent (Chang *et al.*, 2002). Other useful properties include anti-inflammatory, anti-platelet and anti-thrombotic without impairing haemostatic functions (Chang *et al.*, 2007). There have been reports on the anti-bacterial activities of hydroxyl-chavicol (Sharma *et al.*, 2009; Ramji *et al.*, 2002).

Nutritive Value

Though betel leaves are not used as a vegetable, their nutritive value is almost same as any other commonly consumed green leafy vegetables like mint, amaranth etc. The moisture and the protein content of the light and the dark variety are almost the same (87 per cent and 4.5 per cent respectively).

Among the amino acids, the leaves contain leucine–18.3 mg, phenyl alanine–14.2 mg, arginine–2.4 mg, threonine -12 mg, aspartic acid–23 mg, glutamic acid–29.7 mg, valine-3.8 mg, tyrosine-1.2 mg and γ aminobutyric acid–20.2 mg per 100 grams. But lysine, histidine and arginine, occur in traces (CSIR, 1969; Gopalan, 1984; Guha and Jain, 1997). The sugars identified in betel leaves include glucose, fructose, maltose and sucrose. The free reducing sugars in different types of betel leaves from West Bengal ranged from 0.38–0.46 per cent (Airan and Sheth, 1957). The fiber content of the two varieties was somewhat different. The dark variety showed slightly higher value of 2.4 and the light variety was only 1.6 per cent (Subbulakshmi *et al.,* 1988).

Nutritional Composition of Betel Leaves*

Nutrient	Amount (mg/g dry wt.)
Total sugar	47.5
Total proteins	103.4
Total free amino acids	10.4
Carotenoids	1.2

* Subbulakshmi *et al.* (1988).

The Vitamin C content of the leaves is quite high as compared to any dark green leafy vegetables. The leaves also contain good amount of vitamin B complex particularly nicotinic acid. They are rich in calcium, phosphorus and iron. The availability of calcium from the leaf is stated to be poor as it is bound to oxalic acid and the free oxalate present in the leaves may interfere with the absorption of calcium from other foods also. The leaf with proper adjuvant like 'chuna' (lime) is considered as a good and cheap source of dietary calcium. Six leaves with a little bit of slaked lime is said to be comparable to about 300 ml of cow milk particularly for the vitamin and mineral nutrition (Guha, 2006).

Varietal Difference in Selected Micronutrient Composition of Betel Leaves*

Nutrient	Light Variety	Dark Variety
Calcium (mg per cent)	162.8	182.7
Phosphorus (mg per cent)	31.6	36.9
Iron (mg per cent)	1.9	1.4
Vitamin C (mg per cent)	155.0	173.5

* Subbulakshmi *et al.* (1988).

Traditional Therapeutic Uses

The betel leaf plant has many traditional medicinal uses. In Susruta Samhita (1500 BC-500AD), betel leaves are recommended as a digestant, stimulant and carminative; and used in treating cough, cold and anorexia. Malaysians use the leaves for headaches, arthritis and joint pain. In Indonesia it is valued as a natural antibiotic, and taken as tea daily for better health. Tea prepared from the leaves is believed to keep the body free of unpleasant smells of perspiration and menstrual odor. It is also valued for keeping the teeth and gums strong and healthy.

Apart from these, the Chinese folk medicine believes that the betel leaves possess the biological capabilities of de-toxication, anti-oxidation, and anti-mutation. The leaves can serve as a tonic to the brain and liver. There is a belief that the juice of the leaves when dropped in to the eyes can cure night blindness probably because of its high beta-carotene content. The decoction of the leaves is used for healing wounds. The high vitamin C content of the leaves must be responsible for this benefit. Water-soluble extract of the betel leaf has shown to cause humoral immune response stimulation and cellular immune response suppression. The leaves are also useful in elephantiasis of the leg.

Many of these traditional practices and ethno-medicinal claims have been validated clinically for various pharmacological activities (Rai *et al.,* 2011). It is believed to be a mouth freshener, salivary stimulant, digestive, carminative, invigorating, anti phlegmatic, analgesic and antiseptic. It is useful in preventing bleeding, alleviating kapha and is an aphrodisiac and anthelmentic. Pre-clinical experiments have shown that betel leaves possess anti-bacterial, anti-cariogenic, anti-fungal, anti-larval, anti-protozal, anti-filarial, anti-allergic, anti-diabetic, anti-inflammatory, anti-ulcer, anti-fertility, anti-hyper-lipidaemic, anti-platelet, vaso-relaxant, cardio-protective, hepato-protective, and immune-modulatory effects (Kumar *et al.,* 2010; Singh *et al.,* 2009; Norton, 1998; Prabhu *et al.,* 1995; Bhattacharya *et al.,* 2007). The high beta carotene content of some varieties (5760 µg/100 g) has been found to improve the serum retinol levels in pregnant women (Subbulakshmi and Mehra, 1995).

Inserting the stalk of betel leaf dipped in castor oil in the rectum, helps in easy movement of bowels and cures constipation in children. In Thailand and China the roots are crushed and blended with salt to relieve toothache. The essential oil isolated from the leaves is found to be useful in treating respiratory catarrhs and as an anti-septic (Chopra *et al.,* 1982, Satyavati *et al.,* 1987). The oil is an active local stimulant used in the treatment of respiratory catarrhs as a local application or gargle and also an inhalant in diphtheria (Prajapathi, 2003).

Anti-oxidant Property

Betel leaves contain phenolic compounds like hydroxyl-chavicol that are potent anti-oxidants. *Betel* leaf extract showed profound anti-oxidant activity as judged by thiobarbartiuric acid reactive substances (TBARS) and 2, 2 – diphenyl – 1 – picrylhydrazyl (DPPH) scavenging assays. Studies have also shown that the hydro-alcoholic extract of the betel leaf exhibits nitrogen oxide scavenging effect *in-vitro* (Jagetia and Baliga, 2004). Recently, Manigauha *et al.* (2009) have also observed that the methanolic extracts of the betel leaves possess reducing power, DPPH radical and superoxide anion scavenging as well as deoxyribose degradation activities. Jeng *et al.* (2002) found that *Piper Betle* leaf extract is a scavenger of O2(*-) and *OH, and inhibits xanthine oxidase activity and the (*)OH-induced PUC18 DNA breaks.

Bangla variety of betel leaf possessed the best anti-oxidant activity that correlated with the total phenolic content and reducing power of the respective extract (Rathee et al, 2006). The anti-oxidant activities of the three cultivars were in the order of Kauri > Ghanagete > Bagerhati (Dasgupta and De, 2004), but the free radical scavenging ability of commercial betel from Sri Lanka was reported to be higher than that of the Indian cultivars (Arambewela *et al.,* 2006b; 2011).

Interestingly, the anti-oxidant properties of *betel* leaves extracts, including cold ethanolic extract and EO, remained unaltered for a period of 12 months at room temperature (as evaluated by DPPH assay). This supports the potential use of the betel extracts as a natural anti-oxidant in food industry. However, the anti-oxidant property of betel leaves was significantly reduced upon exposure to elevated temperature of 200°C (Arambewela *et al.,* 2011).

Traditionally, betel leaves were used as a preservative in fats and oils. The natural anti-oxidants present in oils and butter is largely eliminated during the refining and hydrogenation processes. Fresh betel leaves are added to butterfat during its conversion into ghee as an anti-oxidant/preservative. In an attempt to introduce betel as a natural anti-oxidant, the cold ethanolic extract of betel leaves was incorporated into fats (cake margarine) and oils (coconut and palm oil) separately and its rancidity was determined in terms of peroxide value (PV). The results showed that PVs were significantly lower in betel treated than in BHT-treated samples (Arambewela *et al.,* 2011).

The phenolic fraction of betel leaves exhibited anti-oxidant activity comparable to that of propyl gallate and butylated hydroxy anisole in groundnut oil. The leaves and their extracts possess anti-oxidant effect on highly unsaturated oil (sardine oil) on storage. The amino acids asparagine, glycine, proline and trytophan present in betel leaves might also act as potent anti-oxidants. Ascorbic acid in the leaves acts as a synergist to the various phenols present in it. It was found that in addition to the anti-oxidant effect on dry cured fish, betel leaves extract was able to check the bacterial spoilage to a certain extent. Addition of betel leaves at 4 and 6 per cent level to palmolein oil stored for 2 months in air tight tins did not change the free fatty acid content showing the anti-oxidant action of the leaves.

According to Bhattacharya *et al.* (2005) the ethonolic extract of *Piper Betle* (PE) effectively prevented gamma-ray induced lipid peroxidation as assessed by measuring thiobarbituric acid reactive substrates, lipid hydroperoxide and conjugated diene. Likewise, it prevented radiation-induced DNA strand breaks in a concentration dependent manner. The radio-protective activity of PE could be attributed to its hydroxyl and superoxide radicals scavenging property along with its lympho-proliferative activity. The radical scavenging capacity of PE was primarily due to its constituent phenolics, which were isolated and identified as chevibetol and allyl-pyrocatechol.

Investigation on the possibility of controlling auto-oxidation and deterioration of lemon grass oil under dark conditions using natural anti-oxidants like betel leaves extract, gooseberry extract, powdered chilli extract and chemical antioxidants such as boric acid, pyrogallol, sodium chloride, hydroxy toluene and ascorbic acid revealed that betel leaves extract offered the highest (83 per cent) protection.

Anti-inflammatory Property

Eugenol, one of the principal constituent of betel leaf has also been shown to possess anti-inflammatory effects in various animal models with various inflamogens (Dohi *et al.,* 1989; Lee *et al.,* 2007). Scientific studies have shown that the ethanolic extract of betel leaf has been reported to possess anti-inflammatory activities in arthritis in rat model. A concentration dependent decrease in the extracellular production of nitric oxide in murine peritoneal macrophages indicated the ability of betel leaf to down-regulate T-helper 1 pro-inflammatory responses (Ganguly *et al.,* 2007).

Immuno-modulation

Many of the disorders today are based on the imbalances of immunological processes. This necessitates the search for newer and safer immune-modulators. The methanolic extract of *Piper betel* L. (MPb) consists of mixture of phenols, flavonoids, tannins and polysaccharides. Piper betel significantly suppressed phyto-haemagglutinin stimulated peripheral blood lymphocyte proliferation in a dose-dependent manner. The decrease in antibody titre and increased suppression of inflammation suggest possible immune-suppressive effect of extract on cellular and humoral response in mice. Thus, the MPb could be explored extensively as a therapeutic agent to treat various immune disorders including auto-immune disorders (Kanjwani *et al.*, 2008).

Crude methanolic extract (100 mg/kg) exhibited a mixed type 1 and type 2 cytokine responses thus suggesting a remarkable immune-modulatory property of betel leaves. The authors opine that induction of differential T-helper cell immune response appears ideal to overcome immune-suppression as observed in case of lymphatic, filarial Brugia malayi infection which may be extended to other infections as well (Singh *et al.,* 2009).

Due to the antioxidant, anti inflammatory and immune modulatory properties, betel leaf exhibits several therapeutic benefits in diseases as discussed below.

Cardiovascular Diseases

The anti-oxidant activity of the leaf extract and its potential to elevate the anti-oxidant status was proved by Choudhary and Kale (2002). The leaf extract was found to diminish the effect of oxidative stress by elevating the level of endogenous defence enzymes, scavenging free radicals and acting as a good immuno-modulatory agent (Kanjwani *et al.,* 2008).

Platelet hyper-activity is important in the pathogenesis of cardiovascular diseases. The betel leaf component Hydroxy-chavicol (HC) tested for its anti-platelet effect by Chang *et al.* (2007) showed that HC is a potent cyclo-oxygenase (COX-1/

COX-2) inhibitor, reactive oxygen species (ROS) scavenger and inhibitor of platelet calcium signaling, thromboxane B(2) production and aggregation. HC could be a potential therapeutic agent for prevention and treatment of atherosclerosis and other cardiovascular diseases through its anti-inflammatory and anti-platelet effects, without affecting haemostatic functions. Lei *et al.* (2003) also found that aqueous components of inflorescence *Piper betle* are potential ROS scavengers and may prevent the platelet aggregation possibly *via* scavenging ROS or inhibition of thromboxane B(2) production. Triterpenes and β-sitosterol isolated from betel laves also have anti-platelet actions (Saeed *et al.,* 1993). Jeng *et al.* (2002) found that *Piper Betle* leaf extract inhibited platelet aggregation via both its antioxidative effects and effects on TXB2 and PGD2 production.

Ma *et al.* (2013) demonstrated that *Piper betle* leaf could activate the reverse cholesterol transport mechanism to enhance the metabolism of the oxidised low density lipoprotein (oxLDL) that could prevent both lipid accumulation and foam cell formation, and further minimize the possible damage of vessels caused by the oxy LDL.

It is known that chewing betel quid with tobacco or without tobacco is a major etiological factor for cardiovascular complications and calcium channel blockers are the major class of drugs prescribed widely for myocardial disturbances. Kumar *et al.* (2009) proved that chewing betel quid without tobacco is cardio-protective, whose activity was further augmented by amlodipine. Histopathological studies also confirmed the biochemical findings while the same with tobacco is cardio-toxic and its effect cannot be reversed using calcium channel blockers.

Diabetes

Betel leaves from Sri Lanka has been found to possess marked hypo-glycemic activity and anti-hyperglycemic activity as judged by improvement in the results of glucose tolerance test and by lowering of the blood glucose levels in rats with streptozotocin (STZ)-induced diabetes, suggesting that the extract has insulin-mimetic activity (Arambewela *et al.,* 2005a).

Oral administration of leaf suspension of *P. betle* (75 and 150 mg/kg of body weight) for 30 days resulted in significant reduction in blood glucose (from 205.00 +/- 10.80 mg/dL to 151.30 +/- 6.53 mg/dL) and glycosylated hemoglobin, and decreased activities of liver glucose-6-phosphatase and fructose-1,6-bisphosphatase, while liver hexokinase increased in STZ induced diabetic rats as compared to untreated diabetic rats. *P. betle* at a dose of 75 mg/kg of body weight exhibited better sugar reduction than 150 mg/kg of body weight. In addition, protection against body weight loss of diabetic animals was also observed. Thus, betel leaf intake influences glucose metabolism beneficially and the effects were comparable to the standard drug glibenclamide (Santhakumari *et al.,* 2006).

Stomach Disorders

Betel leaves act as a digestive, pancreatic lipase stimulant, good remedy against bad breath (Chatterjee and Pakrashi, 1995; Prabhu *et al.,* 1995) and also as a hepato-protective agent (Saravanan *et al.*, 2002) without influencing bile secretion and composition. The Ambadi variety of betel leaf has a positive stimulatory influence on intestinal digestive enzymes, especially lipase, amylase and disaccharidases (Prabhu *et al.*, 1995).

The ethanol extract of the leaf fed at a dose of 150 mg/kg body weight daily for 10 days, after induction of peptic ulcer by NSAID in albino rats, produced significant healing effect (Arambewela *et al.*, 2004). The extract also showed significant *in-vitro* free radical scavenging action. The results suggest that the anti-oxidant (superoxide dismutase and catalase activity, mucus and total gastric tissue sulfhydryl group were increased) or free radical scavenging activity (oxidised lipid and oxidatively modified proteins were reduced to near normalcy) of the plant extract may be responsible for its healing action (Majumdar *et al.*, 2003). Treatment with allylpyrocatechol (APC), the major anti-oxidant constituent of *Piper betel*, (2 mg/kg body wt per day) and misoprostol (1.43 µg/kg body wt per day) for 7 d could effectively heal the stomach ulceration as revealed from the ulcer index and histopathological studies. Compared to the zero day ulcerated group, treatment with APC and misoprostol reduced the ulcer index by 93.4 per cent and 85.4 per cent respectively.

Bhattacharya *et al.* (2007) confirmed the protective activity of allylpyrocatechol (APC), the major anti-oxidant constituent of betel, against the indomethacin-induced stomach ulceration in the rat model and correlated with its anti-oxidative and mucin protecting properties. Treatment with APC (2 mg/kg body wt per day) and misoprostol (1.43 µg/kg body wt per day) for 7 days could effectively heal the stomach ulceration as revealed from the ulcer index and histo-pathological studies.

Cancer

Pioneering studies by Bhide *et al.* (1979) reported for the first time that aqueous extract of betel leaf failed to induce any tumor in both Swiss mice and C17 mice thereby proving that unlike believed betel leaf was not carcinogenic. Interestingly, betel leaves possess cancer-preventive effects and also protects against the carcinogens present in tobacco (Rai *et al.*, 2011; Ko *et al.*, 1995). Betel leaves contain anti-mutagenic and chemo-preventive agents (Shirname *et al.*, 1983), and may inhibit carcinogen-induced oral tumours (Rao, 1984). The genomic damage caused by pan masala was also protected but only after treating the cells for a longer period (Trivedi *et al.*, 1994).

Scientific studies have shown that betel leaf is devoid of mutagenic and carcinogenic effect. Innumerable studies have been performed with the individual constituents of the betel quid and observations have conclusively shown that tobacco (Sundqvist *et al.*, 1989; Boffetta *et al.*, 2008) and areca nut (Bhide *et al.*, 1979; Canniff and Harvey, 1981; Sundqvist *et al.*, 1989; Jin *et al.*, 1996; Wang and Peng, 1996, Jeng *et al.*, 2000; Wang *et al.*, 2003; IARC, 2004; Lee *et al.*, 2005) are both carcinogenic and slaked lime too promotes carcinogenesis (Jeng *et al.*, 1994; Thomas and MacLennan, 1992).

Betel leaf and its constituents decreased the number of papillomas per animal with the maximum protection, considering molar dosage, exhibited by beta-carotene and alpha-tocopherol. Except for beta-carotene, eugenol, hydroxy-chavicol and alpha-tocopherol increased the levels of reduced glutathione in the liver while glutathione S-transferase activity was enhanced by all except eugenol (Bhide *et al.*, 1991b). Water and acetone extracts of betel leaf reduced the mutagenicity of benzo(a)pyrene and dimethyl-benzanthracene. Acetone extract is more potent than water extract in inhibiting mutagenicity of environmental mutagens (Nagabhushan *et al.*, 1987).

The increasing use of panmasala/gutkha (a mix of tobacco and a less moist form of betel quid lacking the betel leaf), seems to be associated with an earlier onset of oral sub mucous fibrosis. Rao as early as in 1984 observed that topical application of betel leaf extract inhibited B(a) P-induced oral tumorigenesis in hamsters. Padma *et al.* (1989a) also showed that betel leaf was effective in preventing tobacco-specific nitrosamines (the N'-nitrosonornicotine and 4-(methyl nitrosamino)-1-(3-pyridyl-1-butanone) induced carcinogenesis of tongue. According to Babu *et al.* (1996), the tobacco content, the absence of betel leaf and its carotenes, and the much higher dry weight of pan masala/gutkha might be responsible the carcinogenesis.

Supplementation of betel leaf extract in drinking water significantly reduced the benzo[a] pyrene-induced fore-stomach neoplasia in a concentration dependent manner in mice (Bhide *et al.*, 1991b).

Young *et al.* (2005) demonstrated that the betel leaf extract was able to increase the sensitivity of Hep G2 cells to cisplatin *via* at least two mechanisms, reducing the expression of MRP2 and inhibiting the activity of total GST and the expression of GSTA. The data of this study support an application of PBL as an additive to reduce drug resistance. Betel leaf extract reduced the tumorigenic effects of NNK (nitrosamine 4-(N-nitrosomethylamino)-1-(3-pyridyl)-1-butanone) by 25 per cent and also inhibited the decrease in levels of vitamin A in liver and plasma induced by NNK. Betel leaf thus has protective effect against the mutagenic, carcinogenic and adverse metabolic effects of NNK in mice (Bhide *et al.*, 1991a).

Subsequent studies have conclusively shown that the betel leaf and some of its phytochemicals also prevented chemical induced cancers in experimental animals. Hydroxychavicol (HC) inhibits 3(H)benzopyrene–DNA interactions (Lahiri and Bhide, 1993), the nitrozation reaction (Nagabhushan *et al.*, 1989) and the DMBA- and tobacco-specific nitrosamine-induced mutations and bone marrow micronucleated cells formation in Swiss mice (Amonkar *et al.*, 1986; 1989; Padma *et al.*, 1989 a, b; Chang *et al.*, 2002).

Hydroxy-chavicol and its analogues induce killing of primary cells in Chronic myelogenous leukemia patients and leukemic cell lines expressing wild type and mutated Bcr-Abl, including the T315I mutation, with minimal toxicity to normal human peripheral blood mononuclear cells (Chakraborty *et al.*, 2012).

Radiotherapy is an integral part of cancer treatment. The leaf extract was also reported to inhibit gamma radiation induced damage in plasmid DNA and lipid peroxidation in rat liver mitochondria under *in-vitro* conditions (Bhattacharya *et al.*, 2005). According to the observations by Verma *et al.* (2010) the treatment with aqueous extract of *Piper betle* leaf (PBL) effectively lowered the radiation induced lipid peroxidation at 24 hrs in all the selected organs with maximum inhibition in thymus. After 48 hrs, lipid peroxidation was maximally inhibited in the group treated with the extract. Frequency of radiation induced micro-nucleated cells declined significantly (34.78 per cent, p < 0.01) at 24 hrs post-irradiation interval by PBL extract administration suggesting that PBL extract has high anti-oxidant potential and relatively non-toxic and thus could be assertively used to mitigate radiotherapy inflicted normal tissues damage and also injuries caused by moderate doses of radiation during unplanned exposures.

Chemo-preventive efficacy of betel leaf extracts and its constituents on 7, 12-dimethybenz (a) anthrocene induced carcinogenesis and their effect on drug detoxification system in mouse skin showed that there was a significant inhibition in tumor formation by about 85 per cent by betel leaf extract, beta carotene and alpha tocopherol in Swiss mice and about 92 per cent at 24 weeks of treatment (Azuine *et al.,* 1991).

The inhibitory effect of oral administration of betel leaf extract or its constituents–beta-carotene and alpha-tocopherol combined with turmeric was higher than that of individual constituents (Azuine and Bhide, 1992). 'Purnark', a mixture of extracts of turmeric, betel leaf and catechu, when tested for its chemo-preventive activity against benzo (a) pyrene-induced DNA damage, it seemed to give 50-60 per cent protection against BP induced sister chromatid exchange and micro-nuclei. Betel leaf extract reduced the tumorigenic effects of NNK (nitrosamine 4-(N-nitroso-methylamino)-1-(3-pyridyl)-1-butanone) by 25 per cent and also inhibited the decrease in levels of vitamin A in liver and plasma induced by NNK. Betel leaf thus has protective effect against the mutagenic, carcinogenic and adverse metabolic effects of NNK in mice (Bhide *et al.,* 1991a). Based on the studies on stomach tumor models and mice, betel leaf extract has been proved to be a promising anti carcinogenic agent in tobacco-induced cancers and also against environmental mutagens.

Aqueous extract of betel leaf was not mutagenic in any of the four strains of *Salmonella typhimurium.* At the same time, when betel nut extract was fed with betel leaf extract, the lung tumorogenisity induced by the nut reduced from 47 per cent to 38 per cent indicating the protective role of betel leaf. Betel leaf water and acetone extracts reduce the mutagenicity of benzo(a)pyrene and dimethyl-benzanthracene. Acetone extract is more potent than water extract in inhibiting mutagenicity of environmental mutagens (Nagabhushan *et al.,* 1987).

Most recently, Paranjpe *et al.* (2013) showed that oral feeding of betel leaf extract (BLE) significantly inhibited the growth of human prostate xenografts implanted in nude mice compared with vehicle-fed controls. They also demonstrated that one fraction in particular, F2, displayed a 3-fold better *in-vitro* efficacy to inhibit proliferation of prostate cancer cells than the parent BLE. Further, the hydroxychavicol containing F2 subfraction was found to be ~8-fold more potent than the F2 subfraction that contained chavibetol, in human prostate cancer PC-3 cells as evaluated by the 3-(4,5-dimethylthiazole-2-yl)-2,5-diphenyl tetrazolium bromide assay.

Liver Diseases

Oral administration of *P. betle* extract (100, 200, or 300 mg/kg body weight) for 30 days significantly decreased aspartate aminotransferase (AST), alanine amino-transferase (ALT), thiobarbituric acid reactive substances (TBARS), and lipid hydro-peroxides in ethanol treated rats. The extract also improved the tissue anti-oxidant status by increasing the levels of nonenzymatic antioxidants (reduced glutathione, vitamin C, and vitamin E) and the activities of free radical-detoxifying enzymes such as superoxide dismutase, catalase, and glutathione peroxidase in liver and kidney of ethanol-treated rats (Saravanan *et al.,* 2002).

D-galactosamine is a well-established hepato-toxicant that induces a diffuse type of liver injury closely resembling human viral hepatitis. D-galactosamine by its property of generating free radicals causes severe damage to the membrane and affects almost all organs of the human body. There was significant improvement in the anti-oxidant status and liver marker enzymes in rats treated with betel leaf extract (200 mg/kg BW) via intra-gastric intubations as compared to the D-galactosamine treated group, demonstrating its hepato-protective and anti-oxidant properties (Pushpavalli *et al.,* 2008). Again Pushpavalli *et al.* (2009) confirmed that administration of betel leaf-extract prevented the increase in the plasma levels of thiobarbituric acid reactive substances (TBARS), lipid hydro-peroxides, and the decrease in vitamin C, vitamin E and reduced glutathione concentrations and brought towards normality. Apart from these, significant decrease in the levels of the VLDL and LDL cholesterol and increase HDL total cholesterol, phospholipids, triglycerides, free fatty acids in the plasma and tissues of liver and kidney. These results suggest that betel leaves could afford a significant anti-oxidant and anti-hyper-lipidaemic effect against D-galactosamine intoxication.

Nervous Disorders

Coadministration of betel leaves along with ethanol resulted in significant reduction of lipid levels (free fatty acids, cholesterol, and phospholipids) and lipid peroxidation markers such as thiobarbituric acid reactive substances and hydro-peroxides. Further, anti-oxidants, like reduced glutathione, vitamin C, vitamin E, superoxide dismutase, catalase, and glutathione peroxidase, were increased in *P. betle* co-administered rats. The higher dose of extract (300 mg kg. (-1) was more effective, and these results indicate the neuro-protective effect of *P. betle* in ethanol-treated rats (Saravanan *et al.,* 2003). The consumption of betel leaves as an infusion was found to have a positive impact in the prevention and treatment of neuro-degenerative diseases as the aqueous extract proved to be cytotoxic to human neuro-blastoma cells at concentrations higher than 500 µg/mL (Ferreres *et al.,* 2014).

Infectious Diseases

The various piper species have been found to have broad spectrum of anti-bacterial activity. The essential oil and leaf extract also showed anti-bacterial, anti-tuberculous and anti-fungal activity. The anti-microbial activity is believed to be due to chavicol, allylpyro-catechol, chavibetol and chavibetol acetate in the essential oil. There have been reports on the anti-bacterial and anti-helminthic activities of hydroxyl-chavicol. The anti-fungal activity of essential oil and leaf extracts of betel leaves especially on aflatoxins is also proved by Chinese researchers (Sharma *et al.,* 2009; Ramji *et al.,* 2002).

Betel leaf chewing strengthens the oral defence system against microbes. There seems to be a synergism in the total salivary peroxidase activity when saliva and leaf extract were mixed (Kumar and Tripati, 2000). Thiocyanate in conjunction with hydrogen peroxide in saliva offers natural defense system against microbial build-up in oral environment. Betel leaf may be contributing a good deal of thiocyanate, which may be one of the factors leading to the increased peroxidase activity.

Nalina and Rahim (2007) found that the crude extract of betel leaves causes plasma cell membrane damage and coagulation of the nucleoid. The leaf extract containing hydroxyl-chavicol, fatty acids (stearic and palmitic) and hydroxy fatty acid esters (stearic, palmitic and myristic) as the main components was found to significantly reduce acid producing properties of the bacteria. It may thus exert anti-cariogenic activities that are related to decrease in acid production and changes in the ultra-structure of *S. mutans*.

Traditional Pan comprising of betel leaf, lime, catechu, 'Gulkand', cardamom and clove tested by Bissa et al, (2007) on oral microorganisms, reduced the growth to 30 per cent. Betel leaf (Meetha) reduced the microflora, approximately 56 per cent as compared to control whereas the microflora reduction was 50 per cent, when betel leaves of 'Lanki' was used. All the experimental materials were found to be effective against bacterial population of mouth cavity. According to the authors, the reduction in oral microbial population is due to synergistic effect of the combination of betel leaf, cardamom and clove as spices are known to affect biological functions and have been traditionally used for many disorders.

The ethanol extract showed activity against *Streptococcus pyogenes*, *E. coli* and *S. aureus*. The MIC values were 1.25×10^3, 5.00×10^3 and 5.00×10^3 µg/mL, respectively. The fungi species Colletotrichum sp., F. oxysporium sp., C. cassicola, Rigidoporous sp. and Phytophthora sp., except Phytophthora sp. showed significant growth inhibition in betel oil (Arambewela *et al.*, 2005b; 2011).

The effect of betel leaves on respiratory infections is scanty. The essential oil of the leaves may be effective in respiratory catarrh, cough, dyspnoea and diphtheria and act as an anti-septic (Chopra *et al.*, 1982, Satyavati *et al.*, 1987).

In the absence of effective and safe treatment for visceral leishmaniasis or Kala-azar–a devastating parasitic disease caused by *Leishmania donovani*–the search for anti-leishmanial agents from natural resources in common use is imperative. Misra *et al.* (2009) found that *P. betle* landrace Bangla Mahoba (PB-BM) was capable of selectively inhibiting both stages of *Leishmania parasites* by accelerating apoptotic events by generation of reactive oxygen species targeting the mitochondria without any cytotoxicity towards macrophages. According to them the anti-leishmanial efficacy of PB-BM methanolic extract mediated through apoptosis is probably due to the higher content of eugenol in the active landrace. Sarkar *et al.* (2008) observed that this leishmanicidal activity of PB was mediated via apoptosis as evidenced by morphological changes, loss of mitochondrial membrane potential, in situ labeling of DNA fragments by terminal deoxy-ribonucleotidyl-transferase-mediated deoxy-uridine triphosphate nick end labeling, and cell-cycle arrest at the sub-G0/G1 phase.

Nair and Chanda (2008) found the methanolic extract of betel leaves as the most active anti-microbial agent compared to the leaves of *Terminalia catappa* L., *Manilkara zapota* L against 10 Gram-positive, 12 Gram-negative bacteria and one fungal strain, *C. tropicalis* using Piperacillin and gentamicin as standards for anti-bacterial assay.

Hydroxy-chavicol from betel leaves exhibited inhibitory effect on fungal species of clinical significance, with the minimum inhibitory concentration (MIC) ranging from 15.62 to 500 µg/ml for yeasts, 125 to 500 µg/ml for *Aspergillus species*, and 7.81 to 62.5 µg/ml for dermatophytes whereas the minimum fungicidal concentrations (MFC) were found to be similar or two fold greater than the MICs. The anti-fungal activity exhibited by this compound supports its use as an anti-fungal agent particularly for treating topical infections, as well as gargle mouthwash against oral Candida infections (Ali *et al.,* 2010). Trakranrungsie *et al.* (2008) tested the crude ethanolic extracts of betel leaves against selected zoonotic dermatophytes (*M. canis*, *M. gypseum* and *T. mentagrophyte*) and the yeast-like *C. albicans* and found to possess effective anti-fungal properties with average IC(50) values ranging from 110.44 to 119.00 µg/ml. Subsequently, 10 per cent *Piper betle* cream formulated and subjected to physical and microbial limit test also supported the traditional wisdom of remedy and suggested a potential value-addition to agricultural products.

The EO from *P. betel* reduced aflatoxin B (1) production in a dose dependent manner and completely inhibited at 0.6 µl/ml. The same exhibited anti-fungal, aflatoxin suppressive and anti-oxidant characters which are desirable for an ideal preservative. Hence, its application as a plant based food additive in protection and enhancement of shelf life of edible commodities during storage and processing is strongly recommended in view of the toxicological implications by synthetic preservatives (Prakash *et al.,* 2010).

Other Medicinal Properties

Cold ethanolic extract of betel leaves, exhibits centrally mediated anti-nociceptive activity against acute pain as evaluated in the hot plate test and in the tail flick test in rats. The anti-nociceptive action of cold ethanolic extract of betel leaves is unlikely to be mediated *via* sedation, but through opioid mechanisms according to Arambewela *et al.* (2006a).

Construction of a stoma is a common procedure in pediatric surgical practice. For care of these stomas, commercially available devices such as ostomy bag, either disposable or of longer duration are usually used. These are expensive, particularly in countries like Bangladesh, and proper-sized ones are not always available. Banu *et al.* (2007) found an alternative for stoma care, betel leaf, which is suitable for Bangladeshis as they are cheap, easy to handle, non-irritant, and non-allergic.

Betel leaves extract significantly inhibited the elevated aspartate aminotransferase and alanine aminotransferase activities caused by carbon tetrachloride intoxication. It also attenuated total glutathione S-transferase (GST) activity and GST alpha isoform activity, and on the other hand, enhanced superoxide dismutase and catalase activities. The histological examination showed that the extract protected liver from the damage induced by carbon tetrachloride by decreasing alpha-smooth muscle actin (alpha-sma) expression, inducing active matrix metalloproteinase-2 expression though Ras/Erk pathway that consequently attenuated the fibrosis of liver (Young *et al.,* 2007).

The extract of the leaves of *Piper betle* possesses potent xanthine oxidase inhibitory activity. Hydroxy-chavicol is a more potent xanthine oxidase inhibitor than allopurinol, which is clinically used for the treatment of hyper-uricemia (Murata *et al.,* 2009).

Panda and Kar (1998) reports that betel leaf can be both stimulatory and inhibitory to thyroid function, particularly for T3 generation and lipid peroxidation in male mice, depending on the amount consumed. While the lowest dose decreased thyroxine (T4) and increased serum triiodothyronine (T3) concentrations, reverse effects were observed at two higher doses. Higher doses also increased LPO with a concomitant decrease in SOD and CAT activities. However, with the lowest dose most of these effects were reversed.

Many investigators studied the anti-fertility effect of betel leaf and therefore its use as an oral contraceptive. In Orissa this slender betel vine and the roots along with black pepper are used to produce sterility in women as it is believed to result in atrophy of the ovaries. It was shown that the betel leaf stalk extract (50 mg/kg body wt) exhibited a mild progestational activity in immature estrogen primed rabbits but some follicle depressant type action was noted as several graffian follicles were seen in their regressive phase when ovaries were sectioned. Thus Tewari *et al.* (1970) proved that betel leaf stalk extract has a definite anti fertility activity. Ratnasooriya et al, (1990); and Ratnasooriya Premakumara, (1997) demonstrated anti fertility activity and anti-motility effects on washed human spermatozoa in male rats. They also noticed that when fed in diestrus phase to female rats, the estrus cycle was not altered nor was any estrogenic or anti estrogenic effect occurred.

The betel leaf can also be used to remove the burnt smell from food items. This will help retain its original taste and aroma.

Betel cream developed for wound healing has been found to be significantly effective on skin rashes by clinical tests on dermatitis patients. Moreover, scientific investigations on betel leaf have led to the development of several other value-added products such as betel toothpaste, mouthwash, face cream, shampoos, instant betel quid, betel pellet, anti-tick lotion and anti-tick powder which improve the industrial prospects and usage of betel leaf (Arambewla *et al.,* 2010).

Safety Concerns

In spite of all the goodness of the leaf being scientifically proved there are a few beliefs on its contra-indications which need to be confirmed by scientific evidences. It has been well established of course, that when betel leaf is combined with tobacco and if consumed in excess it can lead to staining of teeth, heart palpitations and cancer of the mouth. It has also been demonstrated that it is the betel nut in quid and not the betel leaf that is harmful. Rai *et al.* (2011) who reviewed the studies available on the effects of betel leaves have not only confirmed the anti-carcinogenic effects but also reported that betel leaves possess cancer protective effects against the carcinogen present in tobacco. The protective effect of betel leaf extract against the genomic damage caused by pan masala was statistically significant when cells were treated for a long period. Studies conducted in India also reported that ethanolic extract of *P. betle* leaf stalks was nontoxic as judged by hematological, biochemical profiles and enzymatic studies (Arambewela *et al.,* 2003; Sengupta *et al.,* 2000).

Conclusions

Misra *et al.* (2009) emphasizes the need to extend studies related to traditional medicines from bioactive plants below the species level to the gender level for better efficacy and reproducibility. As per the scientific evidence, betel leaves if consumed alone rather than in the form of pan/quid, would be therapeutically beneficial. Moreover, it would be useful if certain recipes are developed incorporating betel leaf.

References

Airan, J.W. and Sheth, A. K. (1957). Chemical composition of leaves of *Piper betle*. *Bombay Univ. J.* 26A: 1–6.

Ali I, Khan FG, Suri KA, Gupta BD, Satti NK, Dutt P, Afrin F, Qazi GN, Khan IA (2010). In-vitro anti-fungal activity of hydroxychavicol isolated from Piper betle L. *Ann. Clin. Microbiol. Antimicrob.* 9: 7.

Amonkar AJ, Nagabhushan M, D'Souza AV, Bhide SV (1986). Hydroxychavicol: a new phenolic anti-mutagen from betel leaf. *Food Chem. Toxicol.* 24 (12): 1321–24.

Amonkar AJ, Padma PR, Bhide SV (1989). Protective effect of hydroxychavicol, a phenolic component of betel leaf, against the tobacco-specific carcinogens. *Mutat. Res.* 210 (2): 249–53.

Arambewela LS, Arawwawala LD, Kumaratunga KG, Dissanayake DS, Ratnasooriya WD, Kumarasingha SP (2011). Investigations on *Piper betle* grown in Sri Lanka. *Pharmacogn. Rev.* (10): 159–63.

Arambewela LSR, Arawwawala LDAM, Ratnasooriya WD (2006a). Anti-nociceptive activities of aqueous and ethanolic extracts of *Piper betle* leaves in rats. *Pharmaceut. Biol.* 43: 766–72.

Arambewela LSR, Arawwawala LDAM, Ratnasooriya WD (2003). Safety evaluation of Sri Lankan *Piper betle* leaf extracts in rats. *J. Trop. Med. Plants.* 4: 195–98.

Arambewela LSR, Arawwawala LDAM, Ratnasooriya WD (2004). Gastro-protective activities of Sri Lankan *Piper betle* leaf extracts in rats. *SLAAS. 60ʰ Annual Session.* p. 117.

Arambewela L, Arawwawala M, Rajapaksa D (2006b). *Piper betle*: A natural anti-oxidant. *Int. J. Food Sci. Technol.* 41: 10–14.

Arambewela LS, Arawwawala LD, Ratnasooriya WD (2005a). Anti-diabetic activities of aqueous and ethanolic extracts of *Piper betle* leaves in rats. *J. Ethnopharmacol.* 102 (2): 239–45.

Arambewela LSR, Arawwawala LDAM, Withanage D, Kulathunga S (2010). Efficacy of betel cream on skin ailments. *J. Complement. Integr. Med.* 7, Article 48.

Arambewela L, Kumaratunga KG, Dias K (2005b). Studies on *Piper betle* of Sri Lanka. *J. Natn. Sci. Foundation. Sri Lanka*. 33: 133–39.

Azuine MA, Bhide SV (1992). Protective single/combined treatment with betel leaf and turmeric against methyl (Acetoxymethyl) nitrosamine-induced hamster oral carcinogenesis. *Cancer*. 28 (3): 412–15.

Azuine MA. Amonkar AJ and Bhide SV. (1991). Chemo preventive efficacy of betel leaf extracts and its constituents on 7,12-dimethybenz (a) anthrocene induced carcinogenesis and their effect on drug detoxification system in mouse skin. *Indian J. Exp. Biol.* 29 (4): 346–51.

Babu S, Bhat RV, Kumar PU, Sesikaran B, Rao KV, Aruna P, Reddy PR (1996). A comparative clinico-pathological study of oral submucous fibrosis in habitual chewers of pan masala and betel quid. *J Toxicol Clin Toxicol*. 34(3):317-22.

Banu T, Talukder R, Chowdhury TK, Hoque M (2007). Betel leaf in stoma care. *J. Pediatr. Surg*. 42 (7): 1263–65.

Bhattacharya S, Banergee D, Bauri AK, Chattopadhyay S, Bandyopadhyay SK (2007). Healing properties of the Piper betle phenol, allylpyrocatechol against indomethacin-induced stomach ulceration and mechanism of action. *World J. Gastroenterol*. 21: 3705–713.

Bhattacharya S, Subramanian M, Roychowdhury S, Bauri AK, Kamat JP, Chattopadhyay S, Bandyopadhyay SK (2005). Radio-protective property of the ethanolic extract of *Piper betel* Leaf. *J. Radiat. Res. (Tokyo)*. 46 (2): 165–71.

Bhide SV, Padma PR, Amonkar AJ.(1991a) Anti-mutagenic and anti-carcinogenic effects of betel leaf extract against the tobacco specific nitrosamine 4-(N-nitrosomethylamino)-1-(3-pyridyl)-1-butanone (NNK). *IARC Sci. Publ*. 105: 520–24.

Bhide SV, Shivapurkar NM, Gothoskar SV, Ranadive KJ (1979). Carcino-genicity of betel quid ingredients: feeding mice with aqueous extract and the polyphenol fraction of betel nut. *Br. J. Cancer*. 40: 922–26.

Bhide SV, Zariwala MB, Amonkar AJ, Azuine MA (1991b) Chemo-preventive efficacy of a betel leaf extract against benzo (a) pyrene-induced fore stomach tumors in mice. *J. Ethnopharmacology*. 34 (2-3): 207–13.

Bissa S, Songara D, Bohra A (2007). Traditions in oral hygiene: Chewing of betel (*Piper betle* L.) leaves. *Current Science*. 92 (1): 26–28.

Boffetta P, Hecht S, Gray N, Gupta P, Straif K (2008). Smokeless tobacco and cancer. *Lancet. Oncol*, 9: 667–75.

Canniff JP, Harvey W (1981). The aetiology of oral submucous fibrosis: the stimulation of collagen synthesis by extracts of areca nut. *Int. J. Oral. Surg*, 10: 163–67.

Chakraborty JB, Mahato SK, Joshi K, Shinde V, Rakshit S, Biswas N, Choudhury Mukherjee I, Mandal L, Ganguly D, Chowdhury AA, Chaudhuri J, Paul K, Pal

BC, Vinayagam J, Pal C, Manna A, Jaisankar P, Chaudhuri U, Konar A, Roy S, Bandyopadhyay S (2012). Hydroxychavicol, a *Piper betle* leaf component, induces apoptosis of CML cells through mitochondrial reactive oxygen species-dependent JNK and endothelial nitric oxide synthase activation and overrides imatinib resistance. *Cancer Sci.* 103 (1): 88–99.

Chang MC, Uang BJ, Tsai CY, Wu HL, Lin BR, Lee CS, Chen YJ, Chang CH, Tsai YL, Kao CJ, Jeng JH. (2007). Hydroxychavicol, a novel betel leaf component, inhibits platelet aggregation by suppression of cyclooxygenase, thromboxane production and calcium mobilization. *Br. J. Pharmacol.* 152 (1): 73–82.

Chang MC, Uang BJ, Wu HL, Lee JJ, Hahn LJ, Jeng JH (2002). Inducing the cell cycle arrest and apoptosis of oral KB carcinoma cells by hydroxychavicol: Roles of glutathione and reactive oxygen species. *Br. J. Pharmacol.* 135: 619–30.

Chatterjee A, Pakrashi SC (1995). *Treatise of Indian Medicinal Plants.* 1: 26–35.

Chopra RN, Chopra IC, Handa KL, Kapur LD (1982). *Chopra's Indigenous Drugs of India*; 2nd ed. AP. New Delhi.

Choudhary D, Kale RK (2002). Anti-oxidant and non-toxic properties of *Piper betle* leaf extract: *in vitro* and *in vivo* studies. *Phytother. Res.* 16 (5): 461–66.

CSIR (1969). *The Wealth of India Council of Scientific and Industrial Research*, New Delhi. 8: 84-94.

Dasgupta N, De B (2004). Anti-oxidant activity of *Piper betle* L. leaf extract *in-vitro.* *Food Chemistry*, 88, 219 – 24.

Dohi T, Terada H, Anamura S, Okamoto H, Tsujimoto A (1989). The anti-inflammatory effects of phenolic dental medicaments as determined by mouse ear edema assay. *Jap. J. Pharmacol*, 49: 535–39.

Ferreres F, Oliveira AP, Gil-Izquierdo A, Valentão P, Andrade PB (2014). *Piper betle* leaves: Profiling phenolic compounds by HPLC/DAD-ESI/MSn and anti-cholinesterase activity. *Phytochem. Anal.* 2014 Apr 14. doi: 10.1002/pca. 2515. [Epub ahead of print].

Ganguly S, Mula S, Chattopadhyay S, Chatterjee M (2007). An ethanol extract of Piper betle Linn. mediates its antiinflammatory activity via down regulation of nitric oxide. *J. Pharma. Pharmacol.* 59: 711–18.

Gopalan C, Ramasastri BV, Balasubramanian SC (1984). *Nutritive Value of Indian Foods*, pp. 108. National Institute of Nutrition (ICMR), Hyderabad, India.

Guha P (1997). "Paan Theke Kutir Silpa Sambhabana" (In Bengali). " *Exploring Betel Leaves for Cottage Industry*", pp. 15–19, In: Krishi, Khadya-O- Gramin Bikash Mela –A Booklet published by the Agricultural and Food Engineering Department, IIT, Kharagpur, India.

Guha P (2006). Betel leaf: The neglected green gold of India. *J. Hum. Ecol.* 19: 87–93.

Guha P, Jain RK (1997). Status report on production, processing and marketing of betel leaf (*Piper betle* L.). Agricultural and Food Engineering Department, IIT, Kharagpur, India.

IARC (2004). Betel-quid and areca nut chewing and some areca-nut derived nitrosamines. In: *IARC Monographs* vol. 85. International Agency for Research on Cancer, WHO.

Jagetia GC, Baliga MS (2004). The evaluation of nitric oxide scavenging activity of certain Indian medicinal plants *in vitro*: a preliminary study. *J. Med. Food*. 7: 343–48.

Jeng JH, Ho YS, Chan CP, *et al.* (2000). Areca nut extract up-regulates prostaglandin production, cyclooxygenase-2 mRNA and protein expression of human oral keratinocytes. *Carcinogenesis*. 21: 1365–70.

Jeng JH, Kuo ML, Hahn LJ, Kuo MY (1994). Genotoxic and nongenotoxic effects of betel quid ingredients on oral mucosal fibroblasts *in vitro*. *J. Dent. Res*. 73: 1043–49.

Jeng JH, Chen SY, Liao CH, Tung YY, Lin BR, Hahn LJ, Chang MC (2002). Modulation of platelet aggregation by areca nut and betel leaf ingredients: roles of reactive oxygen species and cyclooxygenase. *Free Radic. Biol. Med*. 32 (9): 860–71.

Jin YT, Tsai ST, Wong TY, Chen FF, Chen RM (1996). Studies on promoting activity of Taiwan betel quid ingredients in hamster buccal pouch carcinogenesis. *Eur. J. Cancer. B. Oral. Oncol*. 32B: 343–46.

Kanjwani DG, Marathe TP, Chiplunkar SV, Sathaye SS (2008) Evaluation of immunomodulatory activity of methanolic extract of *Piper betel*. Scand. *J. Immunol*. 67 (6): 589–93.

Ko YC, Huang YL, Lee CH, Chen MJ, Lin LM, Tsai CC (1995). Betel quid chewing, cigarette smoking and alcohol consumption related to oral cancer in Taiwan. *J. Oral Pathol. Med*. 24: 450–53.

Kumar N, Misra P, Dube A, *et al.* (2010). *Piper betle* Linn. a maligned Pan-Asiatic plant with an array of pharmacological activities and prospects for drug discovery. *Curr. Sci*. 99: 922–32.

Kumar N, Tripati R (2000). Putative role of bele (*Piper betle* L.) in oral hygiene. *Plant peroxidase Newsl*. 15: 45–48.

Kumar V, Asdaq SM, Asad M (2009). Influence of betel quid on effect of calcium channel blockers on isoprenaline induced myocardial necrosis in mice. *Indian J. Exp. Biol*. 47 (9): 730–36.

Lahiri M, Bhide SV (1993). Effect of four plant phenols, β-carotene and α-tocopherol on 3(H)benzopyrene–DNA interaction *in vitro* in the presence of rat and mouse liver postmitochondrial fraction. *Cancer Lett*. 73: 35–39.

Lee CH, Lee JM, Wu DC, *et al.* (2005). Independent and combined effects of alcohol intake, tobacco smoking and betel quid chewing on the risk of esophageal cancer in Taiwan. *Int. J. Cancer*, 113: 475–82.

Lee YY, Hung SL, Pai SF, Lee YH, Yang SF (2007). Eugenol suppressed the expression of lipo-polysaccharide-induced pro-inflammatory mediators in human macrophages. *J. Endod*. 33, 698 – 702.

Lei D, Chan CP, Wang YJ, Wang TM, Lin BR, Huang CH, Lee JJ, Chen HM, Jeng JH, Chang MC (2003). Anti-oxidative and anti-platelet effects of aqueous inflorescence *Piper betle* extract. *J. Agric. Food Chem.* 51 (7): 2083–88.

Ma GC, Wu PF, Tseng HC, Chyau CC, Lu HC, Chou FP (2013). Inhibitory effect of *Piper betle* leaf extracts on copper-mediated LDL oxidation and oxLDL-induced lipid accumulation via inducing reverse cholesterol transport in macrophages. *Food Chem.* 141 (4): 3703–13.

Maity S (1989). *The Betelvine.* Extension Bulletin. All India Coordinated Research Project on Betelvine, Indian Institute of Horticultural Research, Hessarghatta, Bangalore, India.

Majumdar B, Chaudhuri SG, Ray A, Bandyopadhyay SK (2003). Effect of ethanol extract of *Piper betle* Linn leaf on healing of NSAID-induced experimental ulcer– a novel role of free radical scavenging action. *Indian J. Exp. Biol.* 41 (4): 311–15.

Manigauha A, Ali H, Maheshwari MU (2009). Anti-oxidant activity of ethanolic extract of *Piper betle* leaves. *J. Pharm. Res,* 2: 491–94.

Misra P, Kumar A, Khare P, Gupta S, Kumar N, Dube A (2009). Pro-apoptotic effect of the landrace Bangla Mahoba of *Piper betle* on Leishmania donovani may be due to the high content of eugenol. *J. Med. Microbiol.* 58 (Pt 8): 1058–66.

Murata K, Nakao K, Hirata N, Namba K, Nomi T, Kitamura Y, Moriyama K, Shintani T, Iinuma M, Matsuda H (2009). Hydroxychavicol: a potent xanthine oxidase inhibitor obtained from the leaves of betel, *Piper betle. J. Nat. Med.* 63 (3): 355–59.

Nagabhushan M, Amonkar AJ, D'Souza AV, Bhide SV (1987). Non-mutagenicity of betel leaf and its anti-mutagenic action against environmental mutagens. *Neoplasma.* 34 (2): 159–67.

Nagabhushan M, Amonkar AJ, D'Souza AV, Bhide SV (1989). Hydroxychavicol: a new anti-nitrosating phenolic compound from betel leaf. *Mutagenesis.* 4: 200–204.

Nair R, Chanda S. (2008). Antimicrobial Activity of *Terminalia catappa, Manilkara zapota* and *Piper betel* Leaf Extract. *Indian J Pharm Sci.* 70(3): 390–93.

Nalina T. and Z.H.A. Rahim (2007). The Crude Aqueous Extract of Piper betle L. and its Anti-bacterial Effect Towards *Streptococcus mutans. American Journal of Biotechnology and Biochemistry.* 3 (1): 10–15.

Norton SA (1998). Betel: consumption and consequences. *J. Am. Acad. Dermatol.* 38: 81–88.

Padma PR, Amonkar AJ, Bhide SV (1989 a). Anti-mutagenic effects of betel leaf extract against the mutagenicity of two tobacco-specific N-nitrosamines. *Mutagenesis.* 4: 154–56.

Padma PR, Lalitha VS, Amonkar AJ, Bhide SV (1989b). Anti-carcinogenic effect of betel leaf extract against tobacco carcinogens. *Cancer Lett.* 45 (3): 195 -202.

Panda S, Kar A (1998). Dual role of betel leaf extract on thyroid function in male mice. *Pharmacol. Res.* 38 (6): 493–96.

Paranjpe R, Gundala SR, Lakshminarayana N, Sagwal A, Asif G, Pandey A, Aneja R (2013). *Piper betel* leaf extract: anti-cancer benefits and bio-guided fractionation to identify active principles for prostate cancer management. *Carcinogenesis.* 34 (7): 1558–66.

Prabhu MS, Platel K, Saraswathi G, Srinivasan K (1995). Effect of orally administered betel leaf (*Piper betle* Linn.) on digestive enzymes of pancreas and intestinal mucosa and on bile production in rats. *Indian J. Exp. Biol.* 33 (10): 752–56.

Prajapathi ND (2003). *Agro's Dictionary of Medicinal Plants*, Agrobios, Jodhpur. p. 401.

Prakash B, Shukla R, Singh P, Kumar A, Mishra PK, Dubey NK (2010). Efficacy of chemically characterized *Piper betle* L. essential oil against fungal and aflatoxin contamination of some edible commodities and its anti-oxidant activity. *Int. J. Food Microbiol.* 142 (1-2):114–19.

Pushpavalli G, Veeramani C, Pugalendi KV (2008). Influence of *Piper betle* on hepatic marker enzymes and tissue anti-oxidant status in D-galactosamine-induced hepatotoxic rats. *J. Basic Clin. Physiol. Pharmacol.* 19 (2): 131–50.

Pushpavalli G, Veeramani C, Pugalendi KV (2009). Effect of *Piper betle* on plasma anti-oxidant status and lipid profile against D-galactosamine-induced hepatitis in rats. *Redox. Rep.* 14 (1): 7–12.

Rai MP, Thilakchand KR, Palatty PL, Rao P, Rao S, Bhat HP, Baliga MS (2011). *Piper betel* Linn (betel vine), the maligned Southeast Asian medicinal plant possesses cancer preventive effects: time to reconsider the wronged opinion. *Asian. Pac. J. Cancer. Prev.* 12 (9): 2149–56.

Ramji N, Ramji N, Iyer R, Chandrasekaran S (2002). Phenolic antibacterials from Piper betle in the prevention of halitosis. *J. Ethnopharmacol.* 83: 149–52.

Rao AR (1984). Modifying influences of betel quid ingredients on B(a)P-induced carcinogenesis in the buccal pouch of hamster. *Int. J. Cancer.* 33: 581–86.

Rastogi RP, Mehrotra BN (1993). *Compendium of Indian Medicinal Plants*, Vol. 3. Publications and Information Directorate: New Delhi. pp. 502–03.

Rathee JS, Patro BS, Mula S, Gamre S, Chattopadhyay S (2006). Antioxidant activity of *Piper betel* leaf extract and its constituents. *J. Agric. Food Chem.* 54 (24): 9046–54.

Ratnasooriya WD, Jayawardena KG, Premakumara GA (1990). Anti-motility effects of Piper betel (L) leaf extract on washed human spermatozoa. *J. Natn. Sci. Coun. Sri Lanka.* 18: 53–60.

Ratnasooriya WD, Premakumara GA (1997). *Piper betle* leaves reversibly inhibits fertility of male rats. *Vidyodaya. J. Sci.* 7: 15–21.

Saeed SA, Farnaz S, Simjee RU, Malik A (1993). Triterpenes and B-sitosterol from *Piper betle*: Isolation, anti-platelet and anti-inflammatory effects. *Biochem. Soc. Transact.* 21: 462S.

Samanta C (1994). Paan chaser samasyabali-o-samadhan: Ekti samikkha (In Bengali): "A Report on the Problems and Solutions of Betel Vine Cultivation". A booklet published by Mr. H. R. Adhikari, C-2/16, Karunamoyee, Salt Lake City, Kolkata-64 (WB), India.

Santhakumari P, Prakasam A, Pugalendi KV (2006). Anti-hyperglycemic activity of *Piper betle* leaf on streptozotocin-induced diabetic rats. *J. Med. Food.* 9 (1): 108–12.

Saravanan R, Prakasam A, Ramesh B, Pugalendi KV (2002). Influence of *Piper betle* on hepatic marker enzymes and tissue anti-oxidant status in ethanol-treated Wistar rats. *J. Med. Food.* 5 (4): 197–204.

Saravanan R, Rajendra Prasad N, Pugalendi KV (2003). Effect of *Piper betle* leaf extract on alcoholic toxicity in the rat brain. *J. Med. Food.* 6 (3): 261–65.

Sarkar A, Sen R, Saha P, Ganguly S, Mandal G, Chatterjee M (2008). An ethanolic extract of leaves of *Piper betle* (Paan) Linn mediates its anti-leishmanial activity via apoptosis. *Parasitol. Res.* 102 (6): 1249–55.

Satyavati GV, Raina MK, Sharma M (1987). *Medicinal Plants of India*. Vol. 1. Indian Council of Medical Research, New Delhi, India.

Sengupta A, Adhikary P, Bask BK, Chakrabarti K, Gangopadhyay P, Banerji J, Chatterjee A (2000). Pre–clinical toxicity evaluation of leaf stalk extractive of *Piper betle* Linn. in rodents. *Indian J. Exp. Biol.* 38: 338–42.

Sharma S, Khan IA, Ali I, Ali F, Kumar M, Kumar A, Johri RK, Abdullah ST, Bani S, Pandey A, Suri KA, Gupta BD, Satti NK, Dutt P, Qazi GN (2009). Evaluation of the anti-microbial, anti-oxidant and anti-inflammatory activities of hydroxychavicol for its potential use as an oral care agent. *Antimicrob. Agents Chemother.* 53: 216–22.

Shirname LP, Menon MM, Nair J, bhide SV (1983). Correlation of mutagenicity and tumorigenicity of betel quid and its ingredients. *Nutr. Cancer.* 5 (2):87-91.

Singh M, Shakya S, Soni VK, Dangi A, Kumar N, Bhattacharya SM (2009). The n-hexane and chloroform fractions of *Piper betle* L. trigger different arms of immune responses in BALB/c mice and exhibit anti-filarial activity against human lymphatic filarid Brugia malayi. *Int. Immunopharmacol.* 9: 716–28.

Subbulakshmi G and Mehra M (1995). Effect of betel leaves on the vitamin A status of expectant mothers. *M.Sc. thesis* submitted to the SNDT Women's University, Mumbai, India.

Subbulakshmi, G., Padma R and Jhansi Rani, P. (1988) Nutrient composition of betel leaves (*Piper betle*), their acceptability in Indian dishes. *M.Sc. Thesis* submitted to Satya Sai Institute of Higher Learning, Anantapur, AP. India. 9–15.

Sundqvist K, Liu Y, Nair J, *et al.* (1989). Cytotoxic and genotoxic effects of areca nut-related compounds in cultured human buccal epithelial cells. *Cancer. Res.* 49: 5294–98.

Tewari PV, Chaturvedi C, Dixit SN (1970). Anti-fertility effect of betel leaf stalk– A preliminary experimental study. *J. Res. Indian Med.* 4 (2): 143–51.

Thomas SJ, MacLennan R (1992). Slaked lime and betel nut cancer in Papua New Guinea. *Lancet.* 340: 577–78.

Trakranrungsie N, Chatchawanchonteera A, Khunkitti W (2008). Ethnoveterinary study for anti-dermatophytic activity of Piper betle, *Alpinia galanga* and *Allium ascalonicum* extracts *in vitro. Res. Vet. Sci.* 84 (1): 80–84.

Trivedi AH, Patel RK, Rawal UM, Adhvaryu SG and Balar DB (1994). Evaluation of chemo-preventive effects of betel leaf on the genotoxicity of pan masala. *Neoplasma.* 41 (3): 177–81.

Verma S, Gupta ML, Dutta A, Sankhwar S, Shukla SK, Flora SJ (2010). Modulation of ionizing radiation induced oxidative imbalance by semi-fractionated extract of *Piper betle*: an in vitro and *in-vivo* assessment. Oxid. *Med. Cell. Longev.* 3 (1): 44–52.

Wang CK, Peng CH (1996). The mutagenicities of alkaloids and N-nitrosoguvacoline from betel quid. *Mutat. Res.* 360: 165–71.

Wang LY, You SL, Lu SN, *et al.* (2003). Risk of hepatocellular carcinoma and habits of alcohol drinking, betel quid chewing and cigarette smoking: a cohort of 2416 HBsAg-seropositive and 9421 HBsAg-seronegative male residents in Taiwan. *Cancer–Causes, Control.* 14: 241–50.

Warrier PK, Nambair VPK, Ramankutty C (1995). Ind *ian Medicinal Plants: A Compendium of 500 Species.* Arya Vaidya Sala, Kottakal, Kerala; Orient Longman, India.

Young SC, Wang CJ, Hsu JD, Hsu JL, Chou FP (2005). Increased sensitivity of Hep G2 cells toward the cytotoxicity of cisplatin by the treatment of *piper betel* leaf extract. *Arch. Toxicol.* 9: 1–9.

Young SC, Wang CJ, Lin JJ, Peng PL, Hsu JL, Chou FP (2007). Protection effect of *Piper betel* leaf extract against carbon tetrachloride-induced liver fibrosis in rats. *Arch. Toxicol.* 81 (1): 45–55.

Chapter 4

Bitter Gourd/ Bitter Melon/Karela (*Momordica charantia*)

M. charantia, an edible vegetable belonging to the *Cucurbitaceae* family, is known as a bitter melon or bitter gourd. Bitter gourd is indigenous to South America and Asia, used in native medicines of Asia and Africa, Oriental and Latin American countries.

Chemical Constituents

Bitter gourd contains bitter glycosides, saponins, alkaloids, reducing sugars, phenolics, oils, free acids, polypeptides, sterols, along with 17-amino acids including methionine, a crystalline product named p-insulin and other unspecific bioactive anti-oxidants. Perla and Jayanty (2013) through *in-vitro* experiments showed the presence of high levels of biguanide related compounds in bitter gourd useful in the treatment of type 2 diabetes.

Traditional Therapeutic Uses

Bitter gourd has been traditionally used as folk medicine for several ailments such as diabetes, tumors, bacterial, parasitic and viral infections (Grover and Yadav, 2004). It is also believed to be anti-dotal, anti-pyretic, tonic, appetizing, stomachic, anti-bilious and laxative.

It is one of the most used plants for the treatment of diabetes and related conditions in traditional system of medicine world over (Baby Joseph and Jini, 2013; Leung *et al.,* 2009). Ethno-botanical survey conducted among the Monpa ethnic group, Arunachal

Pradesh, India, reported the use of bitter gourd for treating intestinal worms and diabetes, while in Kurukshetra district, Haryana, India, it is used as a vegetable for treating chronic edema. Use of bitter gourd for diabetes in Bangladesh is still an essential part of public healthcare (Ocvirk *et al.,* 2013).

Nutritional Value per 100 g of Bitter Gourd Pods (Boiled, drained without salt)

Nutrient	Amount	Nutrient	Amount
Energy (Kcal)	19	Vitamin B$_6$ (mg)	0.041
Carbohydrates (g)	4.32	Folate (µg)	51
- Sugars (g)	1.95	Vitamin C (mg)	33
- Dietary fiber (g)	2	Vitamin E (mg)	0.14
Fat (g)	0.18	Vitamin K (µg)	4.8
Protein (g)	0.84	Calcium (mg)	9
Water (g)	93.95	Iron (mg)	0.38
Vitamin A equiv. (µg)	6	Magnesium (mg)	16
Beta-carotene (µg)	68	Manganese (mg)	0.086
Lutein and zeaxanthin (µg)	1323	Phosphorus (mg)	36
Thiamine (mg)	0.05	Potassium (mg)	319
Riboflavin (mg)	0.05	Sodium (mg)	6
Niacin (mg)	0.28	Zinc (mg)	0.77
Pantothenic acid (mg)	0.19		

Source: USDA Nutrient Database.

In spite of the fact that numerous pre-clinical studies have documented the anti-diabetic and hypo-glycaemic effects of bitter gourd through various postulated mechanisms, Leung *et al.* (2009) called for better-designed clinical trials to further elucidate its possible therapeutic effects. The various therapeutic benefits of bitter gourd have been proved through *in-vitro* and *in-vivo* trials as described below.

Anti-oxidant Property

Bitter gourd has been found to possess strong anti-oxidant potential. The hexane and methanol hydrophilic dialyzed extracts of bitter melon showed the highest anti-oxidant activities with the former containing the highest phenolic content (Kenny *et al.,* 2013). As compared to the native polysaccharide from *Momordica charantia* L., sulfated derivatives of polysaccharide exhibited better anti-oxidant activities *in-vitro.* Moreover, high degree of sulfation and moderate molecular weight improved the anti-oxidant activities of polysaccharide (Liu *et al.,* 2014).

Bitter gourd extract along with *Trigonella foenum-graecum* extract significantly increased the activities of key anti-oxidant enzymes such as superoxide dismutase (SOD), catalase (CAT), glutathione-s-transferase (GST) and reduced glutathione (GSH) contents in heart tissue of alloxan induced diabetic rats (Tripathi *et al.,* 2009). The cerebral oxidative stress and damage, and neurological deficits in experimental rats were found to be dose dependently attenuated by pre-treatment with the lyophilized bitter gourd juice (Malik *et al.,* 2011).

Anti-microbial Agent

The broad spectrum anti-viral activity of bitter gourd protein has been proved through inhibition of not only H1N1 and H3N2 but also H5N1 subtype virus, thus revealing its potential to be developed as an effective therapeutic agent against various well known and newly emerging subtypes of influenza A (Pongthanapisith *et al.,* 2013). The mode of anti-viral activity of bitter gourd has been reported by various researchers worldwide.

Thai bitter gourd protein (MRK29) isolated from the ripe fruit and seeds inhibited the HIV-1 reverse transcriptase with 50 per cent IR at the concentration of 18 µg/ml. The MRK29 (concentrated in the 30–60 per cent salt precipitated fraction) exerted 82 per cent reduction of viral core protein p24 expression in HIV-infected cells at the concentration of 0.175 µg/ml thus exhibiting a modulatory role on immune cells through increased TNF activity by 3-fold (Jiratchariyakul *et al.,* 2001).

MAP30 is an anti-viral protein isolated from bitter melon was reported to enhance the anti-viral activity of weak pharmacological HIV antagonists, dexamethasone and indomethacin (Lee-Huang *et al.,* 1995). Bourinbaiar and Lee-Huang (1995) also found that use of MAP30 in combination with low pharmacological doses of dexamethasone and indomethacin improved the efficacy of anti-HIV therapy. Later Fan *et al.* (2009) also reported that the gene encoding MAP30 protein cloned from bitter melon, and the expressed and purified recombinant MAP30 could inhibit the production of HBV dose-dependently. The expression of HBsAg, the surface antigen of the hepatitis B virus, was significantly decreased by MAP30 dose-dependently and time-dependently. Lower dose of MAP30 (8.0 µg/ml) also could inhibit the expression of HBsAg and HBeAg.

Anti-diabetic Agent

Bitter gourd has been widely studied for its beneficial effects on blood glucose levels. It has been found to show anti-hyperglycemic activities when administered in experimental rats (Singh *et al.,* 2008; Malik *et al.,* 2011; Clouatre *et al.,* 2011; Choudhary *et al.,* 2012; Sitasawad *et al.,* 2000; Ahmed *et al.,* 1998; 2001; Sekar *et al.,* 2005; Virdi *et al.,* 2003). The anti-diabetic components in bitter gourd include charantin, vicine, and polypeptide-p. Keller *et al.* (2011) discovered that a saponin-rich fraction isolated from *M. charantia,* stimulates insulin secretion in an *in-vitro* static incubation assay.

The hypo-glycemic efficacy of aqueous seed extract of two varieties, namely a country and a hybrid variety of *M. charantia* clearly proved their anti-diabetic properties. Both the varieties showed safe and significant hypo-glycemic effects which were more pronounced in country variety compared to hybrid variety (Sekar and Subramanian, 2005a). Metabolic and hypo-glycemic effects of bitter gourd extracts have been demonstrated in cell culture, animal, and human studies (Krawinkel and Keding, 2006). Experiments conducted on rats (Leatherdale *et al.,* 1981; Day, 1990; Chaturvedi *et al.,* 2004; Zheng *et al.,* 2005) and maturity onset diabetic patients (Welihinda *et al.,* 1986), showed that bitter gourd improved glucose tolerance without any significant increase in serum insulin response, along with protective effect on development of diabetes induced cataract. The extract reduced the risk of diabetic complications and exerted rapid protective effects against lipid peroxidation by scavenging of free radicals (Sekar and Subramanian, 2005b). Significant improvement in fasting blood glucose, serum insulin, β-cell number and function was observed with administration of bitter gourd extract to the diabetic rat model (Hafizur *et al.,* 2011).

Mice fed bitter gourd diet showed higher insulinogenic index in an *in-vivo*, oral glucose tolerance test. This anti-diabetic activity of bitter gourd could be partially attributed to its stimulatory effect on the GLP-1 (Huang *et al.,* 2013).

There is strong evidence that increased adipose 11β-HSD1 activity may be an important etiological factor in the current obesity and diabetes type 2 epidemics. Blum *et al.* (2012) showed that bitter gourd extract contains at least one ingredient with selective 11β-HSD1 inhibitory activity thus providing an interesting additional explanation for its well-documented anti-diabetic and hypo-glycaemic effects. Out of the three anti-diabetic compounds namely, charantin, momordenol and momordicilin from bitter gourd, momordicilin was found to show minimum binding energy and thus found to be the most active compound in the respective target site (Hazarika *et al.,* 2012). According to Chaturvedi (2012) the effects of bitter gourd including transport of glucose into the cells, transport of fatty acids into the mitochondria, modulation of insulin secretion, and elevation of levels of uncoupling proteins in adipose and skeletal muscles are similar to those of AMP-activated protein kinase (AMPK) and thyroxine. Therefore, it is proposed that effects of bitter gourd on carbohydrate and fat metabolism are through thyroxine and AMPK.

Bitter melon is commonly known as vegetable insulin. In a human study, Baldwa and his team, as early as 1977, administered insulin like compound obtained from the bitter gourd fruit and as well as from tissue culture of the plant. This compound was homologous to animal insulin and showed consistent hypo-glycemic effect on diabetic subjects. It showed the onset of action within 30–60 min. and the peak effect of six hours after the administration of the dose without any hyper-sensitivity reaction among the subjects. It has been demonstrated that bitter melon suppressed postprandial hyper-glycemia by inhibition of alpha-glucosidase activity and that the most beneficial component is present in the fraction obtained from methanolic extract with a molecular weight of less than 1,300 (Uebanso *et al.,* 2007). Among the several fractions isolated from ethanolic extract of *Momordica charantia* fruits, 7β,25-dihydroxycucurbita-5,23(E)-dien-19-al 3-O- β-d-allo-pyranoside showed potent hypo-glycemic activities by glucose uptake assay (Hsiao *et al.,* 2013). In non-insulin dependent diabetic patients the juice of bitter gourd produced significant reduction in fasting as well as post prandial blood sugars after an oral glucose tolerance test (Ahmad *et al,* 1999). Cummings *et al.* (2004) found the bitter melon fruit juice to act like insulin, exert hypo-glycemic effect and to stimulate amino acid uptake into skeletal muscle cells. Whereas, Ahmed *et al.* (2004) observed that bitter melon juice can regulate glucose uptake into jejunum membrane brush border vesicles and stimulate glucose uptake into skeletal muscle cells similar to the response obtained with insulin.

Day *et al.* (1990) reported that in normal mice, an aqueous extract (A) of bitter gourd lowered the glycemic response to both oral and intra-peritoneal glucose, without altering the insulin response. This aqueous extract and the residue after alkaline chloroform extraction reduced the hyper-glycemia in diabetic mice at 1 hour. Material recovered by acid water wash of the chloroform extract remaining after an alkaline water wash produced a more slowly generated hypo-glycemic effect suggesting that orally administered bitter gourd extracts lower glucose concentrations independently of intestinal glucose absorption and involve an extra-pancreatic effect.

The mechanism of action of the bittergourd extract in alloxan-induced diabetic rats seemed to be enhancing insulin secretion by the islets of Langerhans, reducing glycogenesis in liver tissue, enhancing peripheral glucose utilization and increasing serum protein levels. Further, the treatment restored the altered histological architecture of the islets of Langerhans. Hence, the biochemical, pharmacological and histo-pathological profiles of bitter gourd clearly indicate its potential anti-diabetic activity and other beneficial effects in amelioration of diabetes associated complications including hyper-lipidemia and obesity (Fernandes *et al.,* 2007).

Singh and Gupta (2007) observed the presence of small scattered islets among the acinar tissue when experimental diabetic animals were administered the acetone extract of bitter gourd fruit powder. These small scattered islets were suggestive of the neo-formation of islets from pre-existing islet cells, which if proved by further studies will be beneficial for diabetic population. Alleviation of pancreatic damage and an increase in the number of beta cells in neonatal diabetic rats fed with bitter gourd fruit extract was observed by Abdollahi *et al.* (2011). The islet size, total beta cell area and number of β-cells almost doubled in the diabetic rats treated with bitter gourd extract as compared to the untreated diabetic rats, and the extract-treated diabetic rat beta cells were found to be abundant with insulin granules (Hafizur *et al.,* 2011).

Roffey *et al,* (2007) reported that water-soluble component(s) in bitter gourd enhanced the glucose uptake at sub-optimal concentrations of insulin in 3T3-L1 adipocytes, which was accompanied by and might be the result of increased adiponectin secretion from the 3T3-L1 adipocytes. Bitter melon reversed fructose diet-induced hypo-adiponectinemia which provides a therapeutic advantage to insulin resistance in improving insulin sensitivity. Additionally, bitter melon decreased the weights of epididymal and retroperitoneal white adipose tissue. Furthermore, they demonstrated that bitter melon significantly increased the mRNA expression of glucose transporter 4 (GLUT4) in skeletal muscle. The methanol extract of bitter gourd normalized blood glucose level, reduced triglyceride and LDL levels and increased HDL level in diabetic rats fed a high-fat and a low-carbohydrate diet (Chaturvedi *et al.,* 2004). Kumar *et al.* (2009) demonstrated the significance of Glut-4, PPAR gamma and PI3K up-regulation by bitter gourd in augmenting the glucose uptake and homeostasis. However, the animals reverted to a diabetic state once the extract was discontinued (Chaturvedi, 2005). On the contrary, Singh *et al.* (2008) observed that the blood glucose once lowered by the treatment with bitter gourd fruit extract remained static even after discontinuation of the extract for 15 days and the blood sugar never fell below normal values even with a high dose of this extract.

According to Ahmed *et al.* (1998) oral feeding of *M. charantia* fruit juice may have a role in the renewal of beta cells in STZ-diabetic rats or alternately may permit the recovery of partially destroyed beta cells. Sitasawad *et al.* (2000) found that feeding the aqueous juice of bitter gourd fruit resulted in reduction of STZ-induced hyper-glycemia in mice. It also markedly reduced the STZ-induced lipid peroxidation in pancreas of mice, rat insulinoma cells (RIN) and islets. Further, it reduced the STZ-induced apoptosis in RIN cells indicating the mode of protection of bitter gourd juice on RIN cells, islets and pancreatic beta-cells.

Time course experiments performed in rat adipocytes by Yibchok-anun (2006) revealed that *M. charantia* protein extract significantly increased glucose uptake after 4 and 6 h of incubation. Thus, the *M. charantia* protein extract, a slow acting chemical, exerted both insulin secretagogue and insulin-mimetic activities to lower blood glucose concentrations *in-vivo*.

Extracts of fruit pulp, seed, leaves and whole plant of bitter gourd have all shown to have hypo-glycemic effect in various animal models (Sarkar *et al.*, 1996; Kar *et al.*, 2003; Fernandes *et al.*, 2007; Singh *et al.*, 2008). Polypeptide p, isolated from these parts showed significant hypo-glycemic effect when administered subcutaneously to langurs and humans (Khanna *et al.*, 1981). 200 mg/kg ethanolic extract showed an anti-hyperglycemic effect in normal and STZ diabetic rats. This may be because of inhibition of glucose-6-phosphatase besides fructose-1, 6-biphosphatase in the liver and stimulation of hepatic glucose-6-phosphate dehydrogenase activities (Shibib *et al.*, 1993).

Dried bitter gourd powder at 10 per cent level in the diet in streptozotocin induced diabetic rats brought down the fasting blood glucose level by about 30 per cent. In addition, bitter gourd supplementation alleviated the rise in water consumption, urine volume and urine sugar during diabetes by about 30 per cent as compared to the controls. Renal hyper-trophy was higher in diabetic controls and bitter gourd supplementation partially, but effectively prevented it by 38 per cent. Increased glomerular filtration rate in diabetes was significantly reduced by 27 per cent in the experimental group (Shetty *et al.*, 2005a; b).

The aqueous extract (1 g/kg) of *Momordica charantia* seeds (MCSE) significantly lowered the blood glucose level in normal and diabetic mice. Moreover, MCSE primarily regulated the insulin signaling pathway in muscles and adipose tissues, suggesting that MCSE might target insulin receptor (IR), stimulate the IR-downstream pathway, and subsequently display hypo-glycemic activity in mice (Lo *et al.*, 2013).

A combination of extracts of bitter gourd and fenugreek when administered to alloxan induced diabetes rats for 30 days showed improvement in the fasting blood sugars along with protection of the heart, kidneys and liver tissue against oxidative stress (Tripathi and Chandra, 2009; 2010).

The water extracts of β-Glucan purified from oats (OG) and bitter melon (MC) have shown favorable effects on diabetes and its complications. In variable compositions the extracts were administered orally once a day for 28 days following 7 days post streptozotocin (STZ) dosing. The changes of hyper-glycemia, diabetic nephropathy, hepatopathy, and hyper-lipemia observed as the result of STZ-induced diabetes were dramatically decreased in the OG and MC single-dosing group, and all composition groups. Among variable compositions, the OG:MC 1:2 mixed group showed the most synergic effects (Kim *et al.,* 2012).

Diabetes mellitus is associated with an increase in sialic acid concentration along with other complications. Comparison of serum sialic acid concentration of patients, following bitter melon and rosiglitazone treatment revealed that bitter melon could be more effective in the management of diabetes and its related complications as compared to rosiglitazone (Inayat-ur-Rahman *et al.,* 2009). The seed varieties of bitter gourd were more effective than the glibenclamide, a known synthetic drug in normalizing the impaired oxidative stress in streptozotocin induced-diabetic rats (Sekar *et al.,* 2005). Bitter melon significantly reduced fructosamine levels from baseline among patients with type 2 diabetes who received 2,000 mg/day. However, the hypo-glycemic effect of bitter melon was less than metformin (1,000 mg/day) (Fuangchan *et al.,* 2011). Huang *et al.* (2008) showed that bitter melon can reduce insulin resistance as effective as the anti-diabetic drug thiazolidinedione. Similarly, when *Momordica charantia* was compared to metformin or glibenclamide, there was also no significant change in the parameters of glycaemic control (Li *et al.,* 2012).

Bitter melon may have additive effects when taken with other glucose-lowering agents (Basch *et al.,* 2003). Yadav *et al.* (2005) found that the fruit extract and sodium orthovanadate, a well-known insulin mimetic and an anti-diabetic compound exhibited hypo-lipidemic as well as hypo-glycemic effect in diabetic rats and their effect was pronounced when administered in combination. The reduction in blood glucose level and the increase in serum insulin level by bitter melon in the treated diabetic rats were similar to glibenclamide. In addition to insulin therapy and diet, diabetic dogs receiving 200 mg/kg BW of bitter gourd in capsule form orally twice a day for two months showed significant decrease in the serum fructosamine and fasting blood glucose concentrations (Kamoltip *et al.,* 2010).

The fruit extract alleviated pancreatic damage and increased the number of β-cells in the diabetic treated rats suggesting that oral feeding of the fruit extract may have a significant role in the renewal of pancreatic β-cells in the diabetic control group rats (Abdollahi *et al.,* 2011). Similarly, the ethanolic extract of *E. jambolana* seeds, water extract of *M. charantia* fruits, ethanol extract of *G. sylvestre* leaves, and water extract of fenugreek seeds have shown higher hypo-glycemic and anti-hyper-glycemic potential and may be used as complementary medicine to treat the diabetic population by significantly reducing the dosage of standard drug (Yadav *et al.,* 2010). Many more studies have conclusively stated the beneficial effects of bitter melon on diabetes and have also confirmed that it is either equal or better than the standard drugs in its hypo-glycemic effects.

Recently, a novel protein termed as ADMc1 identified and isolated from the seed extract of *M. charantia* was found to show significant anti-hyperglycemic activity in type 1 diabetic rats (Chhabra and Dixit, 2013). Bitter gourd extract has been found to be a potent wound healing agent. It was found to improve as well as accelerate the delayed wound healing process in rats induced by diabetes (Teoh *et al.,* 2009b). Transdermal patches made up of bitter gourd extract showed improvement in blood glucose parameters of experimental rats without any incidence of skin irritation or reaction (Bhujbal *et al.,* 2011).

Bitter gourd showed reno-protective effects in diabetic rats by causing a significant decrease in serum urea and creatinine levels (Abd El Sattar *et al.,* 2006). Diabetic nephropathy was prevented when bitter gourd was administered in combination with oats beta glucan in ratio of 1:2 to diabetic rats (Kim *et al.,* 2012). In streptozotocin induced diabetic rats, bitter gourd alleviated the increase in urine volume and urine sugar by about 30 per cent, prevented renal hypertrophy by 38 per cent, lowered the increased Glomerular Filtration Rate by 27 per cent and reduced fasting blood sugars by 30 per cent (Shetty *et al.,* 2005b). It also showed a beneficial role in controlling glycol-conjugate and heparan sulfate related kidney complications during diabetes, through significant reduction of glycol-conjugates.

Diabetes associated elevation in the activities of enzymes involved in the synthesis and degradation of glycosaminoglycans (GAGs) were significantly lowered by bitter gourd supplementation. GAGs composition revealed decrease in amino sugar and uronic acid contents during diabetes which were reverted with bitter gourd feeding. Decrease in sulfate content in the GAGs during diabetes was also ameliorated by bitter gourd feeding (Kumar *et al.,* 2008). Changes in the kidneys caused due to diabetes viz- thickening of the basement membrane of the Bowman's capsule, edema and hyper cellurarity of the proximal tubules, necrosis and hyaline deposits, were significantly reversed when bitter gourd extract was administered to diabetic rats. This reversal is mainly due to the anti-oxidant potential of bitter gourd (Teoh *et al.,* 2010).

Some new bioactive drugs and isolated compounds such as roseoside, epigallocatechin gallate, beta-pyrazol-1-ylalanine, cinchonain Ib, leucocyandin 3-O-beta-d-galactosyl cellobioside, leucopelargonidin-3- O-alpha-L rhamnoside, glycyrrhetinic acid, dehydrotrametenolic acid, strictinin, isostrictinin, pedunculagin, epicatechin and christinin from plants show significant insulin-mimetic and anti-diabetic activity with more efficacy than conventional hypo-glycaemic agents (Patel *et al.,* 2013). The anti-diabetic activity of medicinal plants is attributed to the presence of polyphenols, flavonoids, terpenoids, coumarins and other constituents which show reduction in blood glucose levels.

Anti-obesity Agent

Bitter gourd juice showed a beneficial effect on reducing the adiposity in humans by inhibiting lipogenesis and stimulating lipolysis (Nerurkar *et al.,* 2010). Bio-actives present in bitter gourd decreased the leptin levels (Shih *et al.,* 2008; Wang *et al.,* 2011) and body weight (Wang *et al.,* 2011), along with a significant reduction in the weight of epididymal white adipose tissue, visceral fat, adipose leptin and resistin mRNA levels in mice fed with a high fat diet (Shih *et al.,* 2008). Chan *et al.* (2005) demonstrated the anti-obesity effects of bitter gourd through its stimulatory action on the lipid oxidative enzyme activities and uncoupling protein expression along with a reduction in visceral fat mass.

Bitter gourd consumption along with high fat diet ameliorates the negative influence of dietary fat on the body composition. In addition, bitter gourd has also been reported to prevent contribution of dietary fat to body weight. Bitter gourd appears to have multiple influences on glucose and lipid metabolism that strongly counteract the untoward effects of a high fat diet. Chen *et al.* (2003) showed for the first time that simultaneous feeding of bitter gourd along with high fat diet results in less weight gain and visceral adiposity than those fed the HF diet alone. The mechanism of action of bitter gourd included significant improvement in the insulin resistance, serum insulin and leptin concentrations along with increased utilization of body fat as indicated by raised serum free fatty acid concentration. Bitter gourd suppresses the visceral fat accumulation and inhibits adipocyte hypertrophy, which may be associated with marked down-regulated expressions of lipogenic genes in the adipose tissue (Huang *et al.,* 2008).

Cardio-protective Effect

Acetone extract of bitter gourd when administered orally to alloxan induced diabetic rats lowered the blood sugar and serum cholesterol levels to normal range after 15 to 30 days without any increase even after 15 days of discontinuation of the extract (Singh *et al.,* 1989; Platel *et al.,* 1993). Bitter gourd also showed a significant decrease in triglyceride (Fernandes *et al.,* 2007; Ahmed *et al.,* 2001), low density lipoprotein and a significant increase in high density lipoprotein level in diabetic rats (Chaturvedi *et al.,* 2004; Fernandes *et al.,* 2007; Chaturvedi, 2005; Chaturvedi and George, 2010). A reduction in serum free fatty acid concentration was noticed with bitter gourd extract in mice fed a high fat diet. The hypo-lipidemic effect of bitter gourd might be due to its role in peroxisome proliferator activated receptor alpha mediated pathways (Shih *et al.,* 2008). Bitter gourd was also found to act in synergy with sodium ortho-vanadate, a hypo-lipidemic drug in experimental rats (Yadav *et al.,* 2005).

A favorable effect on blood pressure regulation was observed in rats (Clouatre *et al.,* 2011). In a human trial, a 3 month supplementation of 4.8 grams of lyophilized wild bitter gourd powder to Taiwanese adults suffering from metabolic syndrome reduced the incidence of the syndrome along with a reduction in the waist circumference, suggesting the role of wild bitter gourd in ameliorating metabolic syndrome (Chung-Huang *et al.,* 2012; Tsai *et al.,* 2012).

Even the high cholesterol diet when supplemented with bitter gourd fruit could bring down the serum cholesterol level significantly as compared to the one without bitter gourd. Significant increase in the levels of fecal total bile acid upon bitter gourd supplementation suggested inhibition of the re-absorption of bile acids into the intestine and thereby increased conversion of cholesterol to bile acids in the liver by cholesterol 7α-hydroxylase up-regulation (Matsui *et al.,* 2013).

Shih *et al.* (2009) demonstrated that bitter melon was effective in ameliorating the fructose diet-induced hyper-glycemia, hyper-leptinemia, hyper-insulinemia, and hyper-triglyceridemia as well as in decreasing the levels of free fatty acid.

Hepato-protective Agent

Oxidative stress along with muco-polysaccharide deposits in diabetes showed involvement of the hepatocytes with features of inflammation in diabetic rats, leading to liver damage. Extract of bitter gourd showed reversal of liver damage in these rats mainly due to its anti-oxidant potential (Teoh *et al.,* 2009a). It also showed hepato-protective effects in diabetic rats by decreasing the levels of alanine amino-transferase and aspartate amino-transferase. It showed a synergistic effect along with oats beta glucan in the ratio of 1:2 in protecting against diabetic hepatopathy (Kim *et al.,* 2012). Apart from decreasing the hepatic 11 beta hydroxysteroid dehydroxygenase (11beta-HSD1) gene expression, which contributed in attenuating diabetic state, bitter melon extract lowered serum triglycerides (TGs) by inhibition of hepatic fatty acid synthesis by dampening sterol response element binding protein 1c and fatty acid synthase mRNA (Shih *et al.,* 2013).

Anti-cancer Agent

Anti-cancer activity of *M. charantia* extracts have been demonstrated by numerous *in-vitro* and *in-vivo* studies. Anti-growth properties of fractionated *M. charantia* whole plant extracts were first reported by West *et al.* (1971). Subsequently, a number of growth inhibitors have been isolated from *M. charantia* seeds and their anti-proliferative activity has been demonstrated in a variety of tumor cell lines (Vesely *et al.,* 1977; Claflin *et al.,* 1978; Barbieri *et al.,* 1979; Akihisa *et al.,* 2007). Takemoto *et al.* (1982a and b) reported that a crude aqueous extract of the bitter melon (but not bitter melon seed) has cytostatic, cytotoxic and anti-viral activities, and is a competitive inhibitor of guanylate cyclase activity. This crude preparation kills human leukemic lymphocytes in a dose-dependent manner while not affecting the viability of normal human lymphocytes (Takemoto *et al.,* 1980). The purified factor from the extract is cytostatic for both BHK-21 and for the IM9 leukemic cell lines for at least 120 h. However, the cytostatic component had no effect on cellular cyclic GMP metabolism (Takemoto *et al.,* 1983).

Several fractions (cucurbitane-type triterpene glycosides) isolated from ethanolic extract of *Momordica charantia* fruits have been shown to be anti-proliferative against MCF-7, WiDr, HEp-2, and Doay human tumor cell lines (Hsiao *et al.,* 2013). Weng *et al.* (2013) reported through Luciferase reporter assays that 3 β,7 β -dihydroxy-25-methoxycucurbita-5,23-diene-19-al (DMC), a cucurbitane-type triterpene isolated from wild bitter gourd, induced apoptotic death in breast cancer cells through peroxisome proliferator-activated receptor (PPAR)-γ activation.

The bitter melon fruit extract and its components have also been shown to be cytotoxic to leukemic lymphocytes and induce anti-tumor activity *in-vivo* (Takemoto *et al.,* 1982; Jilka *et al.,* 1983). The human myeloid HL60 cells treated with the fractionated bitter melon seed extracts differentiated into granulocytic lineage as characterized by NBT staining, CD11b expression, and specific esterase activity suggesting the potential use of the fractionated bitter melon extracts in differentiation therapy for leukemia in combination with other inducers of differentiation (Soundararajan *et al.,* 2012).

Ru *et al.* (2011) observed that oral administration of bitter melon extract inhibits prostate cancer in transgenic adeno-carcinoma of mouse (TRAMP) by interfering with cell-cycle progression and proliferation. The anti-cancer activity in a rat colonic aberrant crypt foci model and a mouse mammary tumor model has also been reported (Kohno *et al.,* 2002; Nagasawa *et al.,* 2002).

The compounds charantagenins D (1) and goyaglycoside d with an -OMe substituent group in the side chain isolated from bitter melon exhibited significant cytotoxic activities against cancer cells. Impressively, the IC(50) values of the new compound 1 to lung cancer cell line A549, glioblastoma cell line U87, and hepatoma carcinoma cell line Hep3B were 1.07, 1.08, and 14.01 µmol/L, respectively, which were much lower than those of other tested compounds (Wang *et al.,* 2012).

Treatment of the human breast cancer bearing SCID mice with *Gelonium* protein of 31 kDa or *Momordica* protein of 30 kDa at 10 µg/injection, totally 10 injections resulted in significant increase in survival, with 20–25 per cent of the mice remaining tumor free for 96 days. Thus, bitter gourd is proved effective against human breast cancer MDA-MB-231 *in-vitro* and *in-vivo* (Lee-Huang *et al.,* 2000). Alpha-Momorcharin, a glycoprotein isolated from seeds of the bitter gourd, inhibited mouse monocyte-macrophage, Balb/c macrophage, JAR (human placental choriocarcinoma) and sarcoma S180 cell lines. Alpha-Momorcharin also enhanced the tumoricidal effect of mouse macrophages on mouse mastocytomal cells (Ng *et al.,* 1994). Another protein (inhibitor) and a hemagglutinin (lectin) purified from the seeds of *Momordica charantia* inhibited protein and subsequently DNA synthesis in normal (mitogen-stimulated) and leukemic human peripheral blood lymphocytes. Both proteins acted on lymphocytes more markedly at concentrations much lower than those required to inhibit protein synthesis in Yoshida ascites cells (Licastro *et al.,* 1980).

Thai bitter gourd fruits contain mono-functional phase II enzyme inducers as well as the ability to reduce Phase I enzyme activities in rat liver with some anti-carcinogens or chemo-preventive agents, and compounds capable of repressing some mono-oxygenases, especially those involved in the metabolic activation of chemical carcinogens (Kusamran *et al.,* 1998a; b).

Trypanosoma brucei is a fatal disease prevalent in sub-Saharan Africa and the few currently available drug treatments are facing problems of toxicity and resistance. Since bitter melon extracts displayed cytotoxic activity towards different cancer cell lines, the trypanocidal activity of bitter melon extract was investigated by Phillips *et al.* (2013). As it is known that agents exhibiting anti-tumour activity also usually inhibit the growth of *T. brucei,* the bloodstream forms of *T. brucei* with extracts prepared from Chinese and Indian bitter melon varieties were used for the treatment. A decrease in cell proliferation revealed that bitter melon is a promising source for trypanocidal agents which could be used in the development of novel anti-sleeping sickness drugs.

Anti-ulcerogenic Agent

The olive oil extract as well as dried and powdered bitter gourd fruits mixed with filtered honey showed significant and dose-dependent anti-ulcerogenic activity against ethanol-induced ulcerogenesis model in rats. The ethanol extract of the fruits showed significant activity against HCl-EtOH induced ulcerogenesis in indomethacin-pretreated rats and diethyl-dithiocarbamate-induced ulcer models (Gürbüz *et al.,* 2000).

Groups of rats were fed the oily extract (5 and 10 mg/kg) of bitter gourd fruits with olive oil (5 and 10 ml/kg) as a vehicle, and compared the effect with distilled water 5 ml/kg and famotidine (40 mg/kg). According to polymorphonuclear leukocytes infiltration, oily extract (10 ml/kg) and vehicle (10 ml/kg) had similar effects to famotidine. The ulcer inhibition rates were in the order of: 98.04 per cent, for oily extract (10 ml/kg) 91.54 per cent for famotidine, 88.02 per cent, for vehicle (olive oil 10 ml/kg) 53.80 per cent for oily extract (5 ml/kg), and 18.40 per cent for vehicle (olive oil 5 ml/kg) (Ozbaki° and Gürsan, 2005).

Bitter gourd extract (100 mg/kg and 500 mg/kg) healed acetic acid induced gastric ulcer in pylorusligated rats. The extract showed significant decrease in ulcer index, total acidity, free acidity and pepsin content, and an increase in gastric mucosal content. The extract also reduced the ulcer index in stress induced, ethanol induced and indomethacin induced gastric ulcers and cysteamine induced duodenal ulcer. Thus, the methanolic extract of bitter gourd fruit increases healing of gastric ulcer and also prevents development of gastric and duodenal ulcers in rats (Alam *et al.,* 2009).

Anti-fertility Agent

The petroleum ether, benzene and alcohol extracts of the seeds of *Momordica charantia* showed anti-spermatogenic activity (decreased number of spermatocytes, spermatids and spermatozoa) in rats at the dose level of 25 mg/100 g body weight for 35 days. Out of the three extracts, the alcohol extract was more potent (Naseem *et al.,* 1998).

Similarly, a dose-related alteration in the cyto-architecture of seminiferous tubules with marked reduction in spermatogenic series was observed. The prostate gland showed dilatation as well as increased intra-luminal secretions with increasing dose indicating that methanolic extract caused reversible histological alterations in the prostate and testes of Sprague-Dawley rats (Boetse *et al.,* 2011).

Other Benefits

Malik *et al.* (2011) reported that bitter gourd has potent neuro-protective activity against global cerebral ischemia-reperfusion induced neuronal injury and consequent neurological deficits in diabetic mice. Snee *et al.* (2011) suggested that incorporating variety of foods along with bitter gourd in commonly consumed recipes can mask bitter taste of bitter melon. Furthermore, providing positive health information can elicit a change in the intent to consume bitter melon-containing dishes despite mixed palatability results.

Safety

Bitter gourd has the potential to become a component of the diet or a dietary supplement for diabetic and pre-diabetic patients. According to Virdi *et al.* (2003) the aqueous extract powder of fresh unripe whole fruits did not show any signs of nephro-toxicity and hepato-toxicity as judged by histological and biochemical parameters. Thus the aqueous extract powder of *M. charantia* appears to be a safe alternative to reduce blood glucose.

Conclusions

Though the therapeutic inhibition of 11β-HSD1 reductase activity has been extensively studied in diabetes, its application in patients with obesity, metabolic syndrome, glaucoma and osteoporosis remains an exciting prospect for research (Tomlinson *et al.,* 2004). It is important to note that the seeds of bitter gourd also seem to possess good therapeutic value indicating a strong need to educate people to consume the bitter melon with the seeds.

References

Abd El Sattar, El Batran S, El-Gengaihi SE, El Shabrawy OA (2006). Some toxicological studies of *Momordica charantia* L. on albino rats in normal and alloxan diabetic rats. *J. Ethnopharmacol.* 108 (2): 236–42.

Abdollahi M, Zuki AB, Goh YM, Rezaeizadeh A, Noordin MM (2011). Effects of *Momordica charantia* on pancreatic histopathological changes associated with streptozotocin-induced diabetes in neonatal rats. *Histol. Histopathol.* 26 (1): 13–21.

Ahmad N, Hassan MR, Halder H, Bennoor KS (1999). Effect of *Momordica charantia* (Karolla) extracts on fasting and postprandial serum glucose levels in NIDDM patients. *Bangladesh Med. Res. Counc. Bull.* 25 (1): 11–13.

Ahmed I, Adeghate E, Cummings E, Sharma AK, Singh J (2004). Beneficial effects and mechanism of action of *Momordica charantia* juice in the treatment of streptozotocin-induced diabetes mellitus in rat. *Mol. Cell. Biochem.* 261 (1-2): 63–70.

Ahmed I, Adeghate E, Sharma AK, Pallot DJ, Singh J (1998). Effects of *Momordica charantia* fruit juice on islet morphology in the pancreas of the streptozotocin-diabetic rat. *Diab. Res. Clin. Pract.* 40 (3): 145–51.

Ahmed I, Lakhani MS, Gillett M, John A, Raza H (2001). Hypo-triglyceridemic and hypo-cholesterolemic effects of anti-diabetic *Momordica charantia* (karela) fruit extract in streptozotocin-induced diabetic rats. *Diab. Res. Clin. Pract.* 51 (3): 155–61.

Akihisa T, Higo N, Tokuda H, *et al.* (2007). Cucurbitane-type triterpenoids from the fruits of *Momordica charantia* and their cancer chemo-preventive effects. *Journal of Natural Products.* 70 (8): 1233–39.

Alam S, Asad M, Asdaq SM, Prasad VS (2009). Anti-ulcer activity of methanolic extract of *Momordica charantia* L. in rats. *J. Ethnopharmacol.* 123 (3): 464–69.

Baby Joseph, Jini D (2013). Anti-diabetic effects of *Momordica charantia* (bitter melon) and its medicinal potency. *Asian Pac J. Trop. Dis.* 3 (2): 93 – 102.

Baldwa VS, Bhandari CM, Pangaria A, Goyal RK (1977). Clinical trial in patients with diabetes mellitus of an insulin-like compound obtained from plant source. *Ups. J. Med. Sci.* 82 (1): 39–41.

Barbieri L, Lorenzoni E, Stirpe F (1979). Inhibition of protein synthesis *in-vitro* by a lectin from *Momordica charantia* and by other haemagglutinins. *Biochemical Journal.* 182 (2): 633–35.

Basch E, Gabardi S, Ulbricht C (2003). Bitter melon (*Momordica charantia*): a review of efficacy and safety. *Am. J. Health Syst. Pharm.* 60 (4): 356–59.

Bhujbal SS, Hadawale SS, Kulkarni PA, Bidkar JS, Thatte VA, Providencia CA, Yeola RR (2011). A novel herbal formulation in the management of diabetes. *Int. J. Pharm. Investig.* 1 (4): 222–26.

Blum A, Loerz C, Martin HJ, Staab-Weijnitz CA, Maser E (2012). *Momordica charantia* extract, a herbal remedy for type 2 diabetes, contains a specific 11 β-hydroxysteroid dehydrogenase type 1 inhibitor. *J. Steroid. Biochem. Mol. Biol.* 128 (1-2): 51–55.

Boetse YO, Ikechukwu DF, Olugbenga OA, Ayodele OA, Caramel NC (2011). Histomorphological alterations in the prostate gland and epithelium of seminiferous tubule of sprague-dawley rats treated with methanolic extract of *Momordica charantia* seeds. *Iran. J. Med. Sci.* 36 (4): 266–72.

Bourinbaiar AS, Lee-Huang S (1995). Potentiation of anti-HIV activity of anti-inflammatory drugs, dexamethasone and indomethacin, by MAP30, the antiviral agent from bitter melon. *Biochem. Biophys. Res. Commun.* 208 (2):

Chhabra G, Dixit A (2013). Structure modeling and antidiabetic activity of a seed protein of *Momordica charantia* in non-obese diabetic (NOD) mice. *Bioinformation.* 9 (15): 766–70.

Chan LL, Chen Q, Go AG, Lam EK, Li ET (2005). Reduced adiposity in bitter melon (*Momordica charantia*)-fed rats is associated with increased lipid oxidative enzyme activities and uncoupling protein expression. *J. Nutr.* 135 (11): 2517–23.

Chaturvedi P (2005). Role of *Momordica charantia* in maintaining the normal levels of lipids and glucose in diabetic rats fed a high-fat and low-carbohydrate diet. *Br. J. Biomed. Sci.* 62 (3): 124–26.

Chaturvedi P (2012). Anti-diabetic potentials of *Momordica charantia:* multiple mechanisms behind the effects. *J. Med. Food.* 15 (2): 101–107.

Chaturvedi P, George S (2010). *Momordica charantia* maintains normal glucose levels and lipid profiles and prevents oxidative stress in diabetic rats subjected to chronic sucrose load. *J. Med. Food.* 13 (3): 520–27.

Chaturvedi P, George S, Milinganyo M, Tripathi YB (2004). Effect of *Momordica charantia* on lipid profile and oral glucose tolerance in diabetic rats. *Phytother. Res.* 18 (11): 954–56.

Chen Q, Chan LL, Li ET (2003). Bitter melon (*Momordica charantia*) reduces adiposity, lowers serum insulin and normalizes glucose tolerance in rats fed a high fat diet. *J. Nutr.* 133 (4): 1088–93.

Choudhary SK, Chhabra G, Sharma D, Vashishta A, Ohri S, Dixit A (2012). Comprehensive evaluation of anti-hyperglycemic activity of fractionated *Momordica charantia* seed extract in alloxan-induced diabetic rats. Evid. Based Complement. *Alternat. Med.* 2012: 293650.

Chung-Huang T, Emily CC, Hsin-Sheng TH, Ching-jang (2012). Wild bitter gourd improves metabolic syndrome: a preliminary dietary supplementation trial. *J. Nutr.* 11: 4–13.

Claflin AJ, Vesely DL, Hudson JL (1978). Inhibition of growth and guanylate cyclase activity of an undifferentiated prostate adenocarcinoma by an extract of the balsam pear (*Momordica charantia* abbreviata) *Proceedings of the National Academy of Sciences of the United States of America.* 75 (2): 989–93.

Clouatre DL, Rao SN, Preuss HG (2011). Bitter melon extracts in diabetic and normal rats favorably influence blood glucose and blood pressure regulation. *J. Med. Food.* 14 (12): 1496–504.

Cummings E, Hundal HS, Wackerhage H, Hope M, Belle M, Adeghate E, Singh J (2004). *Momordica charantia* fruit juice stimulates glucose and amino acid uptakes in L6 myotubes. *Mol. Cell. Biochem.* 261 (1-2): 99–104.

Day C (1990). Hypo-glycemic compounds from plants. In: *New Anti-diabetic Drugs.* CJ Bailey and PR Flatt, editors. Smith-Gordon. London: pp. 267–78.

Day C, Cartwright T, Provost J, Bailey CJ (1990). Hypo-glycaemic effect of *Momordica charantia* extracts. *Planta. Medica.* 56 (5): 426–29.

Fan JM, Zhang Q, Xu J, Zhu S, Ke T, Gao de F, Xu YB (2009). Inhibition on Hepatitis B virus in vitro of recombinant MAP30 from bitter melon. *Mol. Biol. Rep.* 36 (2): 381–88.

Fernandes NP, Lagishetty CV, Panda VS, Naik SR (2007). An experimental evaluation of the anti-diabetic and anti-lipidemic properties of a standardized *Momordica charantia* fruit extract. *BMC Complementary and Alternative Medicine.* 7: 29.

Fuangchan A, Sonthisombat P, Seubnukarn T, Chanouan R, Chotchaisuwat P, Sirigulsatien V, Ingkaninan K, Plianbangchang P, Haines ST (2011). Hypo-glycemic effect of bitter melon compared with metformin in newly diagnosed type 2 diabetes patients. *J. Ethnopharmacol.* 134 (2): 422–28.

Grover JK, Yadav SP (2004). Pharmacological actions and potential uses of *Momordica charantia*: a review. *Journal of Ethnopharmacology.* 93 (1): 123–32.

Gürbüz I, Akyüz C, Ye°ilada E, Sener B (2000). Anti-ulcerogenic effect of *Momordica charantia* L. fruits on various ulcer models in rats. *J. Ethnopharmacol.* 71 (1-2): 77–82.

Hafizur RM, Kabir N, Chishti S (2011). Modulation of pancreatic β-cells in neonatally streptozotocin-induced type 2 diabetic rats by the ethanolic extract of *Momordica charantia* fruit pulp. *Nat. Prod. Res.* 25 (4): 353 – 67.

Hazarika R, Parida P, Neog B, Yadav RN (2012). Binding Energy calculation of GSK-3 protein of human against some anti-diabetic compounds of *Momordica charantia* Linn. (Bitter melon). *Bioinformation.* 8 (6): 251–54.

Hsiao PC, Liaw CC, Hwang SY, Cheng HL, Zhang LJ, Shen CC, Hsu FL, Kuo YH (2013). Anti-proliferative and hypo-glycemic cucurbitane-type glycosides from the fruits of *Momordica charantia. J. Agric. Food Chem*. 61 (12): 2979 – 86.

Huang HL, Hong YW, Wong YH, Chen YN, Chyuan JH, Huang CJ, Chao PM (2008). Bitter melon (*Momordica charantia* L.) inhibits adipocyte hypertrophy and down regulates lipogenic gene expression in adipose tissue of diet-induced obese rats. *Br. J. Nutr*. 99 (2): 230–39.

Huang TN, Lu KN, Pai YP, Chin Hsu, Huang CJ (2013). Role of GLP-1 in the Hypo-glycemic Effects of Wild Bitter Gourd. Evid. Based Complement. *Alternat. Med*. 2013: 625892.

Inayat-ur-Rahman, Malik SA, Bashir M, Khan R, Iqbal M (2009). Serum sialic acid changes in non-insulin-dependant diabetes mellitus (NIDDM) patients following bitter melon (*Momordica charantia*) and *rosiglitazone* (Avandia) treatment. *Phytomedicine*. 16 (5): 401–405.

Jilka C, Strifler B, Fortner GW (1983). *In-vivo* anti-tumor activity of the bitter melon (*Momordica charantia*). *Cancer Research*. 43 (11): 5151–55.

Jiratchariyakul W, Wiwat C, Vongsakul M, Somanabandhu A, Leelamanit W, Fujii I, Suwannaroj N, Ebizuka Y (2001). HIV inhibitor from Thai bitter gourd. *Planta. Med*. 67 (4): 350–53.

Kamoltip T, Pinit P, Kamonwan F, Sirintorn Y (2010). Treatment of canine diabetes mellitus using *Momordica charantia* capsule and a restricted-fat high-fibre diet. *J. Med. Plants. Res*. 4 (21): 2243–51.

Kar A, Choudhary BK, Bandyopadhyay NG (2003). Comparative evaluation of hypo-glycaemic activity of some Indian medicinal plants in alloxan diabetic rats. *Journal of Ethno-pharmacology*. 84 (1): 105–108.

Keller AC, Ma J, Kavalier A, He K, Brillantes AM, Kennelly EJ (2011). Saponins from the traditional medicinal plant *Momordica charantia* stimulate insulin secretion *in-vitro*. *Phytomedicine*. 19 (1): 32–37.

Kenny O, Smyth TJ, Hewage CM, Brunton NP (2013). Anti-oxidant properties and quantitative UPLC-MS analysis of phenolic compounds from extracts of fenugreek (*Trigonella foenum-graecum*) seeds and bitter melon (*Momordica charantia*) fruit. *Food Chem*. 141 (4): 4295–302.

Khanna P, Jain SC, Panagariya A, Dixit VP (1981). Hypo-glycemic activity of polypeptide- p from a plant source. *J. Nat. Prod*. 44: 648–55.

Kim JW, Cho HR, Moon SB, Kim KY, Ku S (2012). Synergic effects of bitter melon and β-Glucan composition on STZ-induced rat diabetes and its complications. *J. Microbiol. Biotechnol*. 22 (1): 147–55.

Kohno H, Suzuki R, Noguchi R, Hosokawa M, Miyashita K, Tanaka T (2002). Dietary conjugated linolenic acid inhibits azoxymethane-induced colonic aberrant crypt foci in rats. *Jpn. J. Cancer Res*. 93: 133–42.

Krawinkel MB, Keding GB (2006). Bitter gourd (*Momordica Charantia*): A dietary approach to hyperglycemia. *Nutr. Rev.* 64 (7 Pt 1): 331–37.

Kumar R, Balaji S, Uma TS, Sehgal PK (2009). Fruit extracts of *Momordica charantia* potentiate glucose uptake and up-regulate Glut-4, PPAR gamma and PI3K. *J. Ethnopharmacol.* 126 (3): 533–37.

Kumar GS, Shetty AK, Salimath PV (2008). Modulatory effect of bitter gourd (Momordica charantia LINN.) on alterations in kidney heparan sulfate in streptozotocin-induced diabetic rats. *J. Ethnopharmacol.* 115 (2): 276–83.

Kusamran WR, Ratanavila A, Tepsuwan A (1998a). Effects of neem flowers, Thai and Chinese bitter gourd fruits and sweet basil leaves on hepatic mono-oxygenases and glutathione S-transferase activities, and *in-vitro* metabolic activation of chemical carcinogens in rats. *Food Chem. Toxicol.* 36 (6): 475–84.

Kusamran WR, Tepsuwan A, Kupradinun P (1998b). Anti-mutagenic and anti-carcinogenic potentials of some Thai vegetables. *Mutat. Res.* 402 (1-2): 247–58.

Leatherdale BA, Panesar RK, Singh G, Atkins TW, Bailey CJ, Bignell AH (1981). Improvement in glucose tolerance due to *Momordica charantia* (karela). *Br. Med. J. (Clin. Res. Ed.).* 282 (6279): 1823–24.

Lee-Huang S, Huang PL, Sun Y, Chen HC, Kung HF, Huang PL, Murphy WJ (2000). Inhibition of MDA-MB-231 human breast tumor xenografts and HER2 expression by anti-tumor agents GAP31 and MAP30. *Anti-cancer Research.* 20 (2A): 653–59.

Lee-Huang S, Huang PL, Chen HC, Bourinbaiar A, Huang HI, Kung HF (1995). Anti-HIV and anti-tumor activities of recombinant MAP30 from bitter melon. *Gene.*161 (2): 151–56.

Leung L, Birtwhistle R, Kotecha J, Hannah S, Cuthbertson S (2009). Anti-diabetic and hypo-glycaemic effects of *Momordica charantia* (bitter melon): a mini review. *Br. J. Nutr.* 102 (12): 1703–708.

Li CJ, Tsang SF, Tsai CH, Tsai HY, Chyuan JH, Hsu HY (2012). *Momordica charantia* Extract Induces Apoptosis in Human Cancer Cells through Caspase- and Mitochondria-Dependent Pathways. Evid. Based Complement. *Alternat. Med.* 2012: 261971.

Licastro F, Franceschi C, Barbieri L, Stirpe F (1980). Toxicity of *Momordica charantia* lectin and inhibitor for human normal and leukaemic lymphocytes. Virchows. *Arch. B Cell. Pathol. Incl. Mol. Pathol.* 33 (3): 257–65.

Liu X, Chen T, Hu Y, Li K, Yan L (2014). Catalytic synthesis and antioxidant activity of sulfated polysaccharide from *Momordica charantia* L. *Biopolymers.* 101 (3): 210–15.

Lo HY, Ho TY, Lin C, Li CC, Hsiang CY (2013). *Momordica charantia* and Its Novel Polypeptide Regulate Glucose Homeostasis in Mice via Binding to Insulin Receptor. *J Agric Food Chem.* 2013 Feb 27. [Epub ahead of print].

Malik ZA, Singh M, Sharma PL (2011). Neuro-protective effect of *Momordica charantia* in global cerebral ischemia and reperfusion induced neuronal damage in diabetic mice. *J. Ethnopharmacol.* 133 (2): 729–34.

Matsui S, Yamane T, Takita T, Oishi Y, Kobayashi-Hattori K (2013). The hypocholesterolemic activity of *Momordica charantia* fruit is mediated by the altered cholesterol- and bile acid-regulating gene expression in rat liver. *Nutr. Res.* 33 (7): 580–85.

Naseem MZ, Patil SR, Patil SR, Ravindra, Patil RS (1998). Anti-spermatogenic and androgenic activities of *Momordica charantia* (Karela) in albino rats. *J. Ethnopharmacol.* 61 (1): 9–16.

Nerurkar PV, Lee YK, Linden EH, Lim S, Pearson L, Frank J, Nerurkar VR (2006). Lipid lowering effects of *Momordica charantia* (Bitter Melon) in HIV-1-protease inhibitor-treated human hepatoma cells, HepG2. *Br. J. Pharmacol.* 148 (8): 1156–64.

Ng TB, Liu WK, Sze SF, Yeung HW (1994). Action of alpha-momorcharin, a ribosome inactivating protein, on cultured tumor cell lines. *Gen Pharmacol.* 25 (1): 75–77.

Nagasawa H, Watanabe K, Inatomi H (2002). Effects of bitter melon (*Momordica charantia* l.) or ginger rhizome (*Zingiber offifinale* rosc) on spontaneous mammary tumorigenesis in SHN mice. *Am. J. Chin. Med.* 30: 195–205.

Ozbaki° Dengiz G, Gürsan N (2005). Effects of *Momordica charantia* L. (Cucurbitaceae) on indomethacin-induced ulcer model in rats. Turk. *J. Gastroenterol.* 16 (2): 85–88.

Ocvirk S, Kistler M, Khan S, Talukder SH, Hauner H (2013). Traditional medicinal plants used for the treatment of diabetes in rural and urban areas of Dhaka, Bangladesh–an ethno-botanical survey. *J. Ethnobiol. Ethnomed.* 9: 43.

Patel D, Prasad S, Kumar R, Hemalatha S (2012). An overview on anti-diabetic medicinal plants having insulin mimetic property. *Asian Pac. J. Trop. Biomed.* 2 (4): 320–30.

Perla V, Jayanty SS (2013). Biguanide related compounds in traditional anti-diabetic functional foods. *Food Chem.* 138 (2-3): 1574–80.

Phillips EA, Sexton DW, Steverding D (2013). Bitter melon extract inhibits proliferation of *Trypanosoma brucei* bloodstream forms in vitro. *Exp Parasitol.* 133 (3): 353–56.

Platel K, Shurpalekar KS, Srinivasan K (1993). Influence of bitter gourd (*Momordica charantia*) on growth and blood constituents in albino rats. *Nahrung.* 37 (2): 156–60.

Pongthanapisith V, Ikuta K, Puthavathana P, Leelamanit W (2013). Anti-viral protein of *Momordica charantia* L. inhibits different subtypes of influenza A. Evid. Based Complement. *Alternat. Med.* 2013: 729081.

Roffey BW, Atwal AS, Johns T, Kubow S (2007). Water extracts from *Momordica charantia* increase glucose uptake and adiponectin secretion in 3T3-L1 adipose cells. *J. Ethnopharmacol.* 112 (1): 77–84.

Ru P, Steele R, Nerurkar PV, Phillips N, Ray RB (2011). Bitter melon extract impairs prostate cancer cell-cycle progression and delays prostatic intraepithelial neoplasia in TRAMP model. *Cancer Prev. Res. (Phila).* 4 (12): 2122–30.

Sarkar S, Pranava M, Marita RA (1996). Demonstration of the hypo-glycemic action of *Momordica charantia* in a validated animal model of diabetes. *Pharmacological Research.* 33 (1): 1–4.

Sekar SD, Subramanian S (2005a). Beneficial effects of *Momordica charantia* seeds in the treatment of STZ-induced diabetes in experimental rats. *Biol. Pharm. Bull.* 28 (6): 978–83.

Sekar SD, Subramanian S (2005b). Anti-oxidant properties of *Momordica Charantia* (bitter gourd) seeds on Streptozotocin induced diabetic rats. *Asia. Pac. J. Clin. Nutr.* 14 (2): 153–58.

Sekar DS, Sivagnanam K, Subramanian S (2005). Anti-diabetic activity of *Momordica charantia* seeds on streptozotocin induced diabetic rats. *Pharmazie.* 60 (5): 383–87.

Shetty AK, Kumar GS, Sambaiah K, Salimath PV (2005a). Effect of bitter gourd (*Momordica charantia*) on glycaemic status in streptozotocin induced diabetic rats. *Plant Foods Hum. Nutr.* 60 (3): 109–12.

Shetty AK, Kumar GS, Salimath PV (2005b). Bitter gourd (*Momordica charantia*) modulates activities of intestinal and renal disaccharidases in streptozotocin-induced diabetic rats. *Mol. Nutr. Food Res.* 49 (8): 791–96.

Shibib BA, Khan LA, Rahman R (1993). Hypo-glycemic activity of *Coccinia indica* and *Momordica charantia* in diabetic rats: depression of the hepatic gluconeogenic enzymes glucose-6-phosphatase and fructose-1, 6-biphosphatase and elevation of liver and red-cell shunt enzyme glucose-6-phosphate dehydrogenase. *Biochem. J.* 292: 267–70.

Shih CC, Lin CH, Lin WL, Wu JB (2009). *Momordica charantia* extract on insulin resistance and the skeletal muscle GLUT4 protein in fructose-fed rats. *J. Ethnopharmacol.* 123 (1): 82–90.

Shih CC, Lin CH, Lin WL (2008). Effects of *Momordica charantia* on insulin resistance and visceral obesity in mice on high-fat diet. *Diabetes Res. Clin. Pract.* 81 (2): 134–43.

Shih CC, Shlau MT, Lin CH, Wu JB (2013). *Momordica charantia* Ameliorates Insulin Resistance and Dyslipidemia with Altered Hepatic Glucose Production and Fatty Acid Synthesis and AMPK Phosphorylation in High-fat-fed Mice. *Phytother Res.* 2013 Apr 23.

Singh N, Gupta M (2007). Regeneration of beta cells in islets of Langerhans of pancreas of alloxan diabetic rats by acetone extract of *Momordica charantia* (Linn.) (bitter gourd) fruits. *Indian J. Exp. Biol.* 45 (12): 1055–62.

Singh N, Gupta M, Sirohi P, Varsha (2008). Effects of alcoholic extract of *Momordica charantia* (Linn.) whole fruit powder on the pancreatic islets of alloxan diabetic albino rats. *J. Environ. Biol.* 29 (1): 101–06.

Singh N, Tyagi SD, Agarwal SC (1989). Effects of long term feeding of acetone extract of *Momordica charantia* (whole fruit powder) on alloxan diabetic albino rats. *Indian J. Physiol. Pharmacol*. 33 (2): 97–100.

Sitasawad SL, Shewade Y, Bhonde R (2000). Role of bitter gourd fruit juice in stz-induced diabetic state *in-vivo* and *in-vitro*. *J. Ethnopharmacol*. 73 (1-2): 71–79.

Snee LS, Nerurkar VR, Dooley DA, Efird JT, Shovic AC, Nerurkar PV (2011). Strategies to improve palatability and increase consumption intentions for *Momordica charantia* (bitter melon): a vegetable commonly used for diabetes management. *Nutr. J*. 10: 78.

Soundararajan R, Prabha P, Rai U, Dixit A (2012). Anti-leukemic Potential of *Momordica charantia* seed extracts on human myeloid leukemic HL60 cells. Evid. Based Complement. *Alternat. Med*. 2012: 732404.

Sridhar MG, Vinayagamoorthi R, Arul Suyambunathan V, Bobby Z, Selvaraj N (2008). Bitter gourd (*Momordica charantia*) improves insulin sensitivity by increasing skeletal muscle insulin-stimulated IRS-1 tyrosine phosphorylation in high-fat-fed rats. *Br. J. Nutr*. 99 (4): 806–812.

Takemoto DJ, Dunford C, McMurray MM (1982). The cytotoxic and cytostatic effects of the bitter melon (*Momordica charantia*) on human lymphocytes. *Toxicon*. 20 (3): 593–99.

Takemoto DJ, Kresie R, Vaughn D (1980). Partial purification and characterization of a guanylate cyclase inhibitor with cytotoxic properties from the bitter melon (*Momordica charantia*). *Biochem. Biophys. Res. Commun*. 94 (1): 332–39.

Teoh SL, Abd Latiff A, Das S (2010). Histological changes in the kidneys of experimental diabetic rats fed with *Momordica charantia* (bitter gourd) extract. Rom. *J. Morphol. Embryol*. 51 (1): 91–95.

Teoh SL, Latiff AA, Das S (2009b). A histological study of the structural changes in the liver of streptozotocin-induced diabetic rats treated with or without *Momordica charantia* (bitter gourd). *Clin. Ter*. 160 (4): 283–86.

Teoh SL, Latiff AA, Das S (2009a). The effect of topical extract of *Momordica charantia* (bitter gourd) on wound healing in non-diabetic rats and in rats with diabetes induced by streptozotocin. *Clin. Exp. Dermatol*. 34 (7): 815–22.

Tomlinson JW, Walker EA, Bujalska IJ, Draper N, Lavery GG, Cooper MS, Hewison M, Stewart PM (2004). 11β-Hydroxysteroid dehydrogenase Type 1: A tissue-specific regulator of glucocorticoid response. *Endocrine Reviews*. 25 (5): 831–66.

Tripathi UN, Chandra D (2009). The plant extracts of *Momordica charantia* and *Trigonella foenum-graecum* have anti-oxidant and anti-hyperglycemic properties for cardiac tissue during diabetes mellitus. *Oxid. Med. Cell. Longev*. 2 (5): 290–96.

Tripathi UN, Chandra D (2010). Anti-hyperglycemic and anti-oxidative effect of aqueous extract of *Momordica charantia* pulp and *Trigonella foenum graecum* seed in alloxan-induced diabetic rats. *Indian J. Biochem. Biophys*. 47 (4): 227–33.

Tsai CH, Chen EC, Tsay HS, Huang CJ (2012). Wild bitter gourd improves metabolic syndrome: a preliminary dietary supplementation trial. *Nutr. J.* 11: 4.

Uebanso T, Arai H, Taketani Y, Fukaya M, Yamamoto H, Mizuno A, Uryu K, Hada T, Takeda E (2007). Extracts of *Momordica charantia* suppress postprandial hyperglycemia in rats. *Journal of Nutritional Science and Vitaminology.* 53 (6): 482–88.

Vesely DL, Graves WR, Lo TM (1977). Isolation of a guanylate cyclase inhibitor from the balsam pear (*Momordica charantia* abreviata). *Biochemical and Biophysical Research Communications.* 77 (4): 1294–99.

Virdi J, Sivakami S, Shahani S, Suthar AC, Banavalikar MM, Biyani MK (2003). Antihyperglycemic effects of three extracts from *Momordica charantia. J. Ethnopharmacol.* 88 (1): 107–111.

Wang X, Sun W, Cao J, Qu H, Bi X, Zhao Y (2012). Structures of new triterpenoids and cytotoxicity activities of the isolated major compounds from the fruit of *Momordica charantia* L. *Journal of Agricultural and Food Chemistry.* 60: 3927–33.

Wang ZQ, Zhang XH, Yu Y, Poulev A, Ribnicky D, Floyd ZE, Cefalu WT (2011). Bioactives from bitter melon enhance insulin signaling and modulate acyl carnitine content in skeletal muscle in high-fat diet-fed mice. *J. Nutr. Biochem.* 22 (11): 1064–73.

Welihinda J, Karunanayake EH, Sheriff MH, Jayasinghe KS (1986). Effect of *Momordica charantia* on the glucose tolerance in maturity onset diabetes. *J. Ethnopharmacol.* 17 (3): 277–82.

Weng JR, Bai LY, Chiu CF, Hu JL, Chiu SJ, Wu CY (2013). Cucurbitane Triterpenoid from *Momordica charantia* Induces Apoptosis and Autophagy in Breast Cancer Cells, in Part, through Peroxisome Proliferator-Activated Receptor γ Activation. Evid. Based Complement. *Alternat. Med.* 2013: 935675.

West ME, Sidrak GH, Street SP (1971). The anti-growth properties of extracts from *Momordica charantia* L. *West Indian Medical Journal.* 20 (1): 25–34.

Yadav M, Lavania A, Tomar R, Prasad GB, Jain S, Yadav H.(2010). Complementary and comparative study on hypoglycemic and anti-hyperglycemic activity of various extracts of *Eugenia jambolana* seed, *Momordica charantia* fruits, *Gymnema sylvestre*, and *Trigonella foenum graecum* seeds in rats. *Appl. Biochem. Biotechnol.* 160 (8): 2388–400.

Yadav UC, Moorthy K, Baquer NZ (2005). Combined treatment of sodium orthovanadate and *Momordica charantia* fruit extract prevents alterations in lipid profile and lipogenic enzymes in alloxan diabetic rats. *Mol. Cell. Biochem.* 268 (1-2): 111–20.

Yibchok-anun S, Adisakwattana S, Yao CY, Sangvanich P, Roengsumran S, Hsu WH (2006). Slow acting protein extract from fruit pulp of *Momordica charantia* with insulin secretagogue and insulinomimetic activities. *Biol. Pharm. Bull.* 29 (6): 1126–31.

Zheng ZX, Teng JY, Liu JY, Qiu JH, Ouyang H, Xue C (2005). The hypo-glycemic effects of crude polysaccharides extract from *Momordica charantia* in mice. *Wei. Sheng. Yan. Jiu.* 34 (3): 361–63. [Article in Chinese].

Chapter–5

Black Pepper
(*Piper nigrum/Piper longum*)

Since ancient times Black pepper is generally considered as "king of spice" world over. It belongs to the family *Piperaceae*. The plant is native to tropical evergreen rain forest of South Indian state, Kerala, from where it spread to rest of the world. The pepper fruit, also known as *peppercorn*, is actually a berry obtained from the plant. Dried ground pepper has been used since antiquity for both its flavour and as a medicine. Currently Vietnam is the world's largest producer and exporter of pepper, producing 34 per cent of the world's *Piper nigrum* crop as of 2008.

In Malayalam, it is called "kuru mulagu" (seed chilli/pepper) and "nalla mulagu" (good chilli/pepper); in Tulu, it is "edde munchi" (good chilli/pepper) and in Hindi it is "kaali mirch" (black chilli/pepper).

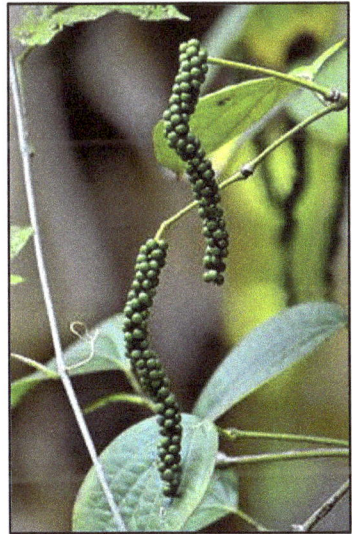

Unripe Drupes of Black Pepper

Black pepper (cooked and dried unripe fruit) is extensively cultivated in south India and other tropical regions. Black pepper is produced from the still-green unripe drupes of the pepper plant. The drupes are cooked briefly in hot water, both to clean them and to prepare them for drying. The heat ruptures cell walls in the pepper, speeding browning enzymes during drying. The drupes are then dried in the sun or by machine for several days, during which the peel around the seed shrinks and darkens into a thin, wrinkled black layer. Once dried, the spice is called black peppercorn. On some estates, the berries are separated from the stem by hand and then sun-dried without the boiling process.

There are different types of black pepper:

- ☆ **Tellicherry Black pepper** from South India–Very bold and pungent aroma. It is considered the best variety cultivated in India.
- ☆ **Sarawak Black** from Borneo and Malaysia–Relatively mild with a fruity bouquet.
- ☆ **Lampong Black** from Sumatra, Indonesia–Mildly pungent and aromatic.
- ☆ **White pepper (unripe fruit seeds)**. White pepper is the seed with the darker coloured skin of the pepper fruit removed by retting. The process involves soaking the fully ripe red pepper berries in water for about a week, rubbing and then drying the seed. Sometimes alternative processes are used for removing the outer layer through mechanical, chemical or biological methods.

Black and White Peppercorns.

- ☆ **Green pepper (dried unripe fruit)**. Green pepper, like black, is made from the unripe drupes. Dried green peppercorns are treated with sulfur dioxide, canning or freeze-drying to retain the green color. Pickled peppercorns, also green, are unripe drupes preserved in brine or vinegar. Fresh, unpreserved green pepper drupes, largely unknown in the West, are used in some Asian cuisines, particularly Thai cuisine. If they are not dried or preserved they decay quickly.
- ☆ **Pink peppercorn** is a dried berry of the shrub *Schinus molle*, commonly known as the Peruvian peppertree.
- ☆ **Orange pepper and red pepper** usually consists of ripe red pepper drupes preserved in brine and vinegar. Ripe red peppercorns can also be dried using the same color-preserving techniques used to produce green pepper.

Pink	Green	Tellicherry Black

Malabar Black	Sarawak White	Muntok White

Nutritional Value Black Pepper

Nutrient	Amount	Nutrient	Amount
Energy (Kcal)	255	Vitamin E-γ (mg)	4.56
Carbohydrates (g)	64.81	Vitamin K (µg)	163.7
Protein (g)	10.95	Calcium (mg)	437
Total Fat (g)	3.26	Copper (mg)	1.127
Cholesterol (mg)	0	Iron (mg)	8.86
Dietary Fiber (g)	26.5	Magnesium (mg)	194
Choline (mg)	11.3	Manganese (mg)	5.625
Folic acid (mcg)	10	Phosphorus (mg)	173
Niacin (mg)	1.142	Zinc (mg)	1.42
Pyridoxine (mg)	0.340	Carotene-β (µg)	156
Riboflavin (mg)	0.240	Carotene-α (µg)	0
Thiamin (mg)	0.109	Crypto-xanthin-β (µg)	48
Vitamin A (IU)	299	Lutein-zeaxanthin (µg)	205
Vitamin C (mg)	21	Lycopene (µg)	6

Source: USDA Nutrient Database.

Chemical and Nutritional Components of Pepper

Pepper essential oil showed the presence of 54 components representing about 96.6 per cent of the total weight. Beta-Caryophylline (29.9 per cent) was the major component along with limonene (13.2 per cent), beta-pinene (7.9 per cent), sabinene (5.9 per cent), and several other minor components. The major component of both ethanol and ethyl acetate oleoresins was found to be piperine (63.9 and 39.0 per cent), with many others in lesser amounts (Kapoor *et al.,* 2009).

The primary phytochemicals isolated from various parts of *P. longum* are piperine, piperlongumine, sylvatin, sesamin, diaeudesmin piperlonguminine, pipermonaline, and piperundecalidine. The alkaloid amides in pepper include Piplartine {5,6-dihydro-1-[1-oxo-3-(3,4,5-trimethoxyphenyl)-2-propenyl]-2(1H)pyridinone} and piperine {1-5-(1,3)-benzodioxol-5-yl)-1-oxo-2,4-pentadienyl]piperidine} (Bezerra *et al.,* 2006).

Black peppercorns contain good amount of minerals like potassium, calcium, zinc, manganese, iron, and magnesium; and anti-oxidant vitamins such as vitamin-C and vitamin-A. They are also rich in flavonoid polyphenolic anti-oxidants like carotenes, crypto-xanthin, zea-xanthin and lycopene.

In the United States alone, the average daily intake of black pepper has been estimated at 359 mg (Makhov *et al.,* 2012).

Therapeutic Properties of *Piper nigum*

Pepper is used in many Asian countries as a stimulant in the treatment of colic, rheumatism, headache, diarrhoea, dysentery, cholera, menstrual pain, flatulence and polyurea. It is also used in folk medicine for stomach disorders, digestive problems, neuralgia and scabies (Singh *et al.,* 2007; 2008). Based on modern cell, animal, and human studies, piperine has been found to have immune-modulatory, anti-oxidant, anti-asthmatic, anti-carcinogenic, anti-inflammatory, anti-ulcer, and anti-amoebic properties (Meghwal and Goswami, 2013). The free-radical scavenging activity of black pepper and its active ingredients might be helpful in chemoprevention and controlling progression of tumor growth. Additionally, piperine assists in cognitive brain functioning, boosts absorption of nutrients and improves gastro-intestinal functionality. Dietary piperine, by favorably stimulating the digestive enzymes of pancreas, enhances the digestive capacity and significantly reduces the gastro-intestinal food transit time (Srinivasan, 2007). It is most commonly used to treat respiratory infections, chronic bronchitis, asthma, cough, paralysis of the tongue, constipation, diarrhea, cholera, chronic malaria, viral hepatitis, stomachache, diseases of the spleen, gonorrhea and tumors (Kumar *et al.,* 2011). Liu *et al.* (2010) demonstrated that black pepper and its constituents like hot pepper, exhibit anti-inflammatory, anti-oxidant and anti-cancer properties.

Anti-oxidant Potential

Piperine has been demonstrated in *in-vitro* studies to protect against oxidative damage by inhibiting or quenching free radicals and reactive oxygen species. Black pepper or piperine treatment has also been evidenced to lower lipid peroxidation *in-vivo* and beneficially influence cellular thiol status, anti-oxidant molecules and anti-oxidant enzymes in a number of experimental situations of oxidative stress (Srinivasan, 2007).

Both water extract and ethanol extract of black pepper exhibited strong total anti-oxidant activity as 75 µg/ml concentration showed 95.5 per cent and 93.3 per cent inhibition on peroxidation of linoleic acid emulsion, respectively while butylated hydroxyl-anisole (BHA) and and butylated hydroxyl-toluene (BHT) and alpha-tocopherol exhibited only 92.1 per cent, 95.0 per cent, and 70.4 per cent inhibition respectively at the same concentration (Gülçin, 2005). The different fractions of petroleum ether extract of *P. nigrum* showed dose dependent free radical scavenging activity. Moreover, the extract scavenged the superoxide radical generated by the Xanthine/Xanthine oxidase system (Singh *et al.,* 2008).

The essential oil and oleoresins evaluated against mustard oil by peroxide, p-anisidine, and thiobarbituric acid showed strong anti-oxidant activity in comparison with BHA and BHT but lower than that of propyl gallate (PG). In addition, their inhibitory action by ferric thiocyanate (FTC) method, scavenging capacity by DPPH (2,2'-diphenyl-1-picrylhydrazyl radical), and reducing power also proved the strong anti-oxidant capacity of both the essential oil and oleoresins of pepper (Kapoor *et al.,* 2009).

Simultaneous supplementation with black pepper or piperine to rats fed with high fat diet lowered TBARS and CD levels and maintained superoxide dismutase (SOD), catalase (CAT), glutathione peroxidase (GPx), glutathione-S-transferase (GST), and reduced glutathione (GSH) levels to near those of control rats confirming that supplementation with black pepper or the active principle of black pepper, piperine, can reduce high-fat diet induced oxidative stress (Vijayakumar *et al.,* 2004) and also improve the anti-oxidant status in high fat diet fed anti-thyroid drug treated rats (Vijayakumar and Nalini, 2006a).

The geno-protective potential of curcumin plus piperine was significantly higher as compared to curcumin alone against BaP induced DNA damage (Sehgal *et al.,* 2011). Piperine as an adjuvant further enhanced the anti-oxidant potential of curcumin on the levels of tissue anti-oxidants like SOD, CAT, glutathione reductase (GR), GPx and GSH in the benzo(a)pyrene (BaP) induced oxidative stress in lungs of male Swiss albino mice (Sehgal *et al.,* 2012a) and clastogenicity (Sehgal *et al.,* 2012b).

Anti-inflammatory Potential

Administration of the methanolic extract of *P. longum* could differentially regulate the level of pro-inflammatory cytokines such as IL-1beta, IL-6, TNF-alpha, GM-CSF. The level of IL-2 and tissue inhibitor of metalloprotease-1 (TIMP-1) was increased significantly when the angiogenesis-induced animals were treated with the extract (Sunila and Kuttan., 2006). The anti-inflammatory, nociceptive, and anti-arthritic effect of piperine, the active phenolic component in black pepper extract has been reported by Bang *et al.* (2009). Piperine inhibited the expression of IL6 and MMP13 and reduced the production of PGE2 in a dose dependant manner at a concentration of 10 µg/ml. In rats, nociceptive and arthritic symptoms were significantly reduced by piperine at days 8 and 4 respectively. Histological staining confirmed that piperine significantly reduced the inflammatory area in the ankle joints.

More recently, Ying *et al.* (2013a) also observed piperine to inhibit the production of PGE2 and NO induced by IL-1β. Piperine significantly decreased the IL-1β-stimulated gene expression and production of MMP-3, MMP-13, iNOS and COX-2 in human OA chondrocytes. Piperine inhibited the IL-1β-mediated activation of NF-κB by suppressing the degradation of its inhibitory protein IκBα in the cytoplasm. It can effectively abrogate the IL-1β-induced over-expression of inflammatory mediators; suggesting that piperine may be a potential agent in the treatment of osteo-arthritis. Ying *et al.* (2013b) demonstrated the anti-inflammatory activity of piperine in RAW264.7 cells too. Ahmed *et al.* (2013) also confirmed that the methanolic extract of *P. nigrum* have potent anti-inflammatory effects against neuro-inflammation characterizing Alzheimer's disease as the extract ameliorated neuro-inflammatory insults in an experimentally induced rat model.

Anti-microbial Activity

Infectious diseases remain a serious health issue in spite of the excellent scientific progress in vaccination and chemotherapy. For various reasons, resistant bacteria against antibiotics have also multiplied and infectious diseases remain one of the leading causes of morbidity worldwide (Ahluwalia and Sharma (2007).

The crude extract from pepper was able to inhibit the growth of twenty nine bacteria within a concentration range of 32 to 1024 µg/mL. As Ampicillin and Norfloxacin used to treat variety of bacterial infections, co-administration of piperine (20 mg/kg) enhanced oral bioavailability of these two drugs as reflected in various pharmacokinetic measurements like Cmax, Tmax, AUC and t(1/2) of the above antibiotics in animal model (Janakiraman and Manavalan, 2008). Noumedem *et al.* (2013) provided promising baseline information for the possible use of P. nigrum, in the control of infections caused by multi-drug resistant (MDR) Gram-negative bacteria and thus its therapeutic use in association with antibiotics to combat MDR pathogens.

The ethanolic extract, hexane fraction, n-butanol soluble fraction at 1000 µg/mL and the chloroform fraction at 500 µg/ml. exhibited *in-vitro* amoebi-cidal action. The ethanolic extract and piperine cured 90 per cent and 40 per cent of rats with caecal amoebiasis, respectively (Ghoshal *et al.,* 1996). Caecal amoebiasis in mice induced by injection of *Entamoeba histolytica trophozoites* directly into the caecum were treated orally with the extract from *Piper longum* fruit, a standard drug (metronidazole), or vehicle p.o. for five consecutive days, beginning 24 h after the infection and examined on the sixth day. At a dose of 1000 mg/kg per day, the extracts of *Piper longum* fruit, had a curative rate of 100 per cent. At a concentration of 500 and 250 mg/kg/day, pepper extract was still effective in 93 and 46 per cent of the cases, respectively as against Metronidazole at a concentration of 125 and of 62.5 mg/kg per day showed a curative rate of 100 and 60 per cent, respectively (Sawangjaroen *et al.,* 2004).

Anti-hyperlipidemic Effect

Piperlonguminine, piperine and pipernonaline, the main anti-hyper-lipidemic constituents of pepper exhibited appreciable hypolipidemic activity *in-vivo,* comparable to that of the standard drug – simvastatin (Jin *et al.,* 2009). Three different piper species (*Piper guineense, Piper nigrum* and *Piper umbellatum*) also prevented the collapse of the anti-oxidant system and the increase of plasma parameters maintaining them towards normality. In addition, the piper species prevented LDL oxidation by increasing the lag time for its oxidation revealing their significant anti-oxidant and anti-atherogenic effect against intoxication by atherogenic diet (Agbor *et al.,* 2012. The methanolic extract of *P. nigrum* fruits has hepato-protective and anti-oxidant effects in rats (Singh *et al.,* 2007).

Black pepper lowers blood lipids *in-vivo* and inhibits cholesterol uptake *in-vitro*, and piperine may mediate these effects. Piperine or black pepper extract (containing the same amount of piperine) dose-dependently reduced cholesterol uptake into Caco-2 cells in a similar manner. Both preparations reduced the membrane levels of NPC1L1 and SR-BI proteins but not their overall cellular expression. Micellar cholesterol solubility of lipid micelles was unaffected except by 1 mg/mL concentration of black pepper extract suggesting that piperine is the active compound in black pepper and reduces cholesterol uptake by internalizing the cholesterol transporter proteins (Duangjai *et al.,* 2013). Supplementation with a dose of 40 mg/kg of piperine with high fat diet significantly reduced not only body weight, triglyceride, total cholesterol, LDL, VLDL, and fat mass, but also increased the HDL levels, with no change in food intake (Shah *et al.,* 2011).

Piperine reduces symptoms of human metabolic syndrome by reducing inflammation and oxidative stress. Diwan *et al.* (2011) reported normalized blood pressure, improved glucose tolerance and reactivity of aortic rings; reduced plasma parameters of oxidative stress and inflammation; attenuated cardiac and hepatic inflammatory cell infiltration and fibrosis; and improved liver function in high carbohydrate, high fat (HCHF) fed rats supplemented with 30 mg of piperine/kg/day. According to Park *et al.* (2012) piperine attenuates fat cell differentiation by down-regulating PPARγ activity as well as suppressing PPARγ expression, thus leading to potential treatment for obesity-related diseases.

Taqvi *et al.* (2008) found piperine to possess a blood pressure-lowering effect mediated possibly through Ca^{2+} channel blockade (CCB), while consistent decrease in BP was restricted by associated vaso-constrictor effect. Additionally, species selectivity was also seen in the CCB effect of piperine. Oral administration of piperine is able to partially prevent the increase of blood pressure caused by chronic L-NAME administration. This effect is probably caused by the blockage of voltage-dependent calcium channels and supported by filamentous actin disassembly (Hlavackova *et al.,* 2010).

Vijayakumar and Nalini (2006a and b) demonstrated that the simultaneous administration of piperine and high fat diet significantly reduced the elevated levels of plasma lipids and lipoproteins levels, except for HDL. Piperine supplementation also improved the plasma levels of apo A-I, T3, T4, testosterone and significantly reduced apo B, TSH, and insulin to near normal levels. These studies provide evidence that piperine possesses thyrogenic activity, thus modulating apo-lipoprotein levels and insulin resistance in HFD-fed rats, opening a new view in the management of dyslipidemia by dietary supplementation with nutrients. Moreover, piperine also inhibits lipid and lipoprotein accumulation by significantly modulating the enzymes of the lipid metabolism, like lecithin-cholesterol acyltransferase and lipoprotien lipase (Vijayakumar and Nalini, 2006b).

Anti-diabetic Effect

Pharmacological inhibition of acyl CoA:diacylglycerol acyltransferase (DGAT, EC 2.3.1.20) has emerged as a potential therapy for the treatment of obesity and type 2 diabetes. A new alkamide named (2E,4Z,8E)-N-[9-(3,4-methylenedioxyphenyl)-2,4,8-nonatrienoyl]piperidine (2), together with four known alkamides: pipernonaline (3), piperrolein B (4), and dehydropipernonaline (5) inhibited DGAT with IC50 values of 29.8 (2), 37.2 (3), 20.1 (4), and 21.2 (5) µM, respectively, indicating that compounds possessing piperidine groups (2-5) can be potential DGAT inhibitors (Lee *et al.*, 2006).

Black pepper consisting of 98 per cent pure piperine for 21 days along with 120 mg coenzyme Q10 in healthy adult male volunteers produced a statistically significant, approximately 30 per cent greater area under the plasma curve than placebo. It is postulated that the bio-enhancing mechanism of piperine to increase plasma levels of supplemental coenzyme Q10 is nonspecific and possibly because piperine in the form of bioperine is termed as a thermo-nutrient (Badmaev *et al.*, 2000).

Piperine treatment of normal rats enhanced hepatic oxidized glutathione (GSSG) concentration by 100 per cent, decreased renal GSH concentration by 35 per cent and renal glutathione reductase activity by 25 per cent when compared to normal controls. Also it reversed the diabetic effects on GSSG concentration in brain, on renal glutathione peroxidase and superoxide dismutase activities, and on cardiac glutathione reductase activity and lipid peroxidation. But it did not reverse the effects of diabetes on hepatic GSH concentrations, lipid peroxidation, or glutathione peroxidase or catalase activities; on renal superoxide dismutase activity; or on cardiac glutathione peroxidase or catalase activities indicating that subacute treatment with piperine for 14 days is only partially effective as an anti-oxidant therapy in diabetes (Rauscher *et al.*, 2000).

Piperine (10 mg/kg) significantly increased the dose-dependent anti-hyper-glycemic activity of nateglinide in alloxan-induced diabetic models, when administered with nateglinide. The synergistic anti-hyperglycemic activity of nateglinide and piperine can be attributed to increased plasma concentration of nateglinide demonstrating that piperine could be used as a potential bio-enhancer along with nateglinide (Sama *et al.*, 2012). A significant blood glucose lowering effect was seen with piperine at dose of 20 mg/kg on day 14 suggesting that sub-acute administration of piperine has statistically significant anti-hyperglycemic activity while acutely it raises blood glucose at high doses of 40 mg/kg (Atal *et al.*, 2012).

Anti-carcinogenic Effect

Sunila and Kuttan (2004) found piperine to be cyto-toxic towards DLA and EAC cells at a concentration of 250 µg/ml. Alcoholic extract and piperine was found to produce cyto-toxicity towards L929 cells in culture at a concentration of 100 and 50 µg/ml, respectively. Administration of alcoholic extract of *Piper longum* as well as piperine-1 inhibited the solid tumor development in mice induced with DLA cells, 2. increased the life span of mice bearing Ehrlich ascites carcinoma tumor by 37.3 and

58.8 per cent, respectively, 3. increased the total WBC count to 142.8 and 138.9 per cent, respectively 4. Significantly enhanced the number of plaque forming cells by 100.3 per cent and 71.4 per cent on 5th day after immunization and 5. increased the bone marrow cellularity and alpha-esterase positive cells. The authors (Sunila and Kuttan (2005) also found the ethanolic extract of *Piper longum* fruits reduced the elevated levels of glutathione pyruvate transaminase (GPT), alkaline phosphatase (ALP), and lipid peroxidation (LPO) in liver and serum of radiation treated animals and increased the reduced glutathione (GSH) production to offer the radio-protection.

The level of IL-2 and tissue inhibitor of metalloprotease-1 (TIMP-1) was increased significantly when the angiogenesis-induced animals were treated with the extract. The extract of *P. longum* at non-toxic concentrations (10 μg/ml, 5 μg/ml, 1 μg/ml) inhibited the VEGF-induced vessel sprouting in rat aortic ring assay and inhibit the VEGF-induced proliferation, cell migration and capillary-like tube formation of primary cultured human endothelial cells (Sunila and Kuttan, 2006). Piperine has also been shown to have anti-invasion activity of B16F-10 melanoma cells (Pradeep and Kuttan, 2004).

Piperine inhibited the proliferation and G(1)/S transition of human umbilical vein endothelial cells (HUVECs) without causing cell death. Piperine also inhibited HUVEC migration and tubule formation *in-vitro*, as well as collagen-induced angiogenic activity by rat aorta explants and breast cancer cell-induced angiogenesis in chick embryos (Doucette *et al.*, 2013).

Both piplartine {5,6-dihydro-1-[1-oxo-3-(3,4,5-trimethoxyphenyl)-2-propenyl]-2(1H)pyridinone} and piperine {1-5-(1,3)-benzodioxol-5-yl)-1-oxo-2,4-pentadienyl]piperidine} isolated from pepper have been reported to show cyto-toxic activity towards several tumor cell lines. Piplartine was more potent than piperine. Piplartine was the most active with IC50 values in the range of 0.7 to 1.7 μg/ml. None of the tested substances induced hemolysis of mouse erythrocytes, suggesting that the cyto-toxicity of piplartine and piperine was not related to membrane damage (Bezerra *et al.*, 2005).

Piperine induced anti-oxidant response element-luciferase and translocated nuclear factor-E2-related factor-2 (Nrf2) to nucleus. It also activated the c-Jun N-terminal kinase (JNK), extracellular signal-regulated kinase and p38 mitogen-activated protein kinase (MAPK) pathways. The JNK pathway played an important role in piperine-induced HO-1 expression. Piperine protected the cells against cisplatin-induced apoptosis. The protective effect of piperine was abrogated by zinc protoporphyrin IX, an HO inhibitor, and antisense oligodeoxy-nucleotides against HO-1 gene. These results demonstrate that the expression of HO-1 by piperine is mediated by both JNK pathway and Nrf2, and the expression inhibits cisplatin-induced apoptosis in HEI-OC1 cells (Choi *et al.*, 2007).

Kim *et al.* (2012) found piperine to dose-dependently decrease phorbol 12-myristate 13-acetate PMA-induced COX-2 expression and PGE(2) production, as well as COX-2 promoter-driven luciferase activity, inhibit PMA-induced NF-κB, C/EBP and c-Jun nuclear translocation and piperine significantly inhibit PMA-induced activation of the Akt and ERK. Thus they demonstrated that piperine effectively attenuates COX-2 production, and provided further insight into the signal transduction pathways involved in the anti-inflammatory effects of piperine.

Administration of piplartine or piperine (50 or 100 mg kg(-1) day(-1) intraperitoneally for 7 days starting 1 day after inoculation) inhibited solid tumor development in mice transplanted with Sarcoma 180 cells by 28.7 and 52.3 per cent; and 55.1 and 56.8 per cent, respectively. Piperine was more toxic to the liver, leading to ballooning degeneration of hepatocytes, accompanied by micro-vesicular steatosis in some areas, than piplartine which, in turn, was more toxic to the kidney, leading to discrete hydropic changes of the proximal tubular and glomerular epithelium and tubular hemorrhage in treated animals (Bezerra *et al.,* 2006).

Leukopenia seen after 5-fluorouracil (5-FU) treatment was reversed by the combined use of piplartine and piperine indicating that piplartine may enhance the therapeutic effectiveness of chemo-therapeutic drugs, and this combination could improve immune-competence hampered by 5-FU (Bezerra *et al.,* 2008a). Mishra *et al.* (2011) isolated new amides from pepper–piperlongimin A (2) [2E-N-isobutyl-hexadecenamide] and piperlongimin B (4) [2E-octadecenoylpiperidine] together with five known compounds inhibited cell proliferation of human leukemia, HL-60 cell lines, and displayed major apoptosis-inducing effects. Piperlonguminine did not show any significant *in-vitro* cytotoxic effect at experimental exposure levels, but showed an *in-vivo* anti-tumor effect. After 7 days of treatment, the inhibition rates were 38.71 per cent and 40.68 per cent at doses of 25 mg kg(-1) and 50 mg kg(-1), respectively (Bezerra *et al.,* 2008b).

Liu *et al.* (2010) observed that the extracts of black pepper at 200 µg/mL and its compounds at 25 µg/mL inhibited LPO by 45–85 per cent, COX enzymes by 31-80 per cent and cancer cells proliferation by 3.5–86.8 per cent. Do *et al.* (2013) reported that piperine strongly inhibited proliferation and induced apoptosis through caspase-3 activation and PARP cleavage. Furthermore, piperine inhibited HER2 gene expression at the transcriptional level. Blockade of ERK1/2 signaling by piperine significantly reduced SREBP-1 and FAS expression. Piperine strongly suppressed EGF-induced MMP-9 expression through inhibition of AP-1 and NF-κB activation by interfering with ERK1/2, p38 MAPK, and Akt signaling pathways resulting in a reduction in migration. Finally, piperine pretreatment enhanced sensitization to paclitaxel killing in HER2-overexpressing breast cancer cells. Wongpa *et al.* (2007) demonstrated that piperine at a dose of 100 mg/kg body weight gave a statistically significant reduction in cyclophosphamide-induced chromosomal aberrations. Thus piperine may have anti-mutagenic potential. Recent studies have demonstrated that piperine inhibit breast cancer by targeting the cancer stem cell renewal properties (Kakarala *et al.,* 2010). Lai *et al.* (2012) also demonstrated that piperine is an effective anti-tumor *compound in-vitro* and *in-vivo* in a 4T1 murine breast cancer model. Thus it has the potential to be developed as a new anti-cancer drug.

Lambert *et al.* (2004) found that 20 µm of piperine inhibited the formation of EGCG-32-glucuronide in human HT-29 colon adenocarcinoma cells by ~50 per cent. Earlier, rodent studies have shown that piperine can inhibit both CYP and UGT activities *in-vivo* (Atal *et al.,* 1985; Reen and Singh, 1991). The anti-oxidant Nacetylcysteine reduced apoptosis in cultures of piperine-treated HRT-18 cells, indicating that piperine-induced cytotoxicity was mediated at least in part by reactive oxygen species. Piperine inhibited the metabolic activity of HRT-18 cells in a dose- and time-dependent fashion, suggesting a cytostatic and/or cytotoxic effect. Similar effects of piperine on rectal cancer cells suggest that this dietary phytochemical may be useful in cancer treatment (Yaffe *et al.,* 2013).

The cyto-protective effect of piperine on B (α)-p (Benzopyrene)-induced experimental lung cancer has been successfully investigated in mice and inferred that piperine could exert its chemopreventive effect by modulating lipid peroxidation and augmenting antioxidant defense system (Selvendiran *et al.,* 2006). Piperine has been reported to reduce cancer incidence in chemical rodent models of lung cancer (Bhardwaj *et al.,* 2002; Pradeep and Kuttan, 2002 and 2004; Selvendiran *et al.,* 2004 and 2006; Shoba *et al.,* 1998).

Prostate cancer is the most common malignancy in men. Samykutty *et al.* (2013) showed that piperine inhibited the proliferation of LNCaP, PC-3, 22RV1 and DU-145 prostate cancer cells in a dose dependent manner. Annexin-V staining demonstrated that piperine treatment induced apoptosis in hormone dependent prostate cancer cells (LNCaP). Using global caspase activation assay, the authors showed that caspase activation in LNCaPμ and PC-3 cells was due to Piperine-induced apoptosis. Dietary consumption of piperine increased the therapeutic efficacy of docetaxel (the mainline treatment approved by the FDA for castration-resistant prostate cancer (CRPC) in a xenograft model without inducing more adverse effects on the treated mice (Makhov *et al.,* 2012).

Piperine offers a promising therapeutic approach to treat cancer as it enhances radio-sensitivity of tumor cells. Cancer cell lines treated with resveratrol and piperine exhibited significantly augmented ionizing radiation-induced apoptosis and loss of mitochondrial membrane potential, presumably through enhanced generation reactive oxygen species. (Tak *et al.,* 2012).

It has been suggested that piperine may also enhance the bioavailability of curcumin if both these chemo-preventive agents are given in combination (Shoba *et al.,* 1998). Li *et al.* (2011) demonstrated that piperine can reverse multidrug resistance (MDR) (which impairs the efficacy of chemotherapy) by multiple mechanisms and it is thus a promising food component for future studies.

Anti-cognitive Impairment and Anti-depressant Effect

Pepper can be considered a potential functional food useful to improve brain function. The anti-depressant-like effect of piperine is mediated *via* the serotonergic system as proved by the enhanced 5-HT(1A) and 5-HT(1B) receptors in both the hippocampus and frontal cortex of mice (Mao *et al.,* 2011a; Mao *et al.,* 2011b). Reduction in lipid peroxidation and acetylcholinesterase enzyme (which reduces cholesterol synthesis) is also proposed as the other possible underlying mechanisms. Moreover, piperine also demonstrated the neurotrophic effect in hippocampus (Chonpathompikunlert *et al.,* 2010).

Daily administration of piperine at the dosage range of 5, 10 and 20mg/kg body weight for 4 weeks exhibited anti-depression like activity and cognitive enhancing effect in Wistar rats (Wattanathorn *et al.*, 2008). Lee *et al.* (2005) suggested that piperine possesses potent anti-depressant-like properties that are mediated in part through the inhibition of monoamine oxidase (MAO) activity, and therefore could be considered a promising pharmaco-therapeutic anti-depressant agent. Al-Baghdadi *et al.* (2012) reported that pepper contains certain other compounds that were selective towards MAO-B and devoid of MAO-A activity with the most potent compound having an IC(50) of 498 nM.

Co-administration of piperine (20 mg/kg; p.o.) with curcumin (100 and 200 mg/kg, p.o.) significantly elevated the protective effect against chronic unpredictable stress induced cognitive impairment and associated oxidative damage in mice as compared to their individual effects suggesting that piperine enhanced the bioavailability of curcumin and potentiated its protective effects *i.e.* anti-immobility, neuro-transmitter enhancing (serotonin and dopamine) and monoamine oxidase inhibitory (MAO-A) effects (Rinwa and Kumar, 2012; Bhutani *et al.*, 2009).

Anti-epileptic Effect

In traditional Chinese medicine, a mixture of radish and pepper is used to treat epilepsy. Appropriate concentration of piperine effectively inhibited the synchronized oscillation of intracellular calcium in rat hippocampal neuronal networks and repressed spontaneous synaptic activities in terms of spontaneous synaptic currents (SSC) and spontaneous excitatory postsynaptic currents (sEPSC). Moreover, pretreatment with piperine exhibited protective effect on glutamate-induced decrease of cell viability and apoptosis of hippocampal neurons. Thus, the neuro-protective effects of piperine might be associated with suppression of synchronization of neuronal networks, presynaptic glutamic acid release, and Ca(2+) overloading (Fu *et al.*, 2010).

Convulsions of E1 mice were completely suppressed by 60 mg/kg of piperine injected intra-peritoneally. The 5-HT level was significantly higher in the cerebral cortex of piperine treated mice than in control mice which may be related directly to the mechanism of inhibition of convulsions by piperine. On the other hand, lower levels of 5-HT were observed in the hippocampus, midbrain and cerebellum. The dopamine level in the piperine treated mice was markedly higher only in the hypothalamus, while the norepinephrine levels were lower in every part of the brain (Mori *et al.*, 1985).

Piperine was shown to significantly block convulsions induced by intra-cerebro-ventricular injection of threshold doses of kainate, but to have no or only slight effects on convulsions induced by L-glutamate, N-methyl-D-aspartate or guanidine-succinate. Piperine suspensions, when injected intra-peritoneally, 1 h before injection of kainate (1 nmol), blocked these convulsions with an ED50 (and 95 per cent confidence interval) of 46 (25-86) mg/kg. (D'Hooge *et al.*, 1996).

Anti-fertility Effect

Interestingly, Piperine when given to pregnant mice from day 2 through 5, day 8 through 12 and day 15 until labor, effectively inhibited implantation, produced abortion and delayed labor without affecting the estrous cycle. Neither uterotropic, anti-estrogenic nor anti-progestational property was observed. Additionally, piperine also inhibited uterine contraction both *in-vivo* and *in-vitro* suggesting that the anti-fertility activity of piperine was not through any hormonal actions or uterotonic activity (Piyachaturawat *et al.,* 1982; 1985).

Piperine (10 and 20 mg/kg b.w.) increased the period of the diestrous phase which seemed to result in decreased mating performance and fertility. Post-partum litter growth was not affected by the piperine treatment. Considerable anti-implantation activity was recorded after five days post-mating with oral treatment of piperine. Intrauterine injection of piperine caused the total absence of implants in either of the uterine horns (16.66 per cent) or one of the horns (33 per cent) of treated females showing that piperine interferes with several crucial reproductive events in a mammalian model (Daware *et al.,* 2000).

Piperine caused a decrease in the activity of anti-oxidant enzymes and sialic acid levels in the epididymis and thereby increased reactive oxygen species levels that could damage the epididymal environment and sperm function. Doses of 10 mg/kg and 100 mg/kg, decreased epididymal sperm count, motility and viability (D'cruz and Mathur, 2005). Further, through immuno-fluorescence studies a dose-dependent increase in caspase 3 and Fas protein in testicular germ cells was found after piperine treatment indicating that piperine induces oxidative stress and thereby triggers apoptosis in the testis, contributing to hampered reproductive functions (D'cruz *et al.,* 2008).

Piper nigrum fruit powder has been shown to possess anti-spermatogenic effect on the male reproductive organs of mice. Testes in treated mice showed non-uniform degenerative changes in the seminiferous tubules. *Piper nigrum* treatment for 20 days did not cause appreciable alterations in histological appearance of the epididymis, while the treatment for 90 days caused detectable alterations in the duct. The treatment also had adverse effects on sperm parameters, levels of sialic acid and fructose, and on litter size. Fifty six days after cessation of treatment, the alterations induced in the reproductive organs recovered to control levels, though the litter size in females impregnated by piper-treated males remained significantly decreased compared to controls (Mishra and Singh, 2009).

Anti-diarrheal Effect

Piperine significantly inhibited diarrhoea produced by the cathartics (castor oil, MgSO4 and arachidonic acid) at a dose of 8 and 32 mg/kg p.o. Inhibition of castor oil induced entero-pooling by piperine suggests its inhibitory effect on prostaglandins thus validating the rationale for its use in traditional anti-diarrhoeal formulations (Bajad *et al.,* 2001).

The cumulative concentrations of the black pepper fruit hot water extract (BPE (0.0625-1 mg mL(-1)) significantly reduced the ileum contractions induced by KCl (60 mM) or carbachol (10 µM). The incubation of the tissue preparation (20 or 30 min) with L-NAME (100 µM), naloxone (1 µM) or propranolol (1 µM) did not reduce the anti-spasmodic effect of BPE on KCl-induced ileum contraction. The extract's spasmolytic effect was attenuated neither by glibenclamide (10 μM) nor by tetra-ethylammonium (1 mM) suggesting that the spasmolytic effect of the BPE on rat ileum was possibly mediated via Ca2+ influx (Naseri and Yahyavi, 2008).

Black Pepper as a Bio-enhancer

Piperine is one of the most promising bioenhancers. By inhibiting the metabolism, piperine improves the bioavailability of drugs like vascicine, sparteine, curcumin, barbiturate and oxyphenylbutazone, zoxazolamine, propranalol and theophylline in animal experiments (Atal *et al.*, 1981, Majumdar *et al.*, 1990a, Shoba *et al.*, 1998, Majumdar *et al.*, 1990b, Majumdar *et al.*, 1999, Bano *et al.*, 1991). Several mechanisms have been postulated for piperine's bioavailability-enhancing effect including the formation of apolar complexes with other compounds, inhibition of efflux transport, and inhibition of gut metabolism (Atal *et al.*, 1985; Khajuria *et al.*, 1998; Bhardwaj *et al.*, 2002).

Peroral administration of piperine at the dose of 112 μg/kg body weight/day to male Wistar rats for 14 consecutive days led to increased intestinal P-glycoprotein (P-gp) levels. (P-glycoprotein is an ATP-binding cassette (ABC) transporter, functions as a biological barrier by extruding toxins and xenobiotics out of cells. In vitro and in vivo studies have demonstrated that P-glycoprotein plays a significant role in drug absorption and disposition (Lin and Yamazaki 2003). However, there was a concomitant reduction in the rodent liver P-gp although the kidney P-gp level was unaffected (Han *et al.*, 2008). The authors suggest that caution should be exercised when piperine is to be co-administered with drugs that are P-gp substrates, particularly for patients whose diet relies heavily on pepper. Aher *et al.* (2009) reported that the P-gp inhibitory activity of the aqueous extract of long pepper fruits was comparable with that of pure piperine and was significantly higher than the alcoholic extract. Pure piperine and the aqueous extract exhibited significant P-gp inhibitory activity compared with control.

Black pepper or piperine is widely used in combination therapy along with drugs in current medical practice. The bioavailability of fexofenadine was increased by approximately 2-folds via the concomitant use of piperine. T(max) tends to be increased which might be attributed to the delayed gastric emptying in the presence of piperine. In contrast, piperine did not alter the intravenous pharmacokinetics of fexofenadine, implying that piperine may increase mainly the gastrointestinal absorption of fexofenadine rather than reducing hepatic extraction (Jin and Han, 2010).

Similarly, enhanced bioavailability and stability of curcumin in the body tissues was evidenced when the same was orally administered concomitantly with piperine. (Suresh and Srinivasan, 2010). Piperine, found in the extract of black pepper, enhances the relative oral bioavailability of curcuminoids by as much as 20-fold in healthy human volunteers (Shoba *et al.*, 1998).

Piperine also enhances the bioavailability of several other compounds including phenytoin, coenzyme Q10, theophylline, and propranolol (Bano *et al.*, 1991; Badmaev *et al.,* 2000; Pattanaik *et al.,* 2006). Piperine also inhibited glucuronidation of (")-epigallocatechin-3-gallate (EGCG) in mice resulting in increased EGCG bioavailability (Lambert *et al.,* 2004).

Other Benefits

Dietary feeding of black pepper caused an increase in bile flow with a concomitant decrease in bile solids – a hydrocholagoguic effect. Cholesterol and bile acid output were not affected by black pepper or piperine at either level irrespective of the mode of administration; in contrast, the secretion of uronic acids in bile was enhanced by both levels of pepper as also of piperine indicating possible excretion of some of the components of black pepper or of piperine as glucuronides (Ganesh Bhat and Chandrasekhara, 1987).

Nasal inhalation of volatile black pepper oil (BPO) for 1 minute shortened latency of the swallowing reflex (LTSR), compared with that of lavender oil and distilled water. Inhalation of BPO activates the insular or orbito-frontal cortex, resulting in improvement of the reflexive swallowing movement and benefit older post stroke patients with dysphagia regardless of their level of consciousness or physical and mental status (Ebihara *et al.,* 2006). Munakata *et al.* (2008) also reported that olfactory stimulation with BPO facilitated oral intake in a subset of patients on long-term enteral nutrition. BPO stimulation may be useful for facilitating oral intake when used in combination with conventional methods.

Piperine exerted a significant protection against tert-butyl hydroperoxide and carbon tetrachloride hepato-toxicity by reducing in-*vitro* and *in-vivo* lipid peroxidation, enzymatic leakage of GPT and AP, and by preventing the depletion of GSH and total thiols in the intoxicated mice. Piperine showed lower hepato-protective potency than silymarin, a known hepato-protective drug (Koul and Kapil, 1993).

Western blotting revealed the down-regulation of SIRT1 protein expression in Daudi cells treated with extracts of black pepper or turmeric. On the other hand, the effect on the SIRT1 gene expression examined by reverse transcription polymerase chain reaction was unaltered (Nishimura *et al.,* 2011).

Piper longum fruits have been traditionally used against snake bites in north-eastern and southern region of India. Ethanolic extract of fruits of *Piper longum* was found to inhibit the venom induced haemorrhage in embryonated fertile chicken eggs, significantly inhibited venom induced lethality, haemorrhage, necrosis, defibrinogenation and inflammatory paw edema in mice in a dose dependent manner and also significantly reduced venom induced mast cell degranulation in rats. Piperine is one of the compounds responsible for the effective venom neutralizing ability of pepper (Shenoy *et al.,* 2013a). Immunization with ethanolic extract of fruits of *Piper longum* and piperine produced a high titre antibody response against Russell's viper venom in mice. The antibodies against the extract and piperine could be useful in anti-venom therapy of Russell's viper bites (Shenoy *et al.,* 2013b).

Conclusions

Thus pepper offers several health benefits in animal as well as human systems. It also potentiates the benefits of other foods such as curcumin, thus suggesting their simultaneous consumption. The potential of pepper as anti-cognitive impairment agent is worth researching further through human trials. Appropriate dosage of piperine for its use as anti-diabetic agent needs to be confirmed. The synergistic interaction of black pepper with different drugs and nutrients suggests a need for controlled randomized trials in human subjects, cohort studies, and meta-analyses with due consideration for many confounding factors affecting the clinical outcome of pharmacokinetic interactions (*e.g.*, dose, dosing, genetic variation and species) in future to help recommend its application in diet-based regimens for therapeutic conditions (Han, 2011; Butt *et al.*, 2013).

References

Agbor GA, Vinson JA, Sortino J, Johnson R (2012). Anti-oxidant and anti-atherogenic activities of three Piper species on atherogenic diet fed hamsters. *Exp. Toxicol. Pathol.* 64 (4): 387 – 91.

Aher S, Biradar S, Gopu CL, Paradkar A (2009). Novel pepper extract for enhanced P-glycoprotein inhibition. *J. Pharm. Pharmacol.* 61 (9): 1179–86.

Ahluwalia G, Sharma SK (2007). Philanthropy and medical science: at last a new dawn for tuberculosis also! *Indian J. Chest Dis. Allied Sci.* 13: 71–73.

Ahmed HH, Salem AM, Sabry GM, Husein AA, Kotob SE (2013). Possible therapeutic uses of *Salvia triloba* and *Piper nigrum* in Alzheimer's disease-induced rats. *J. Med. Food.* 16 (5): 437–46.

Al-Baghdadi OB, Prater NI, Van der Schyf CJ, Geldenhuys WJ (2012). Inhibition of monoamine oxidase by derivatives of piperine, an alkaloid from the pepper plant *Piper nigrum*, for possible use in Parkinson's disease. *Bioorg. Med. Chem. Lett.* 22 (23): 7183 – 88.

Atal CK, Dubey RK, Singh J (1985). Biochemical basis of enhanced drug bioavailability by piperine: evidence that piperine is a potent inhibitor of drug metabolism. *J. Pharmacol. Exp. Ther.* 232: 258–62.

Atal CK, Zutshi U, Rao PG (1981). Scientific evidence on the role of Ayurvedic herbals on bioavailability of drugs. *J. Ethnopharmacol.* 4 (2): 229–32.

Atal S, Agrawal RP, Vyas S, Phadnis P, Rai N (2012). Evaluation of the effect of piperine per se on blood glucose level in alloxan-induced diabetic mice. *Acta. Pol. Pharm.* 69 (5): 965–69.

Badmaev V, Majeed M, Prakash L (2000). Piperine derived from black pepper increases the plasma levels of coenzyme Q10 following oral supplementation. *J. Nutr. Biochem.* 11 (2): 109–113.

Bajad S, Bedi KL, Singla AK, Johri RK (2001). Anti-diarrhoeal activity of piperine in mice. *Planta. Med.* 67 (3): 284–87.

Bang JS, Oh da H, Choi HM, Sur BJ, Lim SJ, Kim JY, Yang HI, Yoo MC, Hahm DH, Kim KS (2009). Anti-inflammatory and antiarthritic effects of piperine in human interleukin 1beta-stimulated fibroblast-like synoviocytes and in rat arthritis models. *Arthritis Res. Ther.* 11 (2): R49.

Bano G, Raina RK, Zutshi U, Bedi KL, Johri RK, Sharma SC (1991). Effect of piperine on bioavailability and pharmacokinetics of propranolol and theophylline in healthy volunteers. *Eur. J. Clin. Pharmacol.* 41 (6): 615–17.

Bezerra DP, Castro FO, Alves AP, Pessoa C, Moraes MO, Silveira ER, Lima MA, Elmiro FJ, Costa-Lotufo LV (2006). *In-vivo* growth-inhibition of Sarcoma 180 by piplartine and piperine, two alkaloid amides from Piper. *Braz J Med Biol Res.* 39 (6): 801–07.

Bezerra DP, de Castro FO, Alves AP, Pessoa C, de Moraes MO, Silveira ER, Lima MA, Elmiro FJ, de Alencar NM, Mesquita RO, Lima MW, Costa-Lotufo LV (2008a). *In-vitro* and *in-vivo* anti-tumor effect of 5-FU combined with piplartine and piperine. *J. Appl. Toxicol.* 28 (2): 156–63.

Bezerra DP, Pessoa C, Moraes MO, Alencar NM, Mesquita RO, Lima MW, Alves AP, Pessoa OD, Chaves JH, Silveira ER, Costa-Lotufo LV (2008b). *In-vivo* growth inhibition of sarcoma 180 by piperlonguminine, an alkaloid amide from the Piper species. *J. Appl. Toxicol.* 28 (5): 599–607.

Bezerra DP, Pessoa C, de Moraes MO, Silveira ER, Lima MA, Elmiro FJ, Costa-Lotufo LV (2005). Anti-proliferative effects of two amides, piperine and piplartine, from Piper species. *Z. Naturforsch. C.* 60 (7-8): 539–43.

Bhardwaj RK, Glaeser H, Becquemont L, Klotz U, Gupta SK, Fromm MF (2002). Piperine, a major constituent of black pepper, inhibits human P-glycoprotein and CYP3A4. *J. Pharmacol. Exp. Ther.* 302:.645–50.

Bhutani MK, Bishnoi M, Kulkarni SK (2009). Anti-depressant like effect of curcumin and its combination with piperine in unpredictable chronic stress-induced behavioral, biochemical and neurochemical changes. *Pharmacol. Biochem. Behav.* 92 (1): 39–43.

Choi BM, Kim SM, Park TK, Li G, Hong SJ, Park R, Chung HT, Kim BR (2007). Piperine protects cisplatin-induced apoptosis via heme oxygenase-1 induction in auditory cells. *J. Nutr. Biochem.* 18 (9): 615–22.

Chonpathompikunlert P, Wattanathorn J, Muchimapura S (2010). Piperine, the main alkaloid of Thai black pepper, protects against neurodegeneration and cognitive impairment in animal model of cognitive deficit like condition of Alzheimer's disease. *Food Chem/Toxicol.* 48 (3): 798–802.

Daware MB, Mujumdar AM, Ghaskadbi S (2000). Reproductive toxicity of piperine in Swiss albino mice. *Planta. Med.* 66 (3): 231–36.

D'cruz SC, Mathur PP (2005). Effect of piperine on the epididymis of adult male rats. *Asian Journal of Andrology.* 7: 363–68.

D'Cruz SC, Vaithinathan S, Saradha B, Mathur PP (2008). Piperine activates testicular apoptosis in adult rats. *J. Biochem. Mol. Toxicol.* 22 (6): 382–88.

D'Hooge R, Pei YQ, Raes A, Lebrun P, van Bogaert PP, de Deyn PP (1996). Anti-convulsant activity of piperine on seizures induced by excitatory amino acid receptor agonists. *Arzneimittelforschung.* 46 (6): 557–60.

Diwan V, Poudyal H, Brown L (2011). Piperine attenuates cardiovascular, Liver and metabolic changes in high carbohydrate, high fat-fed rats. *Cell. Biochem. Biophys.* 2011 Oct 30. [Epub ahead of print].

Do MT, Kim HG, Choi JH, Khanal T, Park BH, Tran TP, Jeong TC, Jeong HG (2013). Anti-tumor efficacy of piperine in the treatment of human HER2-overexpressing breast cancer cells. *Food Chem.* 141 (3): 2591–99.

Doucette CD, Hilchie AL, Liwski R, Hoskin DW (2013). Piperine, a dietary phytochemical, inhibits angiogenesis. *J. Nutr. Biochem.* 24 (1): 231–39.

Duangjai A, Ingkaninan K, Praputbut S, Limpeanchob N (2013). Black pepper and piperine reduce cholesterol uptake and enhance translocation of cholesterol transporter proteins. *J. Nat. Med.* 67 (2): 303–10.

Ebihara T, Ebihara S, Maruyama M, Kobayashi M, Itou A, Arai H, Sasaki H (2006). A randomized trial of olfactory stimulation using black pepper oil in older people with swallowing dysfunction. *J. Am. Geriatr. Soc.* 54 (9): 1401–406.

Fu M, Sun ZH, Zuo HC (2010). Neuro-protective effect of piperine on primarily cultured hippocampal neurons. *Biol. Pharm. Bull.* 33 (4): 598–603.

Ganesh Bhat B, Chandrasekhara N (1987). Effect of black pepper and piperine on bile secretion and composition in rats. *Nahrung.* 31 (9): 913–16.

Ghoshal S, Prasad BN, Lakshmi V (1996). Anti-amoebic activity of *Piper longum* fruits against *Entamoeba histolytica in-vitro* and *in-vivo. J. Ethnopharmacol.* 50 (3): 167–70.

Gülçin I (2005). The anti-oxidant and radical scavenging activities of black pepper (*Piper nigrum*) seeds. *Int. J. Food Sci. Nutr.* 56 (7): 491–99.

Han HK (2011). The effects of black pepper on the intestinal absorption and hepatic metabolism of drugs. Expert Opin. *Drug Metab. Toxicol.* 7 (6): 721–29.

Han Y, Chin Tan TM, Lim LY (2008). *In-vitro* and *in-vivo* evaluation of the effects of piperine on P-gp function and expression. *Toxicol. Appl. Pharmacol.* 230 (3): 283–89.

Hlavackova L, Urbanova A, Ulicna O, Janega P, Cerna A, Babal P (2010). Piperine, active substance of black pepper, alleviates hypertension induced by NO synthase inhibition. *Bratisl. Lek. Listy.* 111 (8): 426–31.

Janakiraman K, Manavalan R (2008). Studies on effect of piperine on oral bioavailability of ampicillin and norfloxacin. *Afr. J. Tradit. Complement. Altern. Med.* 5 (3): 257–62.

Jin MJ, Han HK (2010). Effect of piperine, a major component of black pepper, on the intestinal absorption of fexofenadine and its implication on food-drug interaction. *J. Food Sci.* 75 (3): H93–96.

Jin Z, Borjihan G, Zhao R, Sun Z, Hammond GB, Uryu T (2009). Anti-hyperlipidemic compounds from the fruit of *Piper longum* L. *Phytother. Res.* 23 (8): 1194–96.

Kakarala M, Brenner DE, Korkaya H, Cheng C, Tazi K, Ginestier C, Liu S, Dontu G, Wicha MS (2010). Targeting breast stem cells with the cancer preventive compounds curcumin and piperine. *Breast Cancer. Res. Treat.* 122: 777–85.

Kapoor IP, Singh B, Singh G, De Heluani CS, De Lampasona MP, Catalan CA (2009). Chemistry and *in-vitro* anti-oxidant activity of volatile oil and oleoresins of black pepper (*Piper nigrum*). *J. Agric. Food Chem.* 57 (12): 5358–64.

Khajuria A, Zutshi U, Bedi KL (1998). Permeability characteristics of piperine on oral absorption–an active alkaloid from peppers and a bioavailability enhancer. *Indian J. Exp. Biol.* 36: 46–50.

Kim HG, Han EH, Jang WS, Choi JH, Khanal T, Park BH, Tran TP, Chung YC, Jeong HG (2012). Piperine inhibits PMA-induced cyclooxygenase-2 expression through downregulating NF-κB, C/EBP and AP-1 signaling pathways in murine macrophages. *Food Chem. Toxicol.* 50 (7): 2342–48.

Koul IB, Kapil A (1993). Evaluation of the liver protective potential of piperine, an active principle of black and long peppers. *Planta. Med.* 59 (5): 413–17.

Kumar S, Kamboj J, Suman, Sharma S (2011). Overview for various aspects of the health benefits of *Piper longum* linn. fruit. *J. Acupunct. Meridian Stud.* 4 (2): 134–40.

Lai LH, Fu QH, Liu Y, Jiang K, Guo QM, Chen QY, Yan B, Wang QQ, Shen JG (2012). Piperine suppresses tumor growth and metastasis *in-vitro* and *in-vivo* in a 4T1 murine breast cancer model. *Acta. Pharmacol. Sin.* 33 (4): 523–30.

Lambert JD, Hong J, Kim DH, Mishin VM, Yang CS (2004). Piperine enhances the bioavailability of the tea polyphenol (-)-epigallocatechin-3-gallate in mice. *J. Nutr.* 134: 1948–52.

Lee SA, Hong SS, Han XH, Hwang JS, Oh GJ, Lee KS, Lee MK, Hwang BY, Ro JS (2005). Piperine from the fruits of *Piper longum* with inhibitory effect on monoamine oxidase and anti-depressant-like activity. *Chem. Pharm. Bull. (Tokyo).* 53 (7): 832–35.

Lee SW, Rho MC, Park HR, Choi JH, Kang JY, Lee JW, Kim K, Lee HS, Kim YK (2006). Inhibition of diacylglycerol acyltransferase by alkamides isolated from the fruits of *Piper longum* and *Piper nigrum*. *J. Agric. Food Chem.* 54 (26): 9759–63.

Li S, Lei Y, Jia Y, Li N, Wink M, Ma Y (2011). Piperine, a piperidine alkaloid from *Piper nigrum* re-sensitizes P-gp, MRP1 and BCRP dependent multidrug resistant cancer cells. *Phytomedicine.* 19 (1): 83–87.

Lin JH, Yamazaki M (2003). Role of P-glycoprotein in pharmacokinetics: clinical implications. *Clin. Pharmacokinet.* 42 (1): 59–98.

Liu Y, Yadev VR, Aggarwal BB, Nair MG (2010). Inhibitory effects of black pepper (*Piper nigrum*) extracts and compounds on human tumor cell proliferation, cyclooxygenase enzymes, lipid peroxidation and nuclear transcription factor-kappa-B. *Nat. Prod. Commun.* 5 (8): 1253–57.

Majumdar AM, Dhuley JN, Deshmukh VK, Naik SR (1999). Effect of piperine bioavailability of Oxyphenylbutazone in rats. *Indian Drugs.* 36: 123.

Majumdar AM, Dhuley JN, Deshmukh VK, Raman PH, Naik SR (1990a). Anti-inflammatory activity of piperine. *Japan. J. Med. Sc. Biol.* 43: 95.

Majumdar AM, Dhuley JN, Deshmukh VK, Raman PH, Tharat SL, Naik SR (1990b). Effect of piperine on pentobarbitore induced hypnosis in rats. *Indian J. Exp. Biol.* 28: 486.

Makhov P, Golovine K, Canter D, Kutikov A, Simhan J, *et al.* (2012). Co-administration of piperine and docetaxel results in improved anti-tumor efficacy via inhibition of CYP3A4 activity. *Prostate.* 6: 661–67.

Mao QQ, Huang Z, Ip SP, Xian YF, Che CT (2011b).Role of 5-HT(1A) and 5-HT(1B) receptors in the anti-depressant-like effect of piperine in the forced swim test. *Neurosci. Lett.* 504 (2): 181–84.

Mao QQ, Xian YF, Ip SP, Che CT (2011a). Involvement of serotonergic system in the anti-depressant-like effect of piperine. Prog. Neuro-psychopharmacol. *Biol. Psychiatry.* 35 (4): 1144–47.

Meghwal M, Goswami TK (2013). *Piper nigrum* and piperine: an update. *Phytother. Res.* 27 (8): 1121–30.

Mishra P, Sinha S, Guru SK, Bhushan S, Vishwakarma RA, Ghosal S (2011). Two new amides with cytotoxic activity from the fruits of *Piper longum. J. Asian Nat. Prod. Res.* 13 (2): 143–88.

Mishra RK, Singh SK (2009). Anti-spermatogenic and anti-fertility effects of fruits of *Piper nigrum* L. in mice. *Indian J. Exp. Biol.* 47 (9): 706–14.

Mori A, Kabuto H, Pei YQ (1985). Effects of piperine on convulsions and on brain serotonin and catecholamine levels in E1 mice. *Neurochem. Res.* 10 (9): 1269–75.

Munakata M, Kobayashi K, Niisato-Nezu J, Tanaka S, Kakisaka Y, Ebihara T, Ebihara S, Haginoya K,Tsuchiya S, Onuma A (2008). Olfactory stimulation using black pepper oil facilitates oral feeding in pediatric patients receiving long-term enteral nutrition. *Tohoku J. Exp. Med.* 214 (4): 327–32.

Naseri MK, Yahyavi H (2008). Anti-spasmodic effect of *Piper nigrum* fruit hot water extract on rat ileum. *Pak. J. Biol. Sci.* 11 (11): 1492–96.

Nishimura Y, Kitagishi Y, Yoshida H, Okumura N, Matsuda S (2011). Ethanol extracts of black pepper or turmeric down-regulated SIRT1 protein expression in Daudi culture cells. *Mol. Med. Rep.* 4 (4): 727–30.

Noumedem JA, Mihasan M, Kuiate JR, Stefan M, Cojocaru D, Dzoyem JP, Kuete V (2013). *In-vitro* anti-bacterial and antibiotic-potentiation activities of four edible plants against multidrug-resistant Gram-negative species. *BMC. Complement. Altern. Med.* 13: 190.

Park UH, Jeong HS, Jo EY, Park T, Yoon SK, Kim EJ, Jeong JC, Um SJ (2012). Piperine, a component of black pepper, inhibits adipogenesis by antagonizing PPARγ activity in 3T3-L1 cells. *J. Agric. Food Chem.* 60 (15): 3853–60.

Pattanaik S, Hota D, Prabhakar S, Kharbanda P, Pandhi P (2006). Effect of piperine on the steady-state pharmacokinetics of phenytoin in patients with epilepsy. *Phytother. Res.* 20: 683–86.

Piyachaturawat P, Glinsukon T, Peugvicha P (1982). Postcoital anti-fertility effect of piperine. *Contraception.* 26 (6): 625–33.

Piyachaturawat P, Glinsukon T, Chanjarunee A (1985). Anti-fertility effect of Citrus hystrix DC. *J. Ethnopharmacol.* 13 (1): 105–10.

Pradeep CR, Kuttan G (2002). Effect of piperine on the inhibition of lung metastasis induced B16F-10 melanoma cells in mice. *Clin. Exp. Metastasis.* 19: 703–08.

Pradeep CR, Kuttan G (2004). Piperine is a potent inhibitor of nuclear factor-kappaB (NF-kappaB), c-Fos, CREB, ATF-2 and proinflammatory cytokine gene expression in B16F-10 melanoma cells. *Int. Immunopharmacol.* 4: 1795–803.

Rauscher FM, Sanders RA, Watkins JB 3rd (2000). Effects of piperine on anti-oxidant pathways in tissues from normal and streptozotocin-induced diabetic rats. *J. Biochem. Mol. Toxicol.* 14 (6): 329–34.

Reen RK, Singh J (1991). *In-vitro* and *in-vivo* inhibition of pulmonary cytochrome P450 activities by piperine, a major ingredient of piper species. *Indian J. Exp. Biol.* 29: 568–73.

Rinwa P, Kumar A (2012). Piperine potentiates the protective effects of curcumin against chronic unpredictable stress-induced cognitive impairment and oxidative damage in mice. *Brain Res.* 1488: 38–50.

Sama V, Nadipelli M, Yenumula P, Bommineni MR, Mullangi R (2012). Effect of piperine on anti-hyperglycemic activity and pharmacokinetic profile of nateglinide. *Arzneimittelforschung.* 62 (8): 384–48.

Samykutty A, Shetty AV, Dakshinamoorthy G, Bartik MM, Johnson GL, Webb B, Zheng G, Chen A,Kalyanasundaram R, Munirathinam G (2013). Piperine, a bioactive component of pepper spice exerts therapeutic effects on androgen dependent and androgen independent prostate cancer cells. *PLoS One.* 8 (6): e65889. Print 2013.

Sawangjaroen N, Sawangjaroen K, Poonpanang P (2004). Effects of *Piper longum* fruit, *Piper sarmentosum* root and *Quercus infectoria* nut gall on caecal amoebiasis in mice. *J. Ethnopharmacol.* 91 (2-3): 357–60.

Sehgal A, Kumar M, Jain M, Dhawan DK (2011). Combined effects of curcumin and piperine in ameliorating benzo(a)pyrene induced DNA damage. *Food Chem. Toxicol.* 49 (11): 3002–06.

Sehgal A, Kumar M, Jain M, Dhawan DK (2012a). Synergistic effects of piperine and curcumin in modulating benzo(a)pyrene induced redox imbalance in mice lungs. *Toxicol. Mech. Methods.* 22 (1): 74–80.

Sehgal A, Kumar M, Jain M, Dhawan DK (2012b). Piperine as an adjuvant increases the efficacy of curcumin in mitigating benzo(a)pyrene toxicity. *Hum. Exp. Toxicol.* 31 (5): 473–82.

Selvendiran K, Banu SM, Sakthisekaran D (2004). Protective effect of piperine on benzo(a)pyrene-induced lung carcinogenesis in Swiss albino mice. *Clin. Chim. Acta.* 350: 73–78.

Selvendiran K, Singh JPV, Sakthisekaran D (2006) In vivo effect of piperine on serum and tissue glycoprotein levels in benzo (a) pyrene induced lung carcinogenesis in Swiss albino mice. *Pulm. Pharmacol. Ther.* 19: 107–111.

Shah SS, Shah GB, Singh SD, Gohil PV, Chauhan K, Shah KA, Chorawala M (2011). Effect of piperine in the regulation of obesity-induced dyslipidemia in high-fat diet rats. *Indian J. Pharmacol.* 43 (3): 296–99.

Shenoy PA, Nipate SS, Sonpetkar JM, Salvi NC, Waghmare AB, Chaudhari PD (2013a). Anti-snake venom activities of ethanolic extract of fruits of *Piper longum* L. (Piperaceae) against Russell's viper venom: characterization of piperine as active principle. *J. Ethnopharmacol.* 147 (2): 373–82.

Shenoy PA, Nipate SS, Sonpetkar JM, Salvi NC, Waghmare AB, Chaudhari PD (2013b). Production of high titre antibody response against Russell's viper venom in mice immunized with ethanolic extract of fruits of *Piper longum* L. (Piperaceae) and piperine. Phytomedicine. pii: S0944-7113 (13) 00320–26. [Epub ahead of print].

Shoba G, Joy D, Joseph T, Majeed M, Rajendran R, Srinivas PS (1998). Influence of piperine on the pharmacokinetics of curcumin in animals and human volunteers. *Planta. Med.* 64: 353–56.

Singh R, Singh N, Saini BS, Rao HS (2008). *In-vitro* anti-oxidant activity of pet ether extract of black pepper. *Indian J. Pharmacol.* 40 (4): 147–51.

Singh R, Singh N, Saini BS, Rao HS (2007). Hepato-protective and anti-oxidant properties of methanolic extract of *Piper nigrum* Linn. in rats. *Phcog. Mag.* 3: 251–58.

Srinivasan, K (2007). Black pepper and its pungent principle-piperine: a review of diverse physiological effects. *Crit. Rev. Food Sci. Nutr.* 47 (8): 735–48.

Sunila ES, Kuttan G (2004). Immuno-modulatory and anti-tumor activity of *Piper longum* Linn. and piperine. *J. Ethnopharmacol.* 90 (2-3): 339–46.

Sunila ES, Kuttan G (2005). Protective effect of *Piper longum* fruit ethanolic extract on radiation induced damages in mice: a preliminary study. *Fitoterapia.* 76 (7-8): 649–55.

Sunila ES, Kuttan G (2006). *Piper longum* inhibits VEGF and pro-inflammatory cytokines and tumor-induced angiogenesis in C57BL/6 mice. *Int. Immunopharmacol.* 6 (5): 733–41.

Suresh D, Srinivasan K (2010). Tissue distribution and elimination of capsaicin, piperine and curcumin following oral intake in rats. *Indian J. Med. Res.* 131: 682–91.

Tak JK, Lee JH, Park JW (2012). Resveratrol and piperine enhance radio-sensitivity of tumor cells. *BMB Rep.* 45 (4): 242–46.

Taqvi SI, Shah AJ, Gilani AH (2008). Blood pressure lowering and vasomodulator effects of piperine. *J. Cardiovasc. Pharmacol.* 52 (5): 452–58.

USDA, National Nutrition Data Base for standard reference, Release-26, NDB No-02030.

Vijayakumar RS, Nalini N (2006a). Efficacy of piperine, an alkaloidal constituent from *Piper nigrum* on erythrocyte anti-oxidant status in high fat diet and anti-thyroid drug induced hyper-lipidemic rats. *Cell. Biochem. Funct.* 24 (6): 491–98.

Vijayakumar RS, Nalini N (2006b). Piperine, an active principle from *Piper nigrum,* modulates hormonal and apo-lipoprotein profiles in hyper-lipidemic rats. *J. Basic Clin. Physiol. Pharmacol.* 17 (2): 71–86.

Vijayakumar RS, Surya D, Nalini N (2004). Anti-oxidant efficacy of black pepper (*Piper nigrum* L.) and piperine in rats with high fat diet induced oxidative stress. *Redox. Rep.* 9 (2): 105–110.

Wattanathorn J, Chonpathompikunlert P, Muchimapura S, Priprem A, Tankamnerdthai O (2008). Piperine, the potential functional food for mood and cognitive disorders. *Food Chem. Toxicol.* 46 (9): 3106–10.

Wongpa S, Himakoun L, Soontornchai S, Temcharoen P (2007). Anti-mutagenic effects of piperine on cyclophosphamide-induced chromosome aberrations in rat bone marrow cells. *Asian Pac. J. Cancer Prev.* 8 (4): 623–27.

Yaffe PB, Doucette CD, Walsh M, Hoskin DW (2013). Piperine impairs cell cycle progression and causes reactive oxygen species-dependent apoptosis in rectal cancer cells. *Exp. Mol. Pathol.* 94 (1): 109–14.

Ying X, Chen X, Cheng S, Shen Y, Peng L, Xu HZ (2013a). Piperine inhibits IL-β induced expression of inflammatory mediators in human osteo-arthritis chondrocyte. *Int. Immuno-pharmacol.* 17 (2): 293–99.

Ying X, Yu K, Chen X, Chen H, Hong J, Cheng S, Peng L (2013b). Piperine inhibits LPS induced expression of inflammatory mediators in RAW 264.7 cells. *Cell. Immunol.* 285 (1-2): 49–54.

Chapter 6

Cardamom
(*Elettaria cardamomum*)

Cardamom, known as Queen of Spices, a much coveted most expensive spice, is a tiny, brown seed. It has a pleasant aroma due to its volatile oil; and a characteristic, light pungent taste. It is a dried, unripened fruit pod and is native to the Middle East, North Africa, and Scandinavia, with India being the second largest producer of cardamom. Interestingly, the consumption figures also show that India, UAE and Saudi Arabia account for 60 per cent of the world consumption. Today cardamom is cultivated in India, Nepal, Sri Lanka, Mexico, Thailand and Central America. In India it is widely grown in Kerala.

Traditional Uses

In Europe, cardamom was traditionally used in festive occasions but today it is an ingredient in breads, pastries (particularly in Scandinavian countries) as well as some desserts, much like cinnamon. It has also been known to be used for gin making. In Russia and Germany, it has been used to flavour liquors for over 500 years. In India, it is a well established culinary flavorant in a wide range of sweets and confectionery; flavored milk, icecream, yogurt, 'srikhand' etc. In South Asia, green cardamom is often used in traditional Indian sweets, tea ('masala chai' -spiced tea) and sometimes coffee. Black cardamom is sometimes used in 'garam masala' for curries. It is occasionally used as a garnish in basmati rice and other dishes. The most common use of cardamom is as an excellent mouth freshener. It is also an essential component of the traditional betel quid.

Types/Varieties

Cardamom belongs to ginger family and refers to two main genera:

1. *Elettaria* (commonly called cardamom, green cardamom, or true cardamom) is distributed from India to Malaysia.

2. *Amomum* (commonly known as black cardamom, brown cardamom, Kravan, Java cardamom, Bengal cardamom, Siamese cardamom, white or red cardamom) is distributed mainly in Asia and Australia.

The three Indian varieties of green cardamom are:

1. *Malabar (Nadan/Native)*–As the name suggests, this is the native variety of Kerala. Its long dark-green leaves are lance shaped, and its yellowish or bluish flowers blossom near the ground. The Malabar type, rounded in shape, has a pleasantly mellow flavor generally regarded as superior.

2. *Mysore*–As the name suggests, this is a native variety of Karnataka with pannicles growing vertically upwards, leaves ribbed and three cornered, has a slightly harsher flavor but retains its original green color.

3. *Vazhuka*–A naturally occurring hybrid between Malabar and Mysore varieties. The pannicles grow in between vertically and horizontally.

Chemical and Nutritional Constituents

The cardamom fruit comprises of 70 per cent seeds and 30 per cent skin/peel. The main chemical constituents of the volatile oil of cardamom seeds are 1,8-cineole, alpha-terpinyl acetate, limonene, sabinene and terpineol in the form of formic and acetic acids, alpha-pinene, alpha-terpineol, borneol, linalool, linalyl acetate, myrcene and branched alpha-glucans. Olennikov and Rokhin (2013) have shown the presence of neutral and acidic components in the water-soluble polysaccharides from true cardamom seeds. Three polysaccharides (380, 166, and 27 kDa) have been isolated from the neutral fraction.

Anti-oxidant Property

The hepatic and cardiac anti-oxidant enzyme and lipid conjugated dienes were found to be significantly enhanced while GSH content was markedly restored in rats fed a high fat diet with cinnamon and cardamom. In addition, they partially counteracted the increase in lipid conjugated dienes and hydroperoxides, the primary products of lipid peroxidation exerting anti-oxidant protection (Dhuley, 1999).

Nutritive Value of Cardamom

Nutrient	Amount	Nutrient	Amount
Energy (Kcal)	311.0	Magnesium (mg)	229.0
Protein (g)	10.8	Manganese mg)	28.0
Fat (g)	6.7	Zinc (mg)	7.5
Fibre (g)	28.0	Copper (mg)	0.383
Carbohydrates (g)	68.5	Thiamine (mg)	0.198
Calcium (mg)	383.0	Riboflavin (mg)	0.182
Phosphorus (mg)	178.0	Niacin (mg)	1.102
Iron (mg)	14.0		

Source: USDA Nutrient Database–26.

Anti-microbial Property

The essential oil isolated from Greater cardamom (*A. subulatum* Roxb. Zingiberaceae) was effective against majority of micro-organisms used *viz. B. pumilus, S. aureus, S. epidermidis, P. aeruginosa* and *S. cerevisiae* (Agnihotri and Wakode, 2010).

Cardamom completely inhibited aflatoxins (*A. flavus, A. versicolor* and *P. citrinum*) and sterigmatocystin production (Bokhari, 2007). The extracts are effective against oral pathogenic bacteria like S. mutans and C. albicans (Aneja and Radhika, 2009). The chloroform and acetone extracts of spices–cardamom and cloves exhibited inhibitory activity on growth and toxin elaboration of xerophilic and aflatoxigenic fungi (both tea contaminants), *A. niger* ML01 and *A. flavus* ML02 (Al-Sohaibani *et al.,* 2011).

Anti-toxic Property

The deleterious effects of pan masala on lungs (adeno-carcinoma, edema, and inflammation with increased activity of acid phosphatase, alkaline phosphatase, and lactate dehydrogenase) were seen to be less in cardamom treated group along with significant decrease in the enzymatic activity (Kumari and Dutta, 2013).

Cardamom flavored gum was found to be effective in lessening the nicotine withdrawal symptoms in individuals trying to quit smoking (Cohen *et al.,* 2010).

Anti-inflammatory Property

Cardamom showed a significant immune-suppressive activity proving the folkloric use in treating immune-related disorders in Morooco (Daoudi *et al.,* 2013). Cardamom oil in doses of 175 and 280 µl/kg and indomethacin in a dose of 30 mg/kg revealed excellent anti-inflammatory activity against acute carrageenan-induced planter oedema in male albino rats. In addition, cardamom oil exerted its anti-spasmodic action through muscarinic receptor blockage (al-Zuhair *et al.,* 1996).

Therapeutic Benefits of Cardamom

Cardamom is used as a common folk remedy to treat stomach aches (Kubo *et al.,* 1991). It is used in traditional medicines for colds, bronchitis, fevers, along with liver, kidney and bladder problems. The volatile oil of cardamom is anti-microbial, analgesic and cardio-tonic. Cardamom is an effective remedy for various ailments like anorexia and respiratory problems such as asthma. Being rich in phytochemical cineole, cardamom is responsible for stimulating the digestive system, reducing gastric acidity, relieving flatulence and increasing appetite. It is anti-inflammatory and anti-spasmodic; thus effective for heartburn and gastric troubles.

Gastro-intestinal Disorders

The crude methanolic extract of *Amomum subulatum* (black cardamom) and its fractions, *viz.* essential oil, petroleum ether and ethyl acetate, inhibited gastric lesions induced by ethanol significantly, but not those which were induced by pylorus ligation and aspirin (Jafri *et al.,* 2001). On the contrary, Jamal *et al.* (2006) reported that all fractions of crude methanolic extract (TM), essential oil (EO), petroleum ether soluble (PS) and insoluble (PI) fractions of cardamom significantly inhibited gastric lesions induced by ethanol as well as aspirin. They also reported that the reduction in the lesions by TM was about 70 per cent in the EtOH-induced ulcer model at 500 mg/kg. In the aspirin-induced gastric ulcer, PS fraction showed the best gastro-protective effect, where the inhibition was nearly 100 per cent at 12.5 mg/kg and interestingly, this fraction at the dosage level of ≤12.5mg/kg proved to be more effective than ranitidine at 50mg/kg.

Cancer

The chemo-preventive potential of cardamom was confirmed in mice treated orally with 0.5 mg of cardamom powder in suspension continuously at pre, peri, and post initiational stages of papilloma-genesis by a significant reduction in the values of tumor incidence, tumor burden, and tumor yield; and the cumulative number of papillomas as compared to the control group. Further, reduced glutathione level was significantly elevated and the lipid peroxidation level was significantly decreased in the liver indicating the chemo-preventive potential of cardamom against two-stages of skin cancer (Qiblawi *et al.,* 2012).

Black pepper significantly enhanced T helper (Th)1 cytokine release by splenocytes and suppressed Th2 cytokine release while cardamom significantly suppressed T helper (Th)1 cytokine release by splenocytes and enhanced Th2 cytokine release suggesting that cardamom exerts anti-inflammatory role, and significantly enhanced the cyto-toxic activity of natural killer cells, indicating its potential anti-cancer effects (Majdalawieh and Carr, 2010).

Oral administration of cardamom to 7,12-dimethylbenz[a]anthracene (DMBA)-treated mice up-regulated the phase II detoxification enzymes, such as glutathione-S-transferase and glutathione peroxidase; and upregulated reduced glutathione, glutathione reductase, superoxide dismutase and catalase in mice. In addition, cardamom also blocked NF-κB activation and down-regulated cyclo-oxygenase-2 expression resulting in reduction of both the size and the number of skin papillomas (Das *et al.,* 2012). Oral dose of 0.5 per cent cardamom, in aqueous suspension, fed daily for 8 weeks to male Swiss albino mice with azoxymethane (AOM) induced colonic aberrant crypt foci (ACF) brought significant reduction in the incidence of aberrant crypt foci. This reduction in ACF was accompanied by suppression of cell proliferation and induction of apoptosis (Sengupta *et al.,* 2005).

The inhibitory effect of cinnamon and cardamom on azoxymethane induced colon carcinogenesis has been found to be by virtue of their anti-inflammatory, anti-proliferative and pro-apoptotic activity. They have also shown that the aqueous suspensions of cinnamon and cardamom to enhance the level of detoxifying enzyme (GST activity) with simultaneous decrease in lipid peroxidation levels in the treatment groups when compared to that of the carcinogen control group (Bhattacharjee *et al.,* 2007).

Limonene and cineole demonstrated promising effects against carcinogenesis (Acharya *et al.,* 2010). Cardamom's bioactive principles (eucalyptol, alpha-pinene, beta-pinene, d-limonene and geraniol) revealed pro-apoptopic, anti-inflammatory, anti-proliferative, anti-invasive and anti-angiogenic properties through a dual reverse virtual screening protocol (Bhattacharjee and Chatterjee, 2013).

CVD

Gilani *et al,* (2008) found that cardamom exhibits gut excitatory and inhibitory effects mediated through cholinergic and Ca^{++} antagonist mechanisms respectively and lowers BP *via* combination of both pathways. The diuretic and sedative effects may offer added value in its use in hypertension and epilepsy.

Cardamom powder (3 g) in two divided doses for 12 weeks administered to individuals with primary hypertension of stage 1 significantly decreased systolic, diastolic and mean blood pressure, and in addition significantly increased fibrinolytic activity. Total anti-oxidant status was also significantly increased by 90 per cent (Verma *et al.,* 2009).

Suneetha and Krishnakantha (2005) found the inhibitory effect of cardamom extract on human platelets was dose dependent with concentrations varying between 0.14 and 0.70 mg and time dependent at IC50. An increase in concentration of cardamom decreased the MDA formation significantly. Thus the aqueous extract of cardamom may have component(s), which protect platelets from aggregation and lipid peroxidation.

Conclusion

Cardamom, often used as mouth freshener and fragrance enhancer in cookery, possesses several health benefits as described above proving its potential as a therapeutic food ingredient.

References

Acharya A, Das I, Singh S, Saha T (2010). Chemo-preventive properties of indole-3-carbinol, diindolylmethane and other constituents of cardamom against carcinogenesis. *Recent Pat. Food Nutr. Agric.* 2 (2): 166–77.

Agnihotri S, Wakode S (2010). Anti-microbial activity of essential oil and various extracts of fruits of greater cardamom. *Indian J. Pharm. Sci.* 72 (5): 657–59.

Al-Sohaibani S, Murugan K, Lakshimi G, Anandraj K (2011). Xerophilic aflatoxigenic black tea fungi and their inhibition by *Elettaria cardamomum* and *Syzygium aromaticum* extracts. *Saudi. J. Biol. Sci.* 18 (4): 387–94.

al-Zuhair H, el-Sayeh B, Ameen HA, al-Shoora H (1996). Pharmacological studies of cardamom oil in animals. *Pharmacol. Res.* 34 (1-2): 79–82.

Aneja KR, Radhika J (2009). Anti-microbial activity of *Amomum subulatum* and *Elettaria cardamomum* against dental caries causing micro-organisms. *Ethnobotanical Leaf.* 13: 840–49.

Bhattacharjee B, Chatterjee J (2013). Identification of pro-apoptopic, anti-inflammatory, anti-proliferative, anti-invasive and anti-angiogenic targets of essential oils in cardamom by dual reverse virtual screening and binding pose analysis. *Asian Pac. J. Cancer. Prev.* 14 (6): 3735–42.

Bhattacharjee S, Rana T, Sengupta A (2007). Inhibition of lipid peroxidation and enhancement of GST activity by cardamom and cinnamon during chemically induced colon carcinogenesis in Swiss albino mice. *Asian Pac. J. Cancer Prev.* 8 (4): 578–82.

Bokhari FM (2007). Spices mycobiota and mycotoxins available in Saudi Arabia and their abilities to inhibit growth of some toxigenic fungi. *Mycobiology.* 35 (2): 47–53.

Cohen LM, Collins FL, Jr, Vanderveen JW, Weaver CC (2010). The effect of chewing gum flavor on the negative affect associated with tobacco abstinence among dependent cigarette smokers. *Addict. Behav.* 35: 955–60.

Daoudi A, Aarab L, Abdel-Sattar E (2013). Screening of immune-modulatory activity of total and protein extracts of some Moroccan medicinal plants. *Toxicol. Ind. Health.* 29 (3): 245–53.

Das I, Acharya A, Berry DL, Sen S, Williams E, Permaul E, Sengupta A, Bhattacharya S, Saha T (2012). Antioxidative effects of the spice cardamom against non-melanoma skin cancer by modulating nuclear factor erythroid-2-related factor 2 and NF-κB signalling pathways. *Br. J. Nutr.* 108 (6): 984–97.

Dhuley JN (1999). Anti-oxidant effects of cinnamon (*Cinnamomum verum*) bark and greater cardamom (*Amomum subulatum*) seeds in rats fed high fat diet. *Indian J. Exp. Biol.* 37 (3): 238–42.

Gilani AH, Jabeen Q, Khan AU, Shah AJ (2008). Gut modulatory, blood pressure lowering, diuretic and sedative activities of cardamom. *J. Ethnopharmacol.* 115 (3): 463–72.

Jafri MA, Farah, Javed K, Singh S (2001). Evaluation of the gastric anti-ulcerogenic effect of large cardamom (fruits of *Amomum subulatum* Roxb). *J. Ethnopharmacol.* 75 (2-3): 89–94.

Jamal A, Javed K, Aslam M, Jafri MA (2006). Gastro-protective effect of cardamom, *Elettaria cardamomum* Maton. fruits in rats. *J. Ethnopharmacol.* 103 (2): 149–53.

Kubo I, Himejima M, Muroi H (1991). Anti-microbial activity of flavor components of cardamom *Elettaria cardamomum* (Zingiberaceae) seed. *J. Agric. Food Chem.* 39: 1984–86.

Kumari S, Dutta A (2013). Protective effect of *Eleteria cardamomum* (L.) Maton against Pan masala induced damage in lung of male Swiss mice. Asi*an Pac. J. Trop. Med.* 6 (7): 525–31.

Majdalawieh AF, Carr RI (2010). *In-vitro* investigation of the potential immunomodulatory and anti-cancer activities of black pepper (*Piper nigrum*) and cardamom (*Elettaria cardamomum*). *J. Med.Food.* 13 (2): 371–81.

Olennikov DN, Rokhin AV (2013). Water-soluble glucans from true cardamom (*Elettaria cardamomum* White at Maton) seeds. *Prikl. Biokhim. Mikrobiol.* 49 (2): 197–202. [Article in Russian].

Qiblawi S, Al-Hazimi A, Al-Mogbel M, Hossain A, Bagchi D (2012). Chemo-preventive effects of cardamom (*Elettaria cardamomum* L.) on chemically induced skin carcinogenesis in Swiss albino mice. *J Med Food.* 15 (6): 576 – 80.

Sengupta A, Ghosh S, Bhattacharjee S (2005). Dietary cardamom inhibits the formation of azoxymethane-induced aberrant crypt foci in mice and reduces COX-2 and iNOS expression in the colon. Asian Pac. *J. Cancer Prev.* 6 (2): 118–22.

Suneetha WJ, Krishnakantha TP (2005). Cardamom extract as inhibitor of human platelet aggregation. *Phytother. Res.* 19 (5): 437–40.

USDA, National Nutrition Data Base for Standard Reference. Release-26, NDB No-02006.

Verma SK, Jain V, Katewa SS (2009). Blood pressure lowering, fibrinolysis enhancing and anti-oxidant activities of cardamom (*Elettaria cardamomum*). *Indian J. Biochem. Biophys.* 46 (6): 503–506.

Chapter 7

Fennel Seeds
(*Foeniculum vulgare*)

Fennel is a perennial herb and one of the most well-known Mediterranean aromatic plants belonging to the *Apiaceae* family (formerly the Umbelliferae) (Napoli *et al.,* 2010; Miraldi, 1999). It is a native to Southern Europe and grown extensively all over Europe, Middle-Eastern, China, India, and Turkey. The seeds resemble anise seeds in appearance, light brown in color with fine vertical stripes over their surface. Fennel seeds are commonly used as a spice and flavoring agent; and also as mouth freshener after a meal in both the Indian sub-continent and around the world.

Phytochemical and Nutritional Composition

The total phenolic content of fennel is different between wild and cultivated plants. The cultivated fennel variety had a high content of terpenes such as linool, alpha-terpinol, alpha-terpinyl acetate, thymol, caryophyllene, aromandrene, selinene, farnesene, and cadinene, while the wild laurel had a high content of eugenol and methyl eugenol, vitamin E, and sterols (Conforti *et al.,* 2006). Two compounds initially named as AA and BB were isolated for the first time from the wild fennel and later identified as 3,4-dihydroxy-phenethylalchohol-6-O-caffeoyl-β-D-glucopyranoside and 32,82-binaringenin, respectively (Ghanem *et al.,* 2012).

Despite being widely studied for its essential oils by gas chromatography-mass spectrometry little information is available on the nonvolatile constituents of the fennel (Bilia *et al.,* 2002). Whereas, the most important compounds identified in volatile oil of fennel by gas chromatography-olfactometry were trans-anethole, estragole, fenchone, limonene, alpha-pinene, gamma-terpinene and 1-octen-3-ol (Díaz-Maroto *et al.,* 2005; Aprotosoaie *et al.,* 2008).

The fractionation of an anti-mycobacterial extract of fennel seeds led to the isolation and characterization of 5-hydroxyfurano-coumarin. Major compounds obtained from the active fractions were 1, 3-benzenediol, 1-methoxycyclohexene, o-cymene, sorbic acid, 2-hydroxy-3-methyl-2-cyclopenten-1-one, estragole, limonene-10-ol and 3-methyl-2-cyclopenten-1-one (Esquivel-Ferriño *et al.,* 2012).

Nutritional Value

Nutrient	Amount	Nutrient	Amount
Energy (Kcal)	345	Vitamin C (mg)	21
Carbohydrates (g)	52.3	Sodium (mg)	88
Protein (g)	15.8	Potassium (mg)	1694
Total Fat (g)	14.9	Calcium (mg)	1196
Dietary Fiber (g)	39.8	Copper (mg)	1.07
Niacin (mg)	6.05	Iron (mg)	18.54
Pyridoxine (mg)	0.47	Magnesium (mg)	385
Riboflavin (mg)	0.35	Manganese (mg)	6.53
Thiamin (mg)	0.41	Phosphorus mg)	487
Vitamin A (IU)	135	Zinc (mg)	3.70

Source: USDA Nutrient Database.

Anti-oxidant Property

The possible health protective effects of fennel are generally associated with the anti-oxidant activity of its polyphenolics (Aaby *et al.,* 2005). As mentioned above, wild fennel was found to contain higher total phenolic and total flavonoid content than cultivated varieties (both medicinal and edible fennels) due to which they exhibit a higher free radical scavenging activity (Faudale *et al.,* 2008; Ghanem *et al.,* 2012).

The shoots of fennel plant were found to contain high concentration of tocopherols (34.54 µg/g) and highest phenolic content (65.85 mg/g) compared to leaves, stems, and inflorescences thereby showing the highest radical-scavenging activity and lipid peroxidation inhibition capacity (Barros *et al.,* 2009). They explained that the shoots seem to have the highest radical-scavenging activity and lipid peroxidation inhibition capacity (EC(50) values<1.4 mg/ml), which tallies with its highest content of phenolics (65.85±0.74 mg/g) and ascorbic acid (570.89±0.01 µg/g). The shoots also revealed high concentration of tocopherols (34.54±1.28 µg/g) and were the only part with flavonoids. Therefore, it is highly recommended for diseases induced by oxidative stress such as diabetes, bronchitis, chronic coughs, and for the treatment of kidney stones.

The fennel oil also exhibited anti-oxidant capacities, comparable in some cases to that of alpha-tocopherol and butylated hydroxyl-toluene (BHT), used as reference anti-oxidants. However at higher concentrations, fennel essential oils showed a pro-oxidant activity as assessed by TBARS method. In spite of the hydroxyl radical scavenging capacity of fennel oils (plant and/or fruit oils) being <50 per cent, their ability to inhibit 5-lipoxygenase was evident (Miguel *et al,* 2010).

The essential oil of fennel undergoes rapid oxidation in light but it is lower in the dark. The major component of essential oil of fennel, transanethol, had a lower anti-oxidant activity than essential oil of coriander. Interestingly, the mixture of essential oils from laurel (bay leaves) and coriander could be used to prevent oxidative damage of components of the fennel oil (Misharina and Polshkov, 2005).

Oral administration (200 mg/kg) of fennel significantly increased the plasma superoxide dismutase and catalase activities; and the high density lipoprotein-cholesterol level, whereas the malondialdehyde level was significantly decreased thus supporting its use for relieving inflammation (Choi and Hwang, 2004). Zhang *et al.* (2012) reported that fennel seeds decrease inflammation of liver and prevent progression of the hepatic fibrosis, probably due to its anti-oxidative property.

Fennel seeds displayed a protective effect against ethanol-induced gastric mucosal lesion, which, at least in part, depends upon the reduction in lipid peroxidation and augmentation of the anti-oxidant activity (Birdane *et al.,* 2007).

According to Mohamad *et al.* (2011) methanolic extract of fennel seed could also be used as a safe, effective, and easily accessible source of natural anti-oxidants to improve the oxidative stability of fatty foods during storage.

Anti-microbial Agent

A phenyl propanoid derivative, dillapional (1) was found to be an anti-microbial principle of the stems of fennel seeds with MIC values of 125, 250 and 125 against *B. subtilis, A. niger and C. cladosporioides,* respectively. The essential oil of this plant has been reported to possess anti-bacterial activity against *E. coli, L. monocytogenes, S. typhimurium*, and *S. aureus* (Dadalioglu *et al.,* 2004). Different methods employed for hydro-distillation of fennel seeds affected the essential oil yield and quantitative composition but only slightly altered the anti-fungal activity of the oils against some fungi (Mimica-Dukiæ *et al.,* 2003). Dichloromethane extracts and essential oil from *F. vulgare* showed anti-fungal activity against *C. albicans* (Park and Seong, 2010).

Kwon *et al.* (2002) isolated a coumarin derivative, scopoletin which was found to be marginally anti-microbial while other compounds dillapiol, bergapten, imperatorin and psolaren were found to be inactive.

Kaur and Arora (2009) reported good, strain specific inhibitory activity of fennel seeds against the tested bacteria *viz. E. faecalis, S. aureus, E. coli, K. pneumoniae, K. pneumoniae, P. aeruginosa, P. aeruginosa, S. typhi, S. flexneri.* Aqueous as well as acetone extracts of seeds showed almost comparable anti-bacterial activity, supporting their traditional use against infectious diseases. The presence of various phytochemicals might be responsible for their anti-microbial effects.

Fennel seeds exhibited varietal difference in its anti-microbial property. The essential oil samples from three fennel cultivars displayed different anti-bacterial activity against Gram negative and Gram positive bacteria. Essential oils from theazoricum and dulce cultivars were more effective anti-oxidants than that from the vulgare cultivar (Shahat *et al.*, 2011).

Fennel essential oil among 10 other Turkish spice essential oils showed anti-bacterial activity against one or more of the *Bacillus* species (*B. amyloliquefaciens, B. brevis, B. cereus, B. megaterium, B. subtilis,* and *B.* subtilis var. niger) and were more effective at 1:50 and 1:100 levels (Ozcan *et al.*, 2006). Fennel oil, which contains a high level of trans-anethole, was active against *S. aureus*, with MICs ranging from 64 to 256 µg/ml. Furthermore, fennel oil, when used at sub-inhibitory concentrations, could dose-dependently decrease the expression of *S. aureus* exotoxins, including α-toxin, *Staphylococcal* entero-toxins and toxic shock syndrome toxin 1 (Qiu *et al.*, 2012).

Fennel essential oil exhibited anti-bacterial activity on forty eight isolates of *A. baumannii* suggesting the potential use of the fennel essential oil for the control of multi-drug resistant *A. baumannii* infections (Jazani *et al.*, 2009). Oils from the two samples of *F. vulgare* showed a higher and broader degree of inhibition than that of *C. maritimum* (Ruberto *et al.*, 2000).

Fennel oil samples also showed high activity against *C. albicans* and significant synergistic activity with amoxicillin or tetracycline against *E. coli, Sarcina lutea* and *B. subtilis* strains. The significant anti-bacterial activity of essential oils on the bacterial pathogens of mushrooms appears promising (Lo Cantore *et al.*, 2004; Aprotosoaie *et al.*, 2008). The *in-vitro* anti-microbial assays showed that the essential oil, anethole, and hexane extract were effective against most of the food borne pathogenic, saprophytic, probiotic, and myco-toxigenic micro-organisms tested. (E)-anethole, the main component of Florence fennel essential oil, is responsible for the anti-microbial activity and that the essential oils as well as the hexane extract can be used as a food preservative (Cetin *et al.*, 2010).

Traditional Therapeutic Uses

Fennel is a popular medicinal plant with various pharmacological activities such as anti-oxidant, cytotoxic, anti-inflammatory, anti-microbial, broncho-dilatory, estrogenic, diuretic, lithontripic, galactogogue, emmenagogue, anti-thrombotic, hypo-tensive, gastro-protective, hepato-protective, memory enhancing, and anti-mutagenic activities as mentioned in traditional Iranian medicine and modern phytotherapy (Rahimi and Ardekani, 2013). It has been used in the folk medicine for centuries as a carminative, digestive, lactagogue, and diuretic. Its herbal drug preparations are used for mild spasmodic gastrointestinal complaints (bloating and flatulence) and respiratory disorders. Fennel is also used for catarrh of the upper respiratory tract (Parejo *et al.*, 2004).

CVD

Oral administration of *Foeniculum* extract lowered the systolic blood pressure of spontaneously hyper-tensive but not of normo-tensive Wistar-Kyoto rats and appeared to act mainly as a diuretic and a natri-uretic (El Bardai *et al.*, 2001). Swaminathan *et al.* (2012) confirmed the functional effects of fennel derived-nitrites using *in-vitro* and *ex-vivo* models that describe the promotion of angiogenesis, cell migration, and vaso-relaxation. They also showed that chewing fennel seeds enhanced nitrite content of saliva thus indicating the potential role of fennel derived-nitrites on the vascular system.

The essential oil and anethole tested in rat aorta with or without endothelium, displayed comparable NO-independent vaso-relaxant activity at anti-platelet concentrations which have been proved to be free from cytotoxic effects *in-vitro* (Boskabady *et al.*2004). *In-vivo*, both *F. vulgare* essential oil and anethole orally administered in a sub-acute treatment to mice (30 mg kg (-1) day (-1) for 5 days) showed significant anti-thrombotic activity preventing the paralysis induced by collagen-epinephrine intravenous injection (70 per cent and 83 per cent protection, respectively) (Tognolini *et al.*, 2007).

Cancer

Administration of methanolic extract of fennel seed before irradiation exerted a cyto-protective effect against gamma irradiation, as manifested by a restoration of the MDA level, catalase activity, and GSH content to near-normal levels. Thus, it exhibited an anti-tumor effect by modulating lipid peroxidation and augmenting the anti-oxidant defense system in Ehrlich ascites carcinoma-bearing mice with or without exposure to radiation (Mohamad *et al.*, 2011). The methanolic extract of fennel showed a characteristic mechanism based inactivation on erythromycin N-demethylation mediated by human liver microsomal cytochrome P450 3A4 (CYP3A4). The compound 5-methoxypsoralen showed the strongest inhibition with an IC50 value of 18.3 µM and a mixed type of inhibition. The kinetic parameter for mechanism-based inactivation was characterized by a KI value of 15.0 µM and a kinact value of 0.098 min(-1) (Subehan *et al.*, 2007).

The chemo-preventive potential of the of *F. vulgare* (FV) and *Salvia officinalis* (SO) water infusions were evaluated against chemical carcinogen trichloroacetic acid (TCA)-exposure in rats. The plant infusions sustained the production of anti-oxidant enzymes and reduced malondialdehyde concentration in the tissues thus proving their anti-oxidant properties (Celik and Isik, 2008). Fennel seeds also exhibited a significant reduction in the skin and the fore stomach tumor incidence and tumor multiplicity as compared to the control DMBA-induced papillomagenesis in Swiss albino mice (Singh and Kale, 2008).

Infertility

The fennel seeds have been known to regulate menstruation, alleviate the symptoms of climacteric syndrome and increase libido (Albert-Puleo, 1980). The estrogenic nature of the seed extract was confirmed by Devi *et al.* (1985) as the acetone extracts of fennel seeds at different dose levels (50/ug, 150/ug and 250/ug/100gm body wt.) increased nucleic acids and protein concentration as well as the organ weights in mammary glands and oviducts of castrated rats. Malini *et al.* (1985) also reported that oral administration of the extract for 10 days to female rats, led to vaginal cornification and oestrus cycle. While moderate doses caused increase in weight of mammary glands, higher doses increased the weight of oviduct, endometrium, myometrium, cervix as well as vagina. On the contrary, oral administration of acetone extract of fennel seeds for 15 days to male rats significantly decreased the total protein concentration in testes and vas deferens; and increased in seminal vesicles and prostate gland. There was a decrease in activities of acid and alkaline phosphatase in all these regions, except that alkaline phosphatase was unchanged in vasa confirming the oestrogenic activity of the seed extract. Similar results were reported by Rahimi and Ardekani (2013). In addition they confirmed absence of teratogenicity.

Fennel contains anti-spasmodic anethol agents and may be helpful for management of primary dysmenorrhea as it reduced the severity (Namavar *et al.,* 2003; Bokaie *et al.,* 2013). Fennel seed also possesses properties of emmenogogue and galactogogue (Ostad *et al.,* 2001). Idiopathic hirsutism is defined as the occurrence of excessive male pattern hair growth in women who have a normal ovulatory menstrual cycle and normal levels of serum androgens. It may be a disorder of peripheral androgen metabolism. *F. vulgare* is used as an estrogenic agent. The efficacy of treatment with the cream containing 2 per cent fennel is better than the cream containing 1 per cent fennel and these two were more potent than placebo. The mean values of hair diameter reduction was 7.8 per cent, 18.3 per cent and -0.5 per cent for patients receiving the creams containing 1 per cent, 2 per cent and 0 per cent (placebo) respectively (Javidnia *et al.,* 2003).

Anti-oxidants exert a protective effect on the plasma membrane of frozen boar sperm, thus due to its anti-oxidant property, fennel is found to be useful in cryo-preservation of sperm (Malo *et al.,* 2012).

Other Benefits

The aqueous extract of fennel seeds possesses significant oculo-hypotensive activity, which was found to be comparable to that of timolol though further investigations on possible toxicity, human clinical trials and mechanism of action are suggested (Agarwal *et al.,* 2008).

The aqueous extract of fennel seeds showed statistically significant protective effect against ethanol-induced gastric mucosal lesions in rats with reduction in lipid peroxidation and augmentation in the anti-oxidant activity by significant increase in GSH, nitrite, nitrate, ascorbic acid, retinol and beta-carotene levels (Birdane *et al.,* 2007).

It has been reported that fennel essential oil could be used in pediatric colic and respiratory disorders due to its anti-spasmodic effects (Savino *et al.,* 2005; Ozbek *et al.,* 2003). According to the Wessel criteria the fennel oil emulsion significantly eliminated colic in 65 per cent of infants in the treatment group compared to 23.7 per cent of infants in the control group suggesting that fennel seed oil emulsion is superior to placebo in decreasing intensity of infantile colic (Alexandrovich *et al.,* 2003). The hepato-toxicity produced by acute carbon tetrachloride administration was found to be inhibited by fennel essential oil with evidence of decreased levels of serum aspartate amino-transferase (AST), alanine amino-transferase, alkaline phosphatase (ALP) and bilirubin. The results of this study indicate that fennel essential oil has a potent hepato-protective action against CCl(4)-induced hepatic damage in rats (Ozbek *et al.,* 2003).

Boskabady *et al.* (2004) confirmed the broncho-dilatory effects of ethanol extract and essential oil from *F. vulgare.* A potassium channel opening effect and not calcium channels may contribute to its broncho-dilatory effect on guinea pig tracheal chains.

Methanolic extract of the whole plant of *F. vulgare* administered for eight successive days ameliorated the amnesic effect of scopolamine (0.4 mg/kg) and aging-induced memory deficits in mice. *F. vulgare* extract increased step-down latency and acetyl-cholinesterase inhibition in mice significantly. Hence, *F. vulgare* can be employed in treatment of cognitive disorders such as dementia and Alzheimer's disease (Joshi and Parle, 2006).

Kim *et al.* (2012) observed that the fennel seeds have potential in preventing bone loss in post-menopausal osteoporosis by reducing both osteoclast differentiation and function. Tripathi *et al.* (2013) found pretreatments with fennel essential oil (FEO) significantly inhibited the frequencies of aberrant metaphases, chromosomal aberrations, micronuclei formation, and cyto-toxicity in mouse bone marrow cells induced by cyclo-phosphamide (CP) and also produced a significant reduction of abnormal sperm and antagonized the reduction of CP-induced superoxide dismutase, catalase, and glutathione activities and inhibited increased malondialdehyde content in the liver. Thus FEO inhibited geno-toxicity and oxidative stress induced by CP.

A randomized controlled trial concluded that the phytotherapic compound in fennel has laxative efficacy and is a safe alternative option for the treatment of constipation (Picon *et al.,* 2010).

Further, a topical cream (w/o emulsion) containing extract of fennel showed a significant increase in the texture parameter energy proving the potential anti-aging effects of the formulation (Rasul *et al.,* 2012).

Conclusions

In spite of the high therapeutic potential of fennel seed as evidenced by research, its regular consumption is limited to certain Indian communities mostly as a mouth freshener, and of course as a flavouring agent in some non vegetarian recipes. Hence, being a cost effective spice, the regular and higher consumption of fennel could be popularized.

References

Aaby K, Skerede G, Wrolstad RE (2005). Phenolic composition and anti-oxidant activities in flesh and achenes of strawberries (*Fragaria ananassa*). *J. Agric. Food Chem.* 53: 4032–40.

Agarwal R, Gupta SK, Agrawal SS, Srivastava S, Saxena R (2008). Oculohypotensive effects of *Foeniculum vulgare* in experimental models of glaucoma. *Indian J. Physiol. Pharmacol.* 52 (1): 77–83.

Albert-Puleo M (1980). Fennel and Anise as estrogenic agents. *J. Ethnopharmacol.* 2: 337–44.

Alexandrovich I, Rakovitskaya O, Kolmo E, Sidorova T, Shushunov S (2003). The effect of fennel (*Foeniculum Vulgare*) seed oil emulsion in infantile colic: a randomized, placebo-controlled study. *Altern. Ther. Health Med.* 9 (4): 58–61.

Aprotosoaie AC, Hăncianu M, Poiată A, Tuchilu° C, Spac A, Cioană O, Gille E, Stănescu U (2008). In-vitro anti-microbial activity and chemical composition of the essential oil of *Foeniculum vulgare* Mill. *Rev. Med. Chir. Soc. Med. Nat. Iasi.* 112 (3): 832–36.

Barros L, Heleno SA, Carvalho AM, Ferreira IC (2009). Systematic evaluation of the anti-oxidant potential of different parts of *Foeniculum vulgare* Mill. from Portugal. *Food Chem. Toxicol.* 47 (10): 2458–64.

Bilia AR, Flamini G, Taglioli V, Morelli I, Vincieri FF (2002). GC–MS analysis of essential oil of some commercial Fennel teas. *Food Chem.* 76: 307–10.

Birdane FM, Cemek M, Birdane YO, Gülçin I, Büyükokuroðlu ME (2007). Beneficial effects of *Foeniculum vulgare* on ethanol-induced acute gastric mucosal injury in rats. *World J. Gastroenterol.* 13 (4): 607–11.

Bokaie M, Farajkhoda T, Enjezab B, Khoshbin A, Zarchi Mojgan K (2013). Oral fennel (*Foeniculum vulgare*) drop effect on primary dysmenorrhea: Effectiveness of herbal drug. *Iran. J. Nurs. Midwifery Res.* 18 (2): 128–32.

Boskabady MH, Khatami A, Nazari A (2004). Possible mechanism(s) for relaxant effects of *Foeniculum vulgare* on guinea pig tracheal chains. *Pharmazie.* 59 (7): 561–64.

Celik I, Isik I (2008). Determination of chemopreventive role of *Foeniculum vulgare* and *Salvia officinalis* infusion on trichloroacetic acid-induced increased serum marker enzymes lipid peroxidation and antioxidative defense systems in rats. *Nat. Prod. Res.* 22 (1): 66–75.

Cetin B, Ozer H, Cakir A, Polat T, Dursun A, Mete E, Oztürk E, Ekinci M (2010). Anti-microbial activities of essential oil and hexane extract of Florence fennel [*Foeniculum vulgare* var. *azoricum* (Mill.) Thell.] against food borne micro-organisms. *J. Med. Food.* 13 (1): 196–204.

Choi E, Hwang J (2004). Anti-inflammatory analgesic and anti-oxidant activities of the fruit of *Foeniculum vulgare*. *Fitoterapia* 75, 557–65.

Conforti F, Statti G, Uzunov D, Menichini F (2006). Comparative chemical composition and anti-oxidant activities of wild and cultivated *Laurus nobilis* L. leaves and *Foeniculum vulgare* subsp. *piperitum* (Ucria) coutinho seeds. *Biol. Pharm. Bull.* 29 (10): 2056–64.

Dadalioglu I, Evrendilek GA (2004). Chemical compositions and antibacterial effects of essential oils of Turkish oregano (*Origanum minutiflorum*), bay laurel (*Laurus nobilis*), Spanish lavender (*Lavandula stoechas* L.), and fennel (*Foeniculum vulgare*) on common food borne pathogens. *J. Agric. Food Chem.* 52 (26): 8255–60.

Devi K, Vanithakumari G, Anusya S, Mekala N, Malini T, Elango V (1985). Effect of *Foeniculum vulgare* seed extract on mammary glands and oviducts of ovariectomised rats. *Anc. Sci. Life.* 5 (2): 129–32.

Díaz-Maroto MC, Díaz-Maroto Hidalgo IJ, Sánchez-Palomo E, Pérez-Coello MS (2005). Volatile components and key odorants of fennel (*Foeniculum vulgare* Mill.) and thyme (*Thymus vulgaris* L.) oil extracts obtained by simultaneous distillation-extraction and supercritical fluid extraction. *J. Agric. Food Chem.* 53 (13): 5385–89.

El Bardai S, Lyoussi B, Wibo M, Morel N (2001). Pharmacological evidence of hypotensive activity of *Marrubium vulgare* and *Foeniculum vulgare* in spontaneously hyper-tensive rat. *Clin. Exp. Hypertens.* 23 (4): 329–43.

Esquivel-Ferriño PC, Favela-Hernández JM, Garza-González E, Waksman N, Ríos MY, del Rayo Camacho-Corona M (2012). Anti-mycobacterial activity of constituents from *Foeniculum vulgare* var. dulce grown in Mexico. *Molecules.* 17 (7): 8471–82.

Faudale M, Viladomat F, Bastida J, Poli F, Codina C (2008). Anti-oxidant activity and phenolic composition of wild, edible, and medicinal fennel from different Mediterranean countries. *J. Agric. Food Chem.* 56 (6): 1912–20.

Ghanem MT, Radwan HM, Mahdy el-SM, Elkholy YM, Hassanein HD, Shahat AA (2012). Phenolic compounds from *Foeniculum vulgare* (Subsp. *Piperitum*) (Apiaceae) herb and evaluation of hepato-protective anti-oxidant activity. *Pharmacognosy Res.* 4 (2): 104–08.

Javidnia K, Dastgheib L, Mohammadi Samani S, Nasiri A (2003). Anti-hirsutism activity of Fennel (fruits of *Foeniculum vulgare*) extract. A double-blind placebo controlled study. *Phytomedicine.* 10 (6-7): 455–58.

Jazani NH, Zartoshti M, Babazadeh H, Ali-daiee N, Zarrin S, Hosseini S (2009). Anti-bacterial effects of Iranian fennel essential oil on isolates of *Acinetobacter baumannii*. *Pak. J. Biol. Sci.* 12 (9): 738–41.

Joshi H, Parle M (2006). Cholinergic basis of memory-strengthening effect of *Foeniculum vulgare* Linn. *J. Med. Food.* 9 (3): 413–17.

Kaur GJ and Arora DS (2009). Anti-bacterial and phytochemical screening of Anethum graveolens, *Foeniculum vulgare and Trachyspermum ammi. BMC Complementary and Alternative Medicine.* 9: 30.

Kim TH, Kim HJ, Lee SH, Kim SY (2012). Potent inhibitory effect of *Foeniculum vulgare* Miller extract on osteoclast differentiation and ovariectomy-induced bone loss. *Int. J. Mol. Med.* 29 (6): 1053–59.

Kwon YS, Choi WG, Kim WJ, Kim WK, Kim MJ, Kang WH, Kim CM (2002). Anti-microbial constituents of *Foeniculum vulgare. Arch. Pharm. Res.* 25 (2): 154–57.

Lo-Cantore P, Iacobellis NS, De-Marco A, Capasso F, Senatore F (2004). Anti-bacterial activity of *Coriander sativum L.* and *Foeniculum vulgare* Miller Var. vulgare (Miller) essential oils. *J. Agri. Food Chem.* 52: 7862–66.

Malini T, Vanithakumari G, Megala N, Anusya S, Devi K, Elango V (1985). Effect of *Foeniculum vulgare* Mill. seed extract on the genital organs of male and female rats. *Indian J. Physiol. Pharmacol.* 29 (1): 21–26.

Malo C, Gil L, Cano R, González N, Luño V (2012). Fennel (*Foeniculum vulgare*) provides anti-oxidant protection for boar semen cryo-preservation. *Andrologia.* 44 Suppl. 1: 710–15.

Miguel MG, Cruz C, Faleiro L, Simões MT, Figueiredo AC, Barroso JG, Pedro LG (2010). *Foeniculum vulgare* essential oils: chemical composition, antioxidant and anti-microbial activities. *Nat. Prod. Commun.* 5 (2): 319–28.

Mimica-Dukiæ N, Kujundziæ S, Sokoviæ M, Couladis M (2003). Essential oil composition and antifungal activity of *Foeniculum vulgare* Mill obtained by different distillation conditions. *Phytother. Res.* 17 (4): 368–71.

Miraldi E (1999). Comparison of the essential oils from ten *Foeniculum vulgare* miller samples of fruits of different origin. *Flavour. Fragr. J.* 14: 379–82.

Misharina TA, Polshkov AN (2005). Anti-oxidant properties of essential oils: auto-oxidation of essential oils from laurel and fennel and effects of mixing with essential oil from coriander. *Prikl. Biokhim. Mikrobiol.* 41 (6): 693–702. [Article in Russian].

Mohamad RH, El-Bastawesy AM, Abdel-Monem MG, Noor AM, Al-Mehdar HA, Sharawy SM, El-Merzabani MM (2011). Anti-oxidant and anti-carcinogenic effects of methanolic extract and volatile oil of fennel seeds (*Foeniculum vulgare*). *J. Med. Food.* 14 (9): 986–1001.

Namavar Jahromi B, Tartifizadeh A, Khabnadideh S (2003). Comparison of fennel and mefenamic acid for the treatment of primary dysmenorrhea. *Int. J. Gynaecol. Obstet.* 80 (2): 153–57.

Napoli ME, Curcuruto G, Ruberto G (2010). Screening the essential oil composition of wild Sicilian fennel. *Biochem. Syst. Ecol.* 38: 213–23.

Ostad SN, Soodi M, Shariffzadeh M, Khorshidi N, Marzban H (2001). The effect of fennel essential oil on uterine contraction as a model for dysmenorrhea, pharmacology and toxicology study. *J. Ethnopharmacol.* 76: 299–304.

Ozbek, H. Ugras S. Dulger, H.; Bayram, I.; Tuncer, I.; Ozturk, G (2003). Hepato-protective effect of *Foeniculum vulgare* essential oil. *Fitoterapia.* 74, 317–19.

Ozcan MM, Saðdiç O, Ozkan G (2006). Inhibitory effects of spice essential oils on the growth of *Bacillus* species. *J. Med. Food.* 9 (3): 418–21.

Parejo I, Jauregui O, Snchez-Rabaneda F, Viladomat F, Bastida J, Codina C (2004). Separation and characterization of phenolic compounds in fennel (*Foeniculum vulgare*) using liquid chromatography–negative electrospray ionization tandem mass spectrometry. *J. Agric. Food Chem.* 52: 3679–87.

Park SH, Seong I (2010). Anti-fungal effects of the extracts and essential oils from *Foeniculum vulgare* and *Illicium verum* against *Candida albicans*. *Korean J. Med. Mycol*. 15 (4): 157–64.

Picon PD, Picon RV, Costa AF, Sander GB, Amaral KM, Aboy AL, Henriques AT (2010). Randomized clinical trial of a phytotherapic compound containing *Pimpinella anisum, Foeniculum vulgare, Sambucus nigra,* and *Cassia augustifolia* for chronic constipation. *BMC. Complement. Altern. Med*. 10:17.

Qiu J, Li H, Su H, Dong J, Luo M, Wang J, Leng B, Deng Y, Liu J, Deng X (2012). Chemical composition of fennel essential oil and its impact on *Staphylococcus aureus* exotoxin production. *World J. Microbiol. Biotechnol*. 28 (4): 1399–405.

Rahimi R, Ardekani MR (2013). Medicinal properties of *Foeniculum vulgare* Mill. in traditional Iranian medicine and modern phytotherapy. *Chin. J. Integr. Med*. 19(1): 73–79.

Rasul A, Akhtar N, Khan BA, Mahmood T, Uz Zaman S, Khan HM (2012). Formulation development of a cream containing fennel extract: *in-vivo* evaluation for anti-aging effects. *Pharmazie*. 67 (1): 54–58.

Ruberto G, Baratta MT, Deans SG, Dorman HJD (2000). Anti-oxidant and anti-microbial activity of *Foeniculum vulgare and Crithmum maritimum-* essential oils. *Planta. Medica*. 19: 2943–50.

Savino F, Cresi F, Castagno E, Silvestro L, Oggero R (2005). A randomized double-blind placebo-controlled trial of standardized extract of *Matricariae recutita, Foeniculum vulgare* and *Melissa offinalis* (Colimil) in the treatment of breastfed colicky infants. *Phytother. Res*. 19: 335–40.

Shahat AA, Ibrahim AY, Hendawy SF, Omer EA, Hammouda FM, Abdel-Rahman FH, Saleh MA (2011). Chemical composition, anti-microbial and anti-oxidant activities of essential oils from organically cultivated fennel cultivars. *Molecules*. 16 (2): 1366–77.

Singh B, Kale RK (2008). Chemo-modulatory action of *Foeniculum vulgare* (Fennel) on skin and fore-stomach papillomagenesis, enzymes associated with xenobiotic metabolism and anti-oxidant status in murine model system. *Food Chem. Toxicol*. 46 (12): 3842–50.

Subehan, Zaidi SF, Kadota S, Tezuka Y (2007). Inhibition on human liver cytochrome P450 3A4 by constituents of fennel (*Foeniculum vulgare*): identification and characterization of a mechanism-based inactivator. *J. Agric. Food Chem*. 55 (25): 10162–67.

Swaminathan A, Sridhara SR, Sinha S, Nagarajan S, Balaguru UM, Siamwala JH, Rajendran S, Saran U, Chatterjee S (2012). Nitrites derived from *Foeniculum vulgare* (fennel) seeds promote vascular functions. *J. Food Sci*. 77 (12): H273–79.

Tognolini M, Ballabeni V, Bertoni S, Bruni R, Impicciatore M, Barocelli E (2007). Protective effect of *Foeniculum vulgare* essential oil and anethole in an experimental model of thrombosis. *Pharmacol. Res*. 56 (3): 254–60.

Tripathi P, Tripathi R, Patel RK, Pancholi SS (2013). Investigation of antimutagenic potential of *Foeniculum vulgare* essential oil on cyclophosphamide induced genotoxicity and oxidative stress in mice. *Drug Chem. Toxicol.* 36 (1): 35–41.

USDA, National Nutrition Data Base for standard reference, Release-26, NDB No-02018.

Zhang ZG, Lu XB, Xiao L, Tang L, Zhang LJ, Zhang T, Zhan XY, Ma XM, Zhang YX (2012). Antioxidant effects of the Uygur herb, *Foeniculum Vulgare* Mill, in a rat model of hepatic fibrosis. *Zhonghua. Gan. Zang. Bing. Za. Zhi.* 20 (3): 221–26. [Article in Chinese].

Chapter 8

Grapes and Grape Seeds
(*Vitis vinifera*)

Grapes (genus *Vitis*) are one of the world's largest fruit crops and most commonly consumed in the world both as fresh (table grape) and processed fruits (wine, grape juice, molasses, and raisins) (Percival, 2009; Schamel, 2006).

There are greater than one hundred grape species which are divided into 2 sub-genera: *Euvitis* and *Muscadinia*. Most of the species belong to *Euvitis* sub genera (Vivier and Pretorius, 2002). Several varieties of grapes are used in raisin production including Muscat, Black Corinth, and Sultana (USDA, 2003).

Nutritive Value

Grapes are known to contain substantial amount of simple sugars especially glucose and fructose along with fatty acids (Radler, 1965; Stafford *et al.*, 1974) and amino acids (Bolin and Petrucci, 1985). Wine grapes usually contain ≥19 per cent sugar by fresh weight (Bellincontro *et al.*, 2004).

Grape seeds are waste products of the winery and grape juice industry. These seeds contain lipid, protein, carbohydrates, and 5–8 per cent polyphenols depending on the variety. It is estimated that approximately 60–70 per cent of grape polyphenols exist along with procyanidins in grape seeds (Zhao *et al.*, 1999, Kammerer *et al.*, 2004). Grape seeds are one of the most potent natural anti-oxidants. Polyphenols in grape seeds are mainly flavonoids, including gallic acid, the monomeric flavan-3-ols catechin, epicatechin, gallocatechin, epigallocatechin, and epicatechin 3-O-gallate; and procyanidin dimers, trimers, and more highly polymerized procyanidins (Shi *et al.*, 2003) which are flavan-3-ol derivatives (Prieur *et al.*, 1994). Natural grape seed pro-anthocyanidins are a combination of biologically active polyphenolic flavonoids including oligomeric pro-anthocyanidins which possess a broad spectrum of pharmacological, medicinal, and therapeutic properties (Jayaprakasha *et al.,* 2001; Bagchi *et al.,* 1998; Sato *et al.,* 1999; Ye *et al.,* 1999).

Kamiyama *et al.* (2009) detected the presence of cyanidin-3-glucoside (Cy-3-glc), five anthocyanidins and resveratrol in the skin of Nagano Purple grape (a hybrid created by a cross between Kyoho (*V. Labrusca*) and Rosario Bianco (*V. vinifera*) grapes. The grape skin is concentrated with resveratrols, anthocyanins, and catechins. The Norton grape skin contains 215.6 mg phenolic compounds per gram of the extract (Zhang *et al.,* 2011).

Nutritional Value of Grapes, Red or Green (per 100 g)

Nutrient	Amount	Nutrient	Amount
Energy (kcal)	69	Vitamin C (mg)	3.2
Carbohydrates (g)	18.1	Vitamin E (mg)	0.19
-Sugars (g)	15.48	Vitamin K (µg)	14.6
- Dietary fiber (g)	0.9	Calcium (mg)	10
Thiamine (vit. B_1) (mg)	0.069	Iron (mg)	0.36
Riboflavin (vit. B_2) (mg)	0.07	Magnesium (mg)	7
Niacin (vit. B_3) (mg)	0.188	Manganese (mg)	0.071
Pantothenic acid (B_5) (mg)	0.05	Phosphorus (mg)	20
Vitamin B_6 (mg)	0.086	Potassium (mg)	191
Folate (vit. B_9) (µg)	2	Sodium (mg)	2
Choline (mg)	5.6	Zinc (mg)	0.07
Fluoride (µg)	7.8		

Source: USDA Nutrient Database.

The various phytochemicals reported in raisins include triterpenes (Zhang *et al.,* 2004b), flavonoids, hydroxyl-cinnamic acids (Liggins *et al.,* 2000) and 5-hydroxy-2-furaldehyde (Palma and Tailor, 2001).

The various phytochemicals in grapes and grape products are responsible for their therapeutic benefits which are detailed below.

Anti-oxidant Property

Pakhale *et al.* (2007) reported that all Indian grape cultivars (Bangalore blue, Pandhari sahebi, Sharad seedless and Thompson seedless) and their components (whole grapes, pulp with skin and seeds (with/without seeds) varying in their skin color, seed, and polyphenol content exhibited equally good free radical scavenging activity. But in the Shahani black grapes the activity was found to be in the order of seed methanol extract> skin extract> grape juice> seed hexane extract (Yassa *et al.,* 2008).

Bangalore Blue   **Sharad Seedless**

Grape seed pro-anthocyanidin extract (GSPE) has been shown *in-vitro* and *in-vivo* to possess superior anti-oxidant performance to vitamins C, E and β-carotene (Bagchi *et al.,* 2000). The combination of resveratrol and vitamins C and/or E was more effective in protecting the cell by reducing oxidative stress in PC12 cells induced by addition of Fe^{2+} and t-butyl hydroperoxide than any of these three anti-oxidants alone (Chanvitayapongs *et al.,* 1997).

Dimeric procyanidin B2, one of the main components of grape seed pro-anthocyanidin extract has anti-inflammatory, anti-tumor and cardiovascular protective properties, which are believed at least in part, due to the anti-oxidative effect (Houde *et al.,* 2006; Khanna *et al.,* 2002; Vayalil *et al.,*2004; Chen *et al.,*2006; Mackenzie *et al.,* 2008).

The anti-oxidant activity of samples from grape seed and skin extracts are mainly based on inhibition of lipid peroxidation, whereas the anti-oxidant activity of grape juice is based on free radical scavenging activity (Yassa *et al.,* 2008). Further, peroxyl radical scavenging activity of phenolics present in grape seeds or skins was found to be in decreasing order were resveratrol > catechin > epicatechin = gallocatechin > gallic acid = ellagic acid (Yilmaz and Toledo, 2004).

Dietary intake of grape anti-oxidants helps to prevent lipid oxidation and inhibit the production of reactive oxygen species (ROS). For instance, dietary supplementation of grape seed extract (600 mg/day) for 4 weeks was shown to reduce oxidative stress and improve glutathione (GSH)/oxidized glutathione (GSSG); and total anti-oxidant status (TAOS) in a double-blinded randomized cross-over human trial (Kar *et al.,* 2009). Vinson *et al.* (2001) also demonstrated that long-term grape seed extract supplementation *i.e.* Mega Natural (R) Gold (2 × 300 mg/day) improved plasma anti-oxidant capacity, decreased plasma cholesterol, low-density lipoprotein cholesterol, and high-density lipoprotein cholesterol; and significantly increased triglycerides in the hyper-lipidaemic subjects, but did not change plasma lipids or anti-oxidant capacity in the normo-lipidaemic subjects. Grape seed phenolics have been shown to protect low-density lipoprotein (LDL) against oxidation brought about by Cu^{2+}, oxygen-centered radical-generating AAPH, or peroxynitrite-generating SIN-1 in-vitro systems (Shafiee *et al.*, 2003).

The anti-oxidant abilities of grape seed extract have been well demonstrated in various other systems as well. Aged rats supplemented with grape seed extract (100 mg/kg body weight for 30 days) were reported to have normalized lipid peroxidation and anti-oxidant defenses in the central nervous system (Balu *et al.*, 2005). Procyanidin B4, catechin, gallic acid at low concentration could protect spleen cells from DNA damage induced by hydrogen peroxide (Fan and Lou, 2004). Procyanidin extract from grape seed significantly decreased the damage caused by the geno-toxicity of H_2O_2. Hence, Llópiz *et al.* (2004) pronounced that procyanidins are more effective than the corresponding individual monomers–catechin and epicatechin in preventing DNA lesions in hepato-cytes and that this protection is higher after pre-incubation than after co-incubation. Grape seed extract was also seen to protect the hepatic cellular membrane from oxidative damage and consequently from protein and lipid oxidation. Administration of grape seed extract reversed the increased malondialdehyde levels and the decreased superoxide dismutase and catalase activities in the liver homogenates after radiation therapy. Thus, grape seed extract is a promising therapeutic option in radiation-induced oxidative stress in the rat liver (Cetin *et al.*, 2008).

Resveratrol (3,5,4'-trihydroxystilbene) (RV), a constituent of grape seeds possesses anti-inflammatory and anti-oxidant activities. RV over a non-cytotoxic concentration range inhibited chemotactic and calcium mobilization responses of phagocytic cells to selected chemo-attractants. Further, RV reduced phosphorylation of extra-cellular signal-regulated kinase (ERK1/2) and the activation of nuclear factor NF-kappa B induced by formyl-peptide receptor agonists (Tao *et al.*, 2004).

Since grape skins and seeds are the predominant constituents in the pomace, this biomass is a rich source of phenolic anti-oxidants (Kammerer *et al.*, 2004; Lu *et al.*, 1999) and currently have been recycled as organic fertilizers, manure, and animal feed (Hogan *et al.*, 2010). Red grape pomace was rich in polyphenol compounds with a clear anti-oxidant activity. At very low concentration (20 ppm) of total phenols, the ethanolic extract showed an anti-oxidant activity to be higher than 43 per cent on an average, while at higher concentration (80-160 ppm) all the fractions of the extract had an anti-oxidant activity comparable to that of butylated-hydroxytoluene (Negro *et al.*, 2003). Again, consumption of grapes especially grape pomace with the highest content of flavonoids, beta-carotene, tocopherols and dietary fiber showed the prominent anti-oxidative capacity of inhibiting age-related or Cd-induced increase of lipid peroxidation and DNA damage effectively, promoting liver and red blood cell anti-oxidant enzyme activities (Rho and Kim, 2006).

Grape juice concentrate was able to decrease oxidative DNA damage induced by H_2O_2 in peripheral blood cells, as depicted by the tail moment results. COX-2 expression in the liver did not show statistically significant differences between groups (Aguiar *et al.*, 2011). As compared to grape extracts and juices, red wines possessed the highest anti-oxidant activity which correlated with the amounts of total polyphenols and anthocyanins (Briedis *et al.*, 2003).

Grapes and grape seeds show multiple health benefits in clinical conditions as described below.

Diabetes

Grapes have been found to have a mean glycemic index (GI) and glycemic load (GL) in the low range. The phytochemicals present in Grapes including the stilbene resveratrol, and flavanols (quercetin, catechins, and anthocyanins) have shown potential for reducing hyperglycemia, improving beta-cell function, and protecting against beta-cell loss. Therefore, grapes or grape products may provide health benefits in type 2 diabetics (Zunino, 2009). Though administration of grape seed extract (GSE) and its ethyl acetate/ethanol (EE) fraction did not show any significant effects on body weights and food consumption of diabetic mice, a significant reduction in HbA1c levels (5.7 per cent and 6.1 per cent respectively) was seen at 8 weeks after treatment, revealing their anti-diabetic efficacy (Hwang *et al.,* 2009).

Chronic GSPE treatments in intestinal human cells (Caco-2) showed a decrease of DPP4 activity and gene expression *in-vitro*. Whereas both in healthy and diet-induced obese animals GSPE decreased only intestinal but not plasma DPP4 activity and gene expression. Thus, procyanidin inhibition of intestinal DPP4 activity, either directly and/or via gene expression down-regulation, could be responsible for some of their effects in glucose homeostasis (González-Abuín *et al.,* 2012).

Anti-inflammatory activities of certain grape components may have positive benefits in reducing inflammation-related complications of type 2 diabetes such as cardiovascular disease. Red wines and grape juices contain polyphenolics with anti-oxidant and anti-platelet properties that may be protective against oxidative stress leading to hypertension, insulin resistance, and type 2 diabetes. Banini *et al.* (2006) reported that daily intake of 150 mL of muscadine grape wine or de-alcoholized muscadine grape wine with meals improved several metabolic responses among diabetics. Cai *et al.* (2011) proved that grape seed procyanidin B2 (GSPB2) has therapeutic potential in preventing and treating vascular complications of diabetes mellitus. They observed that it could inhibit AGEs-induced proliferation of vascular smooth muscle cell by affecting the production of ubiquitin COOH-terminal hydrolase 1 (UCH-L1), and the degradation of IκB-α and nuclear translocation of NF-κB in human aortic smooth muscle cells (HASMCs). GSPE significantly decreased aortic pulse wave velocity, blood pressure, and aortic medial thickness, and inhibited the migration of vascular smooth muscle cells. Also GSPE significantly reduced the AGEs and the expression of RAGE in aortas of diabetic rats, thus playing an important role against diabetic macrovascular complications. It was also identified that the anti-oxidants GSPB2 and resveratrol of GSPE protect against AGEs-induced EC apoptosis by inhibiting lactadherin (Li *et al.,* 2009; Li *et al.,* 2011a; b).

Yu *et al.* (2012) reported that milk fat globule epidermal growth factor-8 (MFG-E8) plays an important role in atherogenesis in diabetes through both extracellular signal-regulated kinase and monocyte chemo-attractant protein-1 signaling pathways. GSPB2, significantly inhibited the atherogene changes in arterial wall in db/db mice by down-regulating MFG-E8 expression in aorta as well as its serum concentration. Thus MFG-E8 serum level could be a useful clinical surrogate prognostic marker of atherogenesis in DM patients.

Chis *et al.* (2009) showed that oral administration of GSE (100 mg/kg/day) reduced the levels of lipid peroxides and carbonylated proteins and improved the anti-oxidant activity in plasma and hepatic tissue in diabetic rats. An association between paraoxonase status and GSE supplementation was demonstrated revealing that GSE increased paraoxonase activities especially in diabetic rats. Paraoxonase acts as an anti-oxidant enzyme and protects low-density lipoprotein-cholesterol against oxidation (Kiyici *et al.*, 2010). Oleanolic acid, oleanolic aldehyde isolated from grape skin stimulated insulin production (Zhang *et al.*, 2004b). Treatment with 80 mg/kg of grape seed-derived procyanidins (GSP) alone significantly reduced the index of insulin resistance (HOMA-IR) and increased hepatic glycogen concentration while increasing the consumption of extracellular glucose in HepG2 cells. But simultaneous administration of GSP and gypenosides not only decreased the index of insulin resistance (HOMA-IR index) but reduced serum total cholesterol level, and enhanced glucose tolerance (Zhang *et al.*, 2009b).

Cui *et al.* (2008) demonstrated the protective effect of GSPE against diabetic peripheral neuropathy. The mechanism of action included (a) reduction of advanced glycation end products, (AGEs) and MDA, (b) attenuation of oxidation-associated nerve damage and (c) significant increase in motor nerve conductive velocity, mechanical hyperalgesia and SOD in diabetic rats. In addition, segmental demyelination was decreased and Schwann cells were improved. Chacón *et al.* (2009) found GSPE treatment to show a reduction of IL-6 and MCP-1 expression after an inflammatory stimulus. GSPE stimuli alone modulated adipokine (APM1 and LEP) and cytokine (IL-6 and MCP-1) gene expression. GSPE partially inhibited NF-kappa B translocation to the nucleus in both cell lines and thus enhanced the production of anti-inflammatory adipokine adiponectin suggesting that GSPE has a beneficial effect on low-grade inflammatory diseases such obesity and type 2 diabetes.

According to Liu *et al.* (2006) GSPE protects kidney of diabetic rats, the mechanism might be related with its action in increasing the renal anti-oxidative ability, decreasing the content of NO and the activity of NOS in kidney and serum. Puiggròs *et al.* (2009) observed that the administration of GSPE increased Cu/Zn-SOD activity in both diabetic and non-diabetic rats. *In-vitro* experiments revealed that direct interaction between some small or medium sized GSPE components and the enzyme is responsible for the increase in Cu/Zn-SOD activity.

Accordingly, Mansouri *et al.* (2011) reported that administration of grape seed pro-anthocyanidin extract significantly decreased lipid peroxidation and augmented the activities of anti-oxidant enzymes including catalase, superoxide dismutase and glutathione peroxidase in kidney of rats with diabetic nephropathy and reduced urinary albumin excretion and the kidney weight. However, no change was noticed in the total body weight.

CVD

Epidemiological studies showed an inverse correlation between the intake of dietary flavonoids and death from coronary artery disease (Hertog 1993; 1995a and b; Knekt *et al.,* 1996). This may be explained in part by the anti-oxidant and anti-platelet properties attributed to flavonoids (Beretz *et al.,* 1982, DeWalley *et al.,* 1990). Anti-oxidant properties of grape polyphenols are likely to be central to their mechanism(s) of action, which include cellular signaling and interactions at the genomic level. Leifert and Abeywardena (2008) reviewed some of the evidence favoring the consumption of grape extracts rich in polyphenols in the prevention of cardio-vascular disease and suggest that consumption of grape and grape extracts and/or grape products such as red wine may be beneficial in preventing the development of chronic degenerative diseases such as cardiovascular disease. The anti-oxidant properties of flavonoid-rich grape seed extracts play a potential role in the reversal of vascular disease (Kar *et al.,* 2006). Grape seed extract significantly improved markers of inflammation and glycaemia and a sole marker of oxidative stress over a 4-week period in obese Type 2 diabetic subjects at high risk of cardiovascular events (Kar *et al.,* 2009).

The protective effect of grape juice on CVD has been compared with that of other fruits. Grape juice contains the flavonoids quercetin, kaempferol and myricetin, which are known inhibitors of platelet aggregation *in-vitro*. Orange juice and grapefruit juice, while containing less quercetin than grape juice, primarily contain the flavonoids naringin, luteolin and apigenin glucoside. The flavonoids in grapes were shown *in-vitro* to be good inhibitors of platelet aggregation whereas the flavonoids in oranges and grapefruit to be poor inhibitors of platelet aggregation. It has been shown that 5 ml/kg of red wine or 5-10 ml/kg of purple grape juice but not orange or grapefruit juice inhibited platelet activity, and protected against epinephrine activation of platelets. Red wine and purple grape juice enhanced platelet and endothelial production of nitric oxide (Fitzpatrick *et al.,* 1993; Osman *et al.,* 1998). Purple grape juice was also found to potentiate the anti-oxidant benefit of vitamin E (400 IU/day) against LDL cholesterol oxidation. The flavonoids in purple grape juice and red wine might inhibit the initiation of atherosclerosis. Based on the evidence of anti-platelet and anti-oxidant benefits and improved endothelial function from red wine and purple grape juice, Folts (2002) suggested that moderate amounts of red wine or purple grape juice be included among the 5-7 daily servings of fruits and vegetables per day as recommended by the American Heart Association to help reduce the risk of developing cardiovascular disease.

Bagchi *et al.* (2003) conducted a series of studies and demonstrated the cardio-protective ability of GSPE in animals and humans. GSPE supplementation improved cardiac functional ability including post-ischemic left ventricular function, reduced myocardial infarct size, reduced ventricular fibrillation (VF) and tachycardia and decreased the amount of reactive oxygen species (ROS). This may be at least partially attributed to its ability to block anti-death signaling mediated through the pro-apoptotic transcription factors and genes such as JNK-1 and c-JUN.

Further, GSPE pretreatment significantly inhibited doxorubicin-induced cardio-toxicity as demonstrated by reduced serum creatine kinase activity, DNA damage and histo-pathological changes in the cardiac tissue of mice. Approximately 49 and 63 per cent reduction in foam cells, a biomarker of early stage atherosclerosis, were observed following supplementation of 50 and 100 mg GSPE/kg body weight, respectively. A human clinical trial conducted on hyper-cholesterolemic subjects revealed that GSPE supplementation significantly reduced oxidized LDL, a biomarker of cardiovascular diseases (Bagchi *et al.,* 2003). GSPE offers an alternative approach to therapeutic cardio-protection against Ischemia/reperfusion injury and may offer unique opportunities to improve cardiovascular health by enhancing NO production and increasing Akt-eNOS signaling (Shao *et al.,* 2009).

Interestingly, the components of grape seed and grape skin when present in combination as in red wine grape juice or in a commercial preparation containing both extracts, exhibit a greater anti-platelet effect than when present individually (Shanmuganayagam *et al.,* 2002). Cui *et al.* (2012) proved that GSPE increases endothelial nitric oxide synthase (eNOS) expression by knocking down small interfering RNA (siRNA) in human umbilical vessel cells (HUVECs) *in-vitro,* which was attributed to its transcription factor Krüpple like factor 2 (KLF2) induction. After feeding GSPE for 5 weeks the ouabain induced increase in blood pressure was significantly blocked. Spatial learning (8-arm-radial maze) was also improved by dietary grape seed polyphenols in the spontaneously hypertensive female rats indicating that the decrease in arterial pressure is probably through the anti-oxidant mechanism (Peng *et al.,* 2005).

Consecutive 12-wk administration of tablets containing 200 mg grape seed extract (calculated as pro-anthocyanidin) to 61 healthy subjects with LDL cholesterol levels of 100 to 180 mg/dL, significantly reduced malondialdehyde-modified LDL (MDA-LDL) at 12 weeks, while 400 mg resulted in reduction of MDA-LDL levels at 6 weeks itself as compared to the controls. Thus it is possible to prevent lifestyle-related diseases such as arteriosclerosis with grape seed extract (Sano *et al.,* 2007). Administering 150 and 300 mg per day of GSE (containing powerful vasodilator phenolic compounds) for 4 weeks to subjects with the metabolic syndrome lowered both the systolic and diastolic blood pressure as compared to placebo with no significant changes in serum lipids or blood glucose values (Sivaprakasapillai *et al.,* 2009). Hence the authors suggest that GSE could be used as a nutraceutical in a lifestyle modification programme for patients with the metabolic syndrome.

A gender specific response to the hypo-lipidaemic effect of GSE was noticed. Feeding GSE to familial hyper-lipidemic rabbits had no significant effects in females but was associated with transient less hyper-cholesterolemic response to semi-synthetic diet and with retarded development of aortic atherosclerosis in males as demonstrated by significantly lower cholesterol content in the abdomen (Frederiksen *et al.,* 2007).

Badavi *et al.* (2008) confirmed from his study in male Wistar rats that red grape seed extract has the potential protective effect on lead induced hypertension and heart rate through its anti-oxidant activity. Grape pro-anthocyanidin supplementation significantly reduced arterial pressure in the rats fed the basal and high NaCl diet by 10 mmHg. and 26 mmHg. respectively. Polyphenols from grape seed significantly reduced superoxide production by 23 per cent *in-vitro.*

Kamiyama *et al.* (2009) observed that the anti-oxidant activity of Cy-3-glc identified in the skin of Nagano purple grape significantly inhibited LDL oxidation in the venous blood samples collected from human subjects 1 h after consuming the skins or dried fruits.

Preuss *et al.* (2001) reported that prolonged supplementation of a combination of chromium polynicotinate, grape seed extract and zinc mono-methionine (known to sensitize insulin response and act as anti-oxidant) markedly lowered systolic blood pressures in normo-tensive rats, lessened oxidative damage to fats as indicated by decreased TBARS formation, and lowered HbA1C without showing signs of toxicity. Ingestion of niacin-bound chromium and grape seed pro-anthocyanidin extract has been demonstrated to improve insulin sensitivity and/or ameliorate free radical formation and reduce the signs/symptoms of chronic age-related disorders including syndrome X. These natural strategies possess a highly favourable risk/benefit ratio (Preuss *et al.,* 2002).

Phytoestrogens from grape seeds, pressed out grapes, and fermented grape ridges prevent cholesterol accumulation in cells in post-menopausal women (Nikitina *et al.,* 2006). The major polyphenols in grape seed have been shown to have beneficial health effects in the prevention of dyslipidemia and cardiovascular diseases. Ngamukote *et al.* (2011) showed that in a concentration-dependent manner, gallic acid, catechin, and epicatechin significantly inhibited the binding of bile acids by pancreatic cholesterol esterase, and reduced solubility of cholesterol in micelles which may result in delayed cholesterol absorption.

Short-term ingestion of purple grape juice improved Flow-mediated vaso-dilation and reduced LDL susceptibility to oxidation in coronary artery disease patients. Improved endothelium-dependent vaso-dilation of arterial rings, a phenomenon mediated by the nitric oxide-guanosine 32,52 cyclic mono-phosphate (NO-cGMP) pathway (Kanner *et al.,* 1994; Miyagi *et al.,* 1997) and prevention of LDL oxidation are potential mechanisms for the effect (Stein *et al.,* 1999).

Moderate daily consumption of red wine is a negative risk factor for the development of atherosclerosis and coronary artery disease (CAD), especially in France and other Mediterranean areas where red wine is regularly consumed with meals. Platelets contribute to the development of atherosclerosis, CAD, and acute arterial thrombus formation. The cardio-protective effects of red wine consumption observed in the French and other populations may be attributed in part to the ethanol content of the wine and in part to the anti-oxidant and platelet-inhibitory properties of other naturally occurring compounds in the wine. Because platelet adhesion to damaged endothelium and subsequent platelet aggregation are major steps in both thrombosis and atherogenesis, the long-term inhibition of platelet activity by the consumption of flavonoid-containing foods and beverages may retard atherogenesis and prevent thrombosis on a daily basis (Demrow *et al.,* 1995).

Yamakoshi *et al.* (1999) proved that pro-anthocyanidins, the major polyphenols in red wine, might trap reactive oxygen species in aqueous series such as plasma and interstitial fluid of the arterial wall, thereby inhibiting oxidation of LDL and showing an anti-atherosclerotic activity. Flow-mediated dilatation (FMD) was higher following the consumption of de-alcoholized red wine. The pattern of the response was different between the red wine or de-alcoholized red wine as FMD increased following the ingestion of de-alcoholized red wine, but it decreased after consumption of regular red wine. Fibrinogen concentrations were unaltered (Karatzi *et al.,* 2004). Del Bas *et al.* (2005) demonstrated the beneficial long-term effects associated with moderate red wine consumption. They reported that procyanidins improved the atherosclerotic risk index in the post-prandial state, inducing in the liver the over-expression of CYP7A1 (suggesting an increase of cholesterol elimination *via* bile acids) and small heterodimer partner, a nuclear receptor emerging as a key regulator of lipid homeostasis at the transcriptional level.

The consumption of grape juice containing the inhibitors of platelet aggregation may have some of the protection offered by red wine against the development of coronary artery disease and acute occlusive thrombosis, whereas orange juice or grapefruit juice may be ineffective. Thus, grape juice may be a useful alternative dietary supplement to red wine without the concomitant alcohol intake (Osman *et al.,* 1998). Facino *et al.* (1999) demonstrated in rats that procyanidins supplementation makes the heart less susceptible to ischemia/reperfusion damage which was positively associated the increased plasma anti-oxidant activity.

Al-Awwadi *et al.* (2005) compared the efficacy of three polyphenolic extracts (anthocyanin, galloylated procyanidins and Vitaflavan) in preventing hypertension, cardiac hypertrophy, oxidative stress in the aorta, and increasing the expression of cardiac NAD(P)H oxidase in a model of insulin resistance. The results showed that (a) the anthocyanin treatment prevented hypertension, cardiac hypertrophy, and production of ROS, (b), the galloylated procyanidins treatment prevented insulin resistance, hyper-triglyceridemia, and over-production of ROS but had only minor effects on hyper-tension or hyper-trophy, while (c) Vitaflavan prevented hypertension, cardiac hypertrophy, and over-production of ROS. All polyphenolic treatments prevented the increased expression of the p91phox NADPH oxidase subunit.

Thiruchenduran *et al.* (2011) demonstrated cardio-protective effects of grape seed pro-anthocyanidins against cholesterol-cholic acid diet-induced hyper-cholesterolemia *via* their ability to reduce (directly or indirectly) free radicals in the myocardium. Similarly, Razavi *et al.* (2013) found that consumption of red grape seed extract decreased oxidized low-density lipoprotein particles and has beneficial effects on lipid profile, consequently decreasing the risk of atherosclerosis and cardio-vascular disorders-in mild hyper-lipidemic individuals.

Cancer

Presently, the potential anti-cancer effects of GSE or its pro-anthocyanidin fraction has been in focus. Studies from various scientific groups conclude that both grapes and grape-based products are excellent sources of various anti-cancer agents and their regular consumption should thus be beneficial to the general population (Kaur *et al.,* 2009a; b).

The Potential Mechanisms

The mechanisms of potential anti-cancer effects of grapes include anti-oxidant, anti-inflammatory, and anti-proliferative activities (Seeram *et al.,* 2005) by acting on multiple cellular events associated with tumor initiation, promotion, and progression.

The anti-oxidants in grapes induce cell cycle arrest and apoptosis in cancer cells (Aggarwal *et al.,* 2004) and prevent carcinogenesis and progression of cancer in rodent models (Ebeler *et al.,* 2002; Gravin *et al.,* 2006). Apoptosis can proceed by either the extrinsic pathway (death receptor) or intrinsic pathway (mitochondria) (Igney and Krammer, 2002).

Lazzè *et al.* (2009) observed that the grape extract has a significant anti-proliferative effect in a tumor cell line suggesting its potential as an anti-oxidant chemo-preventive agent. Grape seed procyanidins inhibited cell proliferation but did not appreciably increase lethality. The treatment also induced morphological changes in the cells and the rate of CD-11b and CD-14 cells and NBT reductive activities increased significantly. As anti-oxidant, grape seed procyanidins can induce arrest in the phase G1 and decrease iROS formation (Wang *et al.,* 2012). Further, interestingly, GSPE enhances the viability and growth of normal human gastric mucosal cells and murine macrophage cells (Agarwal *et al.,* 2002; Bagchi *et al.,* 1998).

Irrespective of its source, high doses of GSE induced a significant inhibition on Caco-2 cell growth. Moreover, apoptosis was enhanced through both caspase-dependent and caspase-independent mechanisms, leading to an early release of apoptosis-inducing factor and further, to a dramatic increase in caspase 7 and 3 activity. However, treatment with low and intermediate GSE concentrations (25 and 50 μg/ml) displayed a significant difference in apoptotic rates in different grape cultivars (Dinicola *et al.,* 2010).

Hu and Qin (2006) reported that GSPE does not induce apoptosis in neutrophils because GSPE, as a mitochondrial anti-oxidant, had no effects on neutrophils showing that GSPE induced apoptosis may be cancer specific in AML 14.3D10 cells. Oligomeric pro-anthocyanidins from Thai grape seeds were also found to possess the anti-mutagenic and anti-oxidative effects on DNA damage induced by H_2O_2 in TK6 cells at concentrations of 100, 250, 500, and 1,000 μg/ml by 18.7, 36.4, 30.6, and 60.1 per cent, respectively (Praphasawat *et al.,* 2011).

Three of the major anti-oxidant components found in grapes – resveratrol, catechin, quercetin, and grape seed extract, containing a pro-anthocyanidin B-2-gallate given topically and/or systemically strongly inhibited 7,12-dimethylbenz[a]anthracene (DMBA)-induced epidermal hyperplasia, proliferation, and inflammation. The hydroxylation of 2'-deoxyguanosine was markedly inhibited by topical and dietary administration of test variables, *i.e.,* by approximately 40-70 per cent (Hanausek *et al.,* 2011). GSE exhibited the classical morphological characteristics of cell death under the electron microscope, including cell atrophy, increased vacuoles, crumpled nuclear membrane, chromosome aggregation and apoptosis of PC-3 cells (Shang *et al.,* 2008).

Zhang *et al.* (2009a) reported that F2, an oligomer procyanidin fraction isolated from grape seeds, triggered an original form of cell death in U-87 human glioblastoma cells with a phenotype resembling morphological characteristics of paraptosis. Further study by Zhang *et al.* (2010) complemented their earlier reports on the characterization of F2-induced U-87 cell death and confirmed the mechanism of the action of F2 on glioma.

The anthracycline antibiotic Adriamycin (Doxorubicin, Dox) is a cancer chemotherapeutic agent that interferes with the topoisomerase II enzyme and generates free radicals. Grape seed polyphenol (GSP) inhibited tumor growth by 18.35 per cent and when combined with Adriamycin showed significant inhibition in nude mice. 20 mg/kg GSP could effectively reverse the multi-drug resistance of MCF-7/ADR cells to ADR *in-vivo*, and the rate of inhibition was 54.64 per cent (Zhang *et al.,* 2004a).

Postescu *et al.* (2012) studied the effects of administration of hydro-ethanolic extract of grape-seed Burgund Mare variety (BM) along with Dox (30 min before drug administration) in normal (Hfl-1) and tumor cell lines (HepG2 and Mls). In normal cells, the product statistically decreased cyto-toxicity and markedly inhibited lipid peroxides and protein carbonyl formation, in a dose-dependent relationship. Contrary to this, in tumor cells, the same treatment resulted in a reversed effect, cell death, malondialdehyde and increase in protein carbonyl contents with BM dose enhancement. Treatment with BM extract, prior to subsequent administration of Dox, afforded a differential protection against Dox-negative toxic side effects in normal cells without weakening (even enhancing) Dox's anti-tumor activity.

Ramchandani *et al.* (2008) observed that both seedless and seeded GPEs possessed anti-tumor-promoting activity in target tissues of mice as was evident from their ability to delay tumor formation along with a significant decrease in tumor multiplicity and incidence. Marked and sustained epidermal hyperplasia observed in 7,12-dimethylbenz(a)anthracene-initiated and TPA-promoted mice was greatly reduced on pre-treatment with GPE or catechin. The polyphenolic extracts from Sharad seedless and seeds of Bangalore blue showed the strongest suppressing activity comparable to catechin than the corresponding whole grapes.

High consumption of components of grapes including GSE has been associated with the reduced risk of certain cancers such as breast cancer, prostate, lung, skin, and gastro-intestinal cancer (Falcao *et al.,* 1994; Hertog *et al.,* 1995a and b; Viel *et al.,* 1997, Agarwal *et al.,* 2002; Ye *et al.,* 1999; Zhao *et al.,* 1999) in the in-vitro and in-vivo models (Clifford *et al.,* 1996; Soleas *et al.,* 2002; Kim *et al.,* 2004; Jung *et al.,* 2006; Morré and Morré, 2006).

Anti-tumor-promoting effects of polyphenolic fraction of grape seeds were dose dependent and were evident in terms of a reduction in tumor incidence (35 and 60 per cent inhibition), tumor multiplicity (61 and 83 per cent inhibition) and tumor volume (67 and 87 per cent inhibition) at lower and higher concentrations *i.e.* 0.5 and 1.5 mg, respectively. Procyanidin B5-3'-gallate showed the most potent anti-oxidant activity with an IC(50) of 20 µM in an epidermal lipid peroxidation assay (Zhao *et al.,* 1999).

Cervical Cancer

Grapes and grape extracts on comparison for inhibition of a growth-related and cancer-specific form of cell surface NADH oxidase with protein disulfide-thiol interchange activity designated tNOX from human cervical carcinoma (HeLa) cells and growth of HeLa and mouse mammary 4T1 cells in culture and transplanted tumors in mice revealed grape skin to be much more potent than either grape pulp, juice or seeds (Morré and Morré, 2006).

Liver Cancer

Pre-treatment of hepato-carcinoma cell line HepG2 with GSPE (15 mg/L, 23 h) before HepG2 submission to H_2O_2 (1 mM, 1 h) showed an increase in the mRNA of GPx/GR with respect to the H_2O_2 group, whereas the GSH content was similar to the control group. However, the GPx/GR enzyme activities were not increased. GSPE probably improves the cellular redox status *via* glutathione synthesis pathways instead of regulation of the GPx and/or GR activities protecting against oxidative damage (Puiggros *et al.,* 2005).

Agarwal *et al.* (2007) identified procyanidin B2-3,3'-di-O-gallate as a novel biologically active agent in grape seed extract. Non-esterified B2 exhibited little or no activity, suggesting that the galloyl groups of B2-digallate are primarily responsible for its effects on DU145 cells.

Breast Cancer

GSE is potentially useful in the prevention/treatment of hormone-dependent breast cancer through the inhibition of aromatase activity as well as its expression. Lu *et al.* (2009) provided a novel molecular mechanism underlying the anti-angiogenic action of GSE. They found that GSE inhibited vascular endothelial growth factor messenger RNA (mRNA) and protein expression in U251 human glioma cells and MDA-MB-231 human breast cancer cells and that GSE inhibited vascular endothelial growth factor expression by reducing HIF-1alpha protein synthesis through blocking Akt activation. GSPs suppressed the DNA damage induced by Dox in a dose-dependent manner. The co-treatment with Dox demonstrated that GSPs have some anti-mutagenic activity. However, anti-recombinagenic activity was the major response (de Rezende *et al.,* 2009). In both estrogen receptor-positive MCF-7 and receptor-negative MDA-MB468 cells, a combination of 100 µg/ml GSE with 25-75 nM Dox treatment for 48 h showed a strong synergistic effect [combination index (CI) < 0.5] in cell growth inhibition, but mostly an additive effect (CI approximately 1) in cell death. GSE plus Dox combination showed a very strong and significant G1 arrest in MDA-MB468 cells when compared with Dox alone, however, it was less than that observed with GSE alone. Thus, the synergistic efficacy of GSE and Dox combination for breast cancer treatment, independent of estrogen receptor status of the cancer cell was demonstrated by Sharma *et al.* (2004).

Song *et al.* (2010) verified the value of GSPE in the prevention of human breast cell carcinogenesis induced by repeated exposures to low doses of multiple environmental carcinogens. They concluded that the ability of GSPE to reduce gene expression of cytochrome-P450 enzymes CYP1A1 and CYP1B1, which can bioactivate NNK (tobacco-specific nitrosamine 4-(methylnitrosamino)-1-(3-pyridyl)-1-butanone) and Benzo[a]pyrene (B[a]P), possibly contributes to the preventive mechanism for GSPE in suppression of pre-cancerous cellular carcinogenesis.

The high levels of procyanidin dimers in GSE have been shown to be potent inhibitors of aromatase, the enzyme that converts androgen to estrogen. Kijima *et al.*

(2006) showed that GSE to be potentially useful in the prevention/treatment of hormone-dependent breast cancer through the inhibition of aromatase activity as well as its expression.

Colorectal Cancer

The pro-anthocyanidins in grape seed extract act against colorectal cancer and also protect the liver from oxidative damage following bile duct ligation in rats. This effect possibly involves the inhibition of neutrophil infiltration and lipid peroxidation and thus restoration of oxidant and anti-oxidant status in the tissue (Dulundu *et al.*, 2007).

Adenosine deaminase (ADA), 5' nucleotidase (5'NT) activities increased but xanthine oxidase (XO) decreased in cancerous human colon tissues. Thus, the cancerous tissues may obtain new nucleotides for rapid DNA synthesis through accelerated salvage pathway activity. Black grape extract makes significant inhibition on the ADA and 5'NT activities of cancerous and non cancerous colon tissues, thereby eliminating rapid DNA synthesis, which might be the basis for the beneficial effect of black grape in some kinds of human cancers (Durak *et al.*, 2005).

Grape seed extract exerted both anti-proliferative and apoptotic effects on Caco2 and HCT-8 and inhibits growth of colon cancer cells. In addition, its inhibitive effects were stronger than isolated procyanidins, suggesting a potential additive or synergistic effect among the grape seed components (Dinicola *et al.*, 2012). Radhakrishnan *et al.* (2011) found the combination of resveratrol and grape seed extract to induce more pronounced apoptosis in colon cancer cells, which is strongly correlated with p53 levels and Bax : Bcl-2 ratio.

Velmurugan *et al.* (2010a; b) showed that the chemo-preventive efficacy of grape seed extract is due probably to the targeting of beta-catenin and NF-kappaB signaling in rats proving its potential usefulness for the prevention of human colorectal cancer. Bak *et al.* (2012) found that procyanidins from wild grape seeds can regulate ARE-mediated enzyme expression *via* Nrf2 coupled with P38 and PI3K/Akt pathway in HepG2 cells and could be used as a potential natural chemo-preventive agent.

Feeding GSE to mice at 200 mg/kg dose showed time-dependent inhibition of tumor growth without any toxicity and accounted for 44 per cent decrease in tumor volume per mouse after 8 weeks of treatment. The growth inhibitory and apoptotic effects of GSE against colo-rectal cancer could be mediated *via* an up-regulation of Cip1/p21 (Kaur *et al.*, 2006). Kaur *et al.* (2008) also showed that irrespective of source, GSE strongly inhibits LoVo, HT29, and SW480 cell growth, with a G1 arrest in LoVo and HT29 cells but an S and/or G2/M arrest in SW480 cell cycle progression. GSE also induced Cip/p21 levels along with induction of apoptosis in all 3 cell lines. According to Engelbrecht *et al.* (2007) the increased apoptosis observed in GSPE-treated CaCo2 cells correlated with an attenuation of PI3-kinase (p110 and p85 subunits) and decreased PKB Ser (473) phosphorylation.

Intra-gastric administration of pro-anthocyanidin (PA) of grape seed strongly enhanced the anti-tumour effect of Doxorubicin and completely eliminated myocardial oxidative stress induced by Doxorubicin and immuno-suppression in tumour-bearing mice (Zhang *et al.,* 2005a). The mechanism is attributed to the promotion of DOX-induced apoptosis through increasing intra-cellular DOX, Ca^{2+} and Mg^{2+} concentrations, and reducing pH value and mitochondrial membrane potential (Zhang *et al.,* 2005b).

GSE significantly reduced myelo-peroxidase activity in the proximal jejunum and distal ileum, decreased qualitative histological scores of damage in the proximal jejunum; increased villus height in the proximal jejunum and distal ileum, and attenuated the 5-Fluorouracil induced reduction of mucosal thickness in the jejunum and ileum representing a promising prophylactic adjunct to conventional chemo-therapy for preventing intestinal mucositis (Cheah *et al.,* 2009).

Pancreatic Cancer

Dietary administration of GSPs (0.5 per cent, w/w) as a supplement in AIN76A (Purified rodent diet extensively used in research) diet significantly inhibited the growth of Miapaca-2 pancreatic tumor xenografts grown subcutaneously in athymic nude mice, which was associated with: (i) inhibition of cell proliferation, (ii) induction of apoptosis of tumor cells, (iii) increased expression of Bax, reduced expression of anti-apoptotic proteins and activation of caspase-3-positive cells, and (iv) decreased expression of PI3K and p-Akt in tumor xenograft tissues suggesting that GSPs may have a potential chemo-therapeutic effect on pancreatic cancer cell growth (Prasad *et al.,* 2012).

Prostate Cancer

The anti-cancer efficacy of GSE against hormone-refractory human prostate cancer *via* its anti-proliferative, pro-apoptotic and anti-angiogenic activities in both cell culture and animal models has been described by Singh *et al.* (2004). Veluri *et al.* (2006) identifed gallic acid as one of the major active constituents in GSE. Several procyanidins, especially the gallate esters of dimers and trimers also may be efficacious against prostate cancer.

GSE at 200 mg/kg dosage inhibited prostate cancer growth and progression in transgenic adeno-carcinoma of the prostate (TRAMP) mice, which could be mediated via a strong suppression of cell cycle progression and cell proliferation and an increase in apoptosis (Raina *et al.,* 2007). According to Huang *et al.* (2008) GSE significantly inhibited the growth of PC-3 cells in a concentration- and time-dependent manner, but had only a mild inhibitory effect on the kidney cells. Kim *et al.* (2009) found that piceatannol (trans-3,4,3',5'-tetrahydroxy-stilbene) a polyphenol found in grapes and red wine induces apoptosis *via* the activation of the death receptor and mitochondrial-dependent pathways in prostate cancer cells.

Muscadine grape skin extract significantly inhibited tumor cell growth in all transformed prostate cancer cell lines but not normal prostate epithelial cells exhibiting high rates of apoptosis through targeting of the phosphatidyl-inositol 3-kinase-Akt and mitogen-activated protein kinase survival pathways (Hudson *et al.,* 2007).

Lung Cancer

Melanoma is one of the neoplasias that most frequently metastasize, especially in the lung, representing a challenge in oncology since current treatment is ineffective, and mortality is high. Grape-seed extract and red wine reduced the number of metastatic nodules by 26.07 and 20.81 per cent, respectively. The reduction in the invasion index was 31.65 for grape-seed extract and 17.57 per cent for red wine (Martinez *et al.*, 2005). Later, Singh *et al.* (2011) reported that grape seed proanthocyanidins (GSPs) induce apoptosis of NSCLC cells, A549 and H1299, *in-vitro* which is mediated through increased expression of pro-apoptotic protein Bax, decreased expression of anti-apoptotic proteins Bcl2 and Bcl-xl, disruption of mitochondrial membrane potential, and activation of caspases 9, 3 and ADP-ribose and polymerase (PARP). Thus GSP can be a potential therapeutic agent for the non-small cell lung cancer.

Skin Cancer

Dietary GSPs inhibited the expression of proliferating cell nuclear antigen, cyclin D1, inducible nitric oxide synthase, and cyclooxygenase-2 in the skin showing that GSPs have the ability to protect the skin from the adverse effects of UVB radiation *via* modulation of the MAPK and NF-kappaB signaling pathways and provide a molecular basis for the photo-protective effects of GSPs in an *in-vivo* animal model (Sharma *et al.*, 2007). *In-vitro* treatment of NSCLC cells (A549, H1299, H460, H226, and H157) with GSPs resulted in significant growth inhibition and induction of apoptosis, which were associated with the inhibitory effects of GSPs on the over-expression of COX-2 and prostaglandin (PG) E2 receptors (EP1 and EP4) in these cells (Sharma *et al.*, 2010).

Ultraviolet (UV) radiation-induced immune-suppression has been implicated in skin carcinogenesis. Sharma and Katiyar (2006) observed that grape seed proanthocyanidins (GSPs) modulate UVB-induced immune-suppression thus prevented photo-carcinogenesis in mice. Vaid *et al.* (2010) found that GSPs repaired UV induced CPD(+) cells in xeroderma pigmentosum complementation group A (XPA)-proficient fibroblasts from a healthy individual but did not repair in XPA-deficient fibroblasts from XPA patients. Further, they reported that GSPs prevented ultraviolet B (UVB)-induced immune-suppression through DNA repair-dependent functional activation of dendritic cells in mice (Vaid *et al.*, 2013).

Filip *et al.* (2011) found the anti-oxidant activity of red grape seeds (*Vitis vinifera* L, BM variety) extract was higher than those of *Calluna vulgaris* (Cv) extract as determined using stable free radical DPPH assay and ABTS test. Twenty hours following irradiation BM extract inhibited UVB-induced sunburn cells and CPDs formation. Pre-treatment with Cv and BM extracts resulted in significantly reduced levels of IL-6 and TNF-α compared with UVB alone suggesting that BM extracts might be a potential candidate in preventing the damages induced by UV in skin.

On histological and immune-histochemical examination, skin treated with GSPE before UV radiation showed fewer sunburn cells and mutant p53-positive epidermal cells and more Langerhans cells compared with skin treated with 2-MED UV radiation only (Yuan *et al.*, 2012).

Dietary GSPs prevent photo-carcinogenesis in mice. Meeran and Katiyar (2007) reported that *in-vitro* treatment of human epidermoid carcinoma A431 cells with GSPs inhibited cellular proliferation (13-89 per cent) and induced cell death (1-48 per cent) in a dose (5-100 µg/ml)- and time (24, 48 and 72 h)-dependent manner. GSP-induced inhibition of cell proliferation was associated with an increase in G1-phase arrest at 24 h, which was mediated through the inhibition of cyclin-dependent kinases (Cdk) Cdk2, Cdk4, Cdk6 and cyclins D1, D2 and E, and simultaneous increase in protein expression of cyclin-dependent kinase inhibitors (Cdki), Cip1/p21 and Kip1/p27, and enhanced binding of Cdki-Cdk.

Oral Cancer

Red table grapes (RTG) are potent anti-mutagens that protect DNA from oxidative damage and also cyto-toxic toward the HL60 tumor cell line. The skin, seed and pulp of RTG exerted a desmutagenic effect, with seeds and skin showing the most potent effect. The cyto-toxicity tests using HL60 cells indicated that only skin and pulp fractions are able to inhibit the tumor growth (Anter *et al.*, 2011). Kingsley *et al.* (2010) found that oral cancer proliferation was inhibited by 24 hours in the pro-anthocyanidins concentration range of 50-70 µg/mL with concomitant decreases in mRNA expression of specific cell-cycle regulators, and increases in the expression of apoptosis-specific molecules, such as caspase-2 and caspase-8. The study demonstrated simultaneous, temporal inhibition of cell-cycle signaling pathways with the activation of specific apoptosis-related signaling pathways within oral cancers in response to pro-anthocyanidins. Shrotriya *et al.* (2012) observed that GSE targets both DNA damage and repair and provide mechanistic insights for its efficacy selectively against head and neck squamous cell carcinoma (HNSCC) both in cell culture and mouse xenograft, supporting its translational potential against HNSCC.

Anti-proliferative effectiveness of GSP is closely associated with the p53 status of oral squamous cell carcinoma cells. GSP displays chemo-adjuvant potential *via* cell cycle blockage and apoptotic induction (Lin *et al.,* 2012). Aghbali *et al.* (2013) confirmed the apoptotic potential of GSE on oral squamous cell carcinoma by significant inhibition of cell growth and viability in a dose- and time-dependent manner without inducing damage to non-cancerous cell line human umbilical vein endothelial cells.

Role in Chemo and Radiation Therapy

GSPE attenuated I/R-induced cell death. In the context of GSPE stimulation, Akt may help activate eNOS, leading to protective levels of NO. Pan *et al.* (2012) found grape seed pro-anthocyanidins can enhance the radio-sensitivity of cancer cells *in-vitro*. The mechanism of sensitization effect may be related to the effects of GSPs on oxygen balance and cell cycle.

Matrix metallo-proteinases (MMP) play a crucial role in the development and metastatic spread of cancer. Katiyar (2006) reviewed and discussed the evidences for the beneficial effect of grape seed pro-anthocyanidins, in chemo-prevention of cancer with particular emphasis on the involvement of MMPs in prostate cancer. The author (Katiyar, 2008) also showed that the chemo-preventive effects of dietary GSPs are mediated through the attenuation of (i) UV-induced oxidative stress; (ii) activation of mitogen-activated protein kinases and nuclear factor-kappa B (NF-kappaB) signaling pathways; and (iii) immune-suppression through alterations in immune-regulatory cytokines.

Yousef *et al.* (2009) revealed that GSPE exerts a protective effect by antagonizing cisplatin (the most potent chemotherapeutic anti-tumor drug which induces oxidative stress) toxicity as GSPE treatment reduced cisplatin-induced levels of TBARS in plasma, heart, kidney and liver, total lipid, cholesterol, urea and creatinine, and liver AST and ALT. Moreover, it ameliorated cisplatin-induced decrease in the activities of anti-oxidant enzymes, and GSH, total protein and albumin.

Gastric Ulcers

The gastric protective activity of a series of procyanidins increased with the increasing polymerization of catechin units. Oligomers longer than tetramers showed a strong protective effect against gastric mucosal damage. The mechanism of anti-ulcer activity may be the protection by radical scavenging activity on the stomach surface against radical injury induced by HCl/EtOH solution and the defense action of procyanidins covering the stomach surface by their strong ability to bind protein (Saito *et al.,* 1998). Ethanol-induced gastric ulcers were protected by GSE more effectively than vitamin C or E. The decrease of MDA levels with GSE was greater than with vitamin C and comparable to that achieved with vitamin E. Its protection against aspirin ulcers was comparable with all other treatments (Cuevas *et al.,* 2011).

Other Benefits

Flavanol-rich grapeseed extract significantly decreased ADP-stimulated platelet reactivity at 1, 2, and 6 hours following intake compared to baseline levels in a group of smokers, who were at an elevated risk for vascular disease. Similarly, the supplement decreased epinephrine-stimulated platelet reactivity 2 hours following consumption (Polagruto *et al.,* 2007).

Raisins (dried grapes) contain polyphenols, flavonoids, and high levels of iron that may benefit human health. Compared with commercial bran flakes or raisin bran cereal, a lower plaque pH drop was noted in children who consumed a raisin and bran flake mixture with no sugar. Grape seed extract, high in pro-anthocyanidins, positively affected the *in-vitro* demineralization and/or re-mineralization processes of artificial root caries lesions, suggesting its potential as a promising natural agent for non-invasive root caries therapy (Wu, 2009).

Administration of GSPE remarkably suppressed airway resistance and reduced the total inflammatory cell and eosinophil counts in BALF. Treatment with GSPE significantly enhanced the interferon (IFN)–γ level and decreased interleukin (IL)-4 and IL-13 levels in BALF and total IgE levels in serum. GSPE also attenuated allergen-induced lung eosinophilic inflammation and mucus-producing goblet cells in the airway. The elevated iNOS expression observed in the OVA mice was significantly inhibited by GSPE. Thus GSPE decreases the progression of airway inflammation and hyper-responsiveness by down-regulating the iNOS expression, promising to have a potential in the treatment of allergic asthma (Zhou *et al.,* 2011).

Charradi *et al.* (2012) confirmed that HFD could find some potential application in the treatment of Manganism but GSSE should be used as a safe anti-lipotoxic agent in the prevention and treatment of fat-induced brain injury.

GSE administration decreased urine protein excretion and serum cholesterol and triglyceride levels in rats with nephrosis induced by puromycin-aminonucleoside with no significant changes in serum albumin and creatinine levels (Mattoo and Kovacevic, 2003).

Regular consumption of purple grape juice in the absence of other red, blue, or purple fruits benefited immunity in healthy, middle-aged human subjects (Rowe *et al.,* 2011).

Dosage

Based on animal study the dosage has been translated into human consumption. The human equivalent dose of GSPs was calculated using the following formula (Singh *et al.,* 2011).

$$\text{Human Equivalent Dose (mg/kg)} = \text{Animal dose (mg/kg)} \times \frac{\text{Animal K}_m \text{ factor}}{\text{Human K}_m \text{ factor}}$$

(*Km* factor for mouse =3; *Km* factor for adult human = 37).

Interestingly, recent patents regarding grape polyphenols show a tendency to depend on a minimum use of severe extraction processes and organic solvents. (Gollücke and Ribeiro, 2012).

It is known that high dose of garlic could exert adverse health problems but grape seed and skin extract even at high doses exhibit only beneficial effects. Hamlaoui *et al.* (2012) confirmed that high dose of garlic induced anemia and a pro-oxidative state into erythrocytes characterized by increased malondialdehyde (MDA), carbonyl protein and activities of anti-oxidant enzymes catalase (CAT), peroxidase (POD) and superoxide dismutase (SOD) whereas GSSE treatment counteracted almost all garlic deleterious effects.

Conclusions

The traditional Mediterranean diet, high in polyphenols, derived from vegetables and red wine, suggests that dietary polyphenols are beneficial in reducing the incidence of cardiovascular disease. Specifically, the grape flavonoids in the form of red wine, red grape juice and related preparations prevent cardiovascular risk factors. It is encouraging to note that seeded as well as seedless grapes; the various parts of grapes including skin, pomace and seeds all are highly beneficial. There are no general safety concerns with ingestion of grape products. But according to Huntley (2007) consumption of red wine should be within recommended limits and the high sugar content in grape juice should be considered. Further research is needed on the role of grape flavonoids on neurological disorders, cognition and renal function.

The recent patents emphasize pharmaceutical use of natural grape juice and other polyphenol-rich products. The application of such products in clinical trials as a substitute or co-adjuvant with drugs may be suitable for future research (Gollucke and Ribeiro 2012; Gollucke *et al.*, 2013).

References

Agarwal C, Singh RP, Agarwal R (2002). Grape seed extract induced apoptotic death of human prostate carcinoma DU145 cells via caspases activation accompanied by dissipation of mitochondrial membrane potential and cytochrome c release. *Carcinogenesis*. 23: 1869–76.

Agarwal C, Veluri R, Kaur M, Chou SC, Thompson JA, Agarwal R (2007). Fractionation of high molecular weight tannins in grape seed extract and identification of procyanidin B2-3,3'-di-O-gallate as a major active constituent causing growth inhibition and apoptotic death of DU145 human prostate carcinoma cells. *Carcinogenesis*. 28 (7): 1478–84.

Aggarwal BB, Bhardwaj A, Aggarwal RS, Seeram NP, Shishodia S, Takada Y (2004). Role of resveratrol in prevention and therapy of cancer: preclinical and clinical studies. *Anti-cancer Research*. 24 (5 A): 2783–40.

Aghbali A, Hosseini SV, Delazar A, Gharavi NK, Shahneh FZ, Orangi M, Bandehagh A, Baradaran B (2013). Induction of apoptosis by grape seed extract (*Vitis vinifera*) in oral squamous cell carcinoma. *Bosn. J. Basic Med. Sci.* 13 (3): 186–91.

Aguiar O Jr, Gollücke AP, de Moraes BB, Pasquini G, Catharino RR, Riccio MF, Ihara SS, Ribeiro DA (2011). Grape juice concentrate prevents oxidative DNA damage in peripheral blood cells of rats subjected to a high-cholesterol diet. *Br. J. Nutr.* 105 (5): 694–702.

Al-Awwadi NA, Araiz C, Bornet A, Delbosc S, Cristol JP, Linck N, Azay J, Teissedre PL, Cros G (2005). Extracts enriched in different polyphenol families normalize increased cardiac NADPH oxidase expression while having differential effects on insulin resistance, hypertension, and cardiac hypertrophy in high-fructose-fed rats. *J. Agric. Food Chem.* 53 (1): 151–57.

Anter J, de Abreu-Abreu N, Fernández-Bedmar Z, Villatoro-Pulido M, Alonso-Moraga A, Muñoz-Serrano A (2011). Targets of red grapes: oxidative damage of DNA and leukaemia cells. *Nat. Prod. Commun.* 6 (1): 59–64.

Badavi M, Mehrgerdi FZ, Sarkaki A, Naseri MK, Dianat M (2008). Effect of grape seed extract on lead induced hypertension and heart rate in rat. *Pak. J. Biol. Sci.* 11 (6): 882–87.

Bagchi D, Bagchi M, Stohs S, Das DK, Ray SD, Kuszynsky CA, Joshi SS, Pruess HG (2000). Free radicals and grape seed pro-anthocyanidin extract: importance in human health and disease prevention. *Toxicology*. 148: 187–97.

Bagchi D, Garg A, Krohn RL, Bagchi M, Bagchi DJ, Balmoori J, Stohs SJ (1998). Protective effects of grape seed proanthocyanidins and selected anti-oxidants against TPA-induced hepatic and brain lipid peroxidation and DNA fragmentation, and peritoneal macrophage activation in mice. *General Pharmacology*. 30 (5): 771–76.

Bagchi D, Sen CK, Ray SD, Das DK, Bagchi M, Preuss HG, Vinson JA (2003). Molecular mechanisms of cardioprotection by a novel grape seed proanthocyanidin extract. *Mutat. Res.* 523–24: 87–97.

Bak MJ, Jun M, Jeong WS (2012). Procyanidins from wild grape (Vitis amurensis) seeds regulate ARE-mediated enzyme expression via Nrf2 coupled with p38 and PI3K/Akt pathway in HepG2 cells. *Int. J. Mol. Sci.* 13 (1): 801–818.

Balu M, Sangeetha P, Haripriya D, Panneerselvam C (2005). Rejuvenation of antioxidant system in central nervous system of aged rats by grape seed extract. *Neurosci. Lett.* 383 (3): 295–300.

Banini AE, Boyd LC, Allen JC, Allen HG, Sauls DL (2006). Muscadine grape products intake, diet and blood constituents of non-diabetic and type 2 diabetic subjects. *Nutrition.* 22 (11-12): 1137–45.

Bellincontro A, De Santis D, Botondi R, Villa I, Mencarelli F (2004). Different post-harvest dehydration rates affect quality characteristics and volatile compounds of Malvasia, Trebbiano and Sangiovese grapes for wine production. *Journal of the Science of Food and Agriculture.* 84 (13): 1791–800.

Beretz A, Cazenave JP, Anton R (1982). Inhibition of aggregation and secretion of human platelets by quercetin and other flavonoids: structure-activity relationships. *Agents Actions* 12: 382–87.

Bolin H, Petrucci V (1985). Amino acids in raisins. *J. Food Sci.* 50: 1507–09.

Briedis V, Povilaityte V, Kazlauskas S, Venskutonis PR (2003). Polyphenols and anthocyanins in fruits, grapes juices and wines, and evaluation of their anti-oxidant activity. *Medicina* (Kaunas). 39 Suppl. 2: 104–12. [Article in Lithuanian].

Cai Q, Li BY, Gao HQ, Zhang JH, Wang JF, Yu F, Yin M, Zhang Z (2011). Grape seed procyanidin b2 inhibits human aortic smooth muscle cell proliferation and migration induced by advanced glycation end products. *Biosci. Biotechnol. Biochem.* 75 (9): 1692–97.

Cetin A, Kaynar L, Koçyiðit I, Hacioðlu SK, Saraymen R, Oztürk A, Orhan O, Saðdiç O (2008). The effect of grape seed extract on radiation-induced oxidative stress in the rat liver. *Turk. J. Gastroenterol.* 19 (2): 92–98.

Chacón MR, Ceperuelo-Mallafré V, Maymó-Masip E, Mateo-Sanz JM, Arola L, Guitiérrez C, Fernandez-Real JM, Ardèvol A, Simón I, Vendrell J (2009). Grape-seed procyanidins modulate inflammation on human differentiated adipocytes *in-vitro*. *Cytokine.* 47 (2): 137–42.

Chanvitayapongs S, Draczynska-Lusiak B, Sun AY (1997). Amelioration of oxidative stress by anti-oxidants and resveratrol in PC12 cells. *Neuro. Report.* 8 (6): 1499–502.

Charradi K, Elkahoui S, Karkouch I, Limam F, Hassine FB, Aouani E (2012). Grape seed and skin extract prevents high-fat diet-induced brain lipo-toxicity in rat. *Neurochem. Res.* 37 (9): 2004–13.

Cheah KY, Howarth GS, Yazbeck R, Wright TH, Whitford EJ, Payne C, Butler RN, Bastian SE (2009). Grape seed extract protects IEC-6 cells from chemotherapy-induced cyto-toxicity and improves parameters of small intestinal mucositis in rats with experimentally-induced mucositis. *Cancer Biol. Ther.* 8 (4): 382–90.

Chen DM, Cai X, Kwik-Uribe CL, Zeng R, Zhu XZ (2006). Inhibitory effects of procyanidin B2 dimer on lipid-laden macrophage formation. *J. Cardiovasc. Pharmacol.* 48: 54–70.

Chis IC, Ungureanu MI, Marton A, Simedrea R, Muresan A, Postescu ID, Decea N (2009). Anti-oxidant effects of a grape seed extract in a rat model of diabetes mellitus. *Diab Vasc Dis Res.* 6 (3): 200–4.

Clifford AJ, Ebeler SE, Ebeler JD, Bills ND, Hinrichs SH, Teissedre PL, Waterhouse AL (1996). Delayed tumor onset in transgenic mice fed an amino acid-based diet supplemented with red wine solids. *Am. J. Clin. Nutr.* 64 (5): 748–56.

Cuevas VM, Calzado YR, Guerra YP, Yera AO, Despaigne SJ, Ferreiro RM, Quintana DC (2011). Effects of grape seed extract, vitamin C, and vitamin E on ethanol- and aspirin-induced ulcers. *Adv. Pharmacol. Sci.* 2011: 740687.

Cui XP, Li BY, Gao HQ, Wei N, Wang WL, Lu M (2008). Effects of grape seed pro-anthocyanidin extracts on peripheral nerves in streptozocin-induced diabetic rats. *J. Nutr. Sci. Vitaminol. (Tokyo).* 54 (4): 321–28.

Cui X, Liu X, Feng H, Zhao S, Gao H (2012). Grape seed proanthocyanidin extracts enhance endothelial nitric oxide synthase expression through 5'-AMP activated protein kinase/Surtuin 1-Krüpple like factor 2 pathway and modulates blood pressure in ouabain induced hypertensive rats. *Biol Pharm Bull.* 35 (12): 2192–97.

Del Bas JM, Fernández-Larrea J, Blay M, Ardèvol A, Salvadó MJ, Arola L, Bladé C (2005). Grape seed procyanidins improve atherosclerotic risk index and induce liver CYP7A1 and SHP expression in healthy rats. *FASEB. J.* 19 (3): 479–81.

Demrow HS, Slane PR, Folts JD (1995). Administration of wine and grape juice inhibits *in-vivo* platelet activity and thrombosis in stenosed canine coronary arteries. *Circulation.* 91 (4): 1182–88.

de Rezende AA, Graf U, Guterres Zda R, Kerr WE, Spanó MA (2009). Protective effects of pro-anthocyanidins of grape (*Vitis vinifera* L.) seeds on DNA damage induced by Doxorubicin in somatic cells of Drosophila melanogaster. *Food Chem. Toxicol.* 47 (7): 1466–72.

DeWhalley C V, Rankin SM, Hoult JRS, Jessup W, Leake DS (1990). Flavonoids inhibit the oxidative modification of low density lipoproteins by macrophages. *Biomed. Pharmacol.* 39:1743–50.

Dinicola S, Cucina A, Pasqualato A, D'Anselmi F, Proietti S, Lisi E, Pasqua G, Antonacci D, Bizzarri M (2012). Anti-proliferative and apoptotic effects triggered by grape seed extract (GSE) versus epigallocatechin and procyanidins on colon cancer cell lines. *Int. J. Mol. Sci.* 13 (1): 651–64.

Dinicola S, Cucina A, Pasqualato A, Proietti S, D'Anselmi F, Pasqua G, Santamaria AR, Coluccia P,Laganà A, Antonacci D, Giuliani A, Bizzarri M (2010). Apoptosis-inducing factor and caspase-dependent apoptotic pathways triggered by different grape seed extracts on human colon cancer cell line Caco-2. *Br. J. Nutr.* 104 (6): 824–32.

Dulundu E, Ozel Y, Topaloglu U, Toklu H, Ercan F, Gedik N, Sener G (2007). Grape seed extract reduces oxidative stress and fibrosis in experimental biliary obstruction. *J. Gastroenterol. Hepatol.* 22 (6): 885–92

Durak I, Cetin R, Devrim E, Ergüder IB (2005). Effects of black grape extract on activities of DNA turn-over enzymes in cancerous and non cancerous human colon tissues. *Life Sci.* 76 (25): 2995–3000.

Ebeler SE, Brenneman CA, Kim GS, Jewell WT, Webb MR, Chacon-Rodriguez L, MacDonald EA, Cramer AC, Levi A, Ebeler JD, Islas-Trejo A, Kraus A, Hinrichs SH, Clifford AJ (2002). Dietary catechin delays tumor onset in a transgenic mouse model. *Am. J. Clin. Nutr.* 76 (4): 865–72.

Engelbrecht AM, Mattheyse M, Ellis B, Loos B, Thomas M, Smith R, Peters S, Smith C, Myburgh K (2007). Pro-anthocyanidin from grape seeds inactivates the PI3-kinase/PKB pathway and induces apoptosis in a colon cancer cell line. *Cancer Lett.* 258 (1): 144–53.

Facino RM, Carini M, Aldini G, Berti F, Rossoni G, Bombardelli E, Morazzoni P (1999). Diet enriched with procyanidins enhances anti-oxidant activity and reduces myocardial post-ischaemic damage in rats. *Life Sci.* 64 (8): 627–42.

Falcao JM, Dias JA, Miranda AC, Leitao CN, Lacerda MM, Cayolla Da Motta L (1994). Red wine consumption and gastric cancer in Portugal: a case-control study. *Eur. J. Cancer Prev.* 3 (3): 269–76.

Fan P, Lou H (2004). Effects of polyphenols from grape seeds on oxidative damage to cellular DNA. *Molecular and Cellular Biochemistry.* 267 (1-2): 67–74.

Filip A, Clichici S, Daicoviciu D, Catoi C, Bolfa P, Postescu ID, Gal A, Baldea I, Gherman C, Muresan A (2011). Chemo-preventive effects of *Calluna vulgaris* and *Vitis vinifera* extracts on UVB-induced skin damage in SKH-1 hairless mice. *J. Physiol. Pharmacol.* 62 (3): 385–92.

Fitzpatrick DF, Hirschfield SL, Coffey RG (1993). Endothelium-dependent vaso-relaxing activity of wine and other grape products. *Am. J. Physiol.* 265: H774–78.

Folts JD (2002). Potential health benefits from the flavonoids in grape products on vascular disease. *Adv. Exp. Med. Biol.* 505: 95–111.

Frederiksen H, Mortensen A, Schrøder M, Frandsen H, Bysted A, Knuthsen P, Rasmussen SE (2007). Effects of red grape skin and seed extract supplementation on atherosclerosis in Watanabe heritable hyper-lipidemic rabbits. *Mol. Nutr. Food Res.* 51 (5): 564–71.

Garvin S, Öllinger K, Dabrosin C (2006). Resveratrol induces apoptosis and inhibits angiogenesis in human breast cancer xenografts in-vivo. *Cancer Letters.* 231 (1): 113–22.

Gollucke AP, Aguiar O Jr, Barbisan LF, Ribeiro DA (2013). Use of grape polyphenols against carcinogenesis: putative molecular mechanisms of action using in vitro and in vivo test systems. *J. Med. Food.* 16 (3): 199–205.

Gollücke AP, Ribeiro DA (2012). Use of grape polyphenols for promoting human health: a review of patents. *Recent Pat. Food Nutr. Agric.* 4 (1): 26–30.

González-Abuín N, Martínez-Micaelo N, Blay M, Pujadas G, Garcia-Vallvé S, Pinent M, Ardévol A (2012). Grape seed-derived procyanidins decrease dipeptidyl-peptidase 4 activity and expression. *J. Agric. Food Chem.* 60 (36): 9055–61.

Hamlaoui S, Mokni M, Limam N, Zouaoui K, Ben Rayana MC, Carrier A, Limam F, Amri M, Marzouki L, Aouani E (2012). Protective effect of grape seed and skin extract on garlic-induced erythrocyte oxidative stress. *J. Physiol. Pharmacol.* 63 (4): 381–88.

Hanausek M, Spears E, Walaszek Z, Kowalczyk MC, Kowalczyk P, Wendel C, Slaga TJ (2011). Inhibition of murine skin carcinogenesis by freeze-dried grape powder and other grape-derived major anti-oxidants. *Nutr. Cancer.* 63 (1): 28–38.

Hertog MGL, Feskens EJM, Hollman PCH, Katan MB, Kromhout D (1993). Dietary anti-oxidant flavonoids and risk of coronary heart disease: the Zutphen elderly study. *Lancet* 342: 1007–11.

Hertog MGL, Kromhout D, Aravanis C, Blackburn H, Buzina R, Fidanza F, Giampaoli S, Jansen A, Menotti A, Nedeljkovic S, *et al.* (1995a). Flavonoid intake and long-term risk of coronary heart disease and cancer in the Seven Countries Study. *Archives of Internal Medicine.* 155 (4): 381–86.

Hertog MGL, Kromhout D, Aravanis C, Blackburn H, Buzina R, Fidanza F, Giampaoli S (1995b). Flavonoid intake and long-term risk of coronary heart disease and cancer in the seven countries study. *Arch. Intern. Med.* 155: 381–86.

Hogan S, Zhang L, Li J, Sun S, Canning C, Zhou K (2010). Anti-oxidant rich grape pomace extract suppresses post-prandial hyper-glycemia in diabetic mice by specifically inhibiting alpha-glucosidase. *Nutrition and Metabolism.* 7: 71.

Houde V, Grenier D, Chandad F (2006). Protective effects of grape seed pro-anthocyanidins against oxidative stress induced by lipo-polysaccharides of periodontopathogens. *J. Periodontol.* 77: 1371–79.

Hu H, Qin YM (2006). Grape seed proanthocyanidin extract induced mitochondria-associated apoptosis in human acute myeloid leukaemia 14.3D10 cells. *Chin. Med. J. (Engl).* 119 (5): 417–21.

Huang TT, Shang XJ, Yao GH, Ge JP, Teng WH, Sun Y, Huang YF (2008). Grape seed extract inhibits the growth of prostate cancer PC-3 cells. Zhonghua Nan Ke Xue. 14 (4): 331–33. [Article in Chinese].

Hudson TS, Hartle DK, Hursting SD, Nunez NP, Wang TT, Young HA, Arany P, Green JE (2007). Inhibition of prostate cancer growth by muscadine grape skin extract and resveratrol through distinct mechanisms. *Cancer Res*. 67 (17): 8396–405.

Huntley AL (2007). Grape flavonoids and menopausal health. *Menopause. Int*. 13 (4): 165–69.

Hwang IK, Kim DW, Park JH, Lim SS, Yoo KY, Kwon DY, Kim DW, Moon WK, Won MH (2009). Effects of grape seed extract and its ethylacetate/ethanol fraction on blood glucose levels in a model of type 2 diabetes. *Phytother. Res*. 23 (8): 1182–85.

Igney FH, Krammer PH (2002). Death and anti-death: tumor resistance to apoptosis. *Nat. Rev. Cancer*. 2: 277–88.

Jayaprakasha GK, Singh RP, Sakariah KK (2001). Anti-oxidant activity of grape seed (Vitis vinifera) extracts on peroxidation models in-vitro. *Food Chemistry*. 73 (3): 285–90.

Jung KJ, Wallig MA, Singletary KW (2006). Purple grape juice inhibits 7,12-dimethylbenz[a]anthracene (DMBA)-induced rat mammary tumorigenesis and in-vivo DMBA-DNA adduct formation. *Cancer Letters*. 233 (2): 279 -88.

Kamiyama M, Kishimoto Y, Tani M, Andoh K, Utsunomiya K, Kondo K (2009). Inhibition of low-density lipoprotein oxidation by Nagano purple grape (*Vitis viniferax, Vitis labrusca*). *J. Nutr. Sci. Vitaminol. (Tokyo)*. 55 (6): 471–78.

Kammerer D, Claus A, Carle R, Schieber A (2004). Polyphenol screening of pomace from red and white grape varieties (Vitis vinifera L.) by HPLC-DAD-MS/MS. *J. Agri. Food Chem*. 52 (14): 4360–67.

Kanner J, Frankel E, Granit R, German B, Kinsella JE (1994). Natural anti-oxidants in grapes and wines. *J. Agric. Food Chem*. 42: 64–69.

Kar P, Laight D, Rooprai HK, Shaw KM, Cummings M (2009). Effects of grape seed extract in Type 2 diabetic subjects at high cardiovascular risk: a double blind randomized placebo controlled trial examining metabolic markers, vascular tone, inflammation, oxidative stress and insulin sensitivity. *Diabet. Med. 26* (5): 526 – 31.

Kar P, Laight D, Shaw KM, Cummings MH (2006). Flavonoid-rich grapeseed extracts: a new approach in high cardiovascular risk patients? *Int. J. Clin. Pract*. 60 (11): 1484–92.

Karatzi K, Papamichael C, Aznaouridis K, Karatzis E, Lekakis J, Matsouka C, Boskou G, Chiou A, Sitara M, Feliou G, Kontoyiannis D, Zampelas A, Mavrikakis M (2004). Constituents of red wine other than alcohol improve endothelial function in patients with coronary artery disease. *Coron. Artery Dis. 15* (8) 485–90.

Katiyar SK (2006). Matrix metalloproteinases in cancer metastasis: molecular targets for prostate cancer prevention by green tea polyphenols and grape seed pro-anthocyanidins. *Endocr. Metab. Immune. Disord. Drug. Targets*. 6 (1): 17–24.

Katiyar SK (2008). Grape seed pro-anthocyanidines and skin cancer prevention: inhibition of oxidative stress and protection of immune system. *Mol. Nutr. Food Res.* 52 Suppl 1: S71–76.

Kaur M, Agarwal C, Agarwal R (2009a). Anti-cancer and cancer chemo-preventive potential of grape seed extract and other grape-based products. *J. Nutr.* 139 (9): 1806S–1812S.

Kaur M, Mandair R, Agarwal R, Agarwal C (2008). Grape seed extract induces cell cycle arrest and apoptosis in human colon carcinoma cells. *Nutr. Cancer.* 60 Suppl. 1: 2–11.

Kaur M, Singh RP, Gu M, Agarwal R, Agarwal C (2006). Grape seed extract inhibits *in-vitro* and *in-vivo* growth of human colorectal carcinoma cells. *Clin. Cancer. Res.* 12 (20 Pt 1): 6194–202.

Kaur M, Velmurugan B, Rajamanickam S, Agarwal R, Agarwal C (2009b). Gallic acid, an active constituent of grape seed extract, exhibits anti-proliferative, pro-apoptotic and anti-tumorigenic effects against prostate carcinoma xenograft growth in nude mice. *Pharmaceutical Research.* 26 (9): 2133–40.

Khanna S, Venojarvi M, Roy S, Sharma N, Trikha P, Bagchi D, Bagchi M, Sen CK (2002). Dermal wound healing properties of redox-active grape seed pro-anthocyanidins. *Free Radic. Biol. Med.* 33: 1089–96.

Kijima I, Phung S, Hur G, Kwok SL, Chen S (2006). Grape seed extract is an aromatase inhibitor and a suppressor of aromatase expression. *Cancer. Res.* 66(11): 5960–67.

Kim EJ, Park H, Park SY, Jun JG, Park JH (2009). The grape component piceatannol induces apoptosis in DU145 human prostate cancer cells via the activation of extrinsic and intrinsic pathways. *J. Med. Food.* 12 (5): 943–51.

Kim H, Hall P, Smith M, Kirk M, Prasain JK, Barnes S, Grubbs C (2004). Chemo-prevention by grape seed extract and genistein in carcinogen-induced mammary cancer in rats is diet dependent. *Journal of Nutrition.* 134 (12): 3445S–3452S.

Kingsley K, Jensen D, Toponce R, Dye J, Martin D, Phippen S, Ross D, Halthore VS, O'Malley S (2010). Inhibition of oral cancer growth in vitro is modulated through differential signaling pathways by over-the-counter proanthocyanidin supplements. *J. Diet. Suppl.* 7 (2): 130–44.

Kiyici A, Okudan N, Gökbel H, Belviranli M (2010). The effect of grape seed extracts on serum paraoxonase activities in streptozotocin-induced diabetic rats. *J. Med. Food.* 13 (3): 725–28.

Knekt P, Jarvinen R, Reunanen A, Maatela J (1996). Flavonoid intake and coronary mortality in Finland: a cohort study. *BMJ.* 312 (7029): 478–81.

Lazzè MC, Pizzala R, Gutiérrez Pecharromán FJ, Gatòn Garnica P, Antolín Rodríguez JM, Fabris N, Bianchi L (2009). Grape waste extract obtained by supercritical fluid extraction contains bioactive anti-oxidant molecules and induces anti-proliferative effects in human colon adeno-carcinoma cells. *J. Med. Food.* 12 (3): 561–68.

Leifert WR, Abeywardena MY (2008). Cardio-protective actions of grape polyphenols. *Nutr. Res.* 28 (11): 729–37.

Li BY, Li XL, Cai Q, Gao HQ, Cheng M, Zhang JH, Wang JF, Yu F, Zhou RH (2011a). Induction of lactadherin mediates the apoptosis of endothelial cells in response to advanced glycation end products and protective effects of grape seed procyanidin B2 and resveratrol. *Apoptosis.* 16: 732–45.

Li BY, Li XL, Gao HQ, Zhang JH, Cai Q, Cheng M, Lu M (2011b). Grape seed procyanidin B2 inhibits advanced glycation end products induced endothelial cell apoptosis through regulating GSK3β phosphorylation. *Cell. Biol. Int.* 35: 663–69.

Li XL, Li BY, Gao HQ, Cheng M, Xu L, Li XH, Ma YB (2009). Effects of grape seed pro-anthocyanidin extracts on aortic pulse wave velocity in streptozocin induced diabetic rats. *Biosci. Biotechnol. Biochem.* 73 (6): 1348–54.

Liggins J, Bluck LJ, Runswick S, Atkinson C, Coward WA, Bingham SA (2000). Daidzein and genistein content of fruits and nuts. *J. Nutr. Biochem.* 11: 326–31.

Lin YS, Chen SF, Liu CL, Nieh S (2012). The chemo-adjuvant potential of grape seed procyanidins on p53-related cell death in oral cancer cells. *J. Oral. Pathol. Med.* 41 (4): 322–31.

Liu YN, Shen XN, Yao GY (2006). Effects of grape seed pro-anthocyanidins extracts on experimental diabetic nephropathy in rats]. Wei. Sheng. Yan. Jiu. 35 (6): 703–705. [Article in Chinese].

Llópiz N, Puiggròs F, Céspedes E, Arola L, Ardévol A, Bladé C, Salvadó MJ (2004). Anti-genotoxic effect of grape seed procyanidin extract in Fao cells submitted to oxidative stress. *J. Agric. Food Chem.* 52 (5): 1083–87.

Lu J, Zhang K, Chen S, Wen W (2009). Grape seed extract inhibits VEGF expression via reducing HIF-1alpha protein expression. *Carcinogenesis.* 30 (4): 636–44.

Lu Y, Yeap Foo L (1999). The polyphenol constituents of grape pomace. *Food Chemistry.* 65 (1): 1–8.

Mackenzie GG, Adamo AM, Decker NP, Oteiza PI (2008). Dimeric procyanidin B2 inhibits constitutively active NF-kappaB in Hodgkin's lymphoma cells independently of the presence of IkappaB mutations. *Biochem. Pharmacol.* 75: 1461–71.

Mansouri E, Panahi M, Ghaffari MA, Ghorbani A (2011). Effects of grape seed pro-anthocyanidin extract on oxidative stress induced by diabetes in rat kidney. *Iran. Biomed. J.* 15 (3): 100–06.

Martínez CC, Vicente OV, Yáñez GMJ, García RJM, Canteras JM, Alcaraz BM (2005). Experimental model for treating pulmonary metastatic melanoma usinggrape-seed extract, red wine and ethanol. *Clin. Transl. Oncol.* 7 (3): 115–21. [Article in Spanish].

Mattoo TK, Kovacevic L (2003). Effect of grape seed extract on puromycin-aminonucleoside-induced nephrosis in rats. Pediatr. *Nephrol.* 18 (9): 872–77.

Meeran SM, Katiyar SK (2007). Grape seed pro-anthocyanidins promote apoptosis in human epidermoid carcinoma A431 cells through alterations in Cdki-Cdk-cyclin cascade, and caspase-3 activation via loss of mitochondrial membrane potential. *Exp. Dermatol.* 16 (5): 405–15.

Miyagi Y, Miwa K, Inoue H (1997). Inhibition of human low-density lipoprotein oxidation by flavonoids in red wine and grape juice. *Am. J. Cardiol.* 80: 1627–31.

Morré DM, Morré DJ (2006). Anti-cancer activity of grape and grape skin extracts alone and combined with green tea infusions. *Cancer Letters.* 238 (2): 202–09.

Negro C, Tommasi L, Miceli A (2003). Phenolic compounds and anti-oxidant activity from red grape marc extracts. *Bioresource Technology.* 87 (1): 41–44.

Ngamukote S, Mäkynen K, Thilawech T, Adisakwattana S (2011). Cholesterol-lowering activity of the major polyphenols in grape seed. *Molecules.* 16 (6): 5054–61

Nikitina NA, Sobenin IA, Myasoedova VA, Korennaya VV, Mel'nichenko AA, Khalilov EM, Orekhov AN (2006). Anti-atherogenic effect of grape flavonoids in an *ex-vivo* model. Bull. Exp. Biol. Med. 141 (6): 712–15. [Article in English, Russian].

Osman HE, Maalej N, Shanmuganayagam D, Folts JD (1998). Grape juice but not orange or grapefruit juice inhibits platelet activity in dogs and monkeys. *Nutr.* 128 (12): 2307–12.

Pakhale SS, Karibasappa GS, Ramchandani AG, Bhushan B, Sharma A (2007). Scavenging effect of Indian grape polyphenols on 2,2'-diphenyl-1-picrylhydrazyl (DPPH) radical by electron spin resonance spectrometry. *Indian J. Exp. Biol.* 45 (11): 968–73.

Palma M, Taylor LT (2001). Supercritical fluid extraction of 5-hydroxymethyl-2-furaldehyde from raisins. *J. Agric. Food Chem.* 49: 628–32.

Pan XJ, Wang M, Wang XX, Liu B, Zhang H (2012). Study on the effect of grape seed pro anthocyanidins on increasing the radio sensitivity for X-ray. Zhong. Yao. Cai. 35 (2): 264–69. [Article in Chinese].

Peng N, Clark JT, Prasain J, Kim H, White CR, Wyss JM (2005). Anti-hypertensive and cognitive effects of grape polyphenols in estrogen-depleted, female, spontaneously hyper-tensive rats. *Am. J. Physiol. Regul. Integr. Comp. Physiol.* 289 (3): R771–75.

Percival SS (2009). Grape consumption supports immunity in animals and humans. *Journal of Nutrition.* 139 (9): 1801S–05S.

Polagruto JA, Gross HB, Kamangar F, Kosuna K, Sun B, Fujii H, Keen CL, Hackman RM (2007). Platelet reactivity in male smokers following the acute consumption of a flavanol-rich grape seed extract. *J. Med. Food.* 10 (4): 725–30.

Postescu ID, Chereches G, Tatomir C, Daicoviciu D, Filip GA (2012). Modulation of doxorubicin-induced oxidative stress by a grape (*Vitis vinifera* L.) seed extract in normal and tumor cells. *J. Med. Food.* 15 (7): 639–45.

Praphasawat R, Klungsupya P, Muangman T, Laovitthayanggoon S, Arunpairojana V, Himakoun L (2011). Anti-mutagenicity and anti-oxidative DNA damage properties of oligomeric pro-anthocyanidins from Thai grape seeds in TK6 cells. *Asian Pac. J. Cancer Prev.* 12 (5): 1317–21.

Prasad R, Vaid M, Katiyar SK (2012). Grape pro-anthocyanidin inhibit pancreatic cancer cell growth *in vitro* and *in vivo* through induction of apoptosis and by targeting the PI3K/Akt pathway. *PLoS. One.* 7 (8): e43064.

Preuss HG, Bagchi D, Bagchi M (2002). Protective effects of a novel niacin-bound chromium complex and a grape seed pro-anthocyanidin extract on advancing age and various aspects of syndrome X. Ann. N. Y. Acad. Sci. 957: 250–59.

Preuss HG, Montamarry S, Echard B, Scheckenbach R, Bagchi D (2001). Long-term effects of chromium, grape seed extract, and zinc on various metabolic parameters of rats. *Mol. Cell. Biochem.* 223 (1-2): 95–102.

Prieur C, Rigaud V, Cheynier V, Moutounet M (1994). Oligomeric and polymeric procyanidins from grape seeds. *Phytochemistry.* 36: 781–84.

Puiggros F, Llópiz N, Ardévol A, Bladé C, Arola L, Salvadó MJ (2005). Grape seed procyanidins prevent oxidative injury by modulating the expression of anti-oxidant enzyme systems. *J. Agric. Food Chem.* 53 (15): 6080–86.

Puiggròs F, Sala E, Vaqué M, Ardévol A, Blay M, Fernández-Larrea J, Arola L, Bladé C, Pujadas G, Salvadó MJ (2009). *In-vivo, in-vitro,* and in silico studies of Cu/Zn-superoxide dismutase regulation by molecules in grape seed procyanidin extract. *J. Agric. Food Chem.* 57 (9): 3934–42.

Radhakrishnan S, Reddivari L, Sclafani R, Das UN, Vanamala J (2011). Resveratrol potentiates grape seed extract induced human colon cancer cell apoptosis. *Frontiers in Bioscience.* 3: 1509–23.

Radler F (1965). The surface lipids of fresh and processed raisins. *J. Sci. Food Agric.* 16: 638–43.

Raina K, Singh RP, Agarwal R, Agarwal C (2007). Oral grape seed extract inhibits prostate tumor growth and progression in TRAMP mice. *Cancer Res.* 67 (12): 5976–82.

Ramchandani AG, Karibasappa GS, Pakhale SS (2008). Anti-tumor-promoting effects of polyphenolic extracts from seedless and seeded Indian grapes. *J. Environ. Pathol. Toxicol. Oncol.* 27 (4): 321–31.

Razavi SM, Gholamin S, Eskandari A, Mohsenian N, Ghorbanihaghjo A, Delazar A, Rashtchizadeh N, Keshtkar-Jahromi M, Argani H (2013). Red grape seed extract improves lipid profiles and decreases oxidized low-density lipoprotein in patients with mild hyper-lipidemia. *J. Med. Food.* 16 (3): 255–58.

Rho KA, Kim MK (2006). Effects of different grape formulations on anti-oxidative capacity, lipid peroxidation and oxidative DNA damage in aged rats. *J. Nutr. Sci. Vitaminol. (Tokyo).* 52 (1): 33–46.

Rowe CA, Nantz MP, Nieves C Jr, West RL, Percival SS (2011). Regular consumption of concord grape juice benefits human immunity. *J. Med. Food.* 14 (1-2): 69–78.

Saito M, Hosoyama H, Ariga T, Kataoka S, Yamaji N (1998). Anti-ulcer activity of grape seed extract and procyanidins. *Journal of Agricultural and Food Chemistry.* 46 (4): 1460–64.

Sano A, Uchida R, Saito M, Shioya N, Komori Y, Tho Y, Hashizume N (2007). Beneficial effects of grape seed extract on malondialdehyde-modified LDL. *J. Nutr. Sci. Vitaminol.* (Tokyo). 53 (2): 174–82.

Sato M, Maulik G, Ray PS, Bagchi D, Das DK (1999). Cardioprotective effects of grape seed pro-anthocyanidin against ischemic reperfusion injury. *Journal of Molecular and Cellular Cardiology.* 31 (6): 1289–97.

Schamel G (2006). Geography versus brands in a global wine market. *Agri. business.* 22: 363–74.

Seeram NP, Adams LS, Henning SM, Niu Y, Zhang Y, Nair MG, Heber D (2005). *In*-vitro anti-proliferative, apoptotic and anti-oxidant activities of punicalagin, ellagic acid and a total pomegranate tannin extract are enhanced in combination with other polyphenols as found in pomegranate juice. *J. Nutr. Biochem.* 16 (6): 360–67.

Shafiee M, Carbonneau MA, Urban N, Descomps B, Leger CL (2003). Grape and grape seed extract capacities at protecting LDL against oxidation generated by Cu2+, AAPH or SIN-1 and at decreasing superoxide THP-1 cell production. A comparison to other extracts or compounds. *Free Radical Research.* 37 (5): 573–84.

Shang XJ, Yin HL, Ge JP, Sun Y, Teng WH, Huang YF (2008). Grape seed extract induces morphological changes of prostate cancer PC-3 cells. Zhonghua. Nan. Ke. Xue. 14 (12): 1090–93. [Article in Chinese].

Shanmuganayagam D, Beahm MR, Osman HE, Krueger CG, Reed JD, Folts JD (2002). Grape seed and grape skin extracts elicit a greater anti-platelet effect when used in combination than when used individually in dogs and humans. *J. Nutr.* 132 (12): 3592–98.

Shao ZH, Wojcik KR, Dossumbekova A, Hsu C, Mehendale SR, Li CQ, Qin Y, Sharp WW, Chang WT, Hamann KJ, Yuan CS, Hoek TL (2009). Grape seed pro-anthocyanidins protect cardio-myocytes from ischemia and reperfusion injury *via* Akt-NOS signaling. *J. Cell. Biochem.* 107 (4): 697–705.

Sharma G, Tyagi AK, Singh RP, Chan DC, Agarwal R (2004). Synergistic anti-cancer effects of grape seed extract and conventional cytotoxic agent doxorubicin against human breast carcinoma cells. *Breast Cancer Res. Treat.* 85(1):1-12.

Sharma SD, Katiyar SK (2006). Dietary grape-seed proanthocyanidin inhibition of ultraviolet B-induced immune suppression is associated with induction of IL-12. *Carcinogenesis.* 27 (1): 95–102.

Sharma SD, Meeran SM, Katiyar SK (2007). Dietary grape seed pro-anthocyanidins inhibit UVB-induced oxidative stress and activation of mitogen-activated protein kinases and nuclear factor-kappaB signaling in *in vivo* SKH-1 hairless mice. *Mol. Cancer. Ther.* 6 (3): 995–1005.

Sharma SD, Meeran SM, Katiyar SK (2010). Pro-anthocyanidins inhibit in-vitro and in-vivo growth of human non-small cell lung cancer cells by inhibiting the prostaglandin e2 and prostaglandin e2 receptors. *Mol. Cancer Therapeutics.* 9 (3): 569 – 80.

Shi J, Yu J, Pohorly JE, Kakuda Y (2003). Polyphenolics in grape seeds -biochemistry and functionality. *J. Med. Food.* 6 (4): 291–99.

Shrotriya S, Deep G, Gu M, Kaur M, Jain AK, Inturi S, Agarwal R, Agarwal C (2012). Generation of reactive oxygen species by grape seed extract causes irreparable DNA damage leading to G2/M arrest and apoptosis selectively in head and neck squamous cell carcinoma cells. *Carcinogenesis.* 33 (4): 848–58.

Singh RP, Tyagi AK, Dhanalakshmi S, Agarwal R, Agarwal C (2004). Grape seed extract inhibits advanced human prostate tumor growth and angiogenesis and up-regulates insulin-like growth factor binding protein-3. *Int. J. Cancer.* 108 (5): 733–40.

Singh T, Sharma SD, Katiyar SK (2011). Grape pro-anthocyanidins induce apoptosis by loss of mitochondrial membrane potential of human non-small cell lung cancer cells *in-vitro* and in-vivo. *PLoS. ONE.* 6 (11) e27444.

Sivaprakasapillai B, Edirisinghe I, Randolph J, Steinberg F, Kappagoda T (2009). Effect of grape seed extract on blood pressure in subjects with the metabolic syndrome. *Metabolism.* 58 (12): 1743–46.

Soleas GJ, Grass L, Josephy PD, Goldberg DM, Diamandis EP (2002). A comparison of the anti-carcinogenic properties of four red wine polyphenols. *Clinical Biochemistry.* 35 (2): 119–24.

Song X, Siriwardhana N, Rathore K, Lin D, Wang HC (2010). Grape seed proanthocyanidin suppression of breast cell carcinogenesis induced by chronic exposure to combined 4-(methylnitrosamino)-1-(3-pyridyl)-1-butanone and benzo[a]pyrene. *Mol. Carcinog.* 49 (5): 450–63.

Stafford AE, Fuller G, Bolin HR, MacKey BE (1974). Analysis of fatty acid esters in processed raisins by gas chromatography. *J. Agric. Food Chem.* 22: 478–79.

Stein JH, Keevil JG, Wiebe DA, Aeschlimann S, Folts JD (1999). Purple grape juice improves endothelial function and reduces the susceptibility of LDL cholesterol to oxidation in patients with coronary artery disease. *Circulation.* 100 (10): 1050–55.

Tao HY, Wu CF, Zhou Y, Gong WH, Zhang X, Iribarren P, Zhao YQ, Le YY, Wang JM (2004). The grape component resveratrol interferes with the function of chemoattractant receptors on phagocytic leukocytes. *Cell. Mol. Immunol.* 1 (1): 50–56.

Thiruchenduran M, Vijayan NA, Sawaminathan JK, Devaraj SN (2011). Protective effect of grape seed pro anthocyanidins against cholesterol-cholic acid diet-induced hyper-cholesterolemia in rats. *Cardiovasc. Pathol.* 20 (6): 361–68.

USDA (2003) Agricultural Research Service. Fruit and tree nuts outlook/FTS303. http://www.nal.usda.gov/Briefing/FruitandTreeNuts/fruitnutpdf/Raisins.pdf.

USDA, National Nutrition Data Base for standard reference, Release-26, NDB No-09131.

Vaid M, Sharma SD, Katiyar SK (2010). Pro-anthocyanidins inhibit photo-carcinogenesis through enhancement of DNA repair and xeroderma pigmentosum group A-dependent mechanism. *Cancer. Prev. Res.* (Phila). 3 (12): 1621–29.

Vaid M, Singh T, Prasad R, Elmets CA, Xu H, Katiyar SK (2013). Bioactive grape pro-anthocyanidins enhance immune reactivity in UV-irradiated skin through functional activation of dendritic cells in mice. *Cancer Prev. Res.* (Phila). 6 (3): 242–52.

Vayalil PK, Mittal A, Katiyar SK (2004). Pro-anthocyanidins from grape seeds inhibit expression of matrix metallo-proteinases in human prostate carcinoma cells, which is associated with the inhibition of activation of MAPK and NF kappa B. *Carcinogenesis.* 5: 987–95.

Velmurugan B, Singh RP, Agarwal R, Agarwal C (2010a). Dietary-feeding of grape seed extract prevents azoxymethane-induced colonic aberrant crypt foci formation in fischer 344 rats. *Mol. Carcinogenesis.* 49 (7): 641–52.

Velmurugan B, Singh RP, Kaul N, Agarwal R, Agarwal C (2010b). Dietary feeding of grape seed extract prevents intestinal tumorigenesis in APCmin/+ mice. *Neoplasia.* 12 (1): 95–102.

Veluri R, Singh RP, Liu Z, Thompson JA, Agarwal R, Agarwal C (2006). Fractionation of grape seed extract and identification of gallic acid as one of the major active constituents causing growth inhibition and apoptotic death of DU145 human prostate carcinoma cells. *Carcinogenesis.* 27 (7): 1445–53.

Viel JF, Perarnau JM, Challier B, Faivre-Nappez I (1997). Alcoholic calories, red wine consumption and breast cancer among premenopausal women. *Eur. J. Epidemiol.* 13 (6): 639–43.

Vinson JA, Proch J, Bose P (2001). Mega Natural Gold grape seed extract: in-vitro anti-oxidant and in-vivo human supplementation studies. *Journal of Medicinal Food.* 4 (1): 17–26.

Vivier MA, Pretorius IS (2002). Genetically tailored grape vines for the wine industry. *Trends in Biotechnology.* 20 (11): 472–78.

Wang M, Wang L, Pan XJ, Zhang H (2012). Monocytic differentiation of K562 cells induced by pro-anthocyanidins from grape seeds. *Arch. Pharm. Res.* 35 (1): 129–35.

Wu CD (2009). Grape products and oral health. *J. Nutr.* 139 (9): 1818S–23S.

Yamakoshi J, Kataoka S, Koga T, Ariga T (1999). Pro-anthocyanidin-rich extract from grape seeds attenuates the development of aortic atherosclerosis in cholesterol-fed rabbits. *Atherosclerosis.* 142 (1): 139–49.

Yassa N, Beni HR, Hadjiakhoondi A (2008). Free radical scavenging and lipid peroxidation activity of the Shahani black grape. *Pak. J. Biol. Sci.* 11 (21): 2513–16.

Ye X, Krohn RL, Liu W, Joshi SS, Kuszynski CA, McGinn TR, Bagchi M, Preuss HG, Stohs SJ, Bagchi D (1999). The cytotoxic effects of a novel IH636 grape seed proanthocyanidin extract on cultured human cancer cells. *Molecular and Cellular Biochemistry.* 196 (1-2): 99–108.

Yilmaz Y, Toledo RT (2004). Major flavonoids in grape seeds and skins: anti-oxidant capacity of catechin, epicatechin, and gallic acid. J. Agric. Food Chem. 52 (2): 255–60.

Yousef MI, Saad AA, El-Shennawy LK (2009). Protective effect of grape seed pro-anthocyanidin extract against oxidative stress induced by cisplatin in rats. Food Chem. Toxicol. 47 (6): 1176–83.

Yu F, Li BY, Li XL, Cai Q, Zhang Z, Cheng M, Yin M, Wang JF, Zhang JH, Lu WD, Zhou RH, Gao HQ (2012). Proteomic analysis of aorta and protective effects of grape seed procyanidin B2 in db/db mice reveal a critical role of milk fat globule epidermal growth factor-8 in diabetic arterial damage. PLoS. One. 7 (12): e52541.

Yuan XY, Liu W, Hao JC, Gu WJ, Zhao YS (2012). Topical grape seed pro-anthocyandin extract reduces sunburn cells and mutant p53 positive epidermal cell formation, and prevents depletion of Langerhans cells in an acute sunburn model. Photomed. Laser. Surg. 30 (1): 20–25.

Zhang CJ, Zhou GY, Li L, Ma LL, Gao P, Li H (2004a). Reversal of multidrug resistance of MCF-7/ADR in nude mice by grape seed polyphenol. Zhonghua. Wai. Ke. Za. Zhi. 42 (13): 795–98. [Article in Chinese].

Zhang FJ, Yang JY, Mou YH, Sun BS, Ping YF, Wang JM, Bian XW, Wu CF (2009a). Inhibition of U-87 human glioblastoma cell proliferation and formyl peptide receptor function by oligomer procyanidins (F2) isolated from grape seeds. *Chem. Biol. Interact.* 179 (2-3): 419–29.

Zhang FJ, Yang JY, Mou YH, Sun BS, Wang JM, Wu CF (2010). Oligomer procyanidins from grape seeds induce a paraptosis-like programmed cell death in human glioblastoma U-87 cells. *Pharm. Biol.* 48 (8): 883–90.

Zhang HJ, Ji BP, Chen G, Zhou F, Luo YC, Yu HQ, Gao FY, Zhang ZP, Li HY (2009b). A combination of grape seed-derived procyanidins and gypenosides alleviates insulin resistance in mice and HepG2 cells. *J. Food Sci.* 74 (1): H1 – 7.

Zhang L, Hogan S, Li J, Shi S, Caning C, Zheng SJ, Zhou K (2011). Grape skin extract inhibits mammalian intestinal α-glucosidase activity and suppresses post-prandial glycemic response in streptozocin-treated mice. *Food Chemistry.* 126 (2): 466–71.

Zhang XY, Bai DC, Wu YJ, Li WG, Liu NF (2005b). Pro-anthocyanidin from grape seeds enhances anti-tumor effect of doxorubicin both *in-vitro* and *in-vivo*. *Pharmazie.* 60 (7): 533–38.

Zhang XY, Jayaprakasam B, Seeram NP, Olson LK, DeWitt D, Nair MG (2004b). Insulin secretion and cyclooxygenase enzyme inhibition by cabernet sauvignon grape skin compounds. *J. Agric. Food Chem.* 52: 228–33.

Zhang XY, Li WG, Wu YJ, Gao MT (2005a). Amelioration of doxorubicin-induced myocardial oxidative stress and immune-suppression by grape seed proanthocyanidins in tumour-bearing mice. *J. Pharm. Pharmacol.* 57 (8): 1043–52.

Zhao J, Wang J, Chen Y, Agarwal R (1999). Anti-tumor-promoting activity of a polyphenolic fraction isolated from grape seeds in the mouse skin two-stage initiation-promotion protocol and identification of procyanidin B5-3'-gallate as the most effective anti-oxidant constituent. *Carcinogenesis.* 20 (9): 1737–45.

Zhou DY, Du Q, Li RR, Huang M, Zhang Q, Wei GZ (2011). Grape seed proanthocyanidin extract attenuates airway inflammation and hyperresponsiveness in a murine model of asthma by down-regulating inducible nitric oxide synthase. *Planta. Med.* 77 (14): 1575–81.

Zunino S (2009). Type 2 diabetes and glycemic response to grapes or grape products. *J. Nutr.* 139 (9): 1794S–800S.

Chapter 9
Kokum/Vriksh Amla/ Malabar Amla (*Garcinia indica/ Garcinia cambogia*)

Garcinia Indica, popularly known as "Kokum" in India belongs to the mangosteen family. It is a deep purple color fruit. The fruit is dried and used as a spice. It has a tangy flavor. Its seeds, rind, bark, root, juice, pulp–all parts have been said to possess health benefits.

Kokum is native of the Western Ghats; Konkan, Goa, Karnataka and North Malabar regions of southern India (Khare, 2007); and is used more commonly in Gujarat and Maharashrta. It is a preferred substitute for tamarind in curries and other recipes from the Konkan region.

Kokum.

Chemical and Nutritional Constituents of Kokum

Kokum is low in Calories, contains no saturated fats or cholesterol and is rich in dietary fiber. Fresh fruit is a very good source of B-complex vitamins such as thiamin, niacin and folates. It is rich in vitamin C and other anti-oxidants as well as certain minerals (potassium, manganese and magnesium).

Kokum seed contains 23-26 per cent oil, which remains solid at room temperature and is used in the preparation of chocolates, medicines and cosmetics (Wikipedia, 2013). Kokum oil or kokum butter is light grey or yellowish in color. After refining, the kokum fat is equivalent to vanaspati. Kokum butter is an excellent emollient used by the cosmetic industry for preparations of lotions, creams, lip-balms and soaps. It has relatively high melting point and is considered as one of the most stable butter which does not need any refrigeration.

Gambogic acid (GA) is the principal active ingredient which is a resin obtained from various Garcinia species including *Garcinia hanburyi* (Tenghuang) (Panthong *et al.*, 2007).

Garcinol, a poly-isoprenylated benzophenone purified from kokum fruit rind has an anti-oxidant and anti-ulcer property (Yamaguchi *et al.*, 2000b; Selvi *et al.*, 2003). The fruit rind contains 10.3–12.7 per cent HCA (Jayaprakasha and Sakariah, 2002; Jena *et al.*, 2002). Apart from hydroxyl-citric acid (HCA) and garcinol, kokum contains other compounds including citric acid, malic acid, polyphenols, carbohydrates, anthocyanin pigments and ascorbic acid and a colorless isomer of garcinol, isogarcinol with potential anti-oxidant properties.

Therapeutic Properties

Anti-oxidant

The anti-oxidant activity of aqueous extract of kokum has been reported to be higher than other reported spices and fruits. Garcinol, as a potent free radical scavenger, is able to scavenge both hydro-philic and hydrophobic ions including reactive oxygen species. Orally administered garcinol prevented acute ulceration induced by indomethacin and water immersion stress caused by radical formation in rats (Yamaguchi *et al.*, 2000b).

Garcinol exhibited moderate anti-oxidative activity in the micellar linoleic acid peroxidation system and chelating activity at almost the same level as citrate. Using a hypo-xanthine/xanthine oxidase system, garcinol was shown to retard superoxide anion to nearly the same amount as DL-alpha-tocopherol, an established anti-oxidant, while its ability to quell hydroxyl radicals in the Fenton reaction system was even better than that of alpha-tocopherol. It also showed nearly 3 times greater DPPH (1, 1-diphenyl-2-picrylhydrazyl) free radical scavenging activity than DL-alpha-tocopherol by weight in aqueous ethanol solution. In a phenazine methosulfate/NADH-nitroblue tetrazolium system, garcinol exhibited superoxide anion scavenging activity and suppressed protein glycation in a bovine serum albumin/fructose system. Anti-oxidant and anti-glycation activity of garcinol from the fruit rind has also been reported (Yamaguchi *et al.*, 2000a).

Anti-oxidative and neuro-protective properties of garcinol in rat cortical neuron cultures were reported to be due to prevention of nitric oxide accumulation in lipo-polysaccharide treated astrocytes (Liao *et al.,* 2005).

Anti-microbial

Garcinia indica extract has been shown to possess putative bio-preservative properties as it inhibited *A. flavus* and aflatoxin B_1 production (Selvi *et al.,* 2003). Garcinol may be a viable alternative to conventional antibiotics as it shows anti-bacterial activity against Methicillin-resistant *S. aureus* (Rukachaisirikul *et al.,* 2005) which is comparable to that of the antibiotic Vancomycin (MIC–3-12 μg/mL for garcinol Vs. 6 μg/mL for Vancomycin) (Iinuma *et al.,* 1996).

Anti-bacterial activity of a biflavanoid from the ethyl acetate soluble fraction of methanol extract of stem bark of *G. indica* showed good anti-bacterial activity against *S. aureus* and *S. typhi* at higher concentrations, and moderate activity at lower concentrations but no activity against *E. coli* even at higher concentration. But pro-anthocyanin showed partial anti-bacterial activity against *S. Aureus, S. Typhi* and *E. coli* even at higher concentration (Lakshmi *et al.,* 2011).

Although camboginol, (tri-isoprenylated chalcone) has been shown to exhibit therapeutic activity against Gram-positive and Gram-negative cocci; mycobacteria and fungi, it has been found to be inactive against Gram-negative enteric bacilli, yeasts and viruses (Bakana *et al.,* 1987).

Therapeutic Benefits

Kokum is a traditional home remedy for constipation, heart diseases, dysentery and pains (Kirtikar *et al.,* 1991). It is also normally used to combat digestive problems like lack of appetite, flatulence, acidity; and in the treatment of piles and anal fissures. It is believed to possess anti-helmintic properties.

Extensive research within the last two decades has indicated that phytochemicals present in spices including kokum (gambogic acid) can prevent various chronic illnesses including cancerous, diabetic, cardiovascular, pulmonary, neurological and auto-immune diseases (Aggarwal *et al.,* 2008). Hydroxyl-citric acid is a hypo-cholesterolemic agent and also a potential anti-obesity agent (Heymsfield *et al.,* 1998).

Gambogic acid (GA) has various therapeutic effects, such as anti-inflammatory, analgesic and anti-pyretic (Panthong *et al.,* 2007) as well as anti-cancer activities (Kasibhatla *et al.,* 2005; Gu *et al.,* 2008). *In-vitro* and *in-vivo* studies have demonstrated its cyto-toxicity against a variety of malignant tumors, including glioblastoma, as well as cancers of the breast, lung and liver. GA is currently investigated in clinical trials in China (Wu *et al.,* 2004; Qiang *et al.,* 2008; Qi *et al.,* 2008).

Cancer

Defects of either of the enzymes histone acetyl-transferases (HATs) and deacetylases (HDACs) may lead to several diseases, including cancer. Emerging evidence suggests that garcinol could be useful as an anti-cancer agent, and it is increasingly being realized that garcinol is a pleiotropic agent capable of modulating key regulatory cell signaling pathways (Padhye *et al.,* 2009). Garcinol, a poly-isoprenylated benzophenone derivative from *Garcinia indica* fruit rind, a potent inducer of apoptosis alters (predominantly down-regulates) the global gene expression in HeLa cells both *in-vitro* and *in-vivo* (Balasubramanyam *et al.,* 2004). Saadat and Gupta (2012) also reported in their review artivcle that garcinol displays effective epigenetic influence by inhibiting HAT 300 and by possible post-transcriptional modulation by miRNA profiles involved in carcinogenesis.

Prasad *et al.* (2010) observed that garcinol can potentiate TRAIL-induced apoptosis through up-regulation of death receptors and down-regulation of anti-apoptotic proteins. Administration of garcinol significantly inhibited tumor growth, and western blot analysis of remnant tumor lysates showed reduced signal transducer and activator of transcription-3 (STAT-3) expression and activation. These results suggest that garcinol may have translational potential as chemo-preventive or therapeutic agent against multiple cancers and inhibition of STAT-3 signaling pathway is one of the mechanisms by which garcinol exerts its anti-cancer effects (Ahmad *et al.,* 2012a).

Ingestion of 0.01 per cent and 0.05 per cent dietary garcinol in rat models significantly reduced the formation of colonic aberrant crypt foci in a dose-dependent manner, thus suggesting suppression in cancer development. The ability of garcinol on chemo-prevention is through induction of liver GST and QR, inhibition of O(2)(-) and NO generation and/or suppression of iNOS and COX-2 expression, on colon tumori-genesis (Tanaka *et al.,* 2000). Liao *et al.* (2005) reported that garcinol reduces cell invasion and survival through the inhibition of focal adhesion kinase FAK's downstream signaling.

Cell growth (trypan blue exclusion) was also significantly inhibited by garcinol by induction of apoptosis in a dose- and time-dependent manner. Flow cytometric analysis revealed G0-G1 phase cell cycle arrest in both cell lines. The anti-proliferative, pro-apoptotic, anti-metastatic, and anti-angiogenic effects of garcinol were significantly higher, as compared to untreated cells, in both pancreatic cancer cell types (Parasramka and Gupta, 2011). Oike *et al.* (2012) found that garcinol inhibits repair of double-strand breaks including non-homologous end joining without affecting cell cycle checkpoint. Garcinol could radio-sensitize A549 lung and HeLa cervical carcinoma cells with dose enhancement ratios (at 10 per cent surviving fraction) of 1.6 and 1.5, respectively. Cellular senescence induced by IR was enhanced by garcinol.

Garcinol exhibited dose-dependent cancer cell-specific growth inhibition in both the cell lines (ER-positive MCF-7 and ER-negative MDA-MB-231 cells) with a concomitant induction of apoptosis, and had no effect on non-tumorigenic MCF-10A cells suggesting induction of caspase-mediated apoptosis in highly metastatic MDA-MB-231 cells. Down-regulation of NF-kappaB signaling pathway was observed to be the mechanism of apoptosis-induction. A significant decrease in the colony forming ability of all the cell lines was also observed, suggesting the possible application of this compound against metastatic disease and providing pre-clinical evidence to support the use of garcinol against human prostate (AR-positive LNCaP, AR-positive but androgen non-responsive C4-2B cells and AR-negative PC3 cells) and pancreatic cancer (BxPC-3 cells) (Ahmad *et al.,* 2010; 2011).

In-vitro as well as few *in-vivo* studies proved its potential against several types of cancers including breast, colon, pancreatic, and leukemia. Garcinol mediates its anti-tumor effects in squamous cell carcinoma of the head and neck cells and mouse model through the suppression of multiple pro-inflammatory cascades (Li *et al.,* 2013).

Garcinol was found to inhibit NF-κB, miRNAs, vimentin, and nuclear β-catenin. These novel findings suggest that the anti-cancer activity of garcinol against aggressive breast cancer cells is, in part, due to reversal of epithelial-to-mesenchymal transition phenotype, which is mechanically linked with the deregulation of miR-200s, let-7s, NF-κB, and Wnt signaling pathways (Ahmad *et al.,* 2012b). Nicotine-induced human breast cancer (MDA-MB-231) cell proliferation was inhibited by 1 μM of garcinol through down-regulation of α9-nAChR and cyclin D3 expression. The homeostatic regulation of cyclin D3 has the potential to be a molecular target for anti-tumor chemo-therapeutic or chemo-preventive purposes in clinical breast cancer patients (Chen *et al.,* 2011).

Garcinol and its derivatives, including cambogin, garcim-1 and garcim-2, added 1 h after lipo-polysaccharide (LPS) stimulation, significantly inhibited the release of arachidonic acid and its metabolites in macrophages. Garcinol was very effective, showing >50 per cent inhibition. Similar inhibitory activity was also observed in intestinal cells, HT-29, HCT-116 and IEC-6 cells, showing 40-50 per cent inhibition by 1 μM garcinol. In LPS-stimulated macrophages, garcinol inhibited the phosphorylation of cPLA2 without altering its protein level, and the effect was related to the inhibition of ERK1/2 phosphorylation. Thus it is evident that garcinol modulates arachidonic acid metabolism by blocking the phosphorylation of cPLA2 and decreases iNOS protein level by inhibiting STAT-1 activation. These activities may contribute to the anti-inflammatory and anti-carcinogenic actions of garcinol and its derivatives (Hong *et al.,* 2006). Later it was shown that at high concentration, garcinol and its derivatives can inhibit intestinal cell growth. But surprisingly, at low concentrations, they can stimulate normal as well as cancer cell growth. However, this influence on cell growth could be controlled by anti-oxidant enzymes (Hong *et al.,* 2007).

Several anti-apoptotic Bcl-2-family proteins are known to become pathologically over-expressed in human cancers, conferring apoptosis-resistant phenotypes (Schimmer *et al.,* 2003; Letai *et al.,* 2004; Andersen *et al.,* 2005). Zhai *et al.* (2008) identified gambogic acid as a competitive inhibitor that displaced BH3 peptides from Bfl-1 in a fluorescence polarization assay. Analysis of competition for BH3 peptide binding revealed that gambogic acid inhibits all six human Bcl-2 family proteins to various extents, with Mcl-1 and Bcl-B the most potently inhibited. It was found that suppression of anti-apoptotic Bcl-2 family proteins may be among the cytotoxic mechanisms by which gambogic acid kills tumor cells.

Treatment with garcinol reduced growth factor deprivation-mediated cell death and nuclear import of C/EBPβ levels. Garcinol could promote neurite outgrowth in EGF-responsive neural precursor cells and modulate the extracellular signal-regulated kinase pathway in the enhancement of neuronal survival (Weng *et al.,* 2011).

Cheng *et al.* (2010) reported that garcinol activated not only the death receptor and the mitochondrial apoptosis pathways but also the endoplasmic reticulum stress modulator GADD153. Garcinol treatment led to the accumulation of reactive oxygen species (ROS), increased GADD153 expression, and reduced mitochondrial membrane potential. An increase in the Bax : Bcl-2 ratio resulted in enhanced apoptosis. Caspase-8 and tBid (truncated Bid) expression also increased in a time-dependent manner. The enzymatic activities of caspase-3 and caspase-9 increased approximately 13-fold and 7.8-fold, respectively. In addition, the proteolytic cleavage of poly-(ADP-ribose)-polymerase (PARP) and DNA fragmentation factor-45 (DFF-45) increased in dose- and time-dependent manners.

Parasramka *et al.* (2013) identified garcinol-specific miRNA biomarkers that sensitize pancreatic cancer (PaCa) cells to Gemcitabine treatment, thus attenuating the drug-resistance phenotype prompting further interest in garcinol and Gemcitabine combination strategy as a drug modality to improve treatment outcome in patients diagnosed with PaCa.

Yoshida *et al.* (2005) proved that dietary administration of garcinol inhibited 4-nitroquinoline 1-oxide (4-NQO)-induced tongue carcinogenesis through suppression of increased cell proliferation activity in the target tissues and/or COX-2 expression in the tongue lesions in male rats.

Research has shown that gambogic acid can inhibit growth of a wide variety of tumor cells, including hepatoma, pulmonary carcinoma, gastric cancer, and breast cancer (Guo *et al.*, 2004; Wu *et al.*, 2004; Kasibhatla *et al.*, 2005; Yu *et al.*, 2007). A series of novel derivatives of gambogic acid (GA) *i.e.* compounds 3a, 3e, and 3f displayed potent inhibition of human hepato-cellular carcinoma while the most potent compound 3e did not significantly affect the proliferation of non-tumor liver cells, suggesting that it might selectively inhibit HCC proliferation. Compound 3e induced high frequency of Bel-7402 cell apoptosis (He *et al.*, 2012). Gambogic acid inhibits cell adhesion *via* suppressing integrin β1 abundance and cholesterol content as well as the membrane lipid raft-associated integrin function, thus providing new evidence for the anti-cancer activity of gambogic acid (Li *et al.*, 2011).

Chen *et al.* (2009) and Wang *et al.* (2011) demonstrated that a combination of gambogic acid and magnetic $Fe(3)O(4)$ nano-particles represents a promising approach to the treatment of K562 human leukemia and pancreatic cancer. The mechanisms of the synergistic effect may be due to reduced protein expression of Bcl-2 and enhancement of that of Bax, caspase 9, and caspase 3. Wang and Chen (2012) summarized the multiple functional effects of administering gambogic acid in cancer cells including the induction of apoptosis, the inhibition of proliferation and the prevention of cancer metastasis and tumor angiogenesis.

Gambogic acid also exhibits significant anti-metastatic activities on B16-F10 melanoma cancer cells partially through the inhibition of the cell surface expression of integrin α4 in C57BL/6 mice (Zhao *et al.*, 2008). Wang *et al.* (2008) showed that gambogic acid when combined with 5-fluorouracil (5-FU) induced considerably higher apoptosis rates in BGC-823 human gastric cells and inhibited tumor growth in human xenografts. Furthermore, low concentrations of gambogic acid were found to cause a dramatic increase in docetaxel-induced cytotoxicity in docetaxel-resistant BGC-823/Doc cells (Wang *et al.*, 2008). Later, He *et al.* (2009) and Wang *et al.* (2009) demonstrated that when gambogic acid is combined with celastrol in treating Tca8113 cells, the proliferative inhibition and apoptosis induction are much more visibly increased indicating that the combination can be a promising modality for treating oral squamous cell carcinoma.

Similarly, gambogic acid plus docetaxel produced a synergistic anti-tumor effect in gastro-intestinal cancer cells, suggesting that the drug combination may offer a novel treatment option for patients with gastric and colorectal cancers (Zou *et al.*, 2012). Guizzunti *et al.* (2012) indicated that gambogic acid directly targets the mitochondria to induce the intrinsic pathway of apoptosis, and thus represents a new member of the mitocans.

Gambogic acid inhibited TNF-α-induced invasion of PC3 cells *via* inactivation of the PI3K/Akt and NF-κB signaling pathways, which may offer a novel approach for the treatment of human prostate cancer (Lu *et al.*, 2012).

Obesity

Several studies have documented the weight loss effects of kokum (Heymsfield *et al.*, 1998; Leonhardt and Langhans, 2002; Ohia *et al.*, 2002). Hydroxyl-citric acid (HCA) is a potential anti-obesity agent (Heymsfield *et al.*, 1998). Supplementation in experimental animals showed that HCA can suppress appetite and inhibit body fat biosynthesis. In rat brain cortex, HCA was shown to increase availability of 5-hydroxy-tryptamine or serotonin. This neuro-transmitter is implicated in the appetite regulation and control (Ohio *et al.*, 2001).

Preuss *et al.* (2004) demonstrated the bioavailability and efficacy of Ca2+/K+ bound HCA salt (HCA-SX or Super Citri Max) in weight management. HCA-SX increases serotonin availability, reduces appetite, increases fat oxidation, improves blood lipid levels, reduces body weight, and modulates a number of obesity regulatory genes without affecting the mitochondrial and nuclear proteins required for normal, biochemical and physiological functions (Downs *et al.*, 2005) HCA-SX. Roy *et al.* (2004; 2007) also confirmed its anti-lipolytic and anti-adipogenic effects.

The highest dose of *Garcinia indica* (154 mmol HCA/kg diet) showed significant suppression of epididymal fat accumulation in developing male Zucker obese rats, compared with the other groups (Saito *et al.*, 2005). HCA has been reported to limit the synthesis of fatty acids in the muscles and liver by inhibiting the enzyme ATP-citrate lyase (a key enzyme which facilitates the synthesis of fatty acids, cholesterol and triglycerides). As a citrate cleavage enzyme that may play an essential role in *de-novo* lipo-genesis inhibition, with no further synthesis, the existing fatty acids are gradually used up, resulting in reduction of body fat and body weight. Thus *Garcinia cambogia*-derived HCA could be considered a safe and natural supplement for weight management.

Kim *et al.* (2008) reported that *Garcinia cambogia* extract, apart from reducing the body weight gain, was effective in improving fatty liver, dyslipidemia, hyper-insulinemia, and hyper-leptinemia in high fat diet-induced obese mice. Its anti-obesity effects were associated with modulation of the multiple genes associated with visceral adipogenesis such as PPARγ2, SREBP1c, C/EBPα, and aP2 in the visceral fat tissue of mice.

Diabetes

Aqueous extract of *G. indica* significantly decreased both the fasting and post-prandial blood glucose in type 2 diabetic rats. Further, glutathione levels, known to be depleted due to oxidative stress in the erythrocytes of streptozotocin-induced type-2 diabetic rats, were effectively restored, which is beneficial specially in preventing the risk of developing complications (Kirana and Srinivasan, 2010).

Parkinson's Disease

Parkinson's disease is the second most common age-related neuro-degenerative disorder typified by tremor, rigidity, akinesia and postural instability due in part to the loss of dopamine within the nigro-striatal system. Methanolic extract of *Garcinia indica* had significant preventive effect in biochemical indices (dopamine and its metabolites) and in various behaviour tests, (apomorphine-induced rotational behaviour, stepping test, initiation time, postural balance test, and disengage time) as compared to 6-hydroxydopamine-(6-OHDA) treated rats thus acting as an effective neuro-protective agent for striatal dopaminergic neurons in 6-OHDA lesioned rat model of Parkinson's disease (Antala *et al.,* 2012).

Other Benefits

Oral administration of garcinol prevented acute ulceration in rats, suggesting its potential as an anti-ulcer drug (Yamaguchi *et al.,* 2000a). Addition of *Garcinia* extract to fresh skipjack (dark muscle fish) has been demonstrated to prevent histamine formation by lowering the pH to 3.2-3.6 (Thadhani *et al.,* 2002). Since histamine is known to give rise to allergic reactions, *Garcinia* extracts can potentially find use in anti-allergy medications (Pan *et al.,* 2001).

Aqueous extracts of 800 mg/kg of kokum rind demonstrated greater hepato-protection than 400 mg/kg in ethanol-induced oxidative damage probably due to an augmentation of the endogenous anti-oxidants and inhibition of lipid peroxidation in liver (Panda *et al.,* 2012).

Garcinol is a potent inhibitor of histone acetyltransferases both *in-vitro* and *in-vivo* (Balasubramanyam *et al.,* 2004). Local infusion of garcinol into the rat lateral amygdala (LA) impairs the training and retrieval-related acetylation of histone H3 in the LA and either intra-LA or systemic administration of garcinol within a narrow window after either fear conditioning or fear memory retrieval significantly impairs the consolidation and reconsolidation of a Pavlovian fear memory (a widely studied rodent model of post-traumatic stress disorder) and associated neural plasticity in the LA. Thus, the regulation of chromatin function by garcinol may be useful in the treatment of newly acquired or recently reactivated traumatic memories (Maddox *et al.,* 2013).

Conclusions

The therapeutic potential of Kokum is well studied in cancer than any other clinical condition. Hence it is worthwhile to conduct further research on this non conventional and under-utilized condiment in other clinical conditions as well. Furthermore, at the doses usually administered, no side effects or adverse events have been reported in individuals treated with *G. cambogia* (Márquez *et al.,* 2012).

References

Aggarwal BB, Kunnumakkara AB, Harikumar KB, Tharakan ST, Sung B, Anand P (2008). Potential of spice-derived phytochemicals for cancer prevention. *Planta. Med.* 74 (13): 1560–69.

Ahmad A, Sarkar SH, Aboukameel A, Ali S, Biersack B, Seibt S, Li Y, Bao B, Kong D, Banerjee S, Schobert R, Padhye SB, Sarkar FH (2012a). Anti-cancer action of garcinol *in-vitro* and *in-vivo* is in part mediated through inhibition of STAT-3 signaling. *Carcinogenesis.* 33 (12): 2450–56.

Ahmad A, Sarkar SH, Bitar B, Ali S, Aboukameel A, Sethi S, Li Y, Bao B, Kong D, Banerjee S, Padhye SB,Sarkar FH (2012b).Garcinol regulates EMT and Wnt signaling pathways in vitro and in vivo, leading to anticancer activity against breast cancer cells. *Mol. Cancer Ther.* 11 (10): 2193–201.

Ahmad A, Wang Z, Ali R, Maitah MY, Kong D, Banerjee S, Padhye S, Sarkar FH (2010). Apoptosis-inducing effect of garcinol is mediated by NF-kappaB signaling in breast cancer cells. *J. Cell. Biochem.* 109 (6): 1134–41.

Ahmad A, Wang Z, Wojewoda C, Ali R, Kong D, Maitah MY, Banerjee S, Bao B, Padhye S, Sarkar FH (2011). Garcinol-induced apoptosis in prostate and pancreatic cancer cells is mediated by NF- kappaB signaling. *Front. Biosci. (Elite Ed).* 3: 1483–92.

Andersen MH, Svane IM, Kvistborg P, Nielsen OJ, Balslev E, Reker S, Becker JC, Straten PT (2005). Immuno-genicity of Bcl-2 in patients with cancer. *Blood.* 105: 728–34.

Antala BV, Patel MS, Bhuva SV, Gupta S, Rabadiya S, Lahkar M (2012). Protective effect of methanolic extract of *Garcinia indica* fruits in 6-OHDA rat model of Parkinson's disease. *Indian J. Pharmacol.* 44 (6): 683–87.

Bakana P, Claeys M, Totte J, Pieters LA., VanHoof L, Tamba-Vemba DA, Berghe VD, Vlietinck AJ (1987). Structure and chemo-therapeutical activity of a polyisoprenylated benzophenone from the stem bark of *Garcinia huillensis. Journal of Ethnopharmacology.* 21 (1): 75–84.

Balasubramanyam K, Altaf M, Varier RA, Swaminathan V, Ravindran A, Sadhale PP, Kundu TK (2004) Poly-isoprenylated benzophenone, garcinol, a natural histone acetyl-transferase inhibitor, represses chromatin transcription and alters global gene expression. *J. Biol. Chem.* 279: 33716 -26.

Chen BA, Liang YQ, Wu WW, Cheng J, Xia G, Gao F, Ding J, Gao C, Shao Z, Li G, Chen W, Xu W, Sun X, Liu L, Li X, and Wang X (2009). Synergistic effect of magnetic nanoparticles of Fe_3O_4 with gambogic acid on apoptosis of K562 leukemia cells. *Int. J. Nanomedicine.* 4: 251–59.

Chen CS, Lee CH, Hsieh CD, Ho CT, Pan MH, Huang CS, Tu SH, Wang YJ, Chen LC, Chang YJ, Wei PL, Yang YY, Wu CH, Ho YS (2011). Nicotine-induced human breast cancer cell proliferation attenuated by garcinol through down-regulation of the nicotinic receptor and cyclin D3 proteins. *Breast Cancer Res. Treat.* 125 (1): 73–87.

Cheng AC, Tsai ML, Liu CM, Lee MF, Nagabhushanam K, Ho CT, Pan MH (2010). Garcinol inhibits cell growth in hepato-cellular carcinoma Hep3B cells through induction of ROS-dependent apoptosis. *Food Funct.* 1 (3): 301–307.

Downs BW, Bagchi M, Subbaraju GV, Shara MA, Preuss HG, Bagchi D (2005). Bioefficacy of a novel calcium-potassium salt of (-)-hydroxycitric acid. *Mutat. Res.* 579 (1-2): 149–62.

Gu H, Wang X, Rao S, Wang J, Zhao J, Ren FL, Mu R, Yang Y, Qi Q, Liu W, Lu N, Ling H, You Q, Guo Q (2008). Gambogic acid mediates apoptosis as a p53 inducer through down-regulation of mdm2 in wild-type p53-expressing cancer cells. *Mol. Cancer. Ther.* 7: 3298–305.

Guizzunti G, Batova A, Chantarasriwong O, Dakanali M, Theodorakis EA (2012). Subcellular localization and activity of gambogic acid. *Chem. Biochem.* 2012. Apr 24. doi: 10.1002/cbic.201200065.

Guo QL, You QD, Wu ZQ, Yuan ST, Zhao L (2004). General gambogic acids inhibited growth of human hepatoma SMMC-7721 cells in vitro and in nude mice. *Acta. Pharmacol. Sin.* 25: 769–74.

He D, Xu Q, Yan M, Zhang P, Zhou X, Zhang Z, Duan W, Zhong L, Ye D, Chen W (2009). The NF-kappa B inhibitor, celastrol, could enhance the anti-cancer effect of gambogic acid on oral squamous cell carcinoma. *BMC. Cancer.* 9: 343.

He L, Ling Y, Fu L, Yin D, Wang X, Zhang Y (2012). Synthesis and biological evaluation of novel derivatives of gambogic acid as anti-hepatocellular carcinoma agents. *Bioorg. Med. Chem. Lett.* 22 (1): 289–92.

Heymsfield SB, Allison DB, Vasselli JR, Pietrobelli A, Greenfield D, Nuney C (1998). *Garcinia composia* (Hydroxy-citric acid) as a potential anti-obesity agent a randomized controlled trail. *JAMA.* 280: 1596–600.

Hong J, Kwon SJ, Sang S, Ju J, Zhou JN, Ho CT, Huang MT, Yang CS (2007). Effects of garcinol and its derivatives on intestinal cell growth: Inhibitory effects and autoxidation-dependent growth-stimulatory effects. *Free Radic. Biol. Med.* 42 (8): 1211–21.

Hong J, Sang S, Park HJ, Kwon SJ, Suh N, Huang MT, Ho CT, Yang CS (2006). Modulation of arachidonic acid metabolism and nitric oxide synthesis by garcinol and its derivatives. *Carcinogenesis.* 27 (2): 278–86.

Iinuma M, Tosa H, Tanaka T, Kanamaru S, Asai F, Kobayashi Y, Miyauchi K, Shimano R (1996). Anti-bacterial activity of some Garcinia benzophenone derivatives against methicillin-resistant *Staphylococcus aureus. Biol Pharm Bull.* 19: 311–14.

Jayaprakasha GK, Sakariah KK (2002). Determination of organic acids in leaves and rinds of *Garcinia indica* (Desr.) by LC. *J. Pharm. Biomed. Anal.* 28 (2): 379–84.

Jena BS, Jayaprakasha GK, Singh RP, Sakariah KK (2002). Chemistry and biochemistry of (-)-hydroxycitric acid from *Garcinia. J. Agric. Food Chem*. 50 (1): 10–22.

Kasibhatla S, Jessen KA, Maliartchouk S, Wang JY, English NM, Drewe J, Qiu L, Archer SP, Ponce AE, Sirisoma N, Jiang S, Zhang HZ, Gehlsen KR, Cai SX, Green DR, Tseng B. (2005). A role for transferrin receptor in triggering apoptosis when targeted with gambogic acid. *Proc. Natl. Acad. Sci U. S. A.* 102: 12095–100.

Khare CP (2007). *Indian Medicinal Plants: An Illustrated Dictionary*. 2nd ed. Springer. New York. pp. 278–279.

Kim Keun-Young, Lee HN, Kim YJ, Park T (2008). *Garcinia cambogia* extract ameliorates visceral adiposity in C57BL/6J mice fed on a high-fat diet. *Biosci. Biotechnol. Biochem*. 72 (7), 1772 – 80.

Kirana H, Srinivasan B (2010). Aqueous extract of *Garcinia indica* choisy restores glutathione in type 2 diabetic rats. *J. Young. Pharm*. 2 (3): 265–68.

Kirtikar KR, Basu BD. Guttiferae. In: Blatter E, Caius JF, Mhaskar KS (1991). Eds. *Indian Medicinal Plants*. 2 nd ed. Vol. 1. Prabasi press. Calcutta. p. 263.

Lakshmi C, Kumar KA, Dennis TJ, Kumar TS (2011). Anti-bacterial activity of polyphenols of garcinia indica. *Indian J. Pharm. Sci*. 73 (4): 470–73.

Leonhardt M.,and Langhans W (2002). Hydroxy-citrate has long-term effects on feeding behavior, body weight regain and metabolism after body weight loss in male rats. *J. Nutr*. 132. 1977 – 82.

Letai A, Sorcinelli MD, Beard C, Korsmeyer SJ (2004). Anti-apoptotic BCL-2 is required for maintenance of a model leukemia. *Cancer Cell*. 6: 241–49.

Li C, Lu N, Qi Q, Li F, Ling Y, Chen Y, Qin Y, Li Z, Zhang H, You Q, Guo Q (2011). Gambogic acid inhibits tumor cell adhesion by suppressing integrin β1 and membrane lipid rafts-associated integrin signaling pathway. *Biochem. Pharmacol*. 82 (12): 1873–83.

Li F, Shanmugam MK, Chen L, Chatterjee S, Basha J, Kumar AP, Kundu TK, Sethi G (2013). Garcinol, a poly-isoprenylated benzophenone modulates multiple pro-inflammatory signaling cascades leading to the suppression of growth and survival of head and neck carcinoma. *Cancer Prev. Res. (Phila)*. 6 (8): 843–54.

Liao CH, Sang S, Ho CT, Lin JK (2005). Garcinol modulates tyrosine phosphorylation of FAK and subsequently induces apoptosis through down-regulation of Src, ERK, and Akt survival signaling in human colon cancer cells. *J. Cell. Biochem*. 96 (1): 155–69.

Lü L, Tang D, Wang L, Huang LQ, Jiang GS, Xiao XY, Zeng FQ (2012). Gambogic acid inhibits TNF-α-induced invasion of human prostate cancer PC3 cells in vitro through PI3K/Akt and NF-κB signaling pathways. *Acta Pharmacol. Sin*. 33 (4): 531-41.

Maddox SA, Watts CS, Doyère V, Schafe GE (2013). A naturally-occurring histone acetyl-transferase inhibitor derived from *Garcinia indica* impairs newly acquired and reactivated fear memories. *PLoS. One.* 8 (1): e54463.

Márquez F, Babio N, Bulló M, Salas-Salvadó J (2012). Evaluation of the safety and efficacy of hydroxyl-citric acid or *Garcinia cambogia* extracts in humans. *Crit. Rev. Food Sci. Nutr.* 52 (7): 585–94.

Ohia SE, Opere CA, LeDay AM, Bagchi M., Bagchi D, and Stohs SJ (2002). Safety and mechanism of appetite suppression by a novel hydroxyl-citric acid extract (HCA-SX). *Mol. Cell. Biochem.* 238, 89–103.

Ohia SE, Awe SO, Le Day AM, Opere CA, Bagchi D (2001). Effect of hydroxycitric acid on serotonin release from isolated rat brain cortex. *Research Communications in Molecular Pathology and Pharmacology.* 109 (3-4): 210–16.

Oike T, Ogiwara H, Torikai K, Nakano T, Yokota J, Kohno T (2012). Garcinol, a histone acetyl-transferase inhibitor, radio-sensitizes cancer cells by inhibiting non-homologous end joining. *Int. J. Radiat. Oncol. Biol. Phys.* 84 (3): 815–21.

Padhye S, Ahmad A, Oswal N, Sarkar FH (2009). Emerging role of Garcinol, the antioxidant chalcone from *Garcinia indica* Choisy and its synthetic analogs. *J. Hematol. Oncol.* 2: 38.

Pan MH, Chang WL, Lin-Shiau SY, Ho CT, Lin JK (2001). Induction of apoptosis by garcinol and curcumin through cytochrome c release and activation of caspases in human leukemia HL-60 cells. *J. Agric. Food Chem.* 49: 1464–74.

Panda V, Ashar H, Srinath S (2012). Anti-oxidant and hepatoprotective effect of *Garcinia indica* fruit rind in ethanol-induced hepatic damage in rodents. *Interdiscip Toxicol.* 5 (4): 207–213.

Panthong A, Norkaew P, Kanjanapothi D, Taesotikul T, Anantachoke N, Reutrakul V (2007). Anti-inflammatory, analgesic and anti-pyretic activities of the extract of gamboge from *Garcinia hanburyi* Hook f. *J. Ethnopharmacol.* 111: 335–40.

Parasramka MA, Ali S, Banerjee S, Deryavoush T, Sarkar FH, Gupta S (2013). Garcinol sensitizes human pancreatic adenocarcinoma cells to gemcitabine in association with microRNA signatures. *Mol. Nutr. Food Res.* 57 (2): 235–48.

Parasramka MA, Gupta SV (2011). Garcinol inhibits cell proliferation and promotes apoptosis in pancreatic adeno-carcinoma cells. *Nutrition and Cancer.* 63 (3): 456–65.

Prasad S, Ravindran J, Sung B, Pandey MK, Aggarwal BB (2010). *Garcinol* potentiates TRAIL-induced apoptosis through modulation of death receptors and antiapoptotic proteins. *Mol. Cancer. Ther.* 9 (4): 856–68.

Preuss HG, Rao CV, Garis R, Bramble JD, Ohia SE, Bagchi M, Bagchi D (2004). An overview of the safety and efficacy of a novel, natural (-)-hydroxycitric acid extract (HCA-SX) for weight management. *J. Med.* 35 (1-6): 33–48.

Qi Q, Gu H, Yang Y, Lu N, Zhao J, Liu W, Ling H, You QD, Wang X, Guo Q (2008). Involvement of matrix metalloproteinase 2 and 9 in gambogic acid induced suppression of MDA-MB-435 human breast carcinoma cell lung metastasis. *J. Mol. Med*. 86: 1367–77.

Qiang L, Yang Y, You QD, Ma YJ, Yang L, Nie FF, Gu HY, Zhao L, Lu N, Qi Q, Liu W, Wang XT, Guo QL (2008). Inhibition of glioblastoma growth and angiogenesis by gambogic acid: an *in-vitro* and *in-vivo* study. *Biochem. Pharmacol*. 75: 1083–92.

Roy S, Rink C, Khanna S, Phillips C, Bagchi D, Bagchi M, Sen CK (2004). Body weight and abdominal fat gene expression profile in response to a novel hydroxycitric acid-based dietary supplement. *Gene Expr*. 11 (5-6): 251–62.

Roy S, Shah H, Rink C, Khanna S, Bagchi D, Bagchi M, Sen CK (2007). Transcriptome of primary adipocytes from obese women in response to a novel hydroxycitric acid-based dietary supplement. *DNA Cell. Biol*. 26 (9): 627–39.

Rukachaisirikul V, Naklue W, Sukpondma Y, Phongpaichit S (2005). An antibacterial biphenyl derivative from *Garcinia bancana* MIQ. *Chem. Pharm. Bull. (Tokyo)*. 53 (3): 342–43.

Saadat N, Gupta SV (2012). Potential role of garcinol as an anti-cancer agent. *J. Oncol*. 2012: 647206.

Saito M, Ueno M, Ogino S, Kubo K, Nagata J, Takeuchi M (2005). High dose of *Garcinia cambogia* is effective in suppressing fat accumulation in developing male Zucker obese rats, but highly toxic to the testis. *Food Chem. Toxicol*. 43 (3): 411–19.

Selvi AT, Joseph GS, Jayaprakasha GK (2003). Inhibition of growth and aflatoxin production in Aspegillus of growth and aflatoxin production in *Aspegillus flavus* by *Garcinia indica* extract and its anti-oxidant activity. *Food Microbiol*. 20: 455–60.

Schimmer AD, Munk-Pedersen I, Minden MD, Reed JC (2003). Bcl-2 and apoptosis in chronic lymphocytic leukemia. *Curr. Treat. Options. Oncol*. 4: 211–18.

Tanaka T, Kohno H, Shimada R, Kagami S, Yamaguchi F, Kataoka S, Ariga T, Murakami A, Koshimizu K, Ohigashi H (2000). Prevention of colonic aberrant crypt foci by dietary feeding of garcinol in male F344 rats. *Carcinogenesis*. 21 (6): 1183–89.

Thadhani VM, Jansz ER, Peiris H (2002). Effect of exogenous histidine and *Garcinia cambogia* on histamine formation in skipjack (*Katsuwonus pelamis*) homogenates. *Int. J. Food Sci. Nutr*. 53: 29–34.

Wang X, Chen W (2012). Gambogic acid is a novel anti-cancer agent that inhibits cell proliferation, angiogenesis and metastasis. *Anti-cancer Agents. Med. Chem*. 2012 Feb 17. [Epub ahead of print].

Wang J, Liu W, Zhao Q, Qi Q, Lu N, Yang Y, Nei FF, Rong JJ, You QD, Guo QL (2009). Synergistic effect of 5-fluorouracil with gambogic acid on BGC-823 human gastric carcinoma. *Toxicology*. 256: 135–40.

Wang T, Wei J, Qian X, Ding Y, Yu L, Liu B (2008). Gambogic acid, a potent inhibitor of survivin, reverses docetaxel resistance in gastric cancer cells. *Cancer. Lett*. 262: 214–22.

Wang C, Zhang H, Chen B, Yin H, Wang W (2011). Study of the enhanced anti-cancer efficacy of gambogic acid on Capan-1 pancreatic cancer cells when mediated via magnetic Fe3O4 nanoparticles. *Int. J. Nanomedicine*. 6: 1929–35.

Weng MS, Liao CH, Yu SY, Lin JK (2011). Garcinol promotes neurogenesis in rat cortical progenitor cells through the duration of extracellular signal-regulated kinase signaling. *J. Agric. Food Chem*. 59 (3): 1031–40.

Wikipedia.org/wiki/*Garcinia_indica*. Wikimedia Foundation, Inc. y2013.

Wu ZQ, Guo QL, You QD, Zhao L, Gu HY (2004). Gambogic acid inhibits proliferation of human lung carcinoma SPC-A1 cells *in-vivo* and *in-vitro* and represses telomerase activity and telomerase reverse transcriptase mRNA expression in the cells. *Biol. Pharm. Bull*. 27: 1769–74.

Yamaguchi F, Ariga T, Yoshimura Y, Nakazawa H (2000a). Anti-oxidative and anti-glycation activity of garcinol from *Garcinia indica* fruit rind. *J. Agric. Food Chem*. 48 (2): 180–85.

Yamaguchi F, Saito M, Ariga T, Yoshimura Y, Nakazawa H (2000b). Free radical scavenging activity and anti-ulcer activity of garcinol from *Garcinia indica* fruit rind. *J. Agric. Food Chem*. 48 (6): 2320–25.

Yoshida K, Tanaka T, Hirose Y, Yamaguchi F, Kohno H, Toida M, Hara A, Sugie S, Shibata T, Mori H (2005). Dietary garcinol inhibits 4-nitroquinoline 1-oxide-induced tongue carcinogenesis in rats. *Cancer. Lett*. 221 (1): 29–39.

Yu J, Guo QL, You QD, Zhao L, Gu HY, Yang Y, Zhang HW, Tan Z, Wang X. (2007). Gambogic acid-induced G (2)/M phase cell-cycle arrest via disturbing CDK7-mediated phosphorylation of CDC2/p34 in human gastric carcinoma BGC-823 cells. *Carcinogenesis*. 28: 632–38.

Zhao J, Qi Q, Yang Y, Gu HY, Lu N, Liu W, Wang W, Qiang L, Zhang LB, You QD, Guo QL (2008). Inhibition of alpha (4) integrin mediated adhesion was involved in the reduction of B16-F10 melanoma cells lung colonization in C57BL/6 mice treated with gambogic acid. *Eur. J. Pharmacol*. 589: 127–131.

Zhai D, Jin C, Shiau CW, Kitada S, Satterthwait AC, Reed JC (2008). Gambogic acid is an antagonist of anti-apoptotic Bcl-2 family proteins. *Mol. Cancer Ther*. 7 (6): 1639–46.

Zou Z, Xie L, Wei J, Yu L, Qian X, Chen J, Wang T, Liu B (2012). Synergistic anti-proliferative effects of gambogic acid with docetaxel in gastro-intestinal cancer cell lines. BMC. *Complement. Altern. Med*. 12 (1): 58.

Chapter 10

Pomegranate
(*Punica granatum*)

Pomegranate (*Punica granatum*) is commonly known as the 'nature's power fruit' and 'jewel of winter'. It grows on a deciduous shrub from the *Punicaceae* family (Adhami *et al.,* 2009; Kote *et al.,* 2011). Though pomegranate is cultivated throughout the Middle East, European Mediterranean region, the drier parts of Southeast Asia, northern and tropical Africa, and to some extent the United States, specifically California and Arizona, it is considered native of the Himalayas in Northern India (Viuda-Martos *et al.,* 2010).

Nutritional and Phytochemical Composition

Edible parts of pomegranate fruit (about 50 per cent of total fruit weight) comprise of 80 per cent juice and 20 per cent seeds. Fresh juice contains 85 per cent water, 10 per cent total sugars, 1.5 per cent pectin, ascorbic acid, and polyphenolic flavonoids. In pomegranate juice fructose and glucose are present in similar quantities, calcium is 50 per cent of its ash content, and the principal amino acids are glutamic and aspartic acids.

Pomegranate seeds are a rich source of crude fiber, pectin, and sugars. The fatty acid composition of pomegranate cold pressed seed oil showed punicic acid (65.3 per cent) along with palmitic acid (4.8 per cent), stearic acid (2.3 per cent), oleic acid (6.3 per cent), and linoleic acid (6.6 per cent). While the levels of total soluble solids (TSS) and soluble sugars in the aril juices differ only slightly, those of titratable acidity (TA) and citric acid changed significantly, suggesting that they are the main contributors to taste of the juice. Peel homogenates exhibited lower levels of TSS, TA, soluble sugars and organic acids than aril juices (Cemeroglu *et al.,* 1992; El-Nemr *et al.,* 1990; Dafny-Yalin *et al.,* 2010; El-Shaarawy and Nahpetian, 1983).

Nutritional Value of Pomegranate (per 100 g)

Nutrient	Amount	Nutrient	Amount
Energy (Kcal)	83	Choline (mg)	7.6
Carbohydrates (g)	18.7	Vitamin C (mg)	10.2
Sugars (g)	13.67	Vitamin E (mg)	0.6
Dietary fiber (g)	4	Vitamin K (µg)	16.4
Fat (g)	1.17	Calcium (mg)	10
Protein (g)	1.67	Iron (mg)	0.3
Thiamine (mg)	0.067	Magnesium (mg)	12
Riboflavin (mg)	0.053	Manganese (mg)	0.119
Niacin (mg)	0.293	Phosphorus (mg)	36
Pantothenic acid (mg)	0.377	Potassium (mg)	236
Vitamin B_6 (mg)	0.075	Sodium (mg)	3
Folate (µg)	38	Zinc (mg)	0.35

Source: USDA Nutrient Database.

Pomegranate has been shown to contain 124 different phytochemicals, and some of them act in concert to exert anti-oxidant and anti-inflammatory effects on cancer cells. Ellagitannins are bioactive polyphenols present in pomegranate (Heber, 2011). The soluble polyphenol content in pomegranate juice (PJ) varies within the limits of 0.2–1.0 per cent, depending on variety, and includes mainly anthocyanins (such as cyanidin-3-glucoside, cyanidin-3,5-diglucoside, and delphindin-3-glucoside), catechins, ellagic tannins, gallic and ellagic acids (Narr *et al.,* 1996). Dried pomegranate seeds contain the steroid estrogen estrone (Heftaman and Bennett 1966; Moneam *et al.,* 1988), the isoflavone phytoestrogens genistein and daidzein, and the phytoestrogen coumestrol (Sharaf and Nigm, 1964).

The various therapeutic properties of pomegranate are as follows.

Anti-oxidant

Pomegranate has been shown to possess anti-oxidant and anti-inflammatory properties (Longtin, 2003). Ignarro *et al.* (2006) reported that the potent anti-oxidant activity of pomegranate juice results in marked protection of nitric oxide (NO) against oxidative destruction, thereby resulting in augmentation of the biological actions of NO. Guo *et al.* (2008) demonstrated that daily consumption of pomegranate juice is potentially better than apple juice in improving anti-oxidant function in the elderly. As the plasma ascorbic acid, vitamin E, and reduced glutathione contents did not differ significantly between the 2 groups, the functional components contained in pomegranate juice–the phenolics may account for these observations. The anti-oxidant activity of flavonoids obtained from pomegranate juice was observed to be close to that of butylated hydroxyl-anisole, green tea, and significantly greater than red wine (Gil *et al.,* 2000; Noda *et al.,* 2002). Seeram *et al.* (2008) on comparing the anti-oxidant activity of different beverages, found pomegranate juice to possess the highest anti-oxidant potency composite index and was at least 20 per cent greater than any of the other beverages tested. Anti-oxidant potency, ability to inhibit LDL oxidation, and total polyphenol content were consistent in classifying the anti-oxidant capacity of the polyphenol-rich beverages in the following order: pomegranate juice > red wine > Concord grape juice > blueberry juice > black cherry juice, açaí juice, cranberry juice > orange juice, iced tea beverages, and apple juice. Although *in-vitro* anti-oxidant potency does not prove *in-vivo* biological activity, there is consistent clinical evidence of anti-oxidant potency of both pomegranate juice and red wine.

Pomegranate juice showed protection against protein and DNA oxidation in mice. There was also a significant decrease in GSH and GSSG, without change in the GSH/GSSG ratio. All enzymatic activities (GPx, GST, GR, SOD and catalase) were found to be decreased by treatment with pomegranate juice. In addition, GST and GS transcription were also found to be decreased in treatment group (Faria *et al.,* 2007a.).

The *in-vitro* anti-oxidant activity of pomegranate has been attributed to its high polyphenolic content, specifically punicalagins, punicalins, gallagic acid, and ellagic acid. The ellagitannins did not show significant anti-oxidant activity compared to punicalagin from pomegranate juice. The potential systemic biological effects of pomegranate juice ingestion should be attributed to the colonic microflora metabolites rather than to the polyphenols present in the juice (Cerdá *et al.,* 2004). The above-mentioned phytochemical compounds are metabolized during digestion to ellagic acid and urolithins, suggesting that the bioactive compounds that provide *in-vivo* anti-oxidant activity may not be the same as those present in the whole food (Johanningsmeier and Harris, 2011). Anthocyanins and the unique fatty acid profile of the seed oil may also play a role in pomegranate's health effects. According to El Kar *et al.* (2011) pro-anthocyanins are the principal contributors in the anti-oxidant capacity of pomegranate. They also suggest that the high concentrations of K^+ and Na^+ may play a role in the adaptation of pomegranate to arid environments.

The trend in anti-oxidant activity was pomegranate juice > total pomegranate tannin > punicalagin > ellagic acid. The superior bioactivity of pomegranate juice compared to its purified polyphenols illustrated the multi-factorial effects and chemical synergy of the action of multiple compounds compared to single purified active ingredients (Seeram *et al.,* 2005). Supporting this view, Viladomiu *et al.* (2013) in their review article reported that the synergistic action of the pomegranate constituents appears to be superior when compared to individual constituents.

Tzulker *et al.* (2007) found the homogenates prepared from the whole fruit exhibited an approximately 20-fold higher anti-oxidant activity than the level found in the aril juice. The anti-oxidant level in the homogenates correlated significantly to the content of the four hydrolyzable tannins in which punicalagin is predominant, while no correlation was found to the level of anthocyanins.

The peels of fruit grown in the desert climate exhibited higher anti-oxidant activity, and the levels of total phenolics, including the two hydrolyzable tannins, punicalagin and punicalin, were higher compared to those in the peels of fruit grown in the Mediterranean climate indicating that environmental conditions significantly affect pomegranate fruit quality and health beneficial compounds (Schwartz *et al.,*

2009). The methanol extract of pomegranate peel showed the highest anti-oxidant activity as compared to ethyl acetate and water extracts or the seed extracts. The methanol extract showed 56, 58, and 93.7 per cent inhibition using the thiobarbituric acid method, hydroxyl radical scavenging activity, and LDL oxidation, respectively, at 100 ppm. (Singh *et al.*, 2002). Pre-treatment of the rats with a methanolic extract of pomegranate peel at 50 mg/kg (in terms of catechin equivalents) followed by CCl_4 treatment resulted in preservation of catalase, peroxidase, and SOD to values comparable with control values, whereas lipid peroxidation was brought down by 54 per cent as compared to control. Pomegranate peel extract restored the normal hepatic architecture in CCl_4 treated liver (Chidambara Murthy *et al.*, 2002).

Comparison of the anti-oxidant capacities of fruit juices, peels, and seed oils from 6 Tunisian pomegranate ecotypes revealed that the peels had highest values with 25.63 mmol trolox equivalent/100 g and 22.08 mmol TE/100 g for FRAP and ORAC assay, respectively. The anti-oxidant potency of pomegranate extracts correlated with their phenolic compounds content (Elfalleh *et al.*, 2011). Arils, juice and rinds of pomegranate fruit and their aqueous and ethyl acetate extracts displayed good anti-oxidant activity (Ricci *et al.*, 2006). The relative anti-oxidant potency was found to be in the order of aqueous rind extract > pomegranate juice > arils only juice. On the contrary, arils only juice was capable of preventing the deleterious effects of cytotoxicity, DNA damage and depletion of non-protein sulphydrils (NPSH) pool caused by treatment of cells with H_2O_2, tert-butylhydroperoxide (tB-OOH) or oxidized lipoproteins (Ox-LDL) *via* a mechanism which is likely to involve both direct scavenging of radical species and iron chelation. Surprisingly, arils only juice and pomegranate juice slightly sensitized the cells to the cyto-toxic effects of the three agents, thus revealing that arils only juice, the major and tasty part of pomegranate juice, does not contain ellagic acid and punicalagin (*i.e.* the polyphenols highly represented in rind extract which are reputed to be responsible for the anti-oxidant capacity) in amounts sufficient to exert cyto-protection in oxidatively injured living cells (Sestili *et al.*, 2007).

Interestingly, the anti-oxidant activity was higher in commercial juices that were extracted from whole pomegranates than in experimental juices that were obtained from the arils only (Schubert *et al.*, 1999). Commercially available pomegranate juices tested by the Trolox Equivalent Anti-oxidant Capacity (TEAC) assay showed 18 to 20 TEAC that was three times higher than those of red wine and green tea (6–8 TEAC). Borges *et al.* (2010) found that both the anti-oxidant assay and the HPLC on-line anti-oxidant data demonstrated that the ellagitannins were the major anti-oxidants in the commercial pomegranate juices. Three of the "pure" pomegranate juices had the highest ellagitannin content and the highest anti-oxidant capacity. Fermented juice and cold-pressed pomegranate seeds possess anti-oxidant activity and can reduce prostaglandin and leukotriene formation by inhibition of cyclo-oxygenases and lipoxygenases (Schubert *et al.*, 1999).

Anti-oxidant activities of freeze-dried preparations of pomegranate and its 3 major anthocyanidins (delphinidin, cyanidin, and pelargonidin) evaluated by the method of electron spin resonance technique and spin trapping exhibited scavenging activity against OH and O_2 The anthocyanidins were found to inhibit a Fenton reagent ˙OH generating system possibly by chelating with ferrous ion. Also, anthocyanidins scavenged in a dose-dependent manner, and ID_{50} values of delphinidin, cyanidin, and pelargonidin were 2.4, 22, and 456 μM, respectively. Anthocyanidins inhibited H_2O_2-induced lipid peroxidation in the rat brain homogenates and ID_{50} values of delphinidin, cyanidin, and pelargonidin were 0.7, 3.5, and 85 μM, respectively (Noda *et al.*, 2002). According to Vegara *et al.* (2013) hurdle technology (heating plus refrigeration) may help to reduce anthocyanin degradation in pasteurized pomegranate juice, avoiding a dramatic impact on its colour and preserving the beneficial effects of this specific bioactive compound on human health.

Interestingly, dried pomegranate extract produced by using pomegranate peel and added to commercial tomato juice and orange juice with strawberries increased the anti-oxidant activity of the juices significantly, proportional to the concentrations added. But for greater acceptance in the consumer market, the maximum addition of dried pomegranate peel extract was found to be 0.5 per cent in tomato juice and orange juice with strawberries (Mastrodi *et al.*, 2012).

The effect of pasteurisation showed that total phenols, punicalagins and ellagic acid were not much affected by thermal processing. Total anthocyanin content and anti-oxidant capacity were substantially and significantly influenced by the heat treatment applied. A linear relationship was observed between TEAC values and total anthocyanins, suggesting that they contributed strongly to the anti-oxidant capacity of pomegranate juice (Mena *et al.*, 2013).

Kishore *et al.* (2009) demonstrated the embryo protective nature of pomegranate fruit extract against Adriamycin (ADR)–induced oxidative stress. Fazeli *et al.* (2011) showed that probiotication of aril juices of sweet (SWV) and sour (SV) pomegranate cultivars can add to their beneficial anti-oxidant activities. Probiotication improved the anti-oxidant activity of SWV juice from 74.4 per cent to 91.82 per cent, and SV juice from 82.64 per cent to 97.8 per cent and reducing power of the probioticated pomegranate juices was also much stronger than the non-probioticated juices.

Anti-inflammatory

Ouachrif *et al.* (2012) demonstrated that the extracts obtained from fruit peels of two varieties of pomegranate: Amrouz and Sefri exerted a significant anti-inflammatory effect. In addition, pomegranate seed oil (PSO) and its major component, punicic acid (a conjugated linolenic acid), also showed potent anti-inflammatory and anti-oxidative properties (the two processes strongly involved in osteoporosis establishment) both *in-vitro* and *in-vivo*. PSO inhibited expression of pro-inflammatory factors while stimulating anti-inflammatory ones (Spilmont *et al.*, 2013). Standardized pomegranate rind extract (SPRE) was more active as an anti-inflammatory agent than ellagic acid. The anti-inflammatory and analgesic effects of SPRE were achieved through inhibiting the leukocyte infiltration and modulating the pro-inflammatory cytokines IL-β and TNF-α (Mo *et al.*, 2013a).

Consumption of pomegranate extract has been shown to significantly lower the severity of arthritis, reduce joint infiltration by the inflammatory cells, alleviate the destruction of bone and cartilage and decrease the levels of IL-6 in the joints. In addition, pomegranate extract abrogated multiple signal transduction pathways and downstream mediators implicated in the pathogenesis of rheumatoid arthritis (Shukla *et al.*, 2008). Balbir-Gurman *et al.* (2011) also demonstrated that pomegranate extract significantly reduced the composite Disease Activity Index (DAS28) by 17 per cent, which could be related mostly to a significant reduction in the tender joint count (by 62 per cent) in rheumatoid arthritis patients. These results were associated with a significant reduction in serum oxidative status and a moderate but significant increase in serum high density lipoprotein-associated paraoxonase 1 activity.

Administration of pomegranate juice dose dependently prevented the negative effects of iodo-acetate. Chondrocyte damage was significantly prevented, with proteoglycan being less affected, especially in the groups receiving a high amount of pomegranate juice. No cell proliferation or inflammatory cells were detected in the synovial fluid confirming the chondro-protective effect *in-vivo* (Hadipour-Jahromy *et al.*, 2010).

Dietary polyphenols present in pomegranate, such as ellagitannins and ellagic acid have shown to exert anti-inflammatory and anti-oxidant properties. In summarizing the evidence for the efficacy of pomegranate in coping with inflammatory conditions of the gastro-intestinal tract, Colombo *et al.* (2013) reported that the *in-vivo* studies on the whole fruit or juice, peel, and flowers demonstrated anti-ulcer effect as well as a significant anti-inflammatory activity in the gut.

Pomegranate extract and ellagic acid-enriched pomegranate extract drastically decreased COX-2 and iNOS over expression, reduced MAPKs phosporylation and prevented the nuclear NF-κB translocation. Thus Rosillo *et al.* (2012) showed that dietary supplementation of ellagic acid augments the beneficial effect of pomegranate extract in experimental colitis and may be a novel therapeutic strategy to manage inflammatory bowel disease (IBD). According to Larrosa *et al.* (2010) the main microbiota-derived metabolite urolithin-A could be the most active anti-inflammatory compound from pomegranate ingestion in healthy subjects whereas in colon inflammation, the effects could be due to the non-metabolized ellagitannin-related fraction.

Oral administration of pomegranate extract (PE) decreased reactive oxygen species concentration and acute inflammation in the tympanic membrane after myringotomy (Kahya *et al.*, 2011). Yazici *et al.* (2012) also demonstrated that oral administration of PE afforded statistically significant protection to the cochlea in rats from cisplatin toxicity, and thus its protective effect against cisplatin ototoxicity in rats.

Micro-particles containing pomegranate encapsulated extract were able to inhibit leukocytes' recruitment to broncho-alveolar fluid, especially, eosinophils, decreasing cytokines (IL-1β and IL-5) and protein levels in the lungs. This approach can be used as an alternative/supplementary therapy based on the biological effects of pomegranate for managing inflammatory processes, especially those with pulmonary complications (de Oliveira *et al.*, 2013).

Mo *et al.* (2013b) indicated that topical application of standardized pomegranate rind extracts (SPRE) and ellagic acid is promising for use in the treatment of inflammatory skin disorders. SPRE dose-dependently reduced the ear edema with the maximal inhibition of 79.12 per cent and 73.63 per cent, respectively. Myeloperoxidase (MPO) activity in the mouse ear was also decreased by SPRE and EA up to 69.68 per cent and 68.79 per cent, respectively as against Triamcinolone and diclofenac by 76.66 per cent and 80.14 per cent.

Therapeutic Benefits of Pomegranate

Pomegranate has gained widespread popularity as a functional food and nutraceutical source since ancient times. It is used in Indian Unani medicine for treatment of diabetes mellitus. For the ancient Chinese the seeds symbolized longevity and immortality. In Chinese traditional medicine, the peels of *Punica granatum* L. have been used to treat traumatic hemorrhage, burns and ulcers. In the past decade, numerous studies on the anti-oxidant, anti-carcinogenic, and anti-inflammatory properties of pomegranate constituents have been published, focusing on treatment and prevention of cancer, cardiovascular disease, diabetes, dental conditions, erectile dysfunction, bacterial infections and antibiotic resistance; and ultraviolet radiation-induced skin damage. Other potential applications include infant brain ischemia, male infertility, Alzheimer's disease, arthritis, and obesity (Jurenka, 2008).

Koniæ-Ristiæ *et al.* (2013) showed that bioactive-rich extracts of kale and pomegranate that are consumed as traditional plant foods of Black Sea area countries were effective in modulating platelet function. Providing an up-to-date overview of the chemical constituents, traditional uses, phytochemistry, pharmacology and toxicology of *Punica granatum,* Reddy *et al.* (2007) also reported that the daily intake of pomegranate by-product (POMx) and pomegranate juice as dietary supplements augment the human immune system's anti-oxidant, anti-malarial and anti-microbial capacities. Ismail *et al.* (2012) confirmed that the ethno-pharmacological relevance of pomegranate is fully justified by the most recent findings on the medicinal and nutritional value of the fruit in treating a wide range of human disorders and maladies.

Gastritis

Methanolic extract of pomegranate fruit rind showed a significant reduction in the ulcer index (UI) in ethanol-induced gastritis in rats. At 250 mg/kg and 500 mg/kg, the extract inhibited UI by 21 per cent and 63 per cent. Rats treated with 500 mg/kg of PG extract were also protected from intra-luminal bleeding. *In-vivo* anti-oxidant levels such as superoxide dismutase (SOD), catalase, glutathione (GSH) and glutathione peroxidase (GPx) levels were increased and found more or less equal to the normal values thus revealing that the gastro-protective activity of the extract (Ajaikumar *et al.,* 2005).

Lai *et al.* (2009) reported that tannins in pomegranate play a protective role against gastric ulcer. Its anti-ulcer effect is related to the increased secretion of adherent mucus and free mucus from the stomach wall and the decrease of lipid peroxidation as well as NO levels and to the modulation of both SOD and GSH-PX in gastric mucosa.

Oral administration of standardized aqueous methanolic extract of pomegranate (490 and 980 mg/kg bw) significantly reduced the ulcer lesion index produced by alcohol, indomethacin, and aspirin in rats. In addition to the ulcer lesions, the extracts also reduced gastric volume and total acidity in pylorusligated rats by increasing the pH and mucus secretion. The saponin, tannins, and flavonoids present in the extract are believed to be responsible for the activity (Alam *et al.,* 2010). Pomegranate has a potential for ameliorating dextran sulfate sodium-induced colitis. Its ellagic acid rich fraction may be responsible for this effect. Further, the anti-ulcerative effects may be attributed to mast cell stabilizing, anti-inflammatory and anti-oxidant actions (Singh *et al.,* 2009).

Cardio-protection

Atherosclerosis

The bioactive metabolites (mainly ellagic acid) in pomegranate have a potential role in the regulation of a number of physio-pathological processes involving thrombin (or thrombin-like proteinase) (Cuccioloni *et al.,* 2009).

Pomegranate juice (PJ) is rich in tannins and possesses anti-atherosclerotic properties (Stowe, 2011). A daily consumption of PJ for 3 months by 45 patients with coronary heart disease revealed that the extent of stress-induced ischemia decreased in the pomegranate group, but increased in the control group without changes in cardiac medications, blood sugar, hemoglobin A1c, body weight, or blood pressure, in either group (Sumner *et al.,* 2005). However, consumption of 330 ml/day of pomegranate juice for four weeks did not have any effect on pulse wave velocity and plasma FRAP. But there was a significant fall in systolic blood pressure (-3.14 mmHg), diastolic blood pressure (-2.33 mmHg) and mean arterial pressure (-2.60 mmHg) (Lynn *et al.,* 2012).

Basu and Penugonda (2009) observed pomegranate juice to significantly reduce atherosclerotic lesion areas in immune-deficient mice and intima media thickness in cardiac patients on medications. Further, it decreased lipid peroxidation in patients with type 2 diabetes and systolic blood pressure and serum angiotensin converting enzyme activity in hyper-tensive patients. Pomegranate polyphenols significantly decreased malondialdehyde and hydroxynonenal only in the diabetic group (Basu *et al.,* 2013).

Shema-Didi *et al.* (2012) confirmed that pomegranate juice intake yielded a significant time response reduction in polymorpho-nuclear leukocyte priming, protein oxidation, lipid oxidation, and inflammation biomarker levels resulting in a significantly lower incidence rate of the second hospitalization due to infections. Furthermore, 25 per cent of the patients in the pomegranate juice group had improvement and only 5 per cent progression in the atherosclerotic process, while more than 50 per cent of patients in the placebo group showed progression and none showed any improvement. The authors recommend that prolonged pomegranate juice intake can improve non-traditional CV risk factors, attenuate the progression of the atherosclerotic process, strengthen the innate immunity, and thus reduce morbidity among hemo-dialysis patients.

Incubation of the cell with pomegranate juice inhibited macrophage cholesterol biosynthesis by 50 per cent. But the inhibition was not mediated at the 3-hydroxy-3 methyl-glutaryl coenzyme A reductase level along the biosynthetic pathway. Hence, pomegranate juice mediated suppression of Ox-LDL degradation and of cholesterol biosynthesis in macrophages can lead to reduced cellular cholesterol accumulation and foam cell formation (Fuhrman *et al.,* 2005). PJ consumption decreased cellular cholesterol content by 14-19 per cent, and this could be attributed to a significant inhibition of cholesterol biosynthesis in peritoneal macrophages by 20-32 per cent, and/or to stimulation of HDL-mediated cholesterol efflux from the cells by 22-37 per cent. Similarly, peritoneal macrophages triglyceride content and triglyceride biosynthesis rate were both significantly decreased by 40 per cent after PJ consumption, by 16-27 per cent and by 22-28 per cent, respectively (Rosenblat *et al.,* 2010b). PJ (0-50 µM) significantly and dose-dependently decreased same in J774A.1 macrophages or in C57BL/6 MPM by about 30 per cent (Rosenblat and Aviram (2011). The anti-oxidative characteristics of punicalagin and gallic acid from pomegranate juice could be related, at least in part, to their stimulatory effect on macrophage PON2 expression, a phenomenon which was shown to be associated with activation of the transcription factors PAPR gamma and AP-1 (Shiner *et al.,* 2007).

Supplementation of PJ to mice with advanced atherosclerosis reduced their macrophage oxidative stress, macrophage cholesterol flux and even attenuated the development of atherosclerosis. Moreover, a tannin-fraction isolated from pomegranate juice had a significant anti-atherosclerotic activity (Kaplan *et al.,* 2001). Aviram *et al.* (2000) observed a small but significant 16 per cent lower susceptibility to free radical-induced lipid peroxidation, in comparison to plasma obtained prior to pomegranate juice consumption, as measured by lipid peroxides formation, or by total anti-oxidant status in serum. According to them, pomegranate anti-oxidants are attached to the pomegranate sugars, and hence could be beneficial even to diabetic patients. Further, pomegranate anti-oxidants are unique in their ability to decrease blood pressure, increase the activity of the HDL-associated paraoxonase-1 which breaks down harmful oxidized lipids in lipoproteins, in macrophages, and in atherosclerotic plaques (Aviram *et al.,* 2002; Aviram and Rosenblat, 2013).

Aviram *et al.* (2008) analyzed *in-vivo* and *in-vitro* the anti-atherogenic properties and mechanisms of action of all pomegranate fruit parts–peels (POMxl, POMxp), arils (POMa), seeds (POMo), and flowers (POMf), in comparison to whole fruit juice (PJ). All POM extracts were found to possess anti-oxidative properties *in-vitro*. After consumption of PJ, POMxl, POMxp, POMa, or POMf by mice, the atherosclerotic lesion area was significantly decreased by 44, 38, 39, 6, or 70 per cent, respectively. Pomegranate phenolics (punicalagin, punicalin, gallic acid, and ellagic acid), as well as pomegranate unique complexed sugars, could mimic the anti-atherogenic effects of pomegranate extracts. They concluded that attenuation of atherosclerosis development by some of the POM extracts and, in particular, POMf, could be related to the combined beneficial effects on serum lipids levels and on macrophage atherogenic properties. Whole fruit extract of pomegranate cardio-protective effect against Dox-induced cardio-toxicity in rats. There was significant increase in the level of GSH whereas inhibition of LPO and increase in SOD concentration was not significant in the pomegranate group compared to the Dox group (Hassanpour *et al.,* 2011).

de Nigris *et al.* (2006) observed that pomegranate juice exerts beneficial effects on the evolution of clinical vascular complications, coronary heart disease, and atherogenesis in humans by enhancing the endothelial nitric-oxide synthase bioactivity. On the other hand Davidson *et al.* (2009) reported that in subjects at moderate coronary heart disease risk, pomegranate juice consumption had no significant effect on overall progression rate of carotid intima-media thickness (CIMT) but slowed down the CIMT progression in subjects with increased oxidative stress and disturbances in the TG-rich lipoprotein/HDL axis. Further, the combination of simvastatin with punicalagin or with a phytosterol (β-sitosterol), or with pomegranate juice that contains both of them was also proved to show anti-atherogenic activity (Rosenblat *et al.,* 2013).

PJ or the polyphenol-rich extract from pomegranate fruit (POMx) of 'Wonderful' variety reduced platelet aggregation, calcium mobilization, thromboxane A(2) production, and hydrogen peroxide formation, induced by collagen and arachidonic acid. POMx showed a stronger action in reducing platelet activation (Mattiello *et al.,* 2009). Even a short term consumption of polyphenolic-rich juices from Wonderful-variety increased the serum anti-atherogenic properties through its anti-oxidant effects *in- vivo* (Rosenblat *et al.,* 2010a).

High Blood Pressure

Aviram and Dornfeld (2001) observed a 36 per cent decrease in serum ACE activity and a 5 per cent reduction in systolic blood pressure. Similar dose-dependent inhibitory effect (31 per cent) of pomegranate juice on serum ACE activity was observed *in-vitro* too. Administration of pomegranate juice (PJ) extract (100 mg/kg and 300 mg/kg; p.o. for 4 weeks) reduced the mean arterial blood pressure and vascular reactivity changes to various catecholamines and also reversed the biochemical changes induced by diabetes and Ang II probably by combating the oxidative stress induced by diabetes and Ang II and by inhibiting ACE activity (Mohan *et al,.* 2010a; b). Pre-supplementation with PJ for 30 consecutive days and treated with isoproterenol (IP) on days 29th and 30th showed significantly lesser increase in heart weight, infarct size, plasma marker enzymes, lipid peroxidation, Ca^{+2} ATPase and a significant protective effect in endogenous enzymatic and non-enzymatic anti-oxidants, Na^+-K^+ and Mg^{+2} ATPases compared to IP alone treated group (Jadeja *et al.,* 2010). Pomegranate polyphenols can protect the cardiac function of rats with I/R injury probably in association with their actions in enhancing oxygen free radical scavenging activity and decreasing lipid peroxidative damage of the myocardial tissues (Dong *et al.,* 2012).

Recently, Asgary *et al.* (2013) documented that consumption of pomegranate juice was associated with significant reductions in systolic BP and diastolic BP but not flow-mediated dilatation. Serum levels of vascular endothelial adhesion molecule 1 were significantly reduced by pomegranate juice while those of E-selectin were elevated. However, no significant effect was observed on serum levels of intracellular adhesion molecule-1, high-sensitivity C-reactive protein, lipid profile parameters, apolipo-proteins and IL-6 in any of the study groups.

Consumption of 240 ml per day of natural pomegranate juice once daily for 1 month improved endothelial function in adolescents with metabolic syndrome. Notably in obese individuals, activity of *P. granatum* extract (*in-vitro* and *in-vivo*) gave very promising results. Such beneficial effects should be considered in dietary recommendations for the paediatric age group (Hashemi *et al.,* 2010).

Hepato-protection

Due to its anti-oxidant and anti-fibrotic properties, pomegranate peel extract is of potential therapeutic value in protecting the liver from fibrosis and oxidative injury caused by biliary obstruction. Chronic administration of pomegranate peel extract alleviated the bile duct ligation-induced oxidative injury of the liver and improved the hepatic structure and function (Toklu *et al.,* 2007).

Pomegranate flower extract offered protection against Fe-NTA induced liver injury in rats through its anti-oxidant effect. A dose regimen of 50-150 mg/kg.bw for a week significantly and dose dependently demonstrated upto 60 per cent protection against hepatic lipid peroxidation; and preserved glutathione levels and activities of anti-oxidant enzymes *viz.,* catalase, glutathione peroxidase glutathione reductase and glutathione-S-transferase by up to 36 per cent, 28.5 per cent, 28.7 per cent, 40.2 per cent and 42.5 per cent respectively. The various markers of liver injury *viz.,* serum aspartate amino-transferase, alanine amino-transferase, alkaline phosphatase, bilirubin and albumin were also modulated exhibiting hepato-protective effect (Kaur *et al.,* 2006; Pirinccioglu *et al.,* 2012). Similarly, pomegranate seed extract also was found to ameliorate drug (cisplatin) induced pathological changes in rabbits (Yildirim *et al.,* 2013).

Neuro-protection

Polyphenols have been proved as neuro-protective agents in different model systems. Pomegranates containing very high levels of anti-oxidant polyphenolic substances also showed positive impact on behavior, motor and cognitive skills, and neuro-degeneration in a transgenic mouse model. Mice treated with pomegranate juice learned water maze tasks more quickly and swam faster than controls and had significantly less (approximately 50 per cent) accumulation of soluble Abeta42 and amyloid deposition in the hippocampus as compared to control mice (Hartman *et al.,* 2006).

Kumar *et al.* (2008) reported increased sleeping latency and reduced sleeping time with Pomegranate seed extract (250 and 500 mg/kg) in animal models. Tail suspension test showed a significant decrease in the immobility time, similar to imipramine, a recognized anti-depressant drug in mice; hot-plate and tail-flick test exhibited anti-nociceptive property similar to morphine (a recognized anti-nociceptive agent). The effects were attributed to the anti-oxidant phytochemical profile of pomegranate seed extract. Forouzanfar *et al.* (2013) indicated cyto-protective property in pomegranate extracts (hydro-alcoholic and aqueous) and pomegranate juice under serum/glucose deprived condition in PC12 cells, suggesting that pomegranate has the potential to be used as a new therapeutic strategy for neuro-degenerative disorders.

Under intense oxidative stress Alzheimer's disease is a progressive degenerative brain disorder characterized by neuronal loss, neuro-fibrillary tangles, and the abnormal deposition of senile plaque and amyloid β peptide (Aβ). The ethanol extract of *P. granatum* mitigated HO-induced oxidative stress in PC12 cells. In addition, the extract inhibited neuronal cell death caused by Aβ-induced oxidative stress and Aβ-induced learning and memory deficiency (Choi *et al.*, 2011).

Older subjects with age-associated memory complaints, after 4 weeks of consuming 8 ounces of pomegranate juice showed a significant improvement in the Buschke selective reminding test of verbal memory and a significant increase in plasma trolox-equivalent anti-oxidant capacity (TEAC) and urolithin A-glucuronide. Compared to the placebo group, the pomegranate group had increased fMRI activity during verbal and visual memory tasks (Bookheimer *et al.*, 2013). Pomegranate juice has been found to ameliorate both amyloid-β plaque load as well as associated spatial learning deficits in a mouse model for Alzheimer's disease (Hartman *et al.*, 2006). Treatment with pomegranate seed extract (PGSE) significantly improved passive and active memory impairments of bilateral common carotid arteries occlusion in cerebral hypoxia-ischemia. No toxicity was observed even with high-dose of 800 mg/kg consumption for 14 days by rats (Sarkaki *et al.*, 2013).

Nephro-protection

Drug induced nephro-toxicity is a major complication of gentamicin (GEN), treatment in severe Gram-negative infections. Reactive oxygen species are important mediators of GEN-induced nephrotoxicity. Administration of pomegranate extract (PE) to GEN-treated rats significantly increased the level of GSH, significantly lowered MDA levels in kidney cortex tissue than those given GEN alone. In rats treated with GEN + PE, despite the presence of mild tubular degeneration, tubular necrosis is less severe, and glomeruli maintained a better morphology when compared with the GEN-treated group. Thus, PE prevents kidney damage by decreasing oxidative stress in kidney (Cekmen *et al.*, 2013).

The protective effect of pomegranate juice on ethylene glycol induced crystal deposition in renal tubules, renal toxicity, and inducible nitric oxide synthase (iNOS) and nuclear factor-kappa B activities was demonstrated in rat kidneys by Tugcu *et al.* (2008). Pomegranate juice was able to elevate the anti-oxidant defense system, clean up free radicals, lessen oxidative damages and protect the kidney against carbon tetrachloride-induced toxicity, thus having a potential protective effect (Abdel Moneim and El-Khadragy, 2013).

Cancer

Pomegranates have also been effective as anti-carcinogenic food. The anti-proliferative, anti-invasive, and anti-metastatic effects of the fruit induces apoptosis through modulation of Bcl-2 proteins, increases p21 and p27, and down-regulates cyclin-cdk network. In addition, pomegranate inhibits the activation of inflammatory pathways including, but not limited to, the NFκ-B pathway (Faria and Calhau, 2011).

Punicic acid isolated from pomegranate fruit extracts have the potential effects in inhibiting tumour necrosis factor-α (TNF-α) and inflammatory diseases. Polyphenols obtained from fermented juice at concentrations ranging from 100 to 1,000 µg/ml inhibited aromatase and 17-β-hydroxysteroid dehydrogenase type 1 activity by 60–80 per cent *in-vitro* in human breast cancer cell lines MCF-7 and MB-MDA-231 cells. The seed oil was the most effective showing 90 per cent inhibition of proliferation while fermented pomegranate juice polyphenols consistently showed about twice the anti-proliferative effect compared to fresh pomegranate juice polyphenols (Kim *et al.*, 2002).

Some modifications are needed to get better efficacy of chemo-preventive agents that are currently in use and it is proposed that cancer prevention is more feasible than treatment. With the idea of combining agents competent of multiple targets, George *et al.* (2011) arrived at a novel combination therapy with pomegranate fruit extract and diallyl sulfide. This imparted better suppressive activity than either of the agents alone supporting that chemo-prevention using dietary agents will be more beneficial against cancer.

At a concentration of 50 mg/L pomegranate juice significantly suppressed TNFalpha-induced COX-2 protein expression by 79 per cent, total pomegranate tannin extract (TPT) by 55 per cent, and punicalagin by 48 per cent. Additionally, pomegranate juice reduced phosphorylation of the p65 subunit and binding to the NFkappaB response element 6.4-fold. TPT suppressed NFkappaB binding 10-fold, punicalagin 3.6-fold, whereas ellagic acid was ineffective. Pomegranate juice also abolished TNFalpha-induced AKT activation, needed for NFkappaB activity (Adams *et al.*, 2006). Among the ellagitannins derived compounds, ellagic acid, gallagic acid, and urolithins A and B, urolithin B (UB) was shown to be most effective in inhibiting aromatase activity in a live cell assay. UB also significantly inhibited testosterone-induced MCF-7aro cell proliferation suggesting that pomegranate ellagitannins-derived compounds have potential for the prevention of estrogen-responsive breast cancers (Adams *et al.*, 2010).

Pomegranate juice extract significantly induced apoptosis in all cell lines, including non-tumor control cells, although lymphoid cells and 2 of the myeloid cell lines were more sensitive. Furthermore, the extract induced cell cycle arrest. These results were confirmed by DAPI analysis and viable cell counts using trypan blue exclusion assay (Dahlawi *et al.*, 2012).

Skin Tumors

Afaq *et al.* (2005a; b) provided clear evidence that pomegranate fruit extract (PFE) possesses anti-skin tumor-promoting effects in CD-1 mouse. Because PFE is capable of inhibiting conventional as well as novel biomarkers of TPA-induced tumor promotion, it may possess chemo-preventive activity in a wide range of tumor models. UVA is the major portion (90-99 per cent) of solar radiation reaching the surface of the earth and has been described to lead to formation of benign and malignant tumors primarily through the release of reactive oxygen species and is responsible for immune-suppression, photo-dermatoses, photo-aging and photo-carcinogenesis. Syed *et al.* (2006) observed that pomegranate fruit extract is an effective agent for ameliorating UVA-mediated damages by modulating NF-kappaB and MAPK pathways and protecting against the adverse effects of UVB radiation. Afaq *et al.* (2010) further reported that oral feeding of PFE inhibited UVB-mediated: (1) nuclear translocation of NF-κB; (2) activation of IKKα; and (3) phosphorylation and degradation of IκBα. Thus provided evidence that oral feeding of PFE to mice affords substantial protection from the adverse effects of UVB radiation via modulation in early biomarkers of photo-carcinogenesis and provide suggestion for its photo-chemo-preventive potential.

Prostate Cancer

Anti-cancer effects have been demonstrated in prostate cancer (Faria and Calhau, 2011). Pomegranate polyphenols inhibited gene expression and androgen receptor (AR) most consistently in the LNCaP-AR cell line. Therefore, inhibition by pomegranate polyphenols of gene expression involved in androgen-synthesizing enzymes and the AR may be of particular importance in androgen-independent prostate cancer cells and the subset of human prostate cancers where AR is up-regulated (Hong *et al.,* 2008). The increase in NF-kappa B activity during the transition from androgen dependence to androgen independence in the LAPC4 xenograft model was abrogated by pomegranate extract (Rettig *et al.,* 2008).

The components of pomegranate juice (PJ)–luteolin, ellagic acid, and punicic acid together inhibit growth of hormone-dependent and hormone-refractory prostate cancer cells and inhibit their migration and their chemotaxis toward stromal cell-derived factor 1α (SDF1α), a chemokine that is important in prostate cancer metastasis to the bone (Wang *et al.,* 2011; 2012).

Ellagitannins from pomegranate juice also possess chemo-preventive potential against prostate cancer (PCa). Ellagitannins are not absorbed intact into the blood stream but are hydrolyzed to ellagic acid. They are also metabolized by gut flora into urolithins which are conjugated in the liver and excreted in the urine. These urolithins are also bioactive and inhibit prostate cancer cell growth. Inhibition of Nuclear Factor Kappa-B activation has been shown in prostate cancer cells and in human prostate cancer xenografts in mice (Heber, 2008). Co-treatment with low concentrations of ellagic acid and urolithin A dramatically provide information on pomegranate metabolites for the prevention of PCa recurrence, supporting the role of gut flora-derived metabolites for cancer prevention (Vicinanza *et al.,* 2013). González-Sarrías *et al.* (2010) suggested that urolithin glucuronides and dimethyl ellagic acid may be the molecules responsible for the beneficial effects of PJ against PCa. The *Punica granatum* L. var. *spinosa* extract attenuated the human prostate cell proliferation *in-vitro* possibly by inducing apoptosis (Sineh Sepehr *et al.,* 2012).

Compounds inhibiting CYP1B1 activity are contemplated to exert beneficial effects at three stages of prostate cancer development, that is, initiation, progression, and development of drug resistance. Cellular uptake experiments demonstrated a 5-fold increase in urolithin uptake by 22Rv1 cells. Western blots of the CYP1B1 protein indicated that the urolithins interfered with the expression of CYP1B1 protein. Thus, urolithins were found to display a dual mode mechanism by decreasing CYP1B1 activity and expression (Kasimsetty *et al.* (2009). Kasimsetty *et al.* (2010) indicated that the ellagitannins and urolithins released in the colon upon consumption of pomegranate juice in considerable amounts could potentially curtail the risk of colon cancer development, by inhibiting cell proliferation and inducing apoptosis.

Treatment of LAPC4 prostate cancer cells with 10µg/ml pomegranate extract (POMx), prepared from skin and arils minus seeds and standardized to ellagitannin content (37 per cent punicalagins by HPLC), resulted in inhibition of cell proliferation and induction of apoptosis. Interestingly, co-treatment with POMx and IGFBP-3 revealed synergistic stimulation of apoptosis and additive inhibition of cell growth (Koyama *et al.,* 2010).

Pomegranate fruit extract (PFE), through modulations in the cyclin kinase inhibitor-cyclin-dependent kinase machinery, resulted in inhibition of cell growth followed by apoptosis of highly aggressive human prostate carcinoma PC3 cells. These were associated with alterations in the levels of Bax and Bcl-2 shifting the Bax:Bcl-2 ratio in favor of apoptosis. Further, oral administration of a human acceptable dose of PFE to athymic nude mice implanted with CWR22Rnu1 cells resulted in significant inhibition of tumor growth with concomitant reduction in secretion of prostate-specific antigen in the serum (Malik *et al.,* 2005; Malik and Mukhtar, 2006). Pomegranate ellagitannin content (POMx) inhibited the proliferation of LNCaP and human umbilical vein endothelial cells cells significantly under both normoxic and hypoxic conditions. HIF-1alpha and vascular endothelial growth factor (VEGF) protein levels were also reduced by POMx under hypoxic conditions. POMx decreased prostate cancer xenograft size, tumor vessel density, VEGF peptide levels and HIF-1alpha expression after 4 weeks of treatment in SCID mice (Sartippour *et al.* (2008).

Breast Cancer

Jeune *et al.* (2005) demonstrated that both pomegranate extracts and genistein inhibited the growth of MCF-7 breast cancer cells through induction of apoptosis, with combination treatment being more efficacious than single treatments. PJ or a combination of its components inhibited growth of the breast cancer cells, increased cancer cell adhesion and decreased cancer cell migration but did not affect normal cells. These treatments inhibit chemotaxis of the cancer cells to SDF1α, a chemokine that attracts breast cancer cells also to the bone (Rocha *et al.,* 2012).

Cancer stem cells, which are highly resistant to conventional chemotherapeutic agents, are thought to be the origin of both primary and secondary breast tumors, and thus are a critical target in both breast cancer therapy and prevention. Several mechanistic studies in cell culture and mouse models suggest possible estrogen receptor-mediated and non-estrogen receptor-mediated benefits of pomegranate juice with respect to breast cancer risk. These studies demonstrate that various constituents of pomegranates can inhibit aromatase and 17beta-hydroxysteroid dehydrogenase enzymes or have antiestrogenic activity (Sturgeon and Ronnenberg, 2010). Ellagic acid, ursolic acid and luteolin caused a time- and concentration-dependent reduction of cell proliferation and viability, suggesting that they contribute to the inhibitory effect of PE, while caffeic acid had no effect (Dai *et al.,* 2010). Pomegranate extract (PE) treatment of WA4 cells resulted in an increase in caspase-3 enzyme activity in a time-

and concentration-dependent manner, indicating that this cyto-toxic effect was due to the induction of apoptosis. Methanolic extract of pomegranate fruit peel reduced cell proliferation and induced apoptosis on MCF-7 human breast cancer cells, due to its anti-oxidant effect (Dikmen *et al.*, 2011). According to Banerjee *et al.* (2012) the anti-cancer activities of pomegranate extract in breast cancer cells were due in part to targeting microRNAs155 and 27a. Both pathways play an important role in the proliferative/inflammatory phenotype exhibited by these cell lines. Shirode *et al.* (2013) reported that PE down-regulates homologous recombination (HR) which sensitizes cells to DNA double strand break (DSBs) growth inhibition and apoptosis. Because HR represents a novel target for cancer therapy, down-regulation of HR by PE may be exploited for sensitization of tumors to anti-cancer drugs.

Liver and Kidney Cancer

Liver cancer, predominantly hepato-cellular carcinoma (HCC), represents a complex and fatal malignancy driven primarily by oxidative stress and inflammation. Pomegranate juice was shown to be a potent inhibitor of human CYP2C9. The addition of 25 µl (5 per cent v/v) of pomegranate juice resulted in almost complete inhibition of human CYP2C9 activity. Further, ingestion of pomegranate juice inhibited the intestinal metabolism of tolbutamide without inhibiting the hepatic metabolism in rats (Nagata *et al.*, 2007). Faria *et al.* (2007b) found that pomegranate juice consumption decreased total hepatic cytochrome P450 (CYP) content as well as the expression of CYP1A2 and CYP3A. Prevention of pro-carcinogen activation through CYP activity/ expression inhibition may be involved in pomegranate juice's effect on tumor initiation, promotion, and progression. Yoshimura *et al.* (2013) demonstrated that ellagic acid is a potent suppressor of resistin secretion *in-vivo* and a transcriptional activator of ppara in the liver, suggesting a possibility for improving obesity-induced dyslipidemia and hepatic steatosis in KK-A(y) mice.

Pomegranate seed extract elicited a significant protective effect toward liver and kidney by decreasing the level of lipid peroxidation; elevating the levels of glutathione S-transferase; and increasing the activities of glutathione peroxidase, glutathione S-transferase, and superoxide dismutase (Cayir *et al.*, 2011). Bishayee *et al.*, 2011; 2013; Bhatia *et al.* (2013) provided substantial evidence that suppression of the inflammatory cascade through modulation of NF-κB signaling pathway may represent a novel mechanism of liver tumor inhibitory effects of PE against experimental hepato-carcinogenesis. Their findings underline the importance of simultaneously targeting two inter-connected molecular circuits, namely, Nrf2-mediated redox signaling and NF-κB-regulated inflammatory pathway, by pomegranate phyto-constituents to achieve chemo-prevention of hepato-cellular carcinoma.

Lung Cancer

Treatment of A549 cells with pomegranate fruit extract (PFE) resulted in dose-dependent arrest of cells in G0–G1 phase of the cell cycle. Further PFE treatment resulted in (i) induction of WAF1/p21 and KIP1/p27, (ii) decrease in the protein expressions of cyclins D1, D2 and E, and (iii) decrease in cyclin-dependent kinase (cdk) 2, cdk4 and cdk6 expression. The treatment also inhibited (i) phosphorylation of MAPK proteins, (ii) inhibition of PI3K, (iii) phosphorylation of Akt at Thr308, (iv) NF-kappaB and IKKalpha, (v) degradation and phosphorylation of IkappaBalpha, and (vi) Ki-67 and PCNA. It was also found that the treatment caused inhibition of NF-kappaB DNA-binding activity. Thus, PFE significantly inhibits lung tumorigenesis in A/J mice and merits investigation as a chemo-preventive agent for human lung cancer (Khan *et al.,* 2007a; b).

Diabetes

Punicalagin and ellagic, gallic, oleanolic, ursolic, and uallic acids in pomegranate have been identified as having anti-diabetic actions. The anti-oxidant mechanism has been identified as the key mechanism by which pomegranate fractions affect the lipid peroxidation in type 2 diabetes. Neutralizing the reactive oxygen species, increasing certain anti-oxidant enzyme activities, inducing metal chelation activity, reducing resistin formation, and inhibiting or activating certain transcriptional factors, such as nuclear factor κB and peroxisome proliferator-activated receptor γ are the ways in which pomegranate offers anti-oxidant protection (Banihani *et al.,* 2013). Thus, consumption of pomegranate juice significantly reduce cellular peroxides (by 71 per cent), and increase glutathione levels (by 141 per cent) in the monocytes-derived macrophages of the NIDDM patients. Pomegranate juice provides anti-oxidative effects on serum and macrophages, contributing to attenuation of atherosclerosis development in diabetic patients (Rosenblat *et al.,* 2006).

Ethanolic extract of pomegranate hull and commercial pomegranate hull extract exhibited similar aldose reductase inhibitory activity on peroxidatively damaged liposomal membranes in addition to anti-oxidant activity which was more effective than pomegranate seed extract. Pomegranate extracts are thus presented as bifunctional agents with potential therapeutic use in the prevention of diabetic complications (Karasu *et al.,* 2012).

Pomegranate has been documented in the management of diabetes in Unani and Chinese medicine. Comparing the effects of the extracts of different pomegranate parts, including juice, peels, seeds and flowers, on carbohydrate digestive enzymes (α-amylase and α-glucosidase) *in-vitro,* it has been confirmed that pomegranate flower and peel exhibit anti-hyper-glycaemic effects by the inhibition of carbohydrate digestive enzymes as well as their phenolic content (Kam *et al.,* 2012). Barrett *et al.* (2013) reported that larger and more complex tannins, such as those in pomegranate and cranberry, more effectively inhibited the enzymes than did less polymerized cocoa tannins. Interaction of the tannins with the enzymes was confirmed through calorimetric measurements of changes in enzyme thermal stability. The juice sugar fraction was found to have unique anti-oxidant polyphenols (tannins and anthocyanins), which could be beneficial to control conditions in type 2 diabetes (Banihani *et al.,* 2013).

Rock *et al.* (2008) and Fuhrman *et al.* (2010) observed that consumption of 'Wonderful variety' pomegranate (red fruit skin and dark-red aril color and has semi-hard seeds and a bit sour flavor) juice and polyphenol extract by diabetic patients contribute to stabilization of serum paraoxonase 1, increased association with HDL, and enhanced catalytic activities leading to retardation of atherosclerosis development in diabetic patients.

Obesity

Many studies have explored the effects of the pomegranate in obesity. Positive effects on fat reduction have been shown using the pomegranate fruit and its extracts. The beneficial effects are related to the presence of anthocyanins, tannins, and very high levels of anti-oxidants, including polyphenols and flavonoids (Al-Muammar and Khan 2012). Vroegrijk *et al.* (2011) demonstrated that dietary pomegranate seed oil (1 per cent) ameliorated high-fat diet induced obesity by decreasing body weight, body fat mass and insulin resistance by improving peripheral insulin sensitivity in mice, independent of changes in food intake or energy expenditure within 12 weeks.

Neyrinck *et al.* (2013) through their study on mice, found that pomegranate constitutes a promising food in the control of atherogenic and inflammatory disorders associated with diet-induced obesity as it reduced the serum level of cholesterol (total and LDL) induced by high fat feeding. Further, it counteracted the HF-induced expression of inflammatory markers both in the colon and the visceral adipose tissue. In spite of the poor bioavailability of pomegranate polyphenols, its bifidogenic effect observed after consumption of pomegranate peel extract suggests the involvement of the gut microbiota in the management of host metabolism by polyphenolic compounds present in pomegranate.

Infectious Diseases

Infectious diseases caused by bacteria and fungi are the major cause of morbidity and mortality across the globe. Multi-drug resistance in these pathogens augments the complexity and severity of the diseases. Various studies have shown the role of biofilms in multi-drug resistance, where the pathogen resides inside a protective coat made of extracellular polymeric substances. Methanolic extract of pomegranate was shown to inhibit the formation of biofilms by *S. aureus, E. coli,* and *C. albicans.* Apart from inhibiting the formation of biofilm, pomegranate extract disrupted pre-formed biofilms and inhibited germ tube formation by *C. albicans* (Bakkiyaraj *et al.,* 2013).

Purified pomegranate extract was investigated *in-vitro* for its efficacy against *T. vaginalis* on Diamond media. Infected women (18/20) who accepted to be treated with *P. granatum* juice were completely cured and followed-up for two months. The anti-*T. vaginalis* activity of *P. granatum* extract (*in-vitro and in-vivo*) gave very promising results (El-Sherbini *et al.,* 2010).

Methanolic extracts from peels of seven commercially grown pomegranate cultivars showed strong broad-spectrum activity against Gram-positive (*B. subtilis* and *S. aureus*) and Gram-negative bacteria (*E. coli* and *K. pneumonia*) with the minimum inhibitory concentrations (MIC) ranging from 0.2 to 0.78 mg/ml. Ferrous ion chelating (FIC) and ferric ion reducing anti-oxidant power (FRAP activities as well as tyrosinase-inhibition activities were exhibited by all the peel extracts (Fawole *et al.,* 2012).

Arils from six pomegranate (*Punica granatum* L.) varieties grown in the Mediterranean region of Turkey had anti-microbial effect on seven bacteria (*B.*

megaterium DSM 32, *P. aeruginosa* DSM 9027, *S. aureus* Cowan 1, *C. xerosis* UC 9165, *E. coli* DM, *E. faecalis* A10, *M. luteus* LA 2971), and three fungi (*K. marxianus* A230, *R. rubra* MC12, *C. albicans* ATCC 1023) showing inhibition zones ranging in size from 13 to 26 mm. (Duman *et al.,* 2009). Dried and powdered pomegranate peel displayed anti-fungal property and was found to show the highest inhibition of *C. albicans* as compared to *Acacia nilotica, Cuminum cyminum* and *Foeniculum vulgare* (Pai *et al.,* 2010). The spray-drying process preserved the anti-fungal activity against *Candida albicans* (Endo *et al.,* 2012). The methanolic extract was found to be most effective against all tested microorganisms (*in-vitro*) as compared to petroleum ether, chloroform and water extracts (Prashant *et al.,* 2001).

Pomegranate juice and its polyphenols resulted in titer reductions of foodborne virus surrogates after 1 h exposure, showing promise for use in hurdle technologies and/or for therapeutic or preventive use (Su *et al.,* 2010). The anti-microbial activity of pomegranate juice was dependent on the test organism, which varied to highly susceptible (four Gram-positive species) to unaffected ones (*Salmonella and E. coli* O157:H7). Two Gram-negative species, which were inhibited were *H. pylori* and *V. parahemolyticus* (Haghayeghi *et al.,* 2013).

Methanolic extract of pomegranate (PGME) dramatically enhanced the activity of all antibiotics tested–chloramphenicol, gentamicin, ampicillin, tetracycline, and oxacillin. Synergic activity was detected between PGME and the antibiotics tested, ranging from 38 per cent to 73 per cent, and thus, offers an effective alternative for the extension of lifetime of these antibiotics. The bactericidal activity of PGME in combination with ampicillin revealed that the cell viability is reduced by 99.9 per cent and 72.5 per cent in methicillin-sensitive *S. aureus* and methicillin-resistant *S. aureus* populations respectively. In addition, PGME demonstrated the potential to either inhibit the efflux pump NorA or to enhance the influx of the drug. The detection of *in-vitro* variant colonies of *S. aureus* resistant to PGME was low and they did not survive (Braga *et al.,* 2005).

A synergistic action of a combination of ciprofloxacin and methanolic extract of pomegranate fruit pericarp against Gram-negative bacilli *i.e.* extended-spectrum β-lactamase (ESBL) producing *E. coli and K. pneumoniae*; and metallo-β-lactamase (MBL) producing *P. aeruginosa*, was demonstrated by Dey *et al.* (2012) proving the efficacy of an existing drug being improved with the help of an inexpensive alternative therapy.

Consumption of pomegranate products leads to a significant accumulation of ellagitannins in the large intestines, where they interact with complex gut microflora. Pomegranate by-product (POMx) and major pomegranate polyphenols, (punicalagins) supplementation significantly enhanced the growth of *B. breve* and *B. infantis* The products of the intestinal microbial transformation of pomegranate ellagitannins may account for systemic anti-oxidant effects (Bialonska *et al.,* 2009a; b).

On further research, Bialonska *et al.* (2010) revealed that the POMx exposure enhanced the growth of total bacteria, *Bifido-bacterium* spp. and *Lactobacillus* spp., without influencing the *C. coccoides-Eubacterium rectale* group and the *C. histolyticum* group. In addition, POMx increased the concentrations of short chain fatty acids

(SCFA) *viz.* acetate, propionate and butyrate in the fermentation medium. But punicalagins did not affect the growth of bacteria or production of SCFA suggesting that POMx oligomers, composed of gallic acid, ellagic acid and glucose units, may account for the enhanced growth of probiotic bacteria.

Haidari *et al.* (2009) demonstrated that punicalagin is the effective, anti-influenza component of pomegranate polyphenol extract (PPE). Punicalagin blocked replication of the virus RNA, inhibited agglutination of chicken RBC's by the virus and had virucidal effects. Furthermore, the combination of PPE and oseltamivir synergistically increased the anti-influenza effect of oseltamivir. Thus PPE inhibited the replication of human influenza A/Hong Kong (H3N2) *in-vitro*.

Dental Plaque

Recent studies demonstrate that pomegranates can support oral health and is a successful remedy for strengthening gums and fastening loose teeth through its anti-bacterial and astringent activity. Sastravaha, *et al.* (2005) found that, treatment with the extract of *Punica granatum* significantly improved clinical signs of chronic periodontitis. There was a significant reduction in the number of colony forming units of streptococci (23 per cent) and lactobacilli (46 per cent). The ruby red seeds may be a possible alternative for the treatment of dental plaque bacteria (Kote *et al.,* 2011).

On comparing pomegranate phyto-therapeutic gel with miconazole (Daktarin oral gel) against three standard streptococci strains (mutans ATCC 25175, sanguis ATCC 10577 and mitis ATCC 9811), pomegranate gel showed greater efficiency in inhibiting microbial adherence than the miconazole. It is clear that this phyto-therapeutic agent might be used in the control of adherence of different micro-organisms in the oral cavity (Vasconcelos *et al.,* 2006).

The hydro-alcoholic extract from pomegranate showed that it was very effective against dental plaque microorganisms, decreasing the CFU/ml (the number of colony forming units per milliliter) by 84 per cent. Chlorhexidine, used as standard and positive control showed 79 per cent inhibition, but the distilled water group inhibited only by 11 per cent (Menezes *et al.,* 2006). Treatment with pomegranate extract PomElla dissolved in water changed salivary measures relevant to oral health including gingivitis, reduced total protein (which can correlate with plaque forming bacteria readings), reduced activities of aspartate aminotransferase (an indicator of cell injury), and alpha-glucosidase (a sucrose degrading enzyme), increased activities of the anti-oxidant enzyme ceruloplasmin (which could give better protection against oral oxidant stress) and radical scavenging capacity (DiSilvestro *et al.,* 2009).

A pomegranate-containing mouthrinse was found to have anti-plaque effect. And the extract was effective against *Aggregatibacter actinomycetemcomitans, Porphyromonas gingivalis* and *Prevotella intermedia* strains of periodonto-pathogens at various concentrations *in-vitro* (Bhadbhade *et al.,* 2011).

Nearly every part of the pomegranate plant has been tested for anti-microbial activities, including the fruit juice, peel, arils, flowers, and bark. In some cases the combination of the pomegranate constituents offers the most benefit (Howell and D'Souza, 2013).

Infertility

Traditionally, pomegranate fruit is linked with fertility, birth and eternal life. A significant decrease in malondialdehyde (MDA) level and glutathione (GSH), glutathione peroxidase (GSH-Px) and catalase (CAT) activities, and vitamin C level were observed in rats treated with different doses of pomegranate juice. Pomegranate juice increased epididymal sperm concentration, sperm motility, spermatogenic cell density and diameter of seminiferous tubules and germinal cell layer thickness, and it decreased abnormal sperm rate when compared to the control group suggesting improvement in sperm quality and anti-oxidant activity in rats (Türk *et al.,* 2008).

The deleterious effects of lead acetate (LA) administration on sperm production and reproductive system (by reducing the length of the stages related to spermiation (VII and VIII), onset of mitosis (IX-XI), and epididymal sperm number) has been shown to be reversed by ascorbic acid (AA) which is a strong anti-oxidant in rats. Administration of ethanolic extract of pomegranate (EEP) or AA resulted in longer VIII and IX-XI stages and reduced the deleterious effect of LA on daily sperm production (DSP) and epididymal sperm number. EEP showed an anti-oxidant activity similar to that of AA and so is able to reverse the damage produced by LA on spermatogenesis (Leiva *et al.,* 2011).

Chen *et al.* (2012) found that pomegranate juice reduces placental oxidative stress *in-vivo* and *in-vitro* while limiting stimulus-induced death of human trophoblasts in culture. The polyphenol punicalagin mimics this protective effect and hence speculate that antenatal intake of pomegranate may limit placental injury and thereby may confer protection to the exposed fetus.

Mice treated with pomegranate juice extract showed an increase in bone Ca content. Dietary supplementation with extracts of pomegranate juice, pomegranate husk or a mixture of husk and juice significantly increased embryo femur length and osteo-genesis index. The number of viable mesenchymal cells from fetal limb buds was greater in cultures exposed to the extracts than in control cultures. The number of cartilage nodules and their diameters were greater in extract-treated cell cultures, a finding which reflected increased cell proliferation and differentiation rates suggesting that pomegranate is able to enhance bone formation (Monsefi *et al.,* 2012). Pomegranate seed oil in doses of 0.2 – 0.4 ml into overiectomized mice twice daily for 2 days resulted in cornification of vaginal cells (Sharaf *et al.,* 1964).

Other Benefits

Interestingly, ellagitannin supplementation from Wonderful variety pomegranate extract improved recovery of skeletal muscle strength after eccentric exercise (Trombold *et al.,* 2010). Further, pomegranate juice displayed a mild, acute ergogenic effect in the elbow flexor muscles of resistance trained individuals after eccentric exercise (Trombold *et al.,* 2011).

It has been mentioned that space radiation can have negative effects on cognitive skills as well as physical and mental health. Dulcich and Hartman (2013) showed that proton irradiation, which may be encountered in space–depression-like behaviors and worse coordination and balance can be overcome by the pomegranate diet and can confer protection against some of those effects.

Safety Issues

Histo-pathological analysis of liver and kidney corroborated the absence of toxicity of punicalagin in Sprague-Dawley rats upon repeated oral administration of a 6 per cent punicalagin-containing diet for 37 days. However, taking into account the high punicalagin content of pomegranate-derived foodstuffs, safety evaluation should be also carried out in humans with a lower dose and during a longer period of intake (Cerdá *et al.,* 2003).

Conclusions

Pomegranate juice is being increasingly proposed as a nutritional supplement to prevent atherosclerosis in humans due to its anti-oxidant capacity. Moreover, the clinical studies on suppression of oral bacteria by pomegranate are intriguing. But, most of the studies on the anti-bacterial and anti-viral activities against food-borne pathogens and other infectious organisms are based on *in-vitro* cell-based assays. Hence confirmation of *in-vivo* efficacy through human clinical trials is needed.

References

Abdel Moneim AE, El-Khadragy MF (2013). The potential effects of pomegranate (*Punica granatum*) juice on carbon tetrachloride-induced nephrotoxicity in rats. *J. Physiol. Biochem*. 69 (3): 359–70.

Adams LS, Seeram NP, Aggarwal BB, Takada Y, Sand D, Heber D (2006). Pomegranate juice, total pomegranate ellagitannins, and punicalagin suppress inflammatory cell signaling in colon cancer cells. *J. Agric. Food Chem*. 54 (3): 980–85.

Adams LS, Zhang Y, Seeram NP, Heber D, Chen S (2010). Pomegranate ellagitannin-derived compounds exhibit anti-proliferative and anti-aromatase activity in breast cancer cells *in-vitro*. *Cancer Prev. Res. (Phila)*. 3 (1): 108–113.

Adhami VM, Khan N, Mukhtar H (2009). Cancer chemo-prevention by pomegranate: laboratory and clinical evidence. *Nutr. Cancer*. 61 (6): 811–15.

Afaq F, Khan N, Syed DN, Mukhtar H (2010). Oral feeding of pomegranate fruit extract inhibits early biomarkers of UVB radiation-induced carcinogenesis in SKH-1 hairless mouse epidermis. *Photochem. Photobiol*. 86 (6): 1318–26.

Afaq F, Malik A, Syed D, Maes D, Matsui MS, Mukhtar H (2005a). Pomegranate fruit extract modulates UV-B-mediated phosphorylation of mitogen-activated protein kinases and activation of nuclear factor kappa B in normal human epidermal keratinocytes paragraph sign. *Photochem. Photobiol.*; 81 (1): 38–45.

Afaq F, Saleem M, Krueger C, Reed J, Mukhtar H (2005b). Anthocyanin- and hydrolyzable tannin rich pomegranate fruit extract modulates MAPK and NF-kappaB pathways and inhibits skin tumorigenesis in *CD-1* mice. *Intl. J. of Cancer*. 113: 423–33.

Ajaikumar KB, Asheef M, Babu BH, Padikkala J (2005). The inhibition of gastric mucosal injury by Punica granatum L. (pomegranate) methanolic extract. *Journal of Ethnopharmacology*. 96 (1-2): 171–76.

Alam MS, Alam MA, Ahmad S, Najmi AK, Asif M, Jahangir T (2010). Protective effects of *Punica granatum* in experimentally-induced gastric ulcers. *Toxicol. Mech. Methods*. 20 (9): 572–78.

Al-Muammar MN, Khan F (2012). Obesity: the preventive role of the pomegranate (*Punica granatum*). *Nutrition*. 28 (6): 595–604.

Asgary S, Sahebkar A, Afshani MR, Keshvari M, Haghjooyjavanmard S, Rafieian-Kopaei M (2013). Clinical evaluation of blood pressure lowering, endothelial function improving, hypo-lipidemic and anti-inflammatory effects of pomegranate juice in hypertensive subjects. *Phytother. Res*. 2013 Mar 21. doi: 10.1002/ptr.4977. [Epub ahead of print].

Aviram M, Dornfeld L (2001). Pomegranate juice consumption inhibits serum angiotensin converting enzyme activity and reduces systolic blood pressure. *Atherosclerosis*. 158 (1): 195–98.

Aviram M, Dornfeld L, Kaplan M, Coleman R, Gaitini D, Nitecki S, Hofman A, Rosenblat M, Volkova N, Presser D, Attias J, Hayek T, Fuhrman B (2002). Pomegranate juice flavonoids inhibit low-density lipoprotein oxidation and cardiovascular diseases: studies in atherosclerotic mice and in humans. *Drugs Ex. Clin. Res*. 28: 49–62.

Aviram M, Dornfeld L, Rosenblat M, Volkova N, Kaplan M, Coleman R, Hayek T, Presser D, Fuhrman B (2000). Pomegranate juice consumption reduces oxidative stress, atherogenic modifications to LDL, and platelet aggregation: studies in humans and in atherosclerotic apolipoprotein E-deficient mice. *Am. J. Clin. Nutr*. 71 (5): 1062–76.

Aviram M, Rosenblat M (2013). Pomegranate for your cardio-vascular health. Rambam. *Maimonides. Med. J*. 4 (2): e0013.

Aviram M, Volkova N, Coleman R, Dreher M, Reddy MK, Ferreira D, Rosenblat M (2008). Pomegranate phenolics from the peels, arils, and flowers are anti-atherogenic: studies *in-vivo* in atherosclerotic apolipoprotein e-deficient (E 0) mice and *in-vitro* in cultured macrophages and lipoproteins. *J. Agric. Food Chem*. 56 (3): 1148–57.

Bakkiyaraj D, Nandhini JR, Malathy B, Pandian SK (2013). The anti-biofilm potential of pomegranate (*Punica granatum* L.) extract against human bacterial and fungal pathogens. *Biofouling*. 29 (8): 929–37.

Balbir-Gurman A, Fuhrman B, Braun-Moscovici Y, Markovits D, Aviram M (2011). Consumption of pomegranate decreases serum oxidative stress and reduces disease activity in patients with active rheumatoid arthritis: a pilot study. *Isr. Med. Assoc. J*. 13 (8): 474–79.

Banerjee N, Talcott S, Safe S, Mertens-Talcott SU (2012). Cytotoxicity of pomegranate polyphenolics in breast cancer cells *in-vitro* and *vivo*: potential role of miRNA-

27a and miRNA-155 in cell survival and inflammation. *Breast Cancer Res. Treat.* 136 (1): 21–34.

Banihani S, Swedan S, Alguraan Z (2013). Pomegranate and type 2 diabetes. *Nutr. Res.* 33 (5): 341–48.

Barrett A, Ndou T, Hughey CA, Straut C, Howell A, Dai Z, Kaletunc G (2013). Inhibition of α-amylase and glucoamylase by tannins extracted from cocoa, pomegranates, cranberries, and grapes. J. *Agric. Food Chem.* 61 (7): 1477–86.

Basu A, Newman ED, Bryant AL, Lyons TJ, Betts NM (2013). Pomegranate polyphenols lower lipid peroxidation in adults with type 2 diabetes but have no effects in healthy volunteers: a pilot study. *J. Nutr. Metab.* 2013: 708381.

Basu A, Penugonda K (2009). Pomegranate juice: a heart-healthy fruit juice. *Nutr. Rev.* 67 (1): 49–56.

Bhadbhade SJ, Acharya AB, Rodrigues SV, Thakur SL (2011). The antiplaque efficacy of pomegranate mouthrinse. *Quintessence Int.* 42 (1): 29–36.

Bhatia D, Thoppil RJ, Mandal A, Samtani KA, Darvesh AS, Bishayee A (2013). Pomegranate bioactive constituents suppress cell proliferation and induce apoptosis in an experimental model of hepato-cellular carcinoma: Role of Wnt/ β -catenin signaling pathway. Evid. Based Complement. *Alternat. Med.* 2013: 371813.

Bialonska D, Kasimsetty SG, Khan SI, Ferreira D (2009a). Urolithins, intestinal microbial metabolites of Pomegranate ellagitannins, exhibit potent antioxidant activity in a cell-based assay. *J. Agric. Food Chem.* 57 (21): 10181–86.

Bialonska D, Kasimsetty SG, Schrader KK, Ferreira D (2009b). The effect of pomegranate (*Punica granatum* L.) byproducts and ellagitannins on the growth of human gut bacteria. *J. Agric. Food Chem.* 57 (18): 8344–49.

Bialonska D, Ramnani P, Kasimsetty SG, Muntha KR, Gibson GR, Ferreira D (2010). The influence of pomegranate by-product and punicalagins on selected groups of human intestinal microbiota. *Int. J. Food Microbiol.* 140 (2-3): 175–82.

Bishayee A, Bhatia D, Thoppil RJ, Darvesh AS, Nevo E, Lansky EP (2011). Pomegranate-mediated chemo-prevention of experimental hepato-carcinogenesis involves Nrf2-regulated anti-oxidant mechanisms. *Carcinogenesis.* 32 (6): 888–96.

Bishayee A, Thoppil RJ, Darvesh AS, Ohanyan V, Meszaros JG, Bhatia D (2013). Pomegranate phyto-constituents blunt the inflammatory cascade in a chemically induced rodent model of hepato-cellular carcinogenesis. *J. Nutr. Biochem.* 24 (1): 178–87.

Bookheimer SY, Renner BA, Ekstrom A, Li Z, Henning SM, Brown JA, Jones M, Moody T, Small GW (2013). Pomegranate juice augments memory and FMRI activity in middle-aged and older adults with mild memory complaints. Evid. Based Complement. *Alternat. Med.* 2013: 946298.

Borges G, Mullen W, Crozier A (2010). Comparison of the polyphenolic composition and anti-oxidant activity of European commercial fruit juices. *Food Funct.* 1 (1): 73–83.

Braga LC, Leite AA, Xavier KG, Takahashi JA, Bemquerer MP, Chartone-Souza E, Nascimento AM (2005). Synergic interaction between pomegranate extract and antibiotics against *Staphylococcus aureus. Can. J. Microbiol.* 51 (7): 541–47.

Cayir K, Karadeniz A, Sim°ek N, Yýldýrým S, Karaku° E, Kara A, Akkoyun HT, Sengül E (2011). Pomegranate seed extract attenuates chemotherapy- induced acute nephro-toxicity and hepato-toxicity in rats. *J. Med. Food.* 14 (10): 1254–62.

Cekmen M, Otunctemur A, Ozbek E, Cakir SS, Dursun M, Polat EC, Somay A, Ozbay N (2013). Pomegranate extract attenuates gentamicin-induced nephrotoxicity in rats by reducing oxidative stress. *Ren. Fail.* 35 (2): 268–74.

Cemeroglu B, Artik N, Erbas S (1992). Extraction and composition of pomegranate juice. *Fluess. Obst.* 59: 335–40.

Cerdá B, Cerón JJ, Tomás-Barberán FA, Espín JC (2003). Repeated oral administration of high doses of the pomegranate ellagitannin punicalagin to rats for 37 days is not toxic. *J. Agric. Food Chem.* 51 (11): 3493–501.

Cerdá B, Espín JC, Parra S, Martínez P, Tomás-Barberán FA (2004). The potent in vitro antioxidant ellagitannins from pomegranate juice are metabolised into bioavailable but poor anti-oxidant hydroxy-6H-dibenzopyran-6-one derivatives by the colonic microflora of healthy humans. *Eur. J. Nutr.* 43 (4): 205–220.

Chen B, Tuuli MG, Longtine MS, Shin JS, Lawrence R, Inder T, Michael Nelson D (2012). Pomegranate juice and punicalagin attenuate oxidative stress and apoptosis in human placenta and in human placental trophoblasts. *Am. J. Physiol. Endocrinol. Metab.* 302 (9): E1142–52.

Chidambara Murthy KN, Jayaprakasha GK, Singh RP (2002). Studies on anti-oxidant activity of pomegranate (*Punica granatum*) peel extract using *in-vivo* models. *J. Agric. Food Chem.* 50 (17): 4791–95.

Choi SJ, Lee JH, Heo HJ, Cho HY, Kim HK, Kim CJ, Kim MO, Suh SH, Shin DH (2011). *Punica granatum* protects against oxidative stress in PC12 cells and oxidative stress-induced Alzheimer's symptoms in mice. *J. Med. Food.* 14 (7-8): 695–701.

Colombo E, Sangiovanni E, Dell'agli M (2013). A review on the anti-inflammatory activity of pomegranate in the gastro-intestinal tract. *Evid. Based Complement. Alternat. Med.* 2013: 247145.

Cuccioloni M, Mozzicafreddo M, Sparapani L, Spina M, Eleuteri AM, Fioretti E, Angeletti M (2009). Pomegranate fruit components modulate human thrombin. *Fitoterapia.* 80 (5): 301–305.

Dafny-Yalin M, Glazer I, Bar-Ilan I, Kerem Z, Holland D, Amir R (2010). Color, sugars and organic acids composition in aril juices and peel homogenates prepared from different pomegranate accessions. *J. Agric. Food Chem.* 58 (7): 4342–52.

Dahlawi H, Jordan-Mahy N, Clench MR, Le Maitre CL (2012). Bioactive actions of pomegranate fruit extracts on leukemia cell lines *in-vitro* hold promise for new therapeutic agents for leukemia. *Nutr. Cancer.* 64 (1): 100–110.

Dai Z, Nair V, Khan M, Ciolino HP (2010). Pomegranate extract inhibits the proliferation and viability of MMTV-Wnt-1 mouse mammary cancer stem cells *in-vitro. Oncol. Rep.* 24 (4): 1087–91.

Davidson MH, Maki KC, Dicklin MR, Feinstein SB, Witchger M, Bell M, McGuire DK, Provost JC, Liker H, Aviram M (2009). Effects of consumption of pomegranate juice on carotid intima-media thickness in men and women at moderate risk for coronary heart disease. *Am. J. Cardiol.* 104 (7): 936–42.

de Nigris F, Williams-Ignarro S, Botti C, Sica V, Ignarro LJ, Napoli C (2006). Pomegranate juice reduces oxidized low-density lipoprotein down-regulation of endothelial nitric oxide synthase in human coronary endothelial cells. *Nitric Oxide.* 15 (3): 259–63.

de Oliveira JF, Garreto DV, da Silva MC, Fortes TS, de Oliveira RB, Nascimento FR, Da Costa FB, Grisotto MA, Nicolete R (2013). Therapeutic potential of biodegradable microparticles containing *Punica granatum* L. (pomegranate) in murine model of asthma. *Inflamm. Res.* 62 (11): 971–80.

Dey D, Debnath S, Hazra S, Ghosh S, Ray R, Hazra B (2012). Pomegranate pericarp extract enhances the anti-bacterial activity of ciprofloxacin against extended-spectrum β-lactamase (ESBL) and metallo-β-lactamase (MBL) producing Gram-negative bacilli. *Food Chem. Toxicol.* 50 (12): 4302–309.

Dikmen M, Ozturk N, Ozturk Y (2011). The anti-oxidant potency of *Punica granatum* L. fruit peel reduces cell proliferation and induces apoptosis on breast cancer. *J. Med. Food.* 14 (12): 1638–46.

DiSilvestro RA, DiSilvestro DJ, DiSilvestro D (2009). Pomegranate extract mouth rinsing effects on saliva measures relevant to gingivitis risk. *Phytother. Res.* 23 (8): 1123–27.

Dong S, Tong X, Liu H, Gao Q (2012). Protective effects of pomegranate polyphenols on cardiac function in rats with myocardial ischemia/reperfusion injury. *Nan. Fang. Yi. Ke. Da. Xue. Xue. Bao.* 32 (7): 924–27. [Article in Chinese].

Dulcich MS, Hartman RE (2013). Pomegranate supplementation improves affective and motor behavior in mice after radiation exposure. *Evid. Based Complement. Alternat. Med.* 2013: 940830.

Duman AD, Ozgen M, Dayisoylu KS, Erbil N, Durgac C (2009). Anti-microbial activity of six pomegranate (*Punica granatum* L.) varieties and their relation to some of their pomological and phyto-nutrient characteristics. *Molecules.* 14 (5): 1808–817.

Elfalleh W, Tlili N, Nasri N, Yahia Y, Hannachi H, Chaira N, Ying M, Ferchichi A (2011). Anti-oxidant capacities of phenolic compounds and tocopherols from Tunisian pomegranate (*Punica granatum*) fruits. *J. Food Sci.* 76 (5): C707–713.

El Kar C, Ferchichi A, Attia F, Bouajila J (2011). Pomegranate (*Punica granatum*) juices: chemical composition, micro-nutrient cations, and anti-oxidant capacity. *J. Food Sci.* 76 (6): C795–800.

El-Nemr SE, Ismail IA, Ragab M (1990). Chemical composition of juice and seeds of pomegranate fruit. *Nahrung.* 34: 601–606.

El-Shaarawy MI, Nahpetian A (1983). Studies on pomegranate seed oil. *Fette. Seifen. Anstrichmittel.* 83 (3): 123–26.

El-Sherbini GM, Ibrahim KM, El Sherbiny ET, Abdel-Hady NM, Morsy TA (2010). Efficacy of *Punica granatum* extract on *in-vitro* and *in-vivo* control of *Trichomonas vaginalis.* *J. Egypt. Soc. Parasitol.* 40 (1): 229–44.

Endo EH, Ueda-Nakamura T, Nakamura CV, Filho BP (2012). Activity of spray-dried micro-particles containing pomegranate peel extract against *Candida albicans.* *Molecules.* 17 (9): 10094–107.

Faria A, Calhau C (2011). The bioactivity of pomegranate: impact on health and disease. *Crit. Rev. Food Sci. Nutr.* 51 (7): 626–34.

Faria A, Monteiro R, Azevedo I, Calhau C (2007b). Pomegranate juice effects on cytochrome P450S expression: *in vivo* studies. *J. Med. Food.* 10 (4): 643–49.

Faria A, Monteiro R, Mateus N, Azevedo I, Calhau C (2007a). Effect of pomegranate (*Punica granatum*) juice intake on hepatic oxidative stress. *Eur. J. Nutr.* 46 (5): 271 -8.

Fawole OA, Makunga NP, Opara UL (2012). Anti-bacterial, antioxidant and tyrosinase-inhibition activities of pomegranate fruit peel methanolic extract. BMC. Complement. *Altern. Med.* 12: 200.

Fazeli MR, Bahmani S, Jamalifar H, Samadi N (2011). Effect of pro-biotication on anti-oxidant and anti-bacterial activities of pomegranate juices from sour and sweet cultivars. *Nat. Prod. Res.* 25 (3): 288–97.

Forouzanfar F, Afkhami Goli A, Asadpour E, Ghorbani A, Sadeghnia HR (2013). Protective effect of *Punica granatum* L. against serum/glucose deprivation-Induced PC12 cells injury. Evid. Based Complement. *Alternat. Med.* 2013: 716730.

Fuhrman B, Volkova N, Aviram M (2005). Pomegranate juice inhibits oxidized LDL uptake and cholesterol biosynthesis in macrophages. *J. Nutr. Biochem.* 16 (9): 570–76.

Fuhrman B, Volkova N, Aviram M (2010). Pomegranate juice polyphenols increase recombinant paraoxonase-1 binding to high-density lipoprotein: studies *in-vitro* and in diabetic patients. *Nutrition.* 26 (4): 359–66.

George J, Singh M, Srivastava AK, Bhui K, Shukla Y (2011). Synergistic growth inhibition of mouse skin tumors by pomegranate fruit extract and diallyl sulfide: evidence for inhibition of activated MAPKs/NF-κB and reduced cell proliferation. *Food Chem. Toxicol.* 49 (7): 1511–20.

Gil MI, Tomas-Barberan FA, Hess-Pierce B, Holcroft DM, Kader AA (2000). Anti-oxidant activity of pomegranate juice and its relationship with phenolic composition and processing. *J. Agric. Food Chem.* 48: 4581–89.

González-Sarrías A, Giménez-Bastida JA, García-Conesa MT, Gómez-Sánchez MB, García-Talavera NV, Gil-Izquierdo A, Sánchez-Alvarez C, Fontana-Compiano LO, Morga-Egea JP, Pastor-Quirante FA, Martínez-Díaz F, Tomás-Barberán FA, Espín JC (2010). Occurrence of urolithins, gut microbiota ellagic acid metabolites and proliferation markers expression response in the human prostate gland upon consumption of walnuts and pomegranate juice. *Mol. Nutr. Food Res.* 54 (3): 311–22.

Guo C, Wei J, Yang J, Xu J, Pang W, Jiang Y (2008). Pomegranate juice is potentially better than apple juice in improving anti-oxidant function in elderly subjects. *Nutr. Res.* 28 (2): 72–77.

Hadipour-Jahromy M, Mozaffari-Kermani R (2010). Chondro-protective effects of pomegranate juice on mono-iodoacetate-induced osteo-arthritis of the knee joint of mice. *Phytother. Res.* 24 (2): 182–85.

Haghayeghi K, Shetty K, Labbé R (2013). Inhibition of foodborne pathogens by pomegranate juice. *J. Med. Food.* 16 (5): 467–70.

Haidari M, Ali M, Ward Casscells S 3rd, Madjid M (2009). Pomegranate (*Punica granatum*) purified polyphenol extract inhibits influenza virus and has a synergistic effect with oseltamivir. *Phytomedicine.* 16 (12): 1127–36.

Hartman RE, Shah A, Fagan AM, Schwetye KE, Parsadanian M, Schulman RN, Finn MB, Holtzman DM (2006). Pomegranate juice decreases amyloid load and improves behavior in a mouse model of Alzheimer's disease. *Neurobiol. Dis.* 24 (3): 506–515.

Hashemi M, Kelishadi R, Hashemipour M, Zakerameli A, Khavarian N, Ghatrehsamani S, Poursafa P (2010). Acute and long-term effects of grape and pomegranate juice consumption on vascular reactivity in paediatric metabolic syndrome. *Cardiol. Young.* 20 (1): 73–77.

Hassanpour Fard M, Ghule AE, Bodhankar SL, Dikshit M (2011). Cardio-protective effect of whole fruit extract of pomegranate on doxorubicin-induced toxicity in rat. *Pharm. Biol.* 49 (4): 37 7–82.

Heber D (2008). Multi-targeted therapy of cancer by ellagitannins. *Cancer Lett.* 269 (2): 262–68.

Heber D (2011). Pomegranate Ellagitannins. In: Benzie IFF, Wachtel-Galor S, editors. *Herbal Medicine: Biomolecular and Clinical Aspects.* Chapter 10. 2nd edition. CRC Press. Boca Raton (FL).

Heftaman E, Bennett ST (1966). Identification of estrone in pomegranate seeds. *Phytochemistry.* 5: 1337–39.

Hong MY, Seeram NP, Heber D (2008). Pomegranate polyphenols down-regulate expression of androgen-synthesizing genes in human prostate cancer cells overexpressing the androgen receptor. *J. Nutr. Biochem.* 19 (12): 848–55.

Howell AB, D'Souza DH (2013). The pomegranate: effects on bacteria and viruses that influence human health. Evid. Based Complement. *Alternat. Med.* 2013: 606212.

Ignarro LJ, Byrns RE, Sumi D, de Nigris F, Napoli C (2006). Pomegranate juice protects nitric oxide against oxidative destruction and enhances the biological actions of nitric oxide. *Nitric Oxide*. 15 (2): 93–102.

Ismail T, Sestili P, Akhtar S (2012). Pomegranate peel and fruit extracts: a review of potential anti-inflammatory and anti-infective effects. *J. Ethnopharmacol*. 143 (2): 397–405.

Jadeja RN, Thounaojam MC, Patel DK, Devkar RV, Ramachandran AV (2010). Pomegranate (*Punica granatum* L.) juice supplementation attenuates isoproterenol-induced cardiac necrosis in rats. *Cardiovasc. Toxicol*. 10 (3): 174–80.

Jeune MA, Kumi-Diaka J, Brown J (2005). Anti-cancer activities of pomegranate extracts and genistein in human breast cancer cells. *J. Med. Food*. 8 (4): 469–75.

Johanningsmeier SD, Harris GK (2011). Pomegranate as a functional food and nutraceutical source. *Annu. Rev. Food Sci. Technol*. 2: 181–201.

Jurenka JS (2008). Therapeutic applications of pomegranate (*Punica granatum* L.): a review. *Altern. Med. Rev*. 13 (2): 128–44.

Kahya V, Meric A, Yazici M, Yuksel M, Midi A, Gedikli O (2011). Anti-oxidant effect of pomegranate extract in reducing acute inflammation due to myringotomy. *J. Laryngol. Otol*. 125 (4): 370–75.

Kam A, Li KM, Razmovski-Naumovski V, Nammi S, Shi J, Chan K, Li GQ (2012). A comparative study on the inhibitory effects of different parts and cemical constituents of pomegranate on α-amylase and α-glucosidase. *Phytother. Res*. 2012 Dec 19. doi: 10.1002/ptr.4913. [Epub ahead of print].

Kaplan M, Hayek T, Raz A, Coleman R, Dornfeld L, Vaya J, Aviram M (2001). Pomegranate juice supplementation to atherosclerotic mice reduces acrophage lipid peroxidation, cellular cholesterol accumulation and development of atherosclerosis. *J. Nutr*. 131 (8): 2082–89.

Karasu C, Cumaoðlu A, Gürpinar AR, Kartal M, Kovacikova L, Milackova I, Stefek M (2012). Aldose reductase inhibitory activity and antioxidant capacity of pomegranate extracts. Interdiscip. *Toxicol*. 5 (1): 15–20.

Kasimsetty SG, Bialonska D, Reddy MK, Ma G, Khan SI, Ferreira D (2010). Colon cancer chemopreventive activities of pomegranate ellagitannins and urolithins. *J. Agric. Food Chem*. 58 (4): 2180–87.

Kasimsetty SG, Bialonska D, Reddy MK, Thornton C, Willett KL, Ferreira D (2009). Effects of pomegranate chemical constituents/intestinal microbial metabolites on CYP1B1 in 22Rv1 prostate cancer cells. *J. Agric. Food Chem*. 57 (22): 10636–44.

Kaur G, Jabbar Z, Athar M, Alam MS (2006). Punica granatum (pomegranate) flower extract possesses potent anti-oxidant activity and abrogates Fe-NTA induced hepato-toxicity in mice. *Food Chem. Toxicol*. 44 (7): 984–93.

Khan N, Afaq F, Kweon MH, Kim K, Mukhtar H (2007a). Oral consumption of pomegranate fruit extract inhibits growth and progression of primary lung tumors in mice. *Cancer Res*. 67 (7): 3475–82.

Khan N, Hadi N, Afaq F, Syed DN, Kweon MH, Mukhtar H (2007b). Pomegranate fruit extract inhibits pro-survival pathways in human A549 lung carcinoma cells and tumor growth in athymic nude mice. *Carcinogenesis*. 28 (1): 163–73.

Kim ND, Mehta R, Yu W, Neeman I, Livney T, Amichay A, Poirier D, Nicholls P, Kirby A, Jiang W, Mansel R,Ramachandran C, Rabi T, Kaplan B, Lansky E. (2002). Chemo-preventive and adjuvant therapeutic potential of pomegranate (*Punica granatum*) for human breast cancer. *Breast Cancer Res. Treat*. 71 (3): 203–217.

Kishore RK, Sudhakar D, Parthasarathy PR (2009). Embryo protective effect of pomegranate (*Punica granatum* L.) fruit extract in adriamycin-induced oxidative stress. *Indian J. Biochem. Biophys*. 46 (1): 106–111.

Koniæ-Ristiæ A, Srdiæ-Rajiæ T, Kardum N, Aleksiæ-Velièkoviæ V, Kroon PA, Hollands WJ, Needs PW, Boyko N, Hayran O, Jorjadze M, Glibetiæ M (2013). Effects of bioactive-rich extracts of pomegranate, persimmon, nettle, dill, kale and Sideritis and isolated bioactives on arachidonic acid induced markers of platelet activation and aggregation. *J. Sci. Food Agric*. 2013 Jul 30. doi: 10.1002/ jsfa.6328. [Epub ahead of print].

Kote S, Kote S, Nagesh L (2011). Effect of pomegranate juice on dental plaque microorganisms (*Streptococci* and *Lactobacilli*). *Anc. Sci. Life*. 31 (2): 49–51.

Koyama S, Cobb LJ, Mehta HH, Seeram NP, Heber D, Pantuck AJ, Cohen P (2010). Pomegranate extract induces apoptosis in human prostate cancer cells by modulation of the IGF-IGFBP axis. *Growth Horm. IGF Res*. 20 (1): 55–62.

Kumar S, Maheshwari KK, Singh V (2008). Central nervous system activity of acute administration of ethanol extract of *Punica granatum* L. seeds in mice. *Indian J. Exp. Biol*. 46 (12): 811–16.

Lai S, Zhou Q, Zhang Y, Shang J, Yu T (2009). Effects of pomegranate tannins on experimental gastric damages. *Zhongguo Zhongyao Zazhi*. 34 (10): 1290–94.

Larrosa M, González-Sarrías A, Yáñez-Gascón MJ, Selma MV, Azorín-Ortuño M, Toti S, Tomás-Barberán F,Dolara P, Espín JC (2010). Anti-inflammatory properties of a pomegranate extract and its metabolite urolithin-A in a colitis rat model and the effect of colon inflammation on phenolic metabolism. *J. Nutr. Biochem*. 21 (8): 717 – 25.

Leiva KP, Rubio J, Peralta F, Gonzales GF (2011). Effect of *Punica granatum* (pomegranate) on sperm production in male rats treated with lead acetate. *Toxicol. Mech. Methods*. 21 (6): 495–502.

Longtin R (2003). The pomegranate: nature's power fruit? *J. Natl. Cancer. Inst*. 95: 346–48.

Lynn A, Hamadeh H, Leung WC, Russell JM, Barker ME (2012). Effects of pomegranate juice supplementation on pulse wave velocity and blood pressure in healthy young and middle-aged men and women. *Plant Foods Hum Nutr*. 67 (3): 309–314.

Malik A, Afaq F, Sarfaraz S, Adhami VM, Syed DN, Mukhtar H (2005). Pomegranate fruit juice for chemo-prevention and chemotherapy of prostate cancer. *Proc. Natl. Acad. Sci. U S A.* 102 (41): 14813–18.

Malik A, Mukhtar H (2006). Prostate cancer prevention through pomegranate fruit. *Cell. Cycle.* 5 (4): 371–73.

Mastrodi SJ, Baroni FTR, de Oliveira BF, Dos Santos Dias CT (2012). Increased anti-oxidant content in juice enriched with dried extract of pomegranate (*Punica granatum*) Peel. *Plant. Foods Hum. Nutr.* 67 (1): 39–43.

Mattiello T, Trifirò E, Jotti GS, Pulcinelli FM (2009). Effects of pomegranate juice and extract polyphenols on platelet function. *J. Med. Food.* 12 (2): 334–39.

Mena P, Vegara S, Martí N, García-Viguera C, Saura D, Valero M (2013). Changes on indigenous microbiota, colour, bioactive compounds and anti-oxidant activity of pasteurised pomegranate juice. *Food Chem.* 141 (3): 2122-9.

Menezes SM, Cordeiro LN, Viana GS (2006). *Punica granatum* (pomegranate) extract is active against dental plaque. *J. Herb. Pharmacother.* 6 (2): 79–92.

Mo J, Panichayupakaranant P, Kaewnopparat N, Nitiruangjaras A, Reanmongkol W (2013a). Topical anti-inflammatory and analgesic activities of standardized pomegranate rind extract in comparison with its marker compound ellagic acid *in-vivo. J. Ethnopharmacol.* 148 (3): 901–908.

Mo J, Panichayupakaranant P, Kaewnopparat N, Songkro S, Reanmongkol W (2013b). Topical anti-inflammatory potential of standardized pomegranate rind extract and ellagic acid in contact dermatitis. *Phytother. Res.* 2013 Jul 19. doi: 10.1002/ptr.5039. [Epub ahead of print].

Mohan M, Patankar P, Ghadi P, Kasture S (2010a). Cardio-protective potential of *Punica granatum* extract in isoproterenol-induced myocardial infarction in Wistar rats. *J. Pharmacol. Pharmacother.* 1 (1): 32–37.

Mohan M, Waghulde H, Kasture S (2010b). Effect of pomegranate juice on Angiotensin II-induced hypertension in diabetic Wistar rats. Phytother. Res. 24 Suppl. 2: S233–34.

Moneam NMA, El Sharasky AS, Badreldin MM (1988). Oestrogen content of pomegranate seeds. *J. Chromatogr.* 438 (2): 438–42.

Monsefi M, Parvin F, Talaei-Khozani T (2012). Effects of pomegranate extracts on cartilage, bone and mesenchymal cells of mouse fetuses. *Br. J. Nutr.* 107 (5): 683–90.

Nagata M, Hidaka M, Sekiya H, Kawano Y, Yamasaki K, Okumura M, Arimori K (2007). Effects of pomegranate juice on human cytochrome P450 2C9 and tolbutamide pharmacokinetics in rats. *Drug Metab. Dispos.* 35 (2): 302–05.

Narr Ben C, Ayed N, Metche M (1996). Quantitative determination of the polyphenolic content of pomegranate peel. Z. Lebensm. *Unters. Forsch.* 203: 374–78.

Neyrinck AM, Van Hée VF, Bindels LB, De Backer F, Cani PD, Delzenne NM (2013). Polyphenol-rich extract of pomegranate peel alleviates tissue inflammation and hyper-cholesterolaemia in high-fat diet-induced obese mice: potential implication of the gut microbiota. *Br. J. Nutr.* 109 (5): 802–09.

Noda Y, Kaneyuki T, Mori A, Packer L (2002). Anti-oxidant activities of pomegranate fruit extract and its anthocyanidins: delphinidin, cyanidin, and pelargonidin. *J. Agric. Food Chem.* 50: 166–71.

Ouachrif A, Khalki H, Chaib S, Mountassir M, Aboufatima R, Farouk L, Benharraf A, Chait A (2012). Comparative study of the anti-inflammatory and anti-nociceptive effects of two varieties of *Punica granatum*. *Pharm. Biol.* 50 (4): 429–38.

Pai MB, Prashant GM, Murlikrishna KS, Shivakumar KM, Chandu GN (2010). Anti-fungal efficacy of *Punica granatum, Acacia nilotica, Cuminum cyminum and Foeniculum vulgare* on *Candida albicans*: an *in-vitro* study. *Indian J. Dent. Res.* 21 (3): 334–36.

Pirinccioglu M, Kizil G, Kizil M, Kanay Z, Ketani A (2012). The protective role of pomegranate juice against carbon tetrachloride-induced oxidative stress in rats. *Toxicol. Ind. Health.* 2012 Nov 16. [Epub ahead of print].

Prashant D, Asha MK, Amit A (2001). Anti-bacterial activity of *Punica granatum*. *Fitoterapia.* 72: 171–73.

Reddy MK, Gupta SK, Jacob MR, Khan SI, Ferreira D (2007). Anti-oxidant, anti-malarial and anti-microbial activities of tannin-rich fractions, ellagitannins and phenolic acids from *Punica granatum* L. *Planta. Med.* 73 (5): 461–67.

Rettig MB, Heber D, An J, Seeram NP, Rao JY, Liu H, Klatte T, Belldegrun A, Moro A, Henning SM, Mo D, Aronson WJ, Pantuck A (2008). Pomegranate extract inhibits androgen-independent prostate cancer growth through a nuclear factor-kappaB-dependent mechanism. *Mol. Cancer Ther.* 7 (9): 2662–71.

Ricci D, Giamperi L, Bucchini A, Fraternale D (2006). Anti-oxidant activity of *Punica granatum* fruits. *Fitoterapia.* 77 (4): 310–12.

Rocha A, Wang L, Penichet M, Martins-Green M (2012). Pomegranate juice and specific components inhibit cell and molecular processes critical for metastasis of breast cancer. *Breast Cancer Res. Treat.* 136 (3): 647–58.

Rock W, Rosenblat M, Miller-Lotan R, Levy AP, Elias M, Aviram M (2008). Consumption of wonderful variety pomegranate juice and extract by diabetic patients increases paraoxonase 1 association with high-density lipoprotein and stimulates its catalytic activities. *J. Agric. Food Chem.* 56 (18): 8704–13.

Rosenblat M, Aviram M (2011). Pomegranate juice protects macrophages from triglyceride accumulation: inhibitory effect on DGAT1 activity and on triglyceride biosynthesis. *Ann. Nutr. Metab.* 58 (1): 1–9.

Rosenblat M, Hayek T, Aviram M (2006). Anti-oxidative effects of pomegranate juice (PJ) consumption by diabetic patients on serum and on macrophages. *Atherosclerosis.* 187 (2): 363–71.

Rosenblat M, Volkova N, Attias J, Mahamid R, Aviram M (2010a). Consumption of polyphenolic-rich beverages (mostly pomegranate and black currant juices) by healthy subjects for a short term increased serum antioxidant status, and the serum's ability to attenuate macrophage cholesterol accumulation. *Food Funct.* 1 (1): 99–109.

Rosenblat M, Volkova N, Aviram M (2010b). Pomegranate juice (PJ) consumption anti-oxidative properties on mouse macrophages, but not PJ beneficial effects on macrophage cholesterol and triglyceride metabolism, are mediated via PJ-induced stimulation of macrophage PON2. *Atherosclerosis.* 212 (1): 86–92.

Rosenblat M, Volkova N, Aviram M (2013). Pomegranate phytosterol (β-sitosterol) and polyphenolic antioxidant (punicalagin) addition to statin, significantly protected against macrophage foam cells formation. *Atherosclerosis.* 226 (1): 110–17.

Rosillo MA, Sánchez-Hidalgo M, Cárdeno A, Aparicio-Soto M, Sánchez-Fidalgo S, Villegas I, de la Lastra CA (2012). Dietary supplementation of an ellagic acid-enriched pomegranate extract attenuates chronic colonic inflammation in rats. *Pharmacol. Res.* 66 (3): 235–42.

Sarkaki A, Rezaiei M, Gharib Naseri M, Rafieirad M (2013). Improving active and passive avoidance memories deficits due to permanent cerebral ischemia by pomegranate seed extract in female rats. *Malays. J. Med. Sci.* 20 (2): 25–34.

Sartippour MR, Seeram NP, Rao JY, Moro A, Harris DM, Henning SM, Firouzi A, Rettig MB, Aronson WJ, Pantuck AJ, Heber D (2008). Ellagitannin-rich pomegranate extract inhibits angiogenesis in prostate cancer *in-vitro* and *in-vivo. Int. J. Oncol.* 32 (2): 475–80.

Sastravaha G, Gassmann G, Sangtherapitikul P, *et al.* (2005). Adjunctive periodontal treatment with *Centella asiatica* and *Punica granatum* extracts in supportive periodontal therapy. *J. Int. Acad. Periodontol.* 7 (3): 70–79.

Schubert SY, Lansky EP, Neeman I (1999). Anti-oxidant and eicosanoid enzyme inhibition properties of pomegranate seed oil and fermented juice flavonoids. *J. Ethnopharmacol.* 66: 11–17.

Schwartz E, Tzulker R, Glazer I, Bar-Ya'akov I, Wiesman Z, Tripler E, Bar-Ilan I, Fromm H, Borochov-Neori H, Holland D, Amir R (2009). Environmental conditions affect the color, taste, and anti-oxidant capacity of 11 pomegranate accessions' fruits. *J. Agric. Food Chem.* 57 (19): 9197–209.

Seeram NP, Adams LS, Henning SM, Niu Y, Zhang Y, Nair MG, Heber D (2005). In vitro anti-proliferative, apoptotic and anti-oxidant activities of punicalagin, ellagic acid and a total pomegranate tannin extract are enhanced in combination with other polyphenols as found in pomegranate juice. *J. Nutr. Biochem.* 16 (6): 360–67.

Seeram NP, Aviram M, Zhang Y, Henning SM, Feng L, Dreher M, Heber D (2008). Comparison of anti-oxidant potency of commonly consumed polyphenol-rich beverages in the United States. *J. Agric. Food Chem.* 56 (4): 1415–22.

Sestili P, Martinelli C, Ricci D, Fraternale D, Bucchini A, Giamperi L, Curcio R, Piccoli G, Stocchi V (2007). Cytoprotective effect of preparations from various parts of

Punica granatum L. fruits in oxidatively injured mammalian cells in comparison with their antioxidant capacity in cell free systems. *Pharmacol. Res.* 56 (1): 18–26.

Sharaf A, Nigm SAR (1964). The oestrogenic activity of pomegranate seed oil. *J. Endocrinol.* 29: 91–92.

Shema-Didi L, Sela S, Ore L, Shapiro G, Geron R, Moshe G, Kristal B (2012). One year of pomegranate juice intake decreases oxidative stress, inflammation, and incidence of infections in hemodialysis patients: a randomized placebo-controlled trial. *Free Radic. Biol. Med.* 53 (2): 297–304.

Shiner M, Fuhrman B, Aviram M (2007). Macrophage paraoxonase 2 (PON2) expression is up-regulated by pomegranate juice phenolic anti-oxidants via PPAR gamma and AP-1 pathway activation. *Atherosclerosis.* 195 (2): 313–21.

Shirode AB, Kovvuru P, Chittur SV, Henning SM, Heber D, Reliene R (2013). Antiproliferative effects of pomegranate extract in MCF-7 breast cancer cells are associated with reduced DNA repair gene expression and induction of double strand breaks. *Mol. Carcinog.* 2013. Jan. 28. doi: 10.1002/mc.21995. [Epub ahead of print].

Shukla M, Gupta K, Rasheed Z, Khan KA, Haqqi TM (2008). Consumption of hydrolyzable tannins-rich pomegranate extract suppresses inflammation and joint damage in rheumatoid arthritis. *Nutrition.* 24 (7-8): 733–43.

Sineh Sepehr K, Baradaran B, Mazandarani M, Khori V, Shahneh FZ (2012). Studies on the cytotoxic activities of *Punica granatum* L. var. spinosa (apple punice) extract on prostate cell line by induction of apoptosis. *ISRN pharm.* 2012: 547942.

Singh K, Jaggi AS, Singh N (2009). Exploring the ameliorative potential of *Punica granatum* in dextran sulfate sodium induced ulcerative colitis in mice. *Phytother. Res.* 23 (11): 1565–74.

Singh RP, Chidambara Murthy KN, Jayaprakasha GK (2002). Studies on the antioxidant activity of pomegranate (*Punica granatum*) peel and seed extracts using in vitro models. *J. Agric. Food Chem.* 50 (1): 81–86.

Spilmont M, Léotoing L, Davicco MJ, Lebecque P, Mercier S, Miot-Noirault E, Pilet P, Rios L, Wittrant Y, Coxam V (2013). Pomegranate seed oil prevents bone loss in a mice model of osteoporosis, through osteoblastic stimulation, osteoclastic inhibition and decreased inflammatory status. *J. Nutr. Biochem.* pii: S0955-2863 (13) 00095-8.

Stowe CB (2011). The effects of pomegranate juice consumption on blood pressure and cardiovascular health. Complement. *Ther. Clin. Pract.* 17 (2): 113–15.

Sturgeon SR, Ronnenberg AG (2010). Pomegranate and breast cancer: possible mechanisms of prevention. *Nutr. Rev.* 68 (2): 122–28.

Su X, Sangster MY, D'Souza DH (2010). *In-vitro* effects of pomegranate juice and pomegranate polyphenols on foodborne viral surrogates. *Foodborne Pathog. Dis.* 7 (12): 1473–79.

Sumner MD, Elliott-Eller M, Weidner G, Daubenmier JJ, Chew MH, Marlin R, Raisin CJ, Ornish D (2005). Effects of pomegranate juice consumption on myocardial perfusion in patients with coronary heart disease. *Am. J. Cardiol.* 96 (6): 810–14.

Syed DN, Malik A, Hadi N, Sarfaraz S, Afaq F, Mukhtar H (2006). Photo-chemopreventive effect of pomegranate fruit extract on UVA-mediated activation of cellular pathways in normal human epidermal keratinocytes. *Photochem. Photobiol.* 82 (2): 398–405.

Toklu HZ, Dumlu MU, Sehirli O, Ercan F, Gedik N, Gökmen V, Sener G (2007). Pomegranate peel extract prevents liver fibrosis in biliary-obstructed rats. J. *Pharm. Pharmacol.* 59 (9): 1287–95.

Trombold JR, Barnes JN, Critchley L, Coyle EF (2010). Ellagitannin consumption improves strength recovery 2-3 d after eccentric exercise. *Med. Sci. Sports Exerc.* 42 (3): 493–98.

Trombold JR, Reinfeld AS, Casler JR, Coyle EF (2011). The effect of pomegranate juice supplementation on strength and soreness after eccentric exercise. *J. Strength Cond. Res.* 25 (7): 1782–88.

Tugcu V, Kemahli E, Ozbek E, Arinci YV, Uhri M, Erturkuner P, Metin G, Seckin I, Karaca C, Ipekoglu N, Altug T, Cekmen MB, Tasci AI (2008). Protective effect of a potent anti-oxidant, pomegranate juice, in the kidney of rats with nephrolithiasis induced by ethylene glycol. *J. Endourol.* 22 (12): 2723–31.

Türk G, Sönmez M, Aydin M, Yüce A, Gür S, Yüksel M, Aksu EH, Aksoy H (2008). Effects of pomegranate juice consumption on sperm quality, spermatogenic cell density, antioxidant activity and testosterone level in male rats. *Clin. Nutr.* 27 (2): 289–96.

Tzulker R, Glazer I, Bar-Ilan I, Holland D, Aviram M, Amir R (2007). Anti-oxidant activity, polyphenol content, and related compounds in different fruit juices and homogenates prepared from 29 different pomegranate accessions. *J. Agric. Food Chem.* 55 (23): 9559–70.

USDA, National Nutrition Data Base for standard reference, Release-26, NDB No-09286.

Vasconcelos LC, Sampaio FC, Sampaio MC, Pereira Mdo S, Higino JS, Peixoto MH (2006). Minimum inhibitory concentration of adherence of *Punica granatum* Linn (pomegranate) gel against *S. mutans, S. mitis* and *C. albicans. Braz. Dent. J.* 17 (3): 223–27.

Vegara S, Mena P, Martí N, Saura D, Valero M (2013). Approaches to understanding the contribution of anthocyanins to the anti-oxidant capacity of pasteurized pomegranate juices. *Food Chem.* 141 (3): 1630–36.

Vicinanza R, Zhang Y, Henning SM, Heber D (2013). Pomegranate juice metabolites, ellagic acid and urolithin A, synergistically inhibit androgen-independent prostate cancer cell growth via distinct effects on cell cycle control and apoptosis. Evid. Based Complement. *Alternat. Med.* 2013:247504. doi: 10.1155/2013/247504. Epub 2013 Apr 24.

Viladomiu M, Hontecillas R, Lu P, Bassaganya-Riera J (2013). Preventive and prophylactic mechanisms of action of pomegranate bioactive constituents. Evid. Based Complement. *Alternat. Med.* 2013:789764. doi: 10.1155/2013/789764. Epub 2013 Apr 30.

Viuda-Martos M, Fernández-Lóaez J, Pérez-álvarez JA (2010). Pomegranate and its many functional components as related to human health: a review. *Comprehensive Reviews in Food Science and Food Safety.* 9 (6): 635–54.

Vroegrijk IO, van Diepen JA, van den Berg S, Westbroek I, Keizer H, Gambelli L, Hontecillas R,Bassaganya-Riera J, Zondag GC, Romijn JA, Havekes LM, Voshol PJ (2011). Pomegranate seed oil, a rich source of punicic acid, prevents diet-induced obesity and insulin resistance in mice. *Food Chem Toxicol.* 49 (6): 1426–30.

Wang L, Alcon A, Yuan H, Ho J, Li QJ, Martins-Green M (2011). Cellular and molecular mechanisms of pomegranate juice-induced anti-metastatic effect on prostate cancer cells. *Integr. Biol. (Camb).* 3 (7): 742–54.

Wang L, Ho J, Glackin C, Martins-Green M (2012). Specific pomegranate juice components as potential inhibitors of prostate cancer metastasis. *Transl. Oncol.* 5 (5): 344–55.

Yazici ZM, Meric A, Midi A, Arýnc YV, Kahya V, Hafýz G (2012). Reduction of cisplatin ototoxicity in rats by oral administration of pomegranate extract. *Eur. Arch. Otorhinolaryngol.* 269 (1): 45–52.

Yildirim NC, Kandemir FM, Ceribasi S, Ozkaraca M, Benzer F (2013). Pomegranate seed extract attenuates chemotherapy-induced liver damage in an experimental model of rabbits. *Cell. Mol. Biol.* (Noisy-le-grand). 59 Suppl: OL 1842–47.

Yoshimura Y, Nishii S, Zaima N, Moriyama T, Kawamura Y (2013). Ellagic acid improves hepatic steatosis and serum lipid composition through reduction of serum resistin levels and transcriptional activation of hepatic ppara in obese, diabetic KK-A(y) mice. *Biochem. Biophys. Res. Commun.* 434 (3): 486–91.

Chapter 11
Saffron
(*Crocus sativus*)

Saffron, an expensive spice comes from the French word 'Safran' originally derived from the Latin word Safranum meaning yellow and well known as Red Gold in some countries. The flower of *C. sativa* is light purple, but it is the thread-like reddish-colored stigma of the flower that is valued both as a spice and as a natural colorant. It is widely cultivated in Iran, India and Greece. It takes about 36,000 flowers to yield just 1 lb. of stigmas (Wani *et al.*, 2011). Over, 2,00,000 dried stigmas (obtained from about 70,000 flowers) yield 500 g of pure saffron (USDA, 2009). Commercial

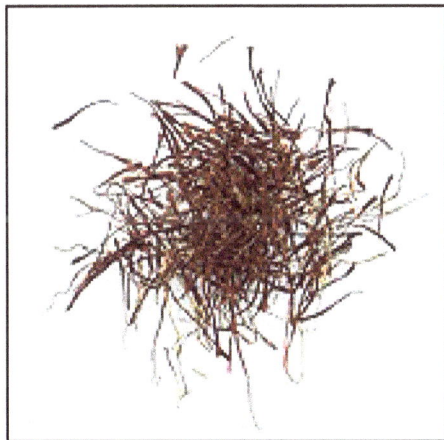

C. sativus Blossom with Crimson Stigmas.

saffron comprises the dried red stigma with a small portion of the yellowish style attached (Zargari, 1990).

Chemical Constituents

Saffron contains more than 150 volatile agents–safranal and aroma-yielding compounds terpenes, terpene alcohol, and their esters, bitter principles (*e.g.*, picrocrocin) and dye materials (*e.g.*, crocetin and its glycoside, crocin) (Ríos *et al.*, 1996). Picrocrocin ($C_{16}H_{26}O_7$) comprising up to 4 per cent of dry saffron is the union of an aldehyde sub-element known as safranal and a carbohydrate. Safranal is less bitter than picrocrocin and comprises up to 70 per cent of dry saffron's volatile fraction. The non-volatile active components in saffron include carotenoids such as zeaxanthin, lycopene, and various α- and β-carotenes (Liakopulou-Kyriakides and Kyriakides, 2002). Crocin (8,8-diapo-8,8-carotenoic acid) is a trans-crocetin di-(β-D-gentiobiosyl) ester and responsible for the color. The Characteristic components of saffron are crocin, picrocrocin and safranal (Srivastava *et al.*, 2010).

The composition of ether fractions of the perianth, stamen, and corm of *C. sativus* is different from each other, but lauric acid, hexadecanoic acid, 4-hydroxydihydro-2(3H)-furanone, and stigmasterol were the common constituents shared by all the three fractions. The ether fraction of stamen displayed the strongest anti-fungal and cyto-toxic activities, whereas ether fractions of both stamen and perianth exhibited significant anti-oxidant activities (Gohari *et al.*, 2013; Zheng *et al.*, 2011).

Montoro *et al.* (2012) observed that the high variety of glycosylated flavonoids found in the metabolic profiles could give value to *C. sativus* petals, stamens and entire flowers. Waste products obtained during saffron production, could represent an interesting source of phenolic compounds, with respect to the high variety of compounds and their free radical scavenging activity.

Anti-oxidant Activity

Recently, identification of new sources of safe natural anti-oxidants for the food industry has been in focus. Saffron exhibited maximum protective effects at a dose of 400mg/kg, which could be due to maintenance of the redox status of the cell reinforcing its role as an anti-oxidant (Sachdeva *et al.*, 2012). Ordoudi *et al.* (2009) reported that saffron extracts exhibit a remarkable intracellular anti-oxidant activity that cannot be revealed using assays repeatedly applied to the evaluation of phenolic-type anti-oxidants.

Gallic acid and pyrogallol, the two active components in the methanol extract of saffron stigma were responsible for its anti-oxidant activity as assessed by the free radical scavenging and ferric reducing power (Karimi *et al.*, 2010). The numerous applications of saffron as an anti-oxidant and anti-cancer agent are due to its secondary metabolites and their derivatives–safranal, crocins, crocetin, dimethyl-crocetin (Kanakis *et al.*, 2007a; 2007b) which have the ability to bind and stabilize the DNA molecule (Kanakis *et al.*, 2007c; 2009). Its capability to suppress the geno-toxicity caused by methyl methane-sulfonate has been demonstrated *in-vivo* (Hosseinzadeh and Sadeghnia, 2007; Hosseinzadeh *et al.*, 2008). The methanol extract of *C. sativus* cultivated in Greece exhibited high anti-oxidant activity which was attributed to crocin and safranal. Crocin showed high radical scavenging activity (50 per cent and 65 per cent for 500 and 1,000 ppm solution in methanol, respectively), followed by safranal (34 per cent for 500 ppm solution) probably due to its high ability to donate a hydrogen atom to the DPPH radical. Thus, Assimopoulou *et al.* (2005) recommended that saffron grown in Greece can be used promisingly in functional foods, drinks, pharmaceutical and cosmetic preparations due to its anti-oxidant and also probably for its anti-aging activity.

In ischemic rats, safranal also exerts a protective activity against oxidative damage in skeletal muscle (Hosseinzadeh *et al.,* 2009) and cerebral tissues (Hosseinzadeh and Sadeghnia, 2005; Hosseinzadeh and Talebzadeh, 2005; Hosseinzadeh *et al.,* 2012). Vakili *et al.* (2012) indicated that crocin has protective effects against ischemic reperfusion injury and cerebral edema in a rat model of stroke. These effects of crocin may have been exerted primarily by suppression of the production of free radicals and increased anti-oxidant enzyme activity.

It has also been reported to have anti-convulsion activity in mice with chronic attacks (Hosseinzadeh and Talebzadeh, 2005). Supplementation with saffron extract protected photo-receptors from death by exposure to bright light (Maccarone *et al.,* 2008) and improved retinal function in early age-related macular degeneration (Falsini *et al.,* 2010). Fernández-Sánchez *et al.* (2012) also demonstrated that administration of safranal to homozygous P23H line-3 rats preserves both photoreceptor morphology and number. Electro-retinographic recordings showed higher a- and b-wave amplitudes under both photopic and scotopic conditions in safranal-treated animals while preserving the capillary network (Ochiai *et al.,* 2007). Thus, a number of studies have proved that safranal is a powerful anti-oxidant that fights oxidative stress in neurons.

Crocin has hepato-protective effects due to its anti-oxidant activity. DZN plus 25 mg crocin decreased significantly malondialdehyde (MDA) level. DZN also induced apoptosis through activation of caspases-9 and -3 and increasing Bax/Bcl-2 ratio. But crocin attenuated the activation of caspases and reduced the Bax/Bcl-2 ratio. Thus, Lari *et al.* (2013a) reported that subacute exposure to DZN induces oxidative stress-mediated apoptosis and crocin may reduce DZN-induced hepato-toxicity.

Goyal *et al.* (2010) observed crocin at the dose of 20mg/kg/day to modulate significantly hemo-dynamic and anti-oxidant derangements in isoproterenol (ISO)-induced MI, which was reconfirmed by histo-pathological and ultra-structural examinations. The effect of crocin at the dose of 20 mg/kg/day was more pronounced than that of 5 and 10 mg/kg/day. They opined from these results that crocin may have cardio-protective effect in ISO-induced cardiac toxicity through modulation of oxidative stress in such a way that maintains the redox status of the cell.

Hepatic ischemia-reperfusion (IR) injury is a common clinical problem and ROS may be a contributing factor for the same. Administration of ethanol extract of saffron attenuated the carbonylation level of several chaperone proteins. IR-induced ER stress and protein ubiquitination is alleviated by saffron extract leading to cell apoptosis (Pan *et al.,* 2013).

Acrylamide (ACR) increases intra-cellular reactive oxygen species (ROS) in cells which play an important role in ACR cyto-toxicity. Pre-treatment of cells with 10–50 µM crocin before ACR treatment significantly attenuated ACR cyto-toxicity in a dose-dependent manner. Crocin inhibited the down-regulation of Bcl-2 and the up-regulation of Bax and decreased apoptosis in treated cells. Also, crocin inhibited ROS generation in cells exposed to ACR (Mehri *et al.,* 2012).

Therapeutic Uses

Saffron is used as an important spice and food colorant in various parts of the world. It possesses anti-oxidative, anti-inflammatory, oxytocic, anti-carcinogenic, exhilarant, anti-depressant, anti-convulsant, anti-tussive and cardiovascular protective effects (Srivastava *et al.,* 2010). In addition, saffron increases the bioavailability and enhances absorption of other drugs (Javadi *et al.,* 2013). In traditional folk medicine, saffron is recommended as an aphrodisiac agent (Madan *et al.,* 1996). In Persian traditional medicine saffron is used for depression (Hosseinzadeh and Younesi, 2002; Karimi *et al.,* 2001; Akhondzadeh *et al.,* 2004; Noorbala *et al.,* 2005). In traditional Indian medicine pistils of saffron are generally used as analgesics and cardio-protective agents, as well as in the treatment of various kinds of mental illnesses and for recovery of ischemia/reperfusion injury and learning and memory in rats. It has anti-cancer and memory enhancing potential (Ríos *et al.,* 1996; Abe and Saito, 2000; Abdullaev, 2002).

Saffron and CVD

Administration of 50 mg saffron dissolved in 100 ml of milk twice a day revealed a significant decrease in susceptibility of lipoprotein oxidation in patients with coronary artery disease indicating its potential as an anti-oxidant (Verma and Bordia, 1998). Both crocetin (a structural component of crocin) and crocin were found to be effective cardio-protective agents. He *et al.* (2007) reported that crocetin could inhibit the formation of atherosclerosis in quails, which might be related to its hypo-lipidemic and anti-oxidative properties. Similarly, Yang *et al.* (2008) demonstrated that crocetin dose-dependently inhibited platelet aggregation induced by ADP, collagen, but not by arachidonic acid. It significantly attenuated dense granule release without altering the platelets adhesion to collagen or cyclic AMP level. Pre-treatment with crocetin partially inhibited Ca^{+2} mobilization *via* reducing both intra-cellular Ca^{+2} release and extra-cellular Ca^{+2} influx.

Crocetin has been found to block inflammatory cascades by inhibiting production of reactive oxygen species and preserving total superoxide dismutase activity to ameliorate the cardiac injury caused by hemorrhage/resuscitation (Yan *et al.,* 2010). Yoshino *et al.* (2011) confirmed that crocetin significantly reduced oxidative stress in the isolated brain as demonstrated by *in-vitro* and *ex-vivo* electron spin resonance analysis. Shen and Qian (2006) found that the cardio-protective effect of crocetin is related to modulation of endogenous anti-oxidant enzymatic activities. Comparison of crocetin with captopril indicated that anti-oxidant activity is an important factor in the therapy of cardiac hypertrophy, but its effects may be limited as an anti-oxidant only.

Crocin was found to reduce the serum TC, TG, LDL-C, inhibit the formation of aortic plaque, reduce malondialdehyde and inhibit the descending of NO in serum. Crocin could inhibit the extracellular Ca^{2+} influx and release of intracellular Ca2+ stores in endoplasmic reticulum (He *et al.,* 2004). It inhibited the proliferation of bovine aortic smooth muscle cells induced by Ox-LDL that plays an important role in the initiation and progression of atherosclerosis (He *et al.,* 2005). The hypo-lipidemic effect of crocin is also attributed to the inhibition of pancreatic lipase, leading to the malabsorption of fat and cholesterol (Sheng *et al.,* 2006).

Both saffron and crocin were effective in decreasing the elevated levels of TG, TC, alkaline phosphatase, aspartate transaminase, alanine aminotransferase, malondialdehyde, glutathione peroxidase enzyme activity, total glutathione and oxidized glutathione and thiobarbituric acid reactive species (TBARS) in serum; and in increasing superoxide dismutase, catalase, ferric reducing/anti-oxidant power, and total sulfhydryl values in liver tissue. But saffron was found to be superior to crocin indicating the involvement of other potential constituents of saffron apart from crocin for its synergistic behavior of quenching the free radicals and ameliorating the damages of hyper-lipidemia (Asdaq and Inamdar, 2010).

Pre-treatment with saffron (20, 40, 80 and 160 mg/kg IP) or safranal (0.025, 0.050, 0.075 ml/kg IP) for 8 days, significantly decreased the serum LDH, CK-MB and myocardial lipid peroxidation. The histological changes revealed reversal of myocardial injury with ISO administration, back to near normal and preservation of tissue architecture with saffron or safranal pre-treatment (Mehdizadeh *et al.,* 2013; Joukar *et al.,* 2010).

Umigai *et al.* (2012) found crocetin to significantly suppress vascular endothelial growth factor (VEGF)-induced tube formation by human umbilical vein endothelial cells (HUVECs) and migration of human retinal micro-vascular endothelial cells (HRMECs). It also significantly inhibited phosphorylation of p38 and protected VE-cadherin expression indicating that crocetin suppresses the VEGF-induced angiogenesis.

Safranal seems to be more important than crocin for lowering blood pressure of rats though the aqueous extract of saffron stigma, safranal and crocin reduced the mean arterial blood pressure in normo-tensive and hyper-tensive anaesthetized rats in a dose-dependent manner. Further, the hypo-tensive properties of the aqueous extract appear to be attributable, in part, to the actions of two major constitutes of this plant, crocin and safranal (Imenshahidi *et al.,* 2010). Significant reduction in systolic blood pressure (SBP) and increase in heart rate (HR) occur in sub-chronic toxicity of diazinon (DZN). Concurrent administration of crocin and DZN restored the reduction of SBP and the elevation of HR induced by DZN in rat. But crocin alone did not have any effect on SBP and HR (Razavi *et al.,* 2013). Administration of 50 mg/g of aqueous extract reduced the rat blood pressure from 133.5 ± 3.9 to 117 ± 2.1 mmHg. This reduction could either be due to the effect of the petals' extracts on the heart itself/ total peripheral resistance, or both. The effect of extracts on peripheral resistance seems to be more important. In the rat isolated vas deferens, contractile responses to electrical field stimulation (EFS) were decreased by the extracts of the petals. The ethanol extract induced greater changes in EFS in the rat isolated *vas deferens* and guinea- pig ileum than the aqueous extract (Fatehi *et al.,* 2003).

Disseminated intra-vascular coagulation (DIC), a life-threatening secondary complication in several diseases, is characterized by large amounts of thrombin that lead to fibrin deposition and micro-thrombus formation throughout the micro-circulation. Recent *in-vitro* studies suggest that crocin, crocetin or safranal, have anti-thrombotic properties, especially anti-Xa activity. Lari *et al.* (2013b) showed that crocetin plays a preventive anti-thrombotic role *in-vivo* and recommended further investigation to confirm the possibility of developing crocetin-based DIC treatment modalities.

Saffron and Osteo-arthritis

Articular cartilage degeneration and inflammation are the hallmark of progressive arthritis and is the leading cause of disability in 10–15 per cent of middle aged individuals across the world. Cartilage and synovium are mainly degraded by either enzymatic or non-enzymatic ways. Matrix metallo-proteinases (MMPs), hyaluronidases and aggrecanases are the enzymatic mediators while inflammatory cytokines and ROS being non-enzymatic mediators. Although several drugs have been used to treat arthritis, numerous reports describe the side effects of these drugs that may turn fatal. Crocin effectively neutralized the augmented serum levels of enzymatic (MMP-13, MMP-3 and MMP-9 and HAases) and non-enzymatic (TNF- α, IL-1β, NF-κB, IL-6, COX-2, PGE(2) and ROS) inflammatory mediators. Further, Crocin re-established the arthritis altered anti-oxidant status of the system (GSH, SOD, CAT and GST). It also protected the bone resorption by inhibiting the elevated levels of bone joint exoglycosidases, cathepsin-D and tartrate resistant acid phosphatases (Hemshekhar *et al.,* 2012).

Interleukin-1β-mediated production of matrix metallo-proteinases (MMPs) plays a pivotal role in the process of osteoarthritis. While IL-1β markedly up-regulated the expression of MMP-1, -3 and -13 in chondrocytes, co-incubation with crocin inhibited this in a dose-dependent manner. Crocin also inhibited IL-1β-induced activation of the nuclear factor kappa B pathway through suppressing degradation of inhibitory-kappa-B-α. Crocin ameliorated cartilage degeneration and expression of the MMP-1, -3 and -13 genes in cartilage significantly. Thus the anti-inflammatory activity of crocin may be of potential value in the prevention and treatment of osteoarthritis (Ding *et al.,* 2013).

Saffron and Cancer

Saffron extract and some of its ingredients displayed a dose-dependent inhibitory activity against different types of human malignant cells *in- vitro*. HeLa cells were more susceptible to saffron than other tested cells. It is suggested that saffron or its carotenoid components might be used as potential cancer chemo-preventive agents (Abdullaev, 2002; Abdullaev *et al.,* 2002; 2003; 2004). The aqueous extract of saffron showed inhibitory effects dose dependently on the growth of both TCC 5637 and normal L929 cell lines (Feizzadeh *et al.,* 2008). According to Nair *et al.* (1995) saffron (dimethyl-crocin) disrupts DNA-protein interactions *e.g.* topo-isomerases II which are important for cellular DNA synthesis. Joukar *et al.* (2013) reported that pre-treatment with saffron, especially at the dosage of 100 mg/kg/day attenuates the susceptibility and incidence of fatal ventricular arrhythmia during the reperfusion period in the rat. This protective effect is apparently mediated through reduction of electrical conductivity and prolonging the action potential duration. Saffron has low biochemical toxic effects on animals (Deng *et al.*, 2002).

It is quite well established that carotenoids possess anti-carcinogenic, anti-mutagenic and immune-modulating effects. Crocin exhibits anti-tumor activity against many human tumors, such as colorectal, pancreatic, and bladder cancer (Bakshi *et al.*, 2010). Chryssanthi *et al.* (2011) showed that crocetin, the main metabolite of crocin, inhibits MDA-MB-231 cell invasiveness *via* down-regulation of MMP expression. Cells treated with crocin exhibited wide cytoplasmic vacuole-like areas, reduced cytoplasm, cell shrinkage and pyknotic nuclei, suggesting apoptosis induction (Escribano *et al.*, 1996). Rezaee *et al.* (2013) also reported that mild cyto-toxic effects on a leukemia cell line exhibited by crocin might be mediated through the increase of DNA fragmentation.

Hoshyar *et al.* (2013) showed the anti-cancer property of crocin against chemical-induced gastric and breast cancer in rats. Crocin revealed a dose- and time-dependent cyto-toxic effect against an AGS cell line, as determined by 3-(4,5-Dimethylthiazol-2-yl)-2,5-diphenyltetrazolium bromide assay. Crocin-induced apoptosis was evidenced by flow cytometry and caspase activity. The increased sub-G1 population and activated caspases in the treated AGS cells confirmed its anti-cancer effect. It has been reported that carotenoids from saffron were effective in inhibiting the proliferation of HL-60 cells (Tarantilis *et al.*, 1994), which may be mediated by the induction of apoptosis and cell cycle arrest and the regulation of Bcl-2 and Bax expression (Sun *et al.*, 2013).

Crocetin is a potential anti-cancer agent, which may be used as a chemo-therapeutic drug or as a chemo-sensitizer for vincristine. Crocetin proves to scavenge free radical and plays an important role in cellular function. Tumor incidence and histopathological studies prove crocetin is a potent anti-tumour agent (Magesh *et al.*, 2006). Proliferation of cervical cancer cell line HeLa, non-small cell lung cancer cell line A549 and ovarian cancer cell line SKOV3 were significantly inhibited by crocetin (60-240 µmol/L) for 48 h in a concentration-dependent manner and at 240 µmol/L significantly induced cell cycle arrest through p53-dependent and independent mechanisms accompanied with p21(WAF1/Cip1) induction (Zhong *et al.*, 2011).

Evidence is now available to indicate that saffron possesses anti-cancer activity against a wide spectrum of tumors, such as leukemia, ovarian carcinoma, colon adeno-carcinoma, rhabdomyosarcoma, papilloma, squamous cell carcinoma, and soft tissue sarcoma. Li *et al.* (2013) demonstrated that a combination of crocin and cisplatin has a strong killing effect on osteosarcoma cells and suppresses the ability of invasion of MG63 and OS732 cells which might be related to up-regulate the expression of caspase-3 and caspase-8.

Bajbouj *et al.* (2012) showed that saffron could induce DNA-damage and apoptosis in HCT116 colorectal cancer cells. However, autophagy delayed the induction of apoptosis in HCT116 p53 cells.

Breast Cancer

Mousavi *et al.* (2009) observed saffron extract to decrease cell viability in MCF-7 cells in a concentration and time dependent manner. Analysis of DNA fragmentation by flow cytometry showed apoptotic cell death in MCF-7 cell treated with saffron extract. Saffron-induced apoptosis could be inhibited by pan-caspase inhibitors, indicating caspase-dependent pathway that was induced by saffronin MCF-7 cells. Bax protein expression was also increased in saffron-treated cells. Thus it was confirmed that saffron exerts pro-apoptotic effects in a breast cancer-derived cell line and could be considered as a potential chemo-therapeutic agent in breast cancer. The same team of researchers (Mousavi *et al.*, 2011) found crocin and its liposomes particularly safranal (Malaekeh-Nikouei *et al.*, 2013) to cause cell death in HeLa and MCF-7 cells, in which liposomal encapsulation improved cyto-toxic effects.

Liver Cancer

Amin *et al.* (2011) provided evidence that saffron exerts a significant chemo-preventive effect against liver cancer in diethyl-nitrosamine (DEN)-induced liver cancer in rats through inhibition of cell proliferation and induction of apoptosis. Saffron protects rat liver from cancer *via* modulating oxidative damage and suppressing inflammatory response. Tavakkol-Afshari *et al.* (2008) also observed that saffron induced a sub-G1 peak in flow cytometry histogram of treated cells indicating apoptotic cell death is involved in saffron toxicity which was independent of ROS production. It was concluded that saffron could cause cell death in HeLa and HepG2 cells, in which apoptosis or programmed cell death plays an important role in hepato-carcinoma. Noureini and Wink (2012) showed that telomerase activity of HepG2 cells decreases after treatment with crocin, which is probably caused by down-regulation of the expression of the catalytic subunit of the enzyme.

Pancreatic Cancer

Dhar *et al.* (2009) indicated that crocetin significantly affects pancreatic cancer growth both *in-vitro* and *in-vivo*. As indicated by Bax/Bcl-2 ratio significantly stimulated both *in-vitro* pancreatic cancer cells and *in-vivo* athymic nude mice tumor apoptosis.

Lung Cancer

Samarghandian *et al.* (2010) observed that although higher concentrations of saffron are safe for L929, the extract exerts pro-apoptotic effects in a lung cancer-derived cell line. But later the authors (Samarghandian *et al.* (2011) ascertained that saffron could decrease the cell viability in the malignant cells and apoptosis of the A549 cells in a concentration and time-dependent manner as determined by flow cytometry histogram of treated cells. Hence saffron could also be considered as a promising chemo-therapeutic agent in treatment of lung cancer in future.

Prostate Cancer

Samarghandian and Shabestari (2013) demonstrated that the prostate cancer cell line to be highly sensitive to safranal-mediated growth inhibition and apoptotic cell death although the molecular mechanisms of safranal action are not clear. Both saffron extract and crocin were effective in prostate cancer. They inhibited cell proliferation and arrested cell cycle progression, by inducing apoptosis in prostate gland. Interestingly, non-malignant cells were not affected (D'Alessandro *et al.,* 2013; Aung *et al.,* 2007). Thus saffron extract, safranal and crocin can be potentially used as chemo-preventive as well as chemo-therapeutic agents for prostate cancer.

Skin Cancer

Saffron given both before and after the induction of skin carcinogenesis exhibited a beneficial effect as per the standard histological examination of skin in mice. Saffron ingestion inhibited the formation of skin papillomas in animals and simultaneously reduced their size in mice if treated early. Inhibition of DMBA-induced skin carcinoma may be due, at least in part, to the induction of cellular defense systems (Das *et al.,* 2010).

Rezaee *et al.* (2013) demonstrated that 48-h crocin treatment at 500 µM, significantly reduced the viability human T-cell leukemia cell line, MOLT-4 exhibiting mild cytotoxic effects on a leukemia cell line which might be mediated through the increase of DNA fragmentation.

Based on the research studies Zhang *et al.* (2013) confirm that saffron possesses free radical-scavenging properties and anti-tumor activities. Significant cancer chemo-preventive effects have been shown in both *in-vitro* and *in-vivo* models. Hence, saffron and its ingredients could be considered as a promising candidate for clinical anti-cancer trials.

Saffron and Diabetes

Crocetin was also demonstrated to alleviate free fatty acid (FFA)-induced insulin insensitivity and dys-regulated mRNA expression of adiponectin, TNF-alpha and leptin in primary cultured rat adipocytes. These findings suggest the possibility of crocetin treatment as a preventive strategy of insulin resistance and related diseases. The favorable impact on adiponectin, TNF-alpha and leptin expression in white adipose tissue may be involved in the improvement of insulin sensitivity observed in crocetin-treated rats (Xi *et al.,* 2007). Saffron, crocin and glutathione pre-treatment reduced glucose toxicity consistent with increased ROS production suggesting that saffron and its carotenoid crocin could be potentially useful in diabetic neuropathy treatment (Mousavi *et al.,* 2010).

Kang *et al.* (2012) elucidated the mechanism of the hypo-glycemic actions of saffron by investigating its signaling pathways associated with glucose metabolism in C(2)C(12) skeletal muscle cells. AMP-activated protein kinase plays a major role in the effects of saffron on glucose uptake and insulin sensitivity in skeletal muscle cells. Thus, the authors provide important insights for the possible mechanism of action of saffron and its potential as a therapeutic agent in diabetic patients.

Shirali *et al.* (2013) reported that crocin significantly decreased the levels of serum glucose, advanced glycation end products, triglyceride, total cholesterol, and low-density lipoprotein; and increased the high-density lipoprotein in the diabetic rats. It was also effective in decreasing HbA1c and micro-albuminuria, as well as insulin resistance in the diabetic rats.

Crocin at a dose of 60 mg/kg was found to significantly reduce the blood glucose level in diabetic animals. In addition, there was a significant increase in TBARS levels and decreased total thiol concentrations in the liver and kidney of diabetic animals. Crocin, at doses of 30 and 60 mg/kg, appears to exert an anti-oxidative activity demonstrated by a lowering of lipid peroxidation levels in these organs (Rajaei *et al.,* 2013).

Saffron in Depression

Depression is a major worldwide health problem. Indeed, by 2020, depressive disorders are estimated to represent the second largest disease burden globally. In spite of a variety of pharmaceutical treatments available for depression, many patients do not tolerate the side effects or respond adequately. Anti-depressant effects of saffron have been demonstrated *in-vivo* by clinical studies (Schmidt *et al.,* 2007). In a double-blind, placebo-controlled and randomized trial, patients exhibiting mild-to-moderate depression receiving capsule of saffron petal (30 mg/day, BD) for 6 weeks, showed a significantly better score on Hamilton Depression Rating Scale than placebo with no side effects thus indicating the efficacy of saffron petal in the treatment of mild-to-moderate depression (Moshiri *et al.,* 2006). The two active ingredients–crocin and safranal have been shown to be responsible for anti-depressant effect of saffron (Bittar *et al.,* 2000).

Further, saffron was found to be effective in relieving depression occurring during pre-menstrual syndrome. A significant difference was observed in the efficacy of saffron on Total Premenstrual Daily Symptoms in cycles 3 and 4 as assessed by Hamilton Depression Rating Scale (Agha-Hosseini *et al.,* 2008).

Wang *et al.* (2010) observed that among the different fractions on the basis of polarity, the petroleum ether fraction and dichloromethane fraction at doses of 150, 300, and 600 mg/kg showed significant anti-depressant-like activities in dose-dependent manner, by means of behavioral models of depression. The authors attributed the effect to crocin in saffron stigma validating its use in traditional medicine. They also suggested that the low polarity parts of *C. sativus* corms should be considered as a new plant material for curing depression.

Saffron for Memory

Recent behavioral and electrophysiological studies have demonstrated that saffron extract affects learning and memory in experimental animals. Saffron extract improved ethanol-induced impairments of learning behaviours in mice by preventing ethanol-induced inhibition of hippocampal long-term potentiation, a form of activity-dependent synaptic plasticity that may underly learning and memory (Hosseinzadeh and Ziaei, 2006;. Sugiura *et al.,* 1995). The effect is attributed to crocin (crocetin di-gentiobiose ester), but not crocetin (Abe and Saito, 2000). In addition, it has been reported that crocin counteracts the ethanol inhibition of NMDA receptor-mediated responses in rat hippocampal neurons (Abe *et al.,* 1998).

Post-training administration of 30 and 60 g/kg of saffron extract and 15 and 30 mg/kg of crocin successfully counteracted extinction of recognition/memory in the normal rat, suggesting that the extract modulates storage and/or retrieval of information. Supporting these observations pre-training treatment as well as crocin treatment (30 mg/kg and to a lesser extent also 15 mg/kg) also significantly antagonized the scopolamine (0.75 mg/kg)-induced performance deficits in the step-through passive avoidance test and (0.2mg/kg)-induced performance deficits in the radial water maze test (Pitsikas and Sakellaridis, 2006; Pitsikas *et al,*. 2007).

Both saffron extract and crocin blocked the ability of chronic stress to impair spatial learning and memory retention. Stressed animals that received saffron extract or crocin had significantly higher levels of lipid peroxidation products, significantly higher activities of anti-oxidant enzymes including glutathione peroxidase, glutathione reductase and superoxide dismutase and lower total anti-oxidant reactivity capacity. Crocin also significantly decreased plasma levels of corticosterone, as measured after the end of stress suggesting its usefulness in pharmacological alleviation of cognitive deficits (Ghadrdoost *et al.,* 2011).

Cerebral ischemia produces brain damage and related behavioral deficits such as memory. All the alterations induced by cerebral ischemia were significantly attenuated by pre-treatment of *Crocus sativus* (100 mg/kg of body weight, p.o.) 7 days before the induction of middle cerebral artery (MCA) occlusion (MCAO) and correlated well with histopathology by decreasing the neuronal cell death following MCAO and reperfusion probably by virtue of its anti-oxidant property (Saleem *et al.,* 2006). Papandreou *et al.* (2011) found both saffron and crocetin provided strong protection in rescuing cell viability (MTT assay), repressing ROS production (DCF assay) and decreasing caspase-3 activation confirming that crocetin is a unique and potent anti-oxidant, capable of mediating the *in-vivo* effects of saffron.

Safranal was found to have some protective effects on different markers of oxidative damage in hippocampal tissue from ischemic rats (Hosseinzadeh and Sadeghnia, 2005; Hosseinzadeh *et al.,* 2005) following quinolinic acid (causing excite-toxicity) administration. This raises the possibility of potential therapeutic application of safranal for preventing and treating neuro-degenerative disorders such as Alzheimer's disease (AD) (Sadeghnia *et al.,* 2013). Safranal reduces the extracellular hippocampal levels of glutamate and aspartate (Hosseinzadeh *et al.,* 2008a). Akhondzadeh *et al.* (2010a; 2010b) demonstrated that supplementation with saffron 30 mg/day (15 mg twice per day) for 16 weeks in forty-six patients with probable AD revealed a significantly better cognitive function (AD assessment scale-cognitive subscale (ADAS-cog), and clinical dementia rating scale-sums of boxes) than placebo.

Different doses of an aqueous solution of crocin or hydro-alcohol extract of saffron administered intra-peritoneally (i.p.) for 5 days after permanent occlusion of the common carotid arteries significantly reduced the escape latency time of spatial learning and memory and traveled distance by crocin (25 mg/kg) and saffron extract (250 mg/kg). The percentages of time spent in the target quadrant, in comparison with the control group (24.16 per cent), increased to 34.25 per cent in the crocin (25 mg/kg) and 34.85 per cent in the saffron extract (250 mg/kg) group thus suggesting that both saffron extract and crocin improve spatial cognitive abilities following chronic cerebral hypo-perfusion and that these effects may be related to the anti-oxidant effects of these compounds (Hosseinzadeh *et al.,* 2005; 2012). According to

Tashakori-Sabzevar *et al.* (2013) crocetin (8 mg/kg) protects cerebro-cortical and hippocampus neurons against ischemia by preventing neuro-pathological alterations in hippocampus thus improving spatial learning memory in rats after chronic cerebral hypo-perfusion. Naghibi *et al.* (2012) demonstrated that the aqueous extract of saffron attenuated morphine-induced memory impairment as judged by the time latency for entering the dark compartment. Purushothuman *et al.* (2013) confirmed that pre-treatment of mice with saffron saved many dopaminergic cells of the substantia nigra pars compacta (SNc) and retina from parkinsonian MPTP (1-methyl-4-phenyl-1,2,3,6-tetrahydropyridine) insult.

Intra-cerebroventricular (ICV) streptozotocin (STZ) has been shown to cause cognitive impairment, associated with free radical generation. Crocin treatment improved cognitive performance and resulted in a significant reduction in MDA levels and elevation in total thiol content and GPx activity. This study demonstrates that crocin may have beneficial effects in the treatment of neuro-degenerative disorders such as Alzheimer's disease (Naghizadeh *et al.*, 2013; Khalili and Hamzeh, 2010). The main therapeutic modality for Alzheimer's disease lies in inhibiting acetylcholine breakdown by acetylcholinesterase (AChE). Saffron has been used in traditional medicine against Alzheimer's disease. *In-vitro* enzymatic and molecular docking studies showed moderate AChE inhibitory activity (up to 30 per cent) by saffron extract. On the basis of the results obtained Geromichalos *et al.* (2012) concluded that safranal interacts only with the binding site of the AChE, but crocin and dimethylcrocetin bind simultaneously to the catalytic and peripheral anionic sites. Thus the previous findings about the beneficial action of saffron against Alzheimer's disease are reinforced.

Aβ is the main constituent of the amyloid plaque found in the brains of patients with Alzheimer's disease. Ghahghaei *et al.* (2013) reported that crocin has the ability to prevent amyloid formation by decreasing the fluorescence intensity. Electron microscopy data also indicated that crocin decreased the amyloid fibril content of Aβ. The ANS-binding assay showed that crocin decreased the hydrophobic area in incubated Aβ. CD spectroscopy results also showed that the peptide undergoes a structural change to α-helical and β-turn. Thus the authors demonstrated that the anti-amyloidogenic effect of crocin is exerted not only by the inhibition of Aβ amyloid formation but also by the disruption of amyloid aggregates.

Aluminium (Al) intake causes memory impairment, decreases of AChE and BuChE activity, activates brain MAO isoforms but inhibits cerebellar MAO-B, elevates brain MDA and reduces GSH content. Although saffron extract co-administration had no effect on cognitive performance of mice, it reversed significantly the Al-induced changes in MAO activity and the levels of MDA and GSH. AChE activity was further significantly decreased in cerebral tissues of Al+saffron group. The biochemical changes support the neuro-protective potential of saffron under toxicity (Linardaki *et al.*, 2013).

Aluminium may affect several enzymes and other biomolecules related to neuro-toxicity and Alzheimer's disease. The promising protective effect of aqueous saffron extract and honey syrup on neuro-toxicity induced by aluminium chloride may be derived from their own anti-oxidant properties. Saffron and honey minimized the toxic effect of aluminium chloride in the liver by alleviating its disruptive effect on the biochemical and molecular levels (Shati and Alamri, 2010). Thus an ameliorative change with saffron extract and honey syrup against aluminium chloride neuro-toxicity was reported by Shati *et al.* (2011).

Crocin and crocetin were shown to be effective in the inhibition of LPS-induced nitric oxide (NO) release from cultured rat brain microglial cells, reducing the LPS-stimulated productions of tumor necrosis factor-α, interleukin-1β, and intracellular reactive oxygen species and also effectively reducing LPS-elicited NF-κB activation thus providing neuro-protection (Nam *et al.,* 2010).

Sciatic nerve crush-injury causes reduction in sciatic functional index (SFI) values, increase in plasma malondialdehyde (MDA) levels and produce Wallerian degeneration in sciatic nerve. Crocin at doses of 20 and 80 mg/kg, accelerated the SFI recovery, decreased MDA levels and reduced Wallerian degeneration severity highlighting the neuro-protective effects afforded by crocin may be due in part to reduction of free radicals-induced toxic effects (Tamaddonfard *et al.,* 2013). Georgiadou *et al.* (2012) reported that the active constituents of *C. Sativus* L. crocins might play a role in compulsive behavior and support a functional interaction between crocins and the serotonergic system.

Saffron water extract and safranal had an important impact on the reduction of both metabolic and behavioral signs of stress in male Wistar rats (Hooshmandi *et al.,* 2011). The aqueous extract of saffron and its constituent crocin reduce side effects of electroshock stress in mice (Halataei *et al.,* 2011).

Amin and Hosseinzadeh (2012) observed that ethanolic and aqueous extracts of saffron as well as safranal could be useful in treatment of different kinds of neuropathic pains and as an adjuvant to conventional medicines.

Macular Degeneration/Cataract

Crocin analogs isolated from saffron significantly increased the blood flow in the retina and choroid and facilitated recovery of retinal function. It could thus find a role in treating ischemic retinopathy and/or age-related macular degeneration (Xuan, 1999). Short-term supplementation of saffron to 25 patients with early age-related macular degeneration (AMD) improved retinal flicker sensitivity, providing important clues that carotenoids may affect AMD in novel and unexpected ways, possibly beyond their anti-oxidant properties (Falsini *et al.,* 2011). Three months of oral supplementation with saffron improved significantly the mean focal electro-retinogram (fERG) amplitude and flicker sensitivity estimate (fERG) sensitivity when compared to baseline values and the changes were stable throughout the follow-up period. The functional effect of saffron supplementation in individual age-related macular degeneration proved to be not related to the major risk genotypes of disease (Marangoni *et al.,* 2013; Piccardi *et al.,* 2012).

Crocetin at a concentration of 3 µM showed the inhibitory effect of 50-60 per cent against tunicamycin- and $H(2)O(2)$-induced cell death and inhibited increase in caspase-3 and -9 activity. Moreover, crocetin inhibited the enzymatic activity of caspase-9 in a cell-free system. Crocetin at 100 mg/kg, p.o. significantly inhibited photo-receptor degeneration and retinal dysfunction and halved the expression of TUNEL-positive cells indicating that crocetin has protective effects against retinal damage *in-vitro* and *in-vivo*, suggesting that the mechanism may inhibit increase in caspase-3 and -9 activities after retinal damage (Yamauchi *et al.,* 2011). Ohno *et al.* (2012) also demonstrated that oral administration of crocetin prevented NMDA-induced retinal damage *via* inhibition of the caspase pathway.

Saffron showed significant protection against selenite-induced cataractogenesis *in-vivo* in Wistar rats, possibly through the reinforcement of anti-oxidant status, reduction of the intensity of lipid peroxidation, protection of the sulfhydryl groups, and inhibition of lysis of the water-soluble fraction of lens proteins (Makri *et al.,* 2013).

Crocin activated the PI3K/AKT pathway in the ganglion cell layer in retinal ischaemia/reperfusion (IR) injury. Intra-vitreal injection of LY294002 blocked the neuro-protective effect of crocin on IR-induced RGC death. Thus crocin prevents retinal IR-induced apoptosis of RGCs by activating the PI3K/AKT signalling pathway (Qi *et al.,* 2013).

Other Benefits

Boskabady *et al.* (2010) indicated an inhibitory effect of aqueous-ethanolic extracts of *Crocus sativus* on histamine H1 receptors on tracheal chains of guinea pigs. Different concentrations of saffron extract and safranal significantly improved lung pathological changes, total and differential white blood cells and serum histamine levels as efficiently as dexamethasone, showing the preventive effect of saffron extract and its constituent safranal on lung inflammation of sensitized guinea pigs. However, safranal was found to be more effective (Boskabady *et al.,* 2012).

Oral supplementation with Satiereal (Inoreal Ltd, Plerin, France), a novel extract of saffron stigma, reduced snacking and enhanced satiety through its suggested mood-improving effect, and thus contribute to weight loss (Gout *et al.,* 2010).

Lower pregnancy rates of in vitro matured oocytes compared to those of in vivo stimulated cycles indicate that optimization of in vitro maturation (IVM) remains a challenge. Reduced developmental competence of in vitro matured oocytes shows that current culture systems for oocyte maturation do not adequately support nuclear and/or cytoplasmic maturation. Treatment with different concentrations of saffron aqueous extract (SAE) revealed significantly higher maturation rate. However, the lower concentrations of SAE (10 and 5 µg/ml) in maturation medium respectively increased the fertilization rate of oocytes and *in-vitro* developmental competence (Tavana *et al.* (2012).

Kawabata *et al.* (2012) reported that dietary crocin suppresses chemically induced colitis and colitis-related colon carcinogenesis in mice, at least partly by inhibiting inflammation and the mRNA expression of certain pro-inflammatory cytokines and inducible inflammatory enzymes.

The efficacy of pollen of saffron extract cream in the treatment of thermal induced burn wounds was compared with silver sulfadiazine (SSD) in rats. The wound size of saffron group was significantly smaller than other groups. Histological comparison showed that saffron significantly increased re-epithelialization in burn wounds, as compared to other cream-treated wounds. Although the exact mechanism of saffron is unclear, anti-inflammatory and antioxidant effects of saffron may have contributed to the healing of burn injuries (Khorasani *et al.,* 2008).

Aqueous and ethanolic extracts of saffrom stigma showed anti-convulsant activity in pentylenetetrazole (PTZ)-induced convulsions and Maximal electro-shock seizure (MES) and this may help control simple and complez seizures (Hosseinzadeh and Khosravan, 2002). Further, safranal (0.15 and 0.35 ml/kg body weight, i.p.) reduced the seizure duration, delayed the onset of tonic convulsions, and protected mice from death. But crocin even at 22 mg/kg, i.p. did not show anti-convulsant activity (Hosseinzadeh and Talebzadeh, 2005b).

The ethanolic extract of saffron (100-800 mg/kg) and safranal (0.25-0.75 ml/kg) reduced the number of cough in guinea pigs. But here again the ethanolic or aqueous extracts of petal and crocin did not show anti-tussive activity (Hosseinzadeh and Ghenaati, 2006).

Saffron, partially due to its constituent safranal, exhibited a potent stimulatory effect on β_2-adrenoreceptors. A possible inhibitory effect of saffron on histamine (H_1) receptors was also suggested (Nemati *et al.,* 2008).

Brown *et al.* (2004) employed medical nutrition therapy as a complementary treatment for psoriasis. Subjects with mild to severe psoriasis consumed along with a diet of fresh fruits and vegetables, small amounts of protein from fish and fowl, fiber supplements, olive oil, saffron tea and slippery elm bark water *Ulmus fulva* daily improved on (a) the Psoriasis Area and Severity Index (PASI) (average pre- and post-test scores were 18.2 and 8.7, respectively), (b) the Psoriasis Severity Scale (PSS) (average pre- and post-test scores were 14.6 and 5.4, respectively), and (c) the lactulose/mannitol test of intestinal permeability (average pre- and post-test scores were 0.066 to 0.026, respectively).

Crocetin administration to experimental animals during resuscitation post hemorrhage increased survival, at least in part by protecting the liver from activation of apoptotic cell death. This agent continues to show promise as a potential treatment strategy for hemorrhagic shock (Yang *et al.,* 2011).

Daily use of 100 mg saffron for 3 weeks increased the IgG level and decreased the IgM level compared with the baseline and placebo (p<0.01), decreased the percentage of basophils and the count of platelets compared with baseline, but increased the percentage of monocytes compared with placebo (p<0.05). However, these parameters returned to the baseline levels after 6weeks. Thus, saffron showed temporary immune-modulatory activities without any adverse effects (Kianbakht and Ghazavi, 2011).

Hosseinzadeh *et al.* (2008b) demonstrated that the aqueous extract of saffron crocin possess aphrodisiac activity too. Saffron treatment dose dependently reduced gentamicin-induced nephrotoxicity (Ajami *et al.,* 2010).

The aqueous extract of *Crocus sativus* stigmas (CSE) and crocin (trans-crocin 4) exhibited genoprotective property by decreasing the methyl methanesulfonate (MMS)-

induced DNA damage in multiple mice organs using the comet assay (Hosseinzadeh *et al.*, 2008a).

El-Beshbishy *et al.* (2012) reported the protective role of crocin in beryllium chloride (BeCl(2) toxicity. BeCl(2) induced oxidation of cellular lipids and proteins. But administration of crocin reduced BeCl(2)-induced oxidative stress combined with initiation of mRNA expression of anti-oxidant genes and restored the hematological and biochemical parameters to near normal levels.

Conclusions

Though very few adverse health effects of saffron have been demonstrated, Poma *et al.* (2012) warns that high doses should be avoided in pregnancy owing to its uterine stimulation activity.

Here in this chapter on saffron, crocin is a natural carotenoid chemical compound found in the flowers crocus. It is the diester formed from the disaccharide gentiobiose and the dicarboxylic acid crocetin.

References

Abe K, Saito H (2000). Effects of saffron extract and its constituent crocin on learning behaviour and long-term potentiation. *Phytother. Res.* 14 (3): 149–52.

Abe K, Sugiura M, Shoyama Y, Saito H (1998). Crocin antagonizes ethanol inhibition of NMDA receptor-mediated responses in rat hippocampal neurons. *Brain Research.* 787 (1): 132–38.

Abdullaev FI (2002). Cancer chemo-preventive and tumori-cidal properties of saffron (*Crocus sativus* L.). *Exp. Biol. and Med.* 227 (1): 20–25.

Abdullaev JF, Caballero-Ortega H, Riverón-Negrete L, Pereda-Miranda R, Rivera-Luna R, Manuel Hernández J, Pérez-López I, Espinosa-Aguirre JJ (2002). *In-vitro* evaluation of the chemo-preventive potential of saffron. *Rev. Invest. Clin.* 54 (5): 430–36. [Article in Spanish].

Abdullaev FI, Espinosa-Aguirre JJ (2004). Biomedical properties of saffron and its potential use in cancer therapy and chemo-prevention trials. *Cancer Detection and Prevention.* 28 (6): 426–32.

Abdullaev FI, Riverón-Negrete L, Caballero-Ortega H, Manuel Hernández J, Pérez-López I, Pereda-Miranda R, *et al.* (2003). Use of *in-vitro* assays to assess the potential anti-genototoxic and cytotoxic effect of saffron (*Crocus sativus*). *Toxicol. In-vitro.* 17: 731–36.

Agha-Hosseini M, Kashani L, Aleyaseen A, Ghoreishi A, Rahmanpour H, Zarrinara AR, Akhondzadeh S (2008). *Crocus sativus* L. (saffron) in the treatment of premenstrual syndrome: a double-blind, randomised and placebo-controlled trial. *BJOG.* 115 (4): 515–19.

Ajami M, Eghtesadi S, Pazoki-Toroudi H, Habibey R, Ebrahimi SA. (2010). Effect of *Crocus sativus* on gentamicin induced nephro-toxicity. *Biol. Res.* 43 (1): 83–90.

Akhondzadeh S, Fallah Pour H, Afkham K, Jamshidi AH, Khalighi-Cigarodi F (2004). Comparison of *Crocus sativus* L. and imipramine in the treatment of mild to moderate depression: a pilot double-blind randomized trial. *BMC. Complement. Altern. Med.* 4: 12.

Akhondzadeh S, Sabet MS, Harirchian MH, Togha M, Cheraghmakani H, Razeghi S, Hejazi SSh, Yousefi MH, Alimardani R, Jamshidi A, Zare F, Moradi A (2010a). Saffron in the treatment of patients with mild to moderate Alzheimer's disease: a 16-week, randomized and placebo-controlled trial. *J. Clin. Pharm. Ther.* 35 (5): 581- 88.

Akhondzadeh S, Shafiee Sabet M, Harirchian MH, Togha M, Cheraghmakani H, Razeghi S, Hejazi SS,Yousefi MH, Alimardani R, Jamshidi A, Rezazadeh SA, Yousefi A, Zare F, Moradi A, Vossoughi A (2010b). A 22-week, multicenter, randomized double-blind controlled trial of *Crocus sativus* in the treatment of mild-to-moderate Alzheimer's disease. *Psychopharmacology.* (Berl). 207 (4): 637– 43.

Amin A, Hamza AA, Bajbouj K, Ashraf SS, Daoud S (2011). Saffron: a potential candidate for a novel anti-cancer drug against hepato-cellular carcinoma. *Hepatology.* 54 (3): 857–67.

Amin B, Hosseinzadeh H (2012). Evaluation of aqueous and ethanolic extracts of saffron, *Crocus sativus* L., and its constituents, safranal and crocin in allodynia and hyperalgesia induced by chronic constriction injury model of neuropathic pain in rats. *Fitoterapia.* 83 (5): 888–95.

Asdaq SM, Inamdar MN (2010). Potential of *Crocus sativus* (saffron) and its constituent, crocin, as hypo-lipidemic and anti-oxidant in rats. *Appl. Biochem. Biotechnol.* 162 (2): 358–72.

Assimopoulou AN, Sinakos Z, Papageorgiou VP (2005). Radical scavenging activity of *Crocus sativus* L. extract and its bioactive constituents. *Phytother. Res.* 19 (11): 997–1000.

Aung HH, Wang CZ, Ni M, *et al.* (2007). Crocin from *Crocus sativus* possesses significant anti-proliferation effects on human colorectal cancer cells. *Experimental Oncology.* 29 (3): 175–80.

Bajbouj K, Schulze-Luehrmann J, Diermeier S, Amin A, Schneider-Stock R (2012). The anti-cancer effect of saffron in two p53 isogenic colorectal cancer cell lines. *BMC. Complement. Altern. Med.* 12: 69.

Bakshi H, Sam S, Rozati R, *et al.* (2010). DNA fragmentation and cell cycle arrest: a hallmark of apoptosis induced by crocin from Kashmiri Saffron in a human pancreatic cancer cell line. *Asian Pacific Journal of Cancer Prevention.* 11 (3): 675– 79.

Bittar M, deSouza MM, Yunus RA, Lento R, Monache FD, Cechinel VF (2000). Anti-nociceptive activity of I-3,II8-binarigenin,a biflavonoid present in plants of the guttiferae. *Planta Med.* 66: 84–86.

Boskabady MH, Ghasemzadeh Rahbardar M, Nemati H, Esmaeilzadeh M (2010). Inhibitory effect of *Crocus sativus* (saffron) on histamine (H1) receptors of guinea pig tracheal chains. *Pharmazie.* 65 (4): 300–305.

Boskabady MH, Tabatabaee A, Byrami G (2012). The effect of the extract of *Crocus sativus* and its constituent safranal, on lung pathology and lung inflammation of ovalbumin sensitized guinea-pigs. *Phytomedicine.* 19 (10): 904–11.

Brown AC, Hairfield M, Richards DG, McMillin DL, Mein EA, Nelson CD (2004). Medical nutrition therapy as a potential complementary treatment for psoriasis: Five case reports. *Altern. Med. Rev.* 9: 297–307.

Chryssanthi DG, Dedes PG, Karamanos NK, Cordopatis P, Lamari FN (2011). Crocetin inhibits invasiveness of MDA-MB-231 breast cancer cells via downregulation of matrix metallo-proteinases. *Planta. Med.* 77 (2): 146–51.

D'Alessandro AM, Mancini A, Lizzi AR, De Simone A, Marroccella CE, Gravina GL, Tatone C, Festuccia C (2013). *Crocus sativus* stigma extract and its major constituent crocin possess significant anti-proliferative properties against human prostate cancer. *Nutr. Cancer.* 65 (6): 930–42.

Das I, Das S, Saha T (2010). Saffron suppresses oxidative stress in DMBA-induced skin carcinoma: A histopathological study. *Acta. Histochem.* 112 (4): 317–27.

Deng Y, Guo ZG, Zeng ZL, Wang Z (2002). Studies on the pharmacological effects of saffron (*Crocus sativus* L.)–a review. Zhongguo. *Zhong. Yao. Za. Zhi.* 27 (8) 565–68. [Article in Chinese].

Dhar A, Mehta S, Dhar G, Dhar K, Banerjee S, Van Veldhuizen P, Campbell DR, Banerjee SK (2009). Crocetin inhibits pancreatic cancer cell proliferation and tumor progression in a xenograft mouse model. *Mol. Cancer. Ther.* 8 (2): 315–23.

Ding Q, Zhong H, Qi Y, Cheng Y, Li W, Yan S, Wang X (2013). Anti-arthritic effects of crocin in interleukin-1β-treated articular chondrocytes and cartilage in a rabbit osteo-arthritic model. *Inflamm. Res.* 62 (1): 17–25.

El-Beshbishy HA, Hassan MH, Aly HA, Doghish AS, Alghaithy AA (2012). Crocin "saffron" protects against beryllium chloride toxicity in rats through diminution of oxidative stress and enhancing gene expression of anti-oxidant enzymes. *Ecotoxicol. Environ. Saf.* 83: 47–54.

Escribano J, Alonso GL, Coca-Prados M, Fernandez JA (1996). Crocin, safranal and picrocrocin from saffron (*Crocus sativus* L.) inhibit the growth of human cancer cells *in-vitro. Cancer Lett.* 100 (1-2): 23–30.

Falsini B, Piccardi M, Minnella A, Savastano C, Capoluongo E, *et al.* (2010). Influence of saffron supplementation on retinal flicker sensitivity in early age-related macular degeneration. *Invest. Ophthalmol. Vis. Sci.* 51 (12): 6118–24.

Fatehi M, Rashidabady T, Fatehi-Hassanabad Z (2003). Effects of Crocus sativus petals' extract on rat blood pressure and on response induced by electrical field stimulation in the rat isolated *vas deferens* and guinea-pig ileum. *J. Ethnopharmacology.* 84: 199–203.

Feizzadeh B, Afshari JT, Rakhshandeh H, Rahimi A, Brook A, Doosti H (2008). Cytotoxic effect of saffron stigma aqueous extract on human transitional cell carcinoma and mouse fibroblast. *Urol. J.* 5 (3): 161–67.

Fernández-Sánchez L, Lax P, Esquiva G, Martín-Nieto J, Pinilla I, Cuenca N (2012). Safranal, a saffron constituent, attenuates retinal degeneration in P23H rats. *PLoS. One.* 7 (8): e43074.

Georgiadou G, Tarantilis PA, Pitsikas N (2012). Effects of the active constituents of *Crocus Sativus* L., crocins, in an animal model of obsessive-compulsive disorder. *Neurosci. Lett.* 528 (1): 27–30.

Geromichalos GD, Lamari FN, Papandreou MA, Trafalis DT, Margarity M, Papageorgiou A, Sinakos Z (2012). Saffron as a source of novel acetylcholinesterase inhibitors: molecular docking and *in-vitro* enzymatic studies. *J. Agric. Food Chem.* 60 (24): 6131–38.

Ghadrdoost B, Vafaei AA, Rashidy-Pour A, Hajisoltani R, Bandegi AR, Motamedi F, Haghighi S, Sameni HR, Pahlvan S (2011). Protective effects of saffron extract and its active constituent crocin against oxidative stress and spatial learning and memory deficits induced by chronic stress in rats. *Eur. J. Pharmacol.* 667 (1-3): 222–29.

Ghahghaei A, Bathaie SZ, Kheirkhah H, Bahraminejad E (2013). The protective effect of crocin on the amyloid fibril formation of Aβ peptide *in-vitro*. *Cell. Mol. Biol. Lett.* 18 (3): 328–39.

Gohari AR, Saeidnia S, Mahmoodabadi MK (2013). An overview on saffron, phytochemicals, and medicinal properties. *Pharmacogn. Rev.* 7 (13): 61–66.

Gout B, Bourges C, Paineau-Dubreuil S (2010). Satiereal, a *Crocus sativus* L extract, reduces snacking and increases satiety in a randomized placebo-controlled study of mildly overweight, healthy women. *Nutr. Res.* 30 (5): 305–13.

Goyal SN, Arora S, Sharma AK, Joshi S, Ray R, Bhatia J, Kumari S, Arya DS (2010). Preventive effect of crocin of *Crocus sativus* on hemodynamic, biochemical, histo-pathological and ultra-stuctural alterations in isoproterenol-induced cardio-toxicity in rats. *Phytomedicine.* 17 (3-4): 227–32.

Halataei BA, Khosravi M, Arbabian S, Sahraei H, Golmanesh L, Zardooz H, Jalili C, Ghoshooni H (2011). Saffron (*Crocus sativus*) aqueous extract and its constituent crocin reduces stress-induced anorexia in mice. *Phytother. Res.* 25 (12): 1833–88.

He SY, Qian ZY, Tang FT (2004). Effect of crocin on intracellular calcium concentration in cultured bovine aortic smooth muscle cells. *Yao. Xue. Xue. Bao.* 39 (10): 778–81.

He SY, Qian ZY, Tang FT, Wen N, Xu GL, Sheng L (2005). Effect of crocin on experimental atherosclerosis in quails and its mechanisms. *Life Sci.* 77 (8): 907–921.

He SY, Qian ZY, Wen N, Tang FT, Xu GL, Zhou CH (2007).Influence of crocetin on experimental atherosclerosis in hyper-lipidamic-diet quails. *Eur. J. Pharmacol.* 554 (2-3): 191–95.

Hemshekhar M, Sebastin Santhosh M, Sunitha K, Thushara RM, Kemparaju K, Rangappa KS, Girish KS (2012). A dietary colorant crocin mitigates arthritis and associated secondary complications by modulating cartilage deteriorating enzymes, inflammatory mediators and anti-oxidant status. *Biochimie.* 94 (12): 2723–33.

Hooshmandi Z, Rohani AH, Eidi A, Fatahi Z, Golmanesh L, Sahraei H (2011). Reduction of metabolic and behavioral signs of acute stress in male Wistar rats by saffron water extract and its constituent safranal. *Pharm Biol.* 49 (9): 947–54.

Hoshyar R, Bathaie SZ, Sadeghizadeh M (2013). Crocin triggers the apoptosis through increasing the Bax/Bcl-2 ratio and caspase activation in human gastric adenocarcinoma, AGS, cells. *DNA Cell. Biol.* 32 (2): 50–57.

Hosseinzadeh H, Abootorabi A, Sadeghnia HR (2008a). Protective effect of *Crocus sativus* stigma extract and crocin (trans-crocin 4) on methyl methanesulfonate-induced DNA damage in mice organs. *DNA Cell. Biol.* 27: 657–64.

Hosseinzadeh H, Ghenaati J (2006). Evaluation of the anti-tussive effect of stigma and petals of saffron (*Crocus sativus*) and its components, safranal and crocin in guinea pigs. *Fitoterapia.* 77: 446–48.

Hosseinzadeh H, Khosravan V (2002). Anti-convulsant effects of aqueous and ethanolic extracts of *Crocus sativus* L. stigma in mice. *Arch. Irn. Med.* 5: 44–47.

Hosseinzadeh H, Modaghegh MH, Saffari Z (2009). *Crocus sativus* L. (Saffron) extract and its active constituents (crocin and safranal) on ischemia-reperfusion in rat skeletal muscle. *Evid. Based Complement. Alternat. Med.* 6: 343–50.

Hosseinzadeh H, Sadeghnia HR, Ghaeni FA, Motamedshariaty VS, Mohajeri SA (2012). Effects of saffron (*Crocus sativus* L.) and its active constituent, crocin, on recognition and spatial memory after chronic cerebral hypoperfusion in rats. *Phytother. Res.* 26 (3): 381–86.

Hosseinzadeh H, Sadeghnia HR, Ziaee T, Danaee A (2005). Protective effect of aqueous saffron extract (*Crocus sativus* L.) and crocin, its active constituent, on renal ischemia-reperfusion-induced oxidative damage in rats. *J. Pharm. Pharm. Sci.* 8 (3): 387–93.

Hosseinzadeh H, Sadeghnia HR (2005). Safranal, a constituent of *Crocus sativus* (saffron), attenuated cerebral ischemia induced oxidative damage in rat hippocampus. *J. Pharm. Pharm. Sci.* 8 (3): 394–99.

Hosseinzadeh H, Sadeghnia HR (2007). Effect of safranal, a constituent of *Crocus sativus* (saffron), on methyl methane-sulfonate (MMS)-induced DNA damage in mouse organs: an alkaline single-cell gel electrophoresis (comet) assay. *DNA Cell. Biol.* 26: 841–46.

Hosseinzadeh H, Talebzadeh F (2005). Anti-convulsant evaluation of safranal and crocin from Crocus sativus in mice. *Fitoterapia.* 76: 722–24.

Hosseinzadeh H, Younesi H (2002). Petal and stigma extracts of *Crocus sativus* L. have antinoceceptive and anti-inflammatory effects in mice. *BMC Pharmacol.* 2: 7.

Hosseinzadeh H, Ziaei T (2006). Effects of Crocus sativus stigma extract and its constituents, crocin and safranal, on intact memory and scopolamine-induced learning deficits in rats performing the Morris water maze task. *Journal of Medicinal Plants.* 5 (19): 40–60.

Hosseinzadeh H, Ziaee T, Sadeghi A (2008b). The effect of saffron, Crocus sativus stigma extract and its constituents, safranal and crocin on sexual behaviours in normal male rats. *Phytomedicine.* 15: 491–95.

Imenshahidi M, Hosseinzadeh H, Javadpour Y (2010). Hypo-tensive effect of aqueous saffron extract (*Crocus sativus* L.) and its constituents, safranal and crocin, in normo-tensive and hyper-tensive rats. *Phytother. Res.* 24 (7): 990–94.

Javadi B, Sahebkar A, Emami SA (2013). A survey on saffron in major islamic traditional medicine books. *Iran. J. Basic Med. Sci. 16* (1): 1–11.

Joukar S, Ghasemipour-Afshar E, Sheibani M, Naghsh N, Bashiri A (2013). Protective effects of saffron (*Crocus sativus*) against lethal ventricular arrhythmias induced by heart reperfusion in rat: a potential anti-arrhythmic agent. *Pharm. Biol.* 51 (7): 836–43.

Joukar S, Najafipour H, Khaksari M, Sepehri G, Shahrokhi N, Dabiri S, Gholamhoseinian A, Hasanzadeh S (2010). The effect of saffron consumption on biochemical and histo-pathological heart indices of rats with myocardial infarction. *Cardiovasc. Toxicol.* 10 (1): 66–71.

Kanakis CD, Tarantilis PA, Pappas C, Bariyanga J, Tajmir-Riahi HA, Polissiou MG (2009). An overview of structural features of DNA and RNA complexes with saffron compounds: Models and anti-oxidant activity. *J. Photochem. Photobiol.* B. 95 (3): 204–212.

Kanakis CD, Tarantilis PA, Tajmir-Riahi HA, Polissiou MG (2007a). Crocetin, dimethylcrocetin, and safranal bind human serum albumin: stability and anti-oxidative properties. *J. Agric. Food Chem.* 55: 970–77.

Kanakis CD, Tarantilis PA, Tajmir-Riahi HA, Polissiou MG (2007b). Interaction of tRNA with Safranal, Crocetin, and Dimethylcrocetin. *J. Biomol. Struct. Dyn.* 24 (6): 537–46.

Kanakis CD, Tarantilis PA, Tajmir-Riahi HA, Polissiou MG (2007c). DNA interaction with saffron's secondary metabolites safranal, crocetin, and dimethylcrocetin. *DNA Cell. Biol.* 26: 63–70.

Kang C, Lee H, Jung ES, Seyedian R, Jo M, Kim J, Kim JS, Kim E (2012). Saffron (*Crocus sativus* L.) increases glucose uptake and insulin sensitivity in muscle cells via multipathway mechanisms. *Food Chem.* 135 (4): 2350–58.

Karimi G, Hosseinzadeh H, Khaleghpanah P (2001). Study of anti-depresant effect of aqueous and ethanolic extract of Crocus sativus in mice. *Iran. J. Basic Med. Sci.* 4: 11–5.

Karimi E, Oskoueian E, Hendra R, Jaafar HZ (2010). Evaluation of *Crocus sativus* L. stigma phenolic and flavonoid compounds and its anti-oxidant activity. *Molecules.* 15 (9): 6244–56.

Kawabata K, Tung NH, Shoyama Y, Sugie S, Mori T, Tanaka T (2012). Dietary Crocin Inhibits Colitis and Colitis-Associated Colorectal Carcinogenesis in Male ICR Mice. *Evid. Based Complement. Alternat. Med.* 2012: 820415.

Khalili M, Hamzeh F (2010). Effects of active constituents of *Crocus sativus* L, crocin on streptozocin-induced model of sporadic Alzheimer's disease in male rats. *Iran. Biomed. J.* 14 (1-2): 59–65.

Khorasani G, Hosseinimehr SJ, Zamani P, Ghasemi M, Ahmadi A (2008). The effect of saffron (*Crocus sativus*) extract for healing of second-degree burn wounds in rats. *Keio. J. Med.* 57 (4): 190–95.

Kianbakht S, Ghazavi A (2011). Immunomodulatory effects of saffron: a randomized double-blind placebo-controlled clinical trial. *Phytother. Res.* 25 (12): 1801–05.

Lari P, Abnous K, Imenshahidi M, Rashedinia M, Razavi M, Hosseinzadeh H (2013a). Evaluation of diazinon-induced hepatotoxicity and protective effects of crocin. *Toxicol. Ind. Health.* 2013 Feb 13. [Epub ahead of print].

Lari P, Abnous K, Imenshahidi M, Rashedinia M, Razavi M, Hosseinzadeh H (2013b). Crocetin administration ameliorates endotoxin-induced disseminated intra-vascular coagulation in rabbits. Blood Coagul. *Fibrinolysis.* 24 (3): 305–310.

Li X, Huang T, Jiang G, Gong W, Qian H, Zou C (2013). Synergistic apoptotic effect of crocin and cisplatin on osteosarcoma cells via caspase induced apoptosis. *Toxicol. Lett.* 221 (3): 197–204.

Liakopulou-Kyriakides M, Kyriakides DA (2002). Crocus sativus: Biological active constituents. Stud. *Nat. Prod. Chem.* 26: 293–312.

Linardaki ZI, Orkoula MG, Kokkosis AG, Lamari FN, Margarity M (2013). Investigation of the neuro-protective action of saffron (*Crocus sativus* L.) in aluminum-exposed adult mice through behavioral and neuro-biochemical assessment. *Food Chem. Toxicol.* 52: 163–70.

Maccarone R, Di Marco S, Bisti S (2008). Saffron supplement maintains morphology and function after exposure to damaging light in mammalian retina. Invest. Ophthalmol. Vis. Sci. 49: 1254–61.

Madan CL, Kapur BM, Gupta US (1996). *Saffron. Econ. Bot.* 20: 377.

Magesh V, Singh JP, Selvendiran K, Ekambaram G, Sakthisekaran D (2006). Anti-tumour activity of crocetin in accordance to tumor incidence, anti-oxidant status, drug metabolizing enzymes and histopathological studies. *Mol. Cell. Biochem.*

287 (1-2): 127–35.

Makri OE, Ferlemi AV, Lamari FN, Georgakopoulos CD (2013). Saffron administration prevents selenite-induced cataractogenesis. *Mol. Vis.* 19: 1188–97.

Malaekeh-Nikouei B, Mousavi SH, Shahsavand S, Mehri S, Nassirli H, Moallem SA (2013). Assessment of Cytotoxic Properties of Safranal and Nano-liposomal Safranal in Various Cancer Cell Lines. *Phytother. Res.* 27 (12): 1868–73.

Marangoni D, Falsini B, Piccardi M, Ambrosio L, Minnella AM, Savastano MC, Bisti S, Maccarone R, Fadda A, Mello E, Concolino P, Capoluongo E (2013). Functional effect of Saffron supplementation and risk genotypes in early age-related macular degeneration: a preliminary report. *J. Transl. Med.* 11 (1): 228.

Mehdizadeh R, Parizadeh MR, Khooei AR, Mehri S, Hosseinzadeh H (2013). Cardio-protective effect of saffron extract and safranal in isoproterenol-induced myocardial infarction in wistar rats. *Iran. J. Basic Med. Sci.* 16 (1): 56–63.

Mehri S, Abnous K, Mousavi SH, Shariaty VM, Hosseinzadeh H (2012). Neuro-protective effect of crocin on acrylamide-induced cytotoxicity in PC12 cells. *Cell. Mol. Neurobiol.* 32 (2): 227–35.

Montoro P, Maldini M, Luciani L, Tuberoso CI, Congiu F, Pizza C (2012). Radical scavenging activity and LC-MS metabolic profiling of petals, stamens, and flowers of *Crocus sativus* L. *J. Food Sci.* 77 (8): C893–900.

Moshiri E, Basti AA, Noorbala AA, Jamshidi AH, Hesameddin Abbasi S, Akhondzadeh S (2006). *Crocus sativus* L. (petal) in the treatment of mild-to-moderate depression: a double-blind, randomized and placebo-controlled trial. *Phytomedicine.* 13 (9 -10): 607–11.

Mousavi SH, Moallem SA, Mehri S, Shahsavand S, Nassirli H, Malaekeh-Nikouei B (2011). Improvement of cytotoxic and apoptogenic properties of crocin in cancer cell lines by its nanoliposomal form. *Pharm. Biol.* 49 (10): 1039–45.

Mousavi SH, Tavakkol-Afshari J, Brook A, Jafari-Anarkooli I (2009). Role of caspases and Bax protein in saffron-induced apoptosis in MCF-7 cells. *Food Chem. Toxicol.* 47 (8): 1909–13.

Mousavi SH, Tayarani NZ, Parsaee H (2010). Protective effect of saffron extract and crocin on reactive oxygen species-mediated high glucose-induced toxicity in PC12 cells. *Cell. Mol. Neurobiol.* 30 (2): 185–91.

Naghibi SM, Hosseini M, Khani F, Rahimi M, Vafaee F, Rakhshandeh H, Aghaie A (2012). Effect of aqueous extract of *Crocus sativus* L. on morphine-induced memory impairment. *Adv. Pharmacol. Sci.* 2012: 494367.

Naghizadeh B, Mansouri MT, Ghorbanzadeh B, Farbood Y, Sarkaki A (2013). Protective effects of oral crocin against intra-cerebroventricular streptozotocin-induced spatial memory deficit and oxidative stress in rats. *Phytomedicine.* 20 (6): 537–42.

Nair SC, Kurumboor SK, Hasegawa JH (1995). Saffron chemo-prevention in biology and medicine: a review. *Cancer Biother.* 10 (4): 257–64.

Nam KN, Park YM, Jung HJ, Lee JY, Min BD, Park SU, Jung WS, Cho KH, Park JH, Kang I, Hong JW, Lee EH (2010). Anti-inflammatory effects of crocin and crocetin in rat brain microglial cells. *Eur. J. Pharmacol.* 648 (1-3): 110–16.

Nemati H, Boskabady MH, Ahmadzadef Vortakolaei H (2008). Stimulatory effects of Crocus sativus (Saffron) on ß2 adreno-receptors of guinea pig tracheal chains. *Phytomedicine.* 15: 1038–45.

Noorbala AA, Akhondzadeh S, Tamacebi-Pour N, Jamshedi AH (2005). Hydro-alcoholic extract of Crocus sativus L. versus fluoxetine in the treatment of mild to moderate depression: a double-blind randomized pilot trial. *J. Ethnopharmacol.* 97: 281–84.

Noureini SK, Wink M (2012). Anti-proliferative effects of crocin in HepG2 cells by telomerase inhibition and hTERT down-regulation. *Asian Pac. J. Cancer Prev.* 13 (5): 2305–309.

Ochiai T, Shimeno H, Mishima K, Iwasaki K, Fujiwara M, Tanaka H, Shoyama Y, Toda A, Eyanagi R, Soeda S (2007). Protective effects of carotenoids from saffron on neuronal injury *in- vitro* and *in-vivo*. *Biochim. Biophys. Acta.* 1770: 578–84.

Ohno Y, Nakanishi T, Umigai N, Tsuruma K, Shimazawa M, Hara H (2012). Oral administration of crocetin prevents inner retinal damage induced by N-methyl-D-aspartate in mice. *Eur. J. Pharmacol.* 690 (1-3): 84–89.

Ordoudi SA, Befani CD, Nenadis N, Koliakos GG, Tsimidou MZ (2009). Further examination of anti-radical properties of *Crocus sativus* stigmas extract rich in crocins. *J. Agric. Food Chem.* 57 (8): 3080–86.

Pan TL, Wu TH, Wang PW, Leu YL, Sintupisut N, Huang CH, Chang FR, Wu YC (2013). Functional proteomics reveals the protective effects of saffron ethanolic extract on hepatic ischemia-reperfusion injury. *Proteomics.* 13 (15): 2297–311.

Papandreou MA, Tsachaki M, Efthimiopoulos S, Cordopatis P, Lamari FN, Margarity M (2011). Memory enhancing effects of saffron in aged mice are correlated with anti-oxidant protection. *Behav. Brain Res.* 219 (2): 197–204.

Piccardi M, Marangoni D, Minnella AM, Savastano MC, Valentini P, Ambrosio L, Capoluongo E, Maccarone R, Bisti S, Falsini B (2012). A longitudinal follow-up study of saffron supplementation in early age-related macular degeneration: sustained benefits to central retinal function. *Evid. Based Complement. Alternat. Med.* 2012: 429124.

Pitsikas N, Sakellaridis N (2006). *Crocus sativus* L. extracts antagonize memory impairments in different behavioural tasks in the rat. *Behav. Brain Res.* 173 (1): 112–15.

Pitsikas N, Zisopoulou S, Tarantilis PA, Kanakis CD, Polissiou MG, Sakellaridis N (2007). Effects of the active constituents of *Crocus sativus* L., crocins on recognition and spatial rats' memory. *Behav. Brain Res.* 183 (2): 141–46.

Poma A, Fontecchio G, Carlucci G, Chichiriccò G (2012). Anti-inflammatory properties of drugs from saffron crocus. Anti-inflamm. Anti-allergy. *Agents Med. Chem.* 11 (1): 37–51.

Purushothuman S, Nandasena C, Peoples CL, El Massri N, Johnstone DM, Mitrofanis J, Stone J (2013). Saffron Pre-Treatment Offers Neuro-protection to Nigral and Retinal Dopaminergic Cells of MPTP-Treated mice. *J. Parkinsons Dis.* 3 (1): 77–83.

Qi Y, Chen L, Zhang L, Liu WB, Chen XY, Yang XG (2013). Crocin prevents retinal ischaemia/reperfusion injury-induced apoptosis in retinal ganglion cells through the PI3K/AKT signalling pathway. *Exp. Eye Res.* 107: 44–51.

Rajaei Z, Hadjzadeh MA, Nemati H, Hosseini M, Ahmadi M, Shafiee S (2013). Anti-hyperglycemic and anti-oxidant activity of crocin in streptozotocin-induced diabetic rats. *J. Med. Food.* 16 (3): 206–210.

Razavi M, Hosseinzadeh H, Abnous K, Motamedshariaty VS, Imenshahidi M (2013). Crocin restores hypotensive effect of subchronic administration of diazinon in rats. Iran. *J. Basic Med. Sci.* 16 (1): 64–72.

Rezaee R, Mahmoudi M, Abnous K, Zamani Taghizadeh Rabe S, Tabasi N, Hashemzaei M, Karimi G (2013). Cyto-toxic effects of crocin on MOLT-4 human leukemia cells. *J. Complement. Integr. Med.* 10. doi: 10.1515/jcim-2013–0011.

Ríos JL, Recio MC, Giner RM, Mánez S (1996). An update review of saffron and its active constituents. *Phytother. Res.* 10 (3): 189–93.

Sachdeva J, Tanwar V, Golechha M, Siddiqui KM, Nag TC, Ray R, Kumari S, Arya DS (2012). *Crocus sativus* L. (saffron) attenuates isoproterenol-induced myocardial injury via preserving cardiac functions and strengthening antioxidant defense system. *Exp. Toxicol. Pathol.* 64 (6): 557–64.

Sadeghnia HR, Kamkar M, Assadpour E, Boroushaki MT, Ghorbani A (2013). Protective effect of safranal, a constituent of *Crocus sativus*, on quinolinic acid-induced oxidative damage in rat hippocampus. *Iran. J. Basic Med. Sci.* 16 (1): 73–82.

Saleem S, Ahmad M, Ahmad AS, Yousuf S, Ansari MA, Khan MB, Ishrat T, Islam F (2006). Effect of Saffron (*Crocus sativus*) on neuro-behavioural and neuro-chemical changes in cerebral ischemia in rats. *J. Med. Food.* 9 (2): 246–53.

Samarghandian S, Boskabady MH, Davoodi S (2010). Use of in vitro assays to assess the potential anti-proliferative and cytotoxic effects of saffron (*Crocus sativus* L.) in human lung cancer cell line. *Pharmacogn. Mag.* 6 (24): 309–314.

Samarghandian S, Shabestari MM (2013). DNA fragmentation and apoptosis induced by safranal in human prostate cancer cell line. *Indian J. Urol.* 29 (3): 177–83.

Samarghandian S, Tavakkol Afshari J, Davoodi S (2011). Suppression of pulmonary tumor promotion and induction of apoptosis by *Crocus sativus* L. extraction. *Appl. Biochem. Biotechnol.* 164 (2): 238–47.

Schmidt M, Betti G, Hensel A (2007). Saffron in phytotherapy: pharmacology and clinical uses. *Wien. Med. Wochenschr.* 157 (13–14): 315–19.

Shati AA, Alamri SA (2010). Role of saffron (*Crocus sativus* L.) and honey syrup on aluminum-induced hepatotoxicity. *Saudi Med. J.* 31 (10): 1106–113.

Shati AA, Elsaid FG, Hafez EE (2011). Biochemical and molecular aspects of aluminium chloride-induced neurotoxicity in mice and the protective role of *Crocus sativus* L. extraction and honey syrup. *Neuroscience*. 175: 66–74.

Shen XC, Qian ZY (2006). Effects of crocetin on antioxidant enzymatic activities in cardiac hypertrophy induced by norepinephrine in rats. *Pharmazie*. 61 (4): 348–52.

Sheng L, Qian Z, Zheng S, Xi L (2006). Mechanism of hypolipidemic effect of crocin in rats: crocin inhibits pancreatic lipase. *Eur J Pharmacol*. 2006 Aug 14: 543 (1-3): 116–22.

Shirali S, Zahra Bathaie S, Nakhjavani M (2013). Effect of crocin on the insulin resistance and lipid profile of streptozotocin-induced diabetic rats. Ph*ytother. Res*. 27 (7): 1042–47.

Srivastava R, Ahmed H, Dixit RK, Dharamveer, Saraf SA (2010). *Crocus sativus* L.: A comprehensive review. *Pharmacogn. Rev*. 4 (8): 200–08.

Sugiura M, Shoyama Y, Saito H, *et al*. (1995). Crocin improves the ethanol-induced impairment of learning behaviors of mice in passive avoidance tasks. *Proceedings of the Japan Academy B*. 71 (10): 319–24.

Sun Y, Xu HJ, Zhao YX, Wang LZ, Sun LR, Wang Z, Sun XF (2013). Crocin exhibits anti-tumor effects on human leukemia HL-60 cells *in-vitro* and *in- vivo*. *Evid. Based Complement. Alternat. Med*. 2013: 690164.

Tamaddonfard E, Farshid AA, Ahmadian E, Hamidhoseyni A (2013). Crocin enhanced functional recovery after sciatic nerve crush injury in rats. *Iran. J. Basic Med. Sci*. 16 (1): 83–90.

Tarantilis PA, Morjani H, Polissiou M, Manfait M (1994). Inhibition of growth and induction of differentiation of promyelocytic leukemia (HL-60) by carotenoids from *Crocus sativus* L. *Anticancer Research A*. 14 (5): 1913–18.

Tashakori-Sabzevar F, Hosseinzadeh H, Motamedshariaty VS, Movassaghi AR, Mohajeri SA (2013). Crocetin attenuates spatial learning dysfunction and hippocampal injury in a model of vascular dementia. *Curr. Neurovasc. Res*. 10 (4): 325–34.

Tavakkol-Afshari J, Brook A, Mousavi SH (2008). Study of cytotoxic and apoptogenic properties of saffron extract in human cancer cell lines. *Food Chem. Toxicol*. 46 (11): 3443–47.

Tavana S, Eimani H, Azarnia M, Shahverdi A, Eftekhari-Yazdi P (2012). Effects of saffron (*Crocus sativus* L.) aqueous extract on *in-vitro* maturation, fertilization and embryo development of mouse oocytes. *Cell. J*. 13 (4): 259–64.

Umigai N, Tanaka J, Tsuruma K, Shimazawa M, Hara H (2012). Crocetin, a carotenoid derivative, inhibits VEGF-induced angiogenesis via suppression of p38 phosphorylation. *Curr. Neurovasc. Res*. 9 (2): 102–09.

USDA (2009). N. The PLANTS database. 2009. h ttp://plants.usda.gov/java.

Vakili A, Einali MR, Bandegi AR (2012). Protective Effect of Crocin against Cerebral Ischemia in a Dose-Dependent Manner in a Rat Model of Ischemic Stroke. J

Stroke Cerebrovasc Dis. 2012 Nov 19. pii: S1052-3057(12)00345-X. [Epub ahead of print].

Verma SK, Bordia A (1998). Anti-oxidant property of saffron in man. *Indian J. Med. Sci.* 52: 205–07.

Wang Y, Han T, Zhu Y, Zheng CJ, Ming QL, Rahman K, Qin LP (2010). Anti-depressant properties of bioactive fractions from the extract of *Crocus sativus* L. *J. Nat. Med.* 64 (1): 24–30.

Wani BA, Hamza AKR, Mohiddin FA (2011). Saffron: A repository of medicinal properties. *J. Med. Plant. Res.* 5: 2131–35.

Xi L, Qian Z, Xu G, Zheng S, Sun S, Wen N, Sheng L, Shi Y, Zhang Y (2007). Beneficial impact of crocetin, a carotenoid from saffron, on insulin sensitivity in fructose-fed rats. *J. Nutr. Biochem.* 18 (1): 64–72.

Xuan B (1999). Effects of crocin analogs on ocular flow and retinal function. *J. Ocul. Pharmacol. Ther.* 15: 143–52.

Yamauchi M, Tsuruma K, Imai S, Nakanishi T, Umigai N, Shimazawa M, Hara H (2011). Crocetin prevents retinal degeneration induced by oxidative and endoplasmic reticulum stresses *via* inhibition of caspase activity. *Eur. J. Pharmacol.* 650 (1): 110–119.

Yan J, Qian Z, Sheng L, Zhao B, Yang L, Ji H, Han X, Zhang R (2010). Effect of crocetin on blood pressure restoration and synthesis of inflammatory mediators in heart after hemorrhagic shock in anesthetized rats. *Shock.* 33 (1): 83–87.

Yang L, Qian Z, Yang Y, Sheng L, Ji H, Zhou C, Kazi HA (2008). Involvement of Ca^{2+} in the inhibition by crocetin of platelet activity and thrombosis formation. *J. Agric. Food Chem.* 56 (20): 9429–33.

Yang R, Vernon K, Thomas A, Morrison D, Qureshi N, Van Way CW 3rd (2011). Crocetin reduces activation of hepatic apoptotic pathways and improves survival in experimental hemorrhagic shock. *J. Parenter. Enteral. Nutr.* 35 (1): 107–13.

Yoshino F, Yoshida A, Umigai N, Kubo K, Lee MC (2011). Crocetin reduces the oxidative stress induced reactive oxygen species in the stroke-prone spontaneously hyper-tensive rats (SHRSPs) brain. *J. Clin. Biochem. Nutr.* 49 (3): 18–87.

Zargari A (1990). *Medicinal Plant.* Tehran University Press. Tehran. p. 574.

Zhang Z, Wang CZ, Wen XD, Shoyama Y, Yuan CS (2013). Role of saffron and its constituents on cancer chemoprevention. *Pharm. Biol.* 51 (7): 920–24.

Zheng CJ, Li L, Ma WH, Han T, Qin LP (2011). Chemical constituents and bioactivities of the liposoluble fraction from different medicinal parts of *Crocus sativus*. *Pharm. Biol.* 49 (7): 756–63.

Zhong YJ, Shi F, Zheng XL, Wang Q, Yang L, Sun H, He F, Zhang L, Lin Y, Qin Y, Liao LC, Wang X (2011). Crocetin induces cyto-toxicity and enhances vincristine-induced cancer cell death via p53-dependent and -independent mechanisms. *Acta. Pharmacol. Sin.* 32 (12): 1529–36.

Chapter 12

Tamarind
(*Tamarindus indica*)

Tamarindus indica L. commonly known as Tamarind is a multipurpose tropical tree. The leaves, flowers, fruits and seeds are all used in the cuisine in many countries. The name tamarind comes from a Persian word "Tamar-I-hind," meaning dates of India. Its name "Amlika" in Sanskrit indicates its ancient presence in the country. It is used as traditional medicine in India, Africa, Pakistan, Bangladesh, Nigeria, and most of the tropical countries.

The tender leaves of tamarind are traditionally used with lentils in Southern India, replacing the tamarind fruit. Apart from the fruit as a condiment, the leaves and flowers are also used in curries, salads, stews and soups (Bhadoriya *et al.,* 2011).

Tamarind   **Seeds**   **Leaves**

Phytochemical Constituents

Phytochemical investigation on *T. indica* revealed the presence of many active constituents, such as phenolic compounds, cardiac glycosides (Rasu *et al.,* 1989), L-(-) mallic acid (Kobayashi *et al.,* 1996), tartaric acid; mucilage and pectin, arabinose, xylose, galactose, glucose, and uronic acid (Ibrahim and Abbas, 1995; Coutino-Rodriguez *et al.,* 2001). The extracts exhibited the presence of secondary metabolites such as alkaloid, flavonoid, saponin and tannin (Mohamad *et al.,* 2012). The pulp has been reported to contain organic acids, such as tartaric acid, acetic acid, citric acid, formic acid, malic acid, succinic acid, some pyrazines (trans-2-hexenal); along with some thiazoles (2-ethylthiazole, 2-methylthiazole) as fragrance contributors (Hänsel *et al.,* 1992). In the leaves of the plant, two triterpenes, lupanone and lupeol were found (Imam *et al.,* 2007). Pumthong (1999) reported that tamarind seed coat was composed of polyphenols including tannins, anthrocyanidin and oligomeric anthrocyanidins. In addition, extracts of tamarind seed coat contains many polyphenols, including catechin, procyanidin B2, epicatechin, procyanidin trimer, procyanidin tetramer, procyanidin pentamer and procyanidin hexamer (Sudjaroen *et al.,* 2005) all of which might influence melanogenic and melanosomogenic activity.

Nutritional Value

The tamarind pulp contains higher moisture than the seed. (Khairunnuur *et al.,* 2009). The ethanolic extract of *T. indica* showed the presence of fatty acids and various essential elements like calcium, cadmium, copper, iron, sodium, manganese, magnesium, potassium, phosphorus, lead, and zinc (Samina *et al.,* 2008). The pulp contains amino acids; invert sugar (25-30 per cent); pectin; protein and fat. Tamarind

Nutritional Value per 100 g

Nutrient	Amount	Nutrient	Amount
Energy (Kcal)	239	Sodium (mg)	28.0
Carbohydrates (g)	62.5	Potassium (mg)	628.0
Protein (g)	2.8	Calcium (mg)	74.0
Fat (g)	0.6	Copper (mg)	0.86
Dietary Fibre (g)	5.1	Iron (mg)	2.8
Folates (µg)	14.0	Magnesium (mg)	92.0
Niacin (mg)	1.94	Phosphorus (mg)	113.0
Pantothenic acid (mg)	0.14	Sodium (mg)	28.0
Pyridoxine (mg)	0.066	Potassium (mg)	628.0
Thiamine (mg)	0.43	Calcium (mg)	74.0
Vitamin C (mg)	3.5	Selenium (µg)	1.3
Vitamin A (IU)	30.0	Zinc (mg)	0.1
Vitamin E (mg)	0.1	β carotene (µg)	18.0
Vitamin K (µg)	2.8		

Source. USDA National Nutrient data base, Release 26.

seed shows a higher percentage of carbohydrate, protein, fat and energy; and higher Ca and vitamin C content than the pulp. The seed polysaccharides were found to contain β-1,4 glucose, xylose (alpha-1,6), galactose, protein, lipids and some keto acids (Hänsel *et al.*, 1992).

Traditional Therapeutic Uses

Many therapeutic effects of the fruit have been described in Ayurveda, which include its usefulness as an infusion in skin ailments such as rashes caused by allergies; in the treatment of burns, scalds and chaffed skin; as a remedy for dysentery and mucous diarrhea; as an appetizer and liver tonic and as a cardiotonic. It is also used in the treatment of bleeding piles and tumors (Shastri, 1956).

It is used traditionally in abdominal pain, diarrhea and dysentery, helminthes infections, wound healing, malaria, fever, constipation, inflammation, cell cyto-toxicity, gonorrhea, and eye diseases. Owing to the numerous phytochemicals present in it, the plant is reported to possess anti-diabetic, anti-microbial, anti-venomic, anti-oxidant, anti-malarial, hepato-protective, anti-asthmatic, laxative, and anti-hyper-lipidemic activity. The fruit of *T. indica* is used traditionally as a laxative, due to the presence of high amounts of malic and tartaric acids (Irvine, 1961). Tamarind fruit pulp is used as a febrifuge and laxative (Norscia and Borgognini-Tarli, 2006; Kheraro and Adam, 1974) throughout the Sahel and Soudan ecological zones. Children in Madagascar are given whole tamarind fruits for breakfast to overcome constipation. The laxative is taken in the form of a sweetmeat (called Bengal by the Wolof people of Senegal) prepared from the unripe fruit of tamarind and sometimes mixed with lime juice or honey (Dalziel, 1937). The leaves are also used to treat throat infections/ cough, fever, intestinal worm infections, urinary problems and liver ailments. Leaves and pulp act as a cholagogue, laxative and anti-congestant; and exhibit anti-oxidant activity in the liver in addition to their blood sugar-reducing properties (El-Siddig *et al.*, 2006). While in West Africa the bark is used to treat diarrhea, in East Africa the leaves are used for this purpose (Havinga *et al.*, 2010).

A macerate of its leaves with potash has been reported in northern Nigeria as a laxative and purgative; and also to relieve abdominal pain (Bhat *et al.*, 1990). In Burkina Faso, the fruits are crushed and soaked for half a day in water with a little salt before consumption. In Ghana, malaria is treated with tamarind leaves (Asase *et al.*, 2005).

Tamarind bark or leaves are most commonly applied, especially in central West Africa, externally on the cuts, wounds, and abscesses, either as a decoction or as a powder or poultice, alone or in combination with other species (Kheraro and Adam, 1974; Diallo *et al.*, 2002; Inngjerdingen *et al.*, 2004).

Anti-oxidant Property

Good anti-oxidant activity (64.5–71.7 per cent) against the linoleic acid emulsion system was shown by the extracts of *T. indica* but the same was lower than butylated hydroxyl anisole and higher than ascorbic acid (Siddhuraju, 2007). The pulp and seed extracted at 100°C/15 minutes showed the highest FRAP value among its groups

(216.17 ± 14.06 μmol (Fe)/g and 659.74 ± 16.40 μmol (Fe)/g respectively) (Khairunnuur *et al.,* 2009). The extract presented radical scavenging ability *in-vitro,* as assessed by the 2,2-diphenyl-1-picrylhydrazyl (DPPH) and superoxide radical assays; and led to decreased lipid peroxidation in serum, as assessed by the thiobarbituric acid reactive substances (TBARS) assay. The extract improved the efficiency of the anti-oxidant defense system, as assessed by the superoxide dismutase, catalase and glutathione peroxidase activities *in-vivo.* Clinically, the anti-oxidant property of tamarind fruit pulp is proved to be responsible for its significant hypo-lipidemic activity in hyper-cholesterolemic hamsters (Martinello *et al.,* 2006).

Tamarind seed coat exhibited anti-oxidant activity when extracted with ethyl acetate and ethanol (Tsuda *et al.,* 1994; Luengthanaphol *et al.,* 2004). Tamarind seed coat of Thai variety showed the most active anti-oxidant property in terms of peroxide value, DPPH free radical scavenging method using ascorbic acid as a standard as well as TBARS method (Luengthanaphol *et al.,* 2004; Vyas *et al.,* 2009; Osawa *et al.,* 1994). Thus, tamarind seed coat, a by-product of the tamarind gum industry, could be used as a safe and low-cost source of anti-oxidant (Ramos *et al.,* 2003).

Anti-microbial Property

T. indica has a broad spectrum anti-bacterial activity. The methanolic extract of the leaves showed inhibitory potentials against *B. pseudomallei* under *in-vitro* conditions. Further animal studies are needed to confirm the role of tamarind in treating melioidosis (Muthu *et al.,* 2005). Acetone and methanol extracts of tamarind seeds were found active against both Gram-positive and Gram-negative organisms. MIC values of potent extracts against susceptible organisms ranged from 53-380 μg/mL. The seeds were found to be bactericidal (Kothari and Seshadri, 2010). Significant anti-microbial activity against *K. pneumoniae* (Vaghasiya and Chanda, 2009) and *S. paratyphi, B. subtilis, S. typhi,* and *S. aureus* (Doughari, 2006) have been reported. T*amarind* flowers also showed anti-microbial activity against *E. coli* and *S. aureus* (Meléndez and Capriles, 2006; Al-Fatimi *et al.,* 2007).

The fruit pulp extracts exhibited a wide spectrum of activity against tested bacterial strains *i.e.* 95.5 per cent, 90.9 per cent and 86.4 per cent by the cold water, hot water and ethanolic extracts respectively (Nwodo *et al.,* 2011a). The fractions and sub-fractions tested separately and in combinations using the agar well diffusion technique showed susceptibility of Gram negative and Gram positive bacterial strains (Nwodo *et al.,* 2011b; Warda *et al.,* 2007). This anti-bacterial activity is said to be indicative of the presence of broad spectrum antibiotic compounds. The leaf extracts were also shown to be anti-fungal and anti-microbial (Doughari *et al.,* 2006; Abubakar *et al.,* 2010).

The potential of using juice of tamarind to reduce *L. monocytogenes* Scott A and *S. Typhimurium* ATCC 14028 populations on raw shrimps after washing and during storage (4 degrees C) was investigated by Norhana *et al.* (2009). Compared to distilled water, tamarind juice significantly reduced naturally occurring aerobic bacteria (0.40-0.70 log cfu/g), *L. monocytogenes* Scott A (0.84-1.58 log cfu/g) and *S. typhimurium* ATCC 14028 (1.03-2.00 log cfu/g) populations in shrimps immediately after washing

(0 day). The control or tamarind-washed shrimps did not differ in sensory acceptability throughout the storage except for acidic or lemony odour at 0 day in the latter.

Anti-inflammatory Property

Tamarind bark is used traditionally in the treatment of pain. Leaf juice with ginger is used in the treatment of bronchitis (Kerharo and Bouquet, 1950). Rimbau *et al.* (1999) reported that the aqueous, ethanol and chloroform extracts of tamarind exhibited anti-inflammatory activity, supporting its use in the traditional medicine of North-African countries.

The dried and powdered bark, added to water was used for the treatment of eye inflammation (Irvine, 1961). The presence of sterols and triterpenes in the extract seems to be responsible for its analgesic activity (Dighe *et al.,* 2009). The aqueous extract of tamarind was reported to possess anti-nociceptive activity at both the peripheral and central levels, which are mediated *via* activation of the opioidergic mechanism (Khalid *et al.,* 2010). Tamarind fruit pulp extract was found to modulate the neutrophil-mediated inflammatory diseases (Paula *et al.,* 2009). Garba *et al.* (2003) reported that tamarind fruit extract significantly increased the bioavailability of Ibuprofen.

In-vitro exposure of RAW 264.7 cells or peritoneal macrophages to 0.2-200 µg/mL of tamarind extract significantly attenuated (as much as 68 per cent) nitric oxide production induced by lipo-polysaccharide (LPS) and interferon gamma (IFN-gamma) in a concentration-dependent manner. *In-vivo* administration of tamarind extract (100-500 mg/kg) to B6C3F1 mice dose-dependently suppressed TPA, LPS and/or IFN-gamma induced production of nitric oxide in isolated mouse peritoneal macrophages in the absence of any effect on body weight (Komutarin *et al.,* 2004).

Tamarind in Fluorosis

Hydro-methanolic (1:1) extract of tamarind fruit pulp was found to reduce fluoride concentration in blood and bone; and enhance urinary fluoride excretion, indicating the fluoride toxicity ameliorative potential of tamarind in rats (Dey *et al.,* 2011). Thus, tamarind intake is likely to help in delaying progression of fluorosis (Khandare *et al.,* 2002, 2004).

Significant improvements in carbohydrate and lipid profiles occurred as evidenced by decreased plasma glucose and lipid levels, lipid peroxidation, increased hepatic glycogen content, hexokinase activity and cholesterol excretion, with simultaneous improvement in anti-oxidant profiles of both hepatic and renal tissues of fluoride-exposed animals (Vasant and Narasimhacharya, 2012).

Sharma *et al.* (2012) reported the protective role of diet supplements including spirulina, tamarind fruit pulp and their combination on toxicity of fluoride (F-) (10 ppm), Al(+3) (3 ppm) and aluminium fluoride (AlF3) (35.4 ppm) in freshwater fish *G. affinis.*

Consumption of aqueous extracts of *T. indica* fruit pulp (100 mg/kg body weight) orally once daily for 90 days lowered plasma fluoride concentrations; and controlled hepatic and renal damage in rabbits receiving fluorinated drinking water (200 mg NaF/Liter water) (Ranjan *et al.,* 2009).

Defluoridation of Water

Defluoridation of water was achieved at pH 7.0 with tamarind seed and the capacity decreased with increase in temperature and particle size. Further, it was found that defluoridation follows first order kinetics and Langmuir adsorption isotherm. Ninety percent desorption was achieved with 0.1 N HCl indicating the involvement of energetic forces such as coulombic interaction in sorption (Murugan and Subramanian, 2006).

Tamarind fruit shells (TIFSs) are naturally calcium rich compounds. They were impregnated with ammonium carbonate and then carbonized, leading to ammonium carbonate activated ACA-TIFS carbon. Virgin fruit shells are useful as adsorbent for the removal of fluoride anions from groundwater. The fluoride removal was greater in groundwater without hydrogen carbonate ions than those containing these ions. The fluoride scavenging ability of TIFS carbons was due to naturally dispersed calcium compounds. X-ray diffraction showed that TIFS carbon contained a mixture of calcium oxalate and calcium carbonate (Sivasankar *et al.,* 2010; 2012).

Obesity

The ethanolic extract of tamarind fruit pulp showed a significant weight-reducing and hypo-lipidemic activity in cafeteria diet and sulpiride-induced obese rats (Jindal *et al.,* 2011) while the aqueous extract decreased the body weight with the concomitant reduction in the levels of plasma total cholesterol, low-density lipoprotein, and triglyceride, and increased high-density lipoprotein (Azman *et al.,* 2012). Moreover, the extract decreased plasma leptin and fatty acid synthase activity; and enhanced the efficiency of the anti-oxidant defense system. The extract also exhibited anti-obesity effects, as indicated by a significant reduction in adipose tissue weights, as well as lowering the degree of hepatic steatosis in the obesity-induced rats. It is evident that the extract possesses hepato-protective activity (Azman *et al.,* 2012).

Diabetes

The aqueous extract of *T. indica* seeds given to mild and severe diabetic rats significantly reduced the blood glucose (Maiti *et al,* 2004) and lipid level (Maiti *et al,* 2005). It was found that the anti-diabetic effect of aqueous tamarind seed extract in STZ-induced diabetes resulted from complex mechanisms of β-cell neogenesis, calcium [Ca^2z] (I) handling, GLUT-2, GLUT-4, and SREBP-1c gene in liver (Sole and Srinivasan, 2012; Sole *et al.,* 2013). Supplementation with the aqueous extract of tamarind seeds ameliorated metabolic syndrome due to the improved insulin action (Shahraki *et al.,* 2011). Aqueous extract of tamarind seeds partially restores pancreatic beta cells and repairs STZ-induced damages in rats (Hamidreza *et al.,* 2010).

CVD

Tamarind extract has a high potential in diminishing the risk of atherosclerosis in humans (Martinello *et al.,* 2006). The hydro-alcoholic extract of tamarind pulp was shown to influence the mediator system of inflammation (Landi Librandi *et al.,* 2007). Dried and pulverized pulp of tamarind fruits at a dose of 15 mg/kg body weight was found to reduce total cholesterol level and LDL-cholesterol level to a significant extent. The fruits exerted no conspicuous effect on body weight and systolic blood pressure, but significantly reduced the diastolic pressure (Iftekhar *et al.,* 2006).

Cancer

T. indica has shown to have an effect on the cellular system. The methanolic extract of the fruit L-(-)-Di-n-butyl maleate exhibited a pronounced cyto-toxicity against sea urchin embryo cells but L-(-)-Di-n-pentyl maleate was a stronger inhibitor (Kobayashi *et al.,* 1996). Tamarind seed extract decreased both oxidative stress markers and, although there was no statistical difference, it delayed the progress of renal cell carcinoma and decreased its incidence (21 per cent) establishing its reno-protective effects (Vargas-Olvera *et al.,* 2012).

Immuno-modulators are well known for their anti-tumor activity and polysaccharides are very proficient immune-modulators. Polysaccharide PST001 isolated from the seed kernel of tamarind has immune-modulatory and tumor inhibitory activities as shown by the increase in total WBC, CD4(+) T-cell population, and bone marrow cellularity. Hence it has the potential to be developed as an anti-cancer agent *i.e.* as a sole agent or as an adjuvant to other chemo-therapeutic drugs (Aravind *et al.,* 2009; 2012; Rolando and Valente, 2007; Sreelekha *et al.,* 1993). Nie and Deters (2013) showed that tamarind xyloglucans promote skin regeneration by a direct influence on cell proliferation and migration.

Other Benefits

The folkloric use of tamarind in the management of diarrhea is well justified. The extract prepared from the dried ripe tamarind fruit (5.0 mg/ml and 10.0 mg/ml) with the seeds intact produced dose-dependent relaxing effects on spontaneous rabbit's jejunum preparations through calcium channel blockade (Ali and Shah, 2010).

A significant hepato-regenerative effect was observed with the aqueous extracts of different parts of *T. indica,* such as fruits, leaves (350 mg/kg p.o.), and unroasted seeds (700 mg/kg p.o.) (Pimple, 2007). The methanolic extract of tamarind leaves exhibited significant anti-histaminic, adaptogenic, and mast cell stabilizing activity in laboratory animals (Tayade *et al.,* 2009). The leaf extract demonstrated promising anti-apoptotic hepato-protective effects in rats (Ghoneim and Eldahshan, 2012).

Significant wound healing efficiency of tamarind seed is exhibited by all the extracts (phosphate buffer saline, water, methanol and ethanol). Phosphate buffer saline induced complete wound healing in shortest period (10 days) while water extract, methanol extract and Solcoseryl ointment treatment induced complete wound healing in 11 days and control groups without any treatment took 14 days (Mohamad *et al.,* 2012).

Tamarind fruit extract significantly increased the bioavailability of aspirin as there was a statistically significant increase in the plasma levels of aspirin and salicylic acid, respectively, when the meal containing tamarind fruit extract was administered with the aspirin tablets than when taken under fasting state or without the fruit extract (Mustapha *et al.*, 1996).

Conclusions

Tamarind, a commonly used souring agent in cookery has been proved to possess innumerable health benefits. Being a less expensive condiment, the therapeutic potential of various parts of tamarind especially leaves could be further explored and popularized.

References

Abubakar MG, Yerima MB, Zahriya AG, Ukwuani AN (2010). Acute toxicity and anti-fungal studies of ethanolic leaves, stem and pulp extract of *Tamarindus indica*. *Res. J. Pharm. Biol. Chem. Sci.* 1 (4): 104–11.

Al-Fatimi M, Wurster M, Schroder G, Lindequist U (2007). Anti-oxidant, anti-microbial and cyto-toxic activities of selected medicinal plants from Yemen. *J. Ethnopharmacol.* 111: 657–66.

Ali N, Shah SW (2010). Spasmolytic activity of fruits of *Tamarindus indica* L. *J. Young Pharmacists*. 2: 261–64.

Aravind SR, Joseph MM, Varghese S, Balaram P, Sreelekha TT (2012). Anti-tumor and immune-potentiating activity of polysaccharide PST001 isolated from the seed kernel of *Tamarindus indica*: an *in-vivo* study in mice. *Scientific World Journal*. 2012: 361382.

Aravind SR, Vijayan KK, Prabha B, Sreelekha TT (2009). Apoptotic evaluation of polysaccharide PST001 in tumor cells. In: *Proceedings of the 21st Kerala Science Congress*. pp. 553–56.

Asase A, Oteng-Yeboah AA, Odamtten GT, Simmonds MSJ (2005). Ethnobotanical study of some Ghanaian anti-malarial plants. *J. Ethnopharmacol.* 99: 273–79.

Azman KF, Amom Z, Azlan A, Esa NM, Ali RM, Shah ZM, Kadir KK (2012). Anti-obesity effect of *Tamarindus indica* L. pulp aqueous extract in high-fat diet-induced obese rats. *J. Nat. Med.* 66 (2): 333–42.

Bhadoriya SS, Ganeshpurkar A, Narwaria J, Rai G, Jain AP (2011). *Tamarindus indica*: Extent of explored potential. *Pharmacogn. Rev.* 5 (9): 73 – 81.

Bhat RB, Eterjere EO, Oladipo VT (1990). Ethnobotanical studies from Central Nigeria. *Econ. Bot.* 44: 382–90.

Coutino-Rodriguez R, Hernandez-Cruz P, Gillis-Rios H (2001). Lectins in fruits having gastro-intestinal activity and their participation in the hem-agglutinating property of *Escherichia Coli* 0157. *Arch. Med. Res.* 32: 251–59.

Dalziel JM (1937). *The Useful Plants of West Tropical Africa*. Crown Agents for Overseas Governments and Administrations. London. p. 612.

Dey S, Swarup D, Saxena A, Dan A (2011). *In-vivo* efficacy of tamarind (*Tamarindus indica*) fruit extract on experimental fluoride exposure in rats. *Res. Vet. Sci.* 91 (3): 422–25.

Diallo D, Sogn C, Samaké FB, Paulsen BS, Michaelsen TE, Keita A (2002). Wound healing plants in Mali, the Bamako region. An ethnobotanical survey and complement fixation of water extracts from selected plants. *Pharm. Biol.* 40: 117–28.

Dighe NS, Pattan1 SR, Nirmal SA, Kalkotwar RS, Gaware VM, Hole MB (2009). Analgesic activity of *Tamarindus indica. Res. J. Pharmacogn. Phytochem.* 1: 69–71.

Doughari JH (2006). Anti-microbial activity of *Tamarindus indica* Linn. *Trop. J. Pharm. Res.* 5 (2): 597–603.

El-Siddig K, Gunasena HPM, Prasada BA, Pushpkumara DKNG, Ramana KVR, Vijayanand P, *et al.* (2006). Tamarind: *Tamarindus indica* L. Southampton Centre for Underutilized Crops. Southampton.

Garba M, Yakasai IA, Bakare MT, Munir HY (2003). Effect of *Tamarindus indica.* L on the bioavailability of ibuprofen in healthy human volunteers. *Eur. J. Drug Metab. Pharmacokinet.* 28 (3): 179–84.

Ghoneim AI, Eldahshan OA (2012). Anti-apoptotic effects of tamarind leaves against ethanol-induced rat liver injury. *J. Pharm. Pharmacol.* 64 (3): 430–38.

Hamidreza H, Heidari Z, Shahraki M, Moudi B (2010). A stereological study of effects of aqueous extract of *Tamarindus indica seeds* on pancreatic islets in streptozotocin-induced diabetic rats. *Pak. J. Pharm. Sci.* 23 (4): 427–34.

Hänsel R, Keller K, Rimpler H, Schneider G, editors. 5th Ed. (1992). Springer Verlag. Berlin. Hagers Handbuch der Pharmzeutischen Praxis; p. 893.

Havinga RM, Hartl A, Putscher J, Prehsler S, Buchmann C, Vogl CR (2010). *Tamarindus indica* L. (Fabaceae): patterns of use in traditional African medicine. *J. Ethnopharmacol.* 127 (3): 573–88.

Ibrahim E, Abbas SA (1995). Chemical and biological evaluation of *Tamarindus indica* L. growing in Sudan. *Acta. Hortic.* 390: 51–57.

Iftekhar AS, Rayhan I, Quadur MA, Akhteruzzaman SF, Hasnat A (2006). Effect of *Tamarindus indica* fruits on blood pressure and lipid-profile in human model–an *in-vivo* approach. *J. Pharm. Sci.* 19: 125–29.

Imam S, Azhar I, Hasan MM (2007). Two triterpenes lupanone and lupanol isolated and identified from *Tamarindus indica. Pak. J. Pharm. Sci.* 20: 125–27.

Inngjerdingen K, Nergard CS, Diallo D, Mounkoro PP, Paulsen BS (2004). An ethnopharmacological survey of plants used for wound healing in Dogonland Mali, West Africa. *J. Ethnopharmacol.* 92: 233–44.

Irvine FR (1961). *Woody Plants of Ghana.* Oxford University Press. London.

Jindal V, Dhingra D, Sharma S, Parle M, Harna RK (2011). Hypolipidemic and weight reducing activity of the ethanolic extract of *Tamarindus indica* fruit pulp in cafeteria diet- and sulpiride-induced obese rats. *J. Pharmacol. Pharmacother*. 2 (2): 80–84.

Khairunnuur FA Jr, Zulkhairi A, Azrina A, Moklas MM, Khairullizam S, Zamree MS, Shahidan MA (2009). Nutritional composition, *in-vitro* anti-oxidant activity and artemia salina L. lethality of pulp and seed of *Tamarindus indica* L. extracts. *Malays. J. Nutr*. 15 (1): 65–75.

Khalid S, Shaik Mossadeq WM, Israf DA, Hashim P, Rejab S, Shaberi AM Mohamad AS, Zakaria ZA,Sulaiman MR (2010). *In-vivo* analgesic effect of aqueous extract of *Tamarindus indica* L. fruits. *Med. Princ. Pract*. 19: 255–59.

Khandare AL, Kumar P U, Shanker RG, Venkaiah K, Lakshmaiah N (2004). Additional beneficial effect of tamarind ingestion over defluoridated water supply to adolescent boys in a fluorotic area. *Nutrition*. 20 (5): 433–36.

Khandare AL, Rao GS, Lakshmaiah N (2002). Effect of tamarind ingestion on fluoride excretion in humans. *Eur. J. Clin. Nutr*. 56 (1): 82–85.

Kheraro J, Adam JG (1974). La pharmacopée sénégalaise traditionnelle, plantes médicinales et Toxiques. Vigot *et* Frères. Paris. p. 1011.

Kerharo J, Bouquet A (1950). Plantes Médicinales *et* Toxiques *de la* Côte d2Ivoire *et* Haute-Volta. Vigot Freres. Paris. p. 297.

Kobayashi A, Adenan ML, Kajiyama SI, Kanzaki H, Kawazu K (1996). A cytotoxic principle of *Tamarindus indica*, di-n-butyl malate and the structure-activity relationship of its analogues. *J. Bio. Sci*. 51: 233–42.

Komutarin T, Azadi S, Butterworth L, Keil D, Chitsomboon B, Suttajit M, Meade BJ (2004). Extract of the seed coat of *Tamarindus indica* inhibits nitric oxide production by murine macrophages *in-vitro* and *in-vivo*. *Food Chem. Toxicol*. 42 (4): 649–58.

Kothari V, Seshadri S (2010). *In-vitro* anti-bacterial activity in seed extracts of *Manilkara zapota, Anona squamosa*, and *Tamarindus indica*. *Biol. Res*. 43 (2): 165–68.

Landi Librandi AP, Chrysóstomo TN, Azzolini AE, Recchia CG, Uyemura SA, de Assis-Pandochi AI (2007). Effect of the extract of Tamarind fruit of the complement system: Studies *in-vitro* and in hamsters submitted to a cholesterol-enriched diet. *Food Chem. Toxicol*. 45: 1487–95.

Luengthanaphol S, Mongkholkhajornsilp D, Douglas S, Douglas PL, Pengsopa L, Pongamphai S (2004). Extraction of anti-oxidants from sweet Thai Tamarind seed coat: Preliminary experiments. *J. Food Eng*. 63: 247–52.

Maiti R, Jana D, Das UK, Ghosh D (2004). Anti-diabetic effect of aqueous extract of seed of *Tamarindus indica* in streptozotocin-induced diabetic rats. *J. Ethnopharmacol*. 92: 85–91.

Maiti R, Das UK, Ghosh D (2005). Attenuation of hyper-glycemia and hyper-lipidemia in streptozotocin-induced diabetic rats by aqueous extract of seeds of *Tamarindus indica*. *Biol. Pharm. Bull*. 28: 1172–76.

Martinello F, Soaresh SM, Franco JJ, Santos AC, Sugohara A, Garcia SB, *et al.* (2006). Hypo-lipemic and anti-oxidant activities from *Tamarindus indica* pulp fruit extract in hyper-cholesterolemic hamsters. *Food Chem Toxicol.* 44: 810–18.

Melendez PA, Capriles VA (2006). Anti-bacterial properties of tropical plant fruit from Puerto Rico. *Phytomedicine.* 13: 272–76.

Mohamad Yusof bin, Akram HB, Bero DN (2012). Tamarind Seed Extract Enhances Epidermal Wound Healing. *Int. J. Biol.* 4 (1): 81–88.

Murugan M, Subramanian E (2006). Studies on defluoridation of water by tamarind seed, an unconventional biosorbent. *J. Water Health.* 4 (4): 453–61.

Mustapha A, Yakasai IA, Aguye IA (1996). Effect of *Tamarindus indica* L. on the bioavailability of aspirin in healthy human volunteers. *Eur. J. Drug Metab. Pharmacokinet.* 21 (3): 223–26.

Muthu SE, Nandakumar S, Rao UA (2005). The effect of methanolic extract of *Tamarindus indica* Linn. on the growth of clinical isolates of *Burkholderia pseudomallei. Indian J. Med. Res.* 122 (6): 525–28.

Nie W, Deters AM (2013). Tamarind seed Xyloglucans promote proliferation and migration of human skin cells through internalization *via* stimulation of pro-proliferative signal transduction pathways. *Dermatol. Res. Pract.* 2013: 359756.

Norhana MN, Azman MN, Poole SE, Deeth HC, Dykes GA (2009). Effects of bilimbi (*Averrhoa bilimbi* L.) and tamarind (*Tamarindus indica* L.) juice on *Listeria monocytogenes* Scott A and *Salmonella Typhimurium* ATCC 14028 and the sensory properties of raw shrimps. *Int. J. Food Microbiol.* 136 (1): 88–94.

Norscia I, Borgognini-Tarli SM (2006). Ethnobotanical reputation of plant species from two forests of Madagascar: A preliminary investigation. *S. Afr. J. Bot.* 72: 656–60.

Nwodo UU, Iroegbu CU, Ngene AA, Chigor VN, Okoh AI (2011a). Effects of fractionation and combinatorial evaluation of *Tamarindus indica* fractions for anti-bacterial activity. *Molecules.* 16 (6): 4818–27.

Nwodo UU, Obiiyeke GE, Chigor VN, Okoh AI (2011b). Assessment of *Tamarindus indica* Extracts for Antibacterial Activity. *Int. J. Mol. Sci.* 12 (10): 6385–96.

Osawa T, Tsuda T, Watanabe M, Ohshima K, Yamamoto A (1994). Anti-oxidative components isolated from the seeds of Tamarind (*Tamarindus indica* L). *J. Agric. Food Chem.* 42: 2671–74.

Paula FS, Kabeya LM, Kanashiro A, de Figueiredo AS, Azzolini AE, Uyemura SA, Lucisano-Valim YM (2009). Modulation of human neutrophil oxidative metabolism and degranulation by extract of *Tamarindus indica* L. fruit pulp. *Food Chem. Toxicol.* 47 (1): 163–70.

Pimple BP, Kadam PV, Badgujar NS, Bafna AR, Patil MJ (2007). Protective effect of *Tamarindus indica* Linn. against paracetamol-induced hepato-toxicity in rats. *Indian J. Pharm. Sci.* 69: 827–31.

Pumthong, G, (1999). Anti-oxidative activity of polyphenolic compounds extracted from seed coat of *Tamarindus indica* Linn. *Ph.D. Thesis*, Chiang Mai University, Thailand.

Ramos A, Visozo A, Piloto J, Garcia A, Rodriguez CA, Rivero R (2003). Screening of anti-mutagenicity via anti-oxidant activity in Cuban medicinal plants. *J. Ethnopharmacol.* 87: 241- 46.

Ranjan R, Swarup D, Patra RC, Chandra V (2009). *Tamarindus indica* L. and *Moringa oleifera* M. extract administration ameliorates fluoride toxicity in rabbits. *Indian J. Exp. Biol.* 47 (11): 900–05.

Rasu N, Saleem B, Nawaz R (1989). Preliminary screening of four common Plants of family *Caesalpiniacae. Pak. J. Pharm. Sci.* 2: 55–57.

Rimbau V, Cerdan C, Vila R, Iglesias J (1999). Anti-inflammatory activity of some extracts from plants used in the traditional medicine of north-African countries (II). *Phytother. Res.* 13 (2): 128–32.

Rolando M, Valente C (2007). Establishing the tolerability and performance of tamarind seed polysaccharide (TSP) in treating dry eye syndrome: results of a clinical study. *BMC Ophthalmology.* 7: article 5.

Samina KK, Shaikh W, Shahzadi S, Kazi TG, Usmanghani K, Kabir A, *et al.* (2008). Chemical constituents of *Tamarindus indica.* Medicinal plant in Sindh. Pak. *J. Bot.* 40: 2553–59.

Shahraki MR, Harati M, Shahraki AR (2011). Prevention of high fructose-induced metabolic syndrome in male wistar rats by aqueous extract of (*Tamarindus indica*) seed. *Acta. Med. Iran.* 49 (5): 277–83.

Sharma KP, Upreti N, Sharma S, Sharma S (2012). Protective effect of spirulina and tamarind fruit pulp diet supplement in fish (*Gambusia affinis Baird and Girard*) exposed to sub-lethal concentration of fluoride, aluminum and aluminum fluoride. *Indian J. Exp. Biol.* 50 (12): 897 – 903.

Shastri BN (1956). *The Wealth of India – A Dictionary of Indian Raw Materials and Industrial Products.* CSIR. New Delhi. India. pp. 101–03.

Siddhuraju P (2007). Anti-oxidant activity of polyphenolic compounds extracted from defatted raw and dry heated *Tamarindus indica* seed coat. *LWT Food Sci. Tech.* 40: 982–90.

Sivasankar V, Rajkumar S, Murugesh S, Darchen A (2012). Tamarind (*Tamarindus indica*) fruit shell carbon: A calcium-rich promising adsorbent for fluoride removal from groundwater. *J. Hazard. Mater.* 225–26: 164–72.

Sivasankar V, Ramachandramoorthy T, Chandramohan A (2010). Fluoride removal from water using activated and MnO_2-coated Tamarind Fruit (*Tamarindus indica*) shell: batch and column studies. *J. Hazard. Mater.* 177 (1-3): 719–29.

Sole SS, Srinivasan BP (2012). Aqueous extract of tamarind seeds selectively increases glucose transporter-2, glucose transporter-4, and islets' intracellular calcium levels and stimulates β-cell proliferation resulting in improved glucose homeostasis in rats with streptozotocin-induced diabetes mellitus. *Nutr. Res.* 32 (8): 626–36.

Sole SS, Srinivasan BP, Akarte AS (2013). Anti-inflammatory action of Tamarind seeds reduces hyperglycemic excursion by repressing pancreatic β-cell damage and normalizing SREBP-1c concentration. *Pharm. Biol.* 51 (3): 350–60.

Sreelekha TT, Vijayakumar T, Ankanthil R, Vijayan KK, Nair MK (1993). Immuno-modulatory effects of a polysaccharide from *Tamarindus indica. Anti-Cancer Drugs.* 4 (2): 209–12.

Sudjaroen Y, Haubner R, Wurtele G, Hull WE, Erben G, Spiegelhalder B, Changbumrung S, Bartsch H, Owen RW (2005). Isolation and structure elucidation of phenolic anti-oxidants from Tamarind (*Tamarindus indica* L.) seeds and pericarp. *Food Chem. Toxicol.* 43: 1673–82.

Tayade PM, Ghaisas MM, Jagtap SA, Dongre SH (2009). Anti-asthmatic activity of methanolic extract of leaves of *Tamarindus Indica* Linn. J. Pharm. Res. 2: 944–47.

Tsuda T, Osawa T, Watanabe M, Ohshima K, Yamamoto A (1994). Anti-oxidative components isolated from the seeds of tamarind (*Tamarindus indica* L). *Journal of Agriculture and Food Chemistry.* 42 (12): 2671–74.

USDA, National Nutrition Data Base for standard reference. Release-26, NDB No-09322.

Vaghasiya Y, Chanda S (2009). Screening of some traditionally used Indian plants for anti-bacterial activity against *Klebsiella pneumonia. J. Herb. Med. Toxicol. 3*: 161–64.

Vargas-Olvera CY, Sánchez-González DJ, Solano JD, Aguilar-Alonso FA, Montalvo-Muñoz F, Martínez-Martínez CM, Medina-Campos ON, Ibarra-Rubio ME (2012). Characterization of N-diethylnitrosamine-initiated and ferric nitrilo-triacetate-promoted renal cell carcinoma experimental model and effect of a tamarind seed extract against acute nephro-toxicity and carcinogenesis. *Mol. Cell. Biochem.* 369 (1-2): 105–17.

Vasant RA, Narasimhacharya AV (2012). Ameliorative effect of tamarind leaf on fluoride-induced metabolic alterations. *Environ. Health. Prev. Med.* 17 (6): 484–93.

Vyas N, Gavatia NP, Gupta B, Tailing M (2009). Anti-oxidant potential of *Tamarindus indica* seed coat. *J. Pharm. Res.* 2: 1705–06.

Warda S, Gadir A, Mohamed F, Bakhiet AO (2007). Anti-bacterial activity of *Tamarindus indica* fruit and *Piper nigrum. Res. J. Microbiol.* 2: 824–30.

Chapter 13

Tomato
(*Solanum lycopersicum/*
Lycopersicon esculentum)

Tomato is one of the fruits/vegetables most commonly consumed all over the world. For a long time tomatoes were known by the name *Lycopersicon esculentum*, but recent work by scientists has shown that they are originally part of the genus Solanum. Epidemiological studies revealed that diets rich in fruits and vegetables can prevent CVD probably due to their bioactive compounds as well as their anti-oxidant properties (Palomo *et al.,* 2009) and anti-platelet activities (Fuentes *et al.,* 2012).

Nutritional and Chemical Constituents of Tomato

Tomatoes and tomato products are rich sources of folate, vitamin C, and potassium (Beecher, 1998). Tomatoes provide an optimal mix of dietary anti-oxidants that may be responsible for the reported health benefits of its consumption. Lycopene is the pigment principally responsible for the characteristic deep-red color of ripe tomato fruits and tomato products. It is an acyclic, biologically active carotenoid. Gupta *et al.* (2013) indicated that average lycopene content of the tomato was 11.6 – 14.0 mg/kg tomato weight.

A newly developed non-genetically modified purple tomato V118 possessed additional phytochemicals like anthocyanins, which can potentially have added health benefits (Li *et al.,* 2011). Naringenin (42,5,7-trihydroxyflavanone), one of the most abundant flavonoids in citrus, grapefruits and tomatoes, has been reported to exhibit interesting pharmacological activities, such as anti-oxidation, anti-inflammation and anti-tumour (Park *et al.,* 2012). The resveratrol concentration in the tomato skin was 18.4 ±1.6 μg/g dry weight. No stilbenes were detected in the flesh of tomato fruit (Ragab *et al.,* 2006).

Nutritional Value of Raw Red Tomatoes per 100 g

Nutrient	Amount	Nutrient	Amount
Energy (kcal)	18	Niacin (mg)	0.594
Carbohydrates (g)	3.9	Vitamin B$_6$ (mg)	0.08
- Sugars (g)	2.6	Vitamin C (mg)	14
- Dietary fiber (g)	1.2	Vitamin E (mg)	0.54
Protein (g)	0.9	Vitamin K (μg)	7.9
Water (g)	94.5	Magnesium (mg)	11
Vitamin A equiv. (μg)	42 5	Manganese (mg)	0.114
beta-carotene (μg)	449	Phosphorus (mg)	24
lutein and zeaxanthin (μg)	123	Potassium (mg)	237
Thiamine (mg)	0.037	Lycopene (μg)	2573

Source: USDA Nutrient Database.

Tomato Products

The bioavailability of lycopene varied significantly depending on the administered matrix. Unlike many other natural compounds, lycopene is generally stable to processing when present in the plant tissue matrix (Story *et al.,* 2010). Lycopene from tomato oleoresin capsules and tomato juice was better absorbed from the intestine than lycopene from raw tomatoes (Böhm and Bitsch, 1999). Lycopene bioavailability from processed tomato products (*e.g.* tomato paste, tomato sauce etc.) especially those cooked in oil is higher than from unprocessed fresh tomatoes (Stahl W, Sies, 1992; Gartner *et al.,* 1997; Shi, Le Maguer, 2000). Thus, the bioavailability of lycopene in processed tomato products is higher than in unprocessed fresh tomatoes. It breaks down the cell walls, thus weakening the bond between lycopene and tissue matrix and making lycopene more accessible and enhancing the cis-isomerization. Frozen foods and heat-sterilized foods exhibit excellent lycopene stability throughout their normal temperature storage shelf life. Thermal processing (bleaching, retorting, and freezing processes) generally cause some loss of lycopene in tomato-based foods. Heat induces isomerization of the all-trans to cis forms (Shi and Le Maguer, 2000). Cooked tomato products may be preferable to the raw vegetable or juices derived from tomatoes bearing on absorption of the active principles.

Optimally, absorption of lycopene, a highly lipid-soluble chemical, is improved in the presence of a small, but essential amount of oil or fat (Weisburger, 1998). Homogenization, heat treatment, and the incorporation of oil in processed tomato products (tomato paste, tomato sauce, or pizza cooked in oil) leads to increased lycopene bioavailability while some of the same processes cause significant loss of other nutrients. Nutrient content is also affected by variety and maturity of tomatoes (Gartner *et al.,* 1997; Willcox *et al.,* 2003).

Technological processing, packaging materials, and storage conditions have an impact on the nutritional quality of tomato products by affecting the stability of anti-oxidant nutrients to different extents. But, the total lycopene, total phenolic compounds, and total flavonoids remained almost stable during storage for 12 months, regardless of the packaging material used, indicating that tomato juice maintains its nutritional value in terms of anti-oxidant composition during their shelf life. It is the lipophilic total anti-oxidant activity that remained substantially stable throughout the storage trial while the hydrophilic total anti-oxidant activity paralleled the losses in ascorbic acid content (García-Alonso *et al.,* 2009).

Tomato as Anti-oxidant

Lycopene is one of the most potent anti-oxidants, (Miller *et al.,* 1996; Mortensen and Skibsted, 1997; Woodall *et al.,* 1997; DiMascio *et al.,* 1989) with a singlet-oxygen-quenching ability which is twice as high as that of β-carotene and 10 times higher than that of α-tocopherol (DiMascio *et al.,* 1989).

Supplementation with lycopene in the form of juice or capsules for 4 months significantly increased total anti-oxidant capacity and decreased lipid peroxidation, protein oxidation, and bone resorption (Mackinnon *et al.*, 2011a). Walfisch *et al.* (2003) showed that tomato-oleoresin supplementation increases lycopene concentrations in serum, adipose tissue and skin. Tomato product consumption can affect not only the lycopene status, but also that of other anti-oxidant micro-constituents–beta-carotene and lutein (Tyssandier *et al.,* 2004; Riso *et al.,* 2004; *Takeoka et al., 2001*). Lycopene, and beta-carotene, are apparently the main tomato micro-constituents responsible for the effect of tomato products on anti-oxidant status. Consumption of tomato products increased the lycopene concentration and vitamin C concentrations in plasma and lymphocytes, while reducing oxidative stress effectively (Riso *et al.,* 2004; Rao, 2004). Sakamoto *et al.* (1994) found the serum lycopene level to increase >3 times following the ingestion of 2 or 3 cans of tomato juice. Although the β-carotene content of the tomato juice was about 1/13th that of lycopene, the β-carotene level in serum was about double in the subjects who ingested 3 cans daily, suggesting that continued ingestion of tomato juice raises the serum levels of lycopene and β-carotene.

Thus a regular intake of small amounts of tomato products can offer increased cell protection from DNA damage induced by oxidant species. This effect may also originate from the synergism of different anti-oxidants present in tomatoes. Tomato products reduced the oxidative stress induced by post-prandial lipidemia and associated inflammatory response (Burton-Freeman *et al.,* 2012) by inhibiting iNOS proteins and mRNA expressions in mouse macrophage cell lines. Furthermore, cyclooxygenase-2 (COX-2) protein and mRNA expression were not affected by treatment with lycopene (Rafi *et al.,* 2007).

Tomato peels had the highest anti-oxidant activity, both, as measured by the FRAP (46.9 ± 0.9 μmol Fe(+2)/g), and the DPPH assays (97.4 ± 0.2 per cent, 1000 μg/mL). Pomace extracts showed the highest anti-platelet activity induced by ADP, collagen, TRAP-6, and arachidonic acid. The maturation stage of tomato fruit affected the anti-oxidant property (Fuentes *et al.,* 2013). Significant anti-oxidant activity was found in both the hexane (containing lycopene) and the methanol fraction, which contained the phenolic anti-oxidants caffeic and chlorogenic acid suggesting that in addition to lycopene, polyphenols in tomatoes may also be important in conferring protective anti-oxidative effects (Takeoka *et al.*, 2001). This explains the report by Gitenay *et al.* (2007) that tomatoes, containing or not containing lycopene, have a higher potential than lycopene to attenuate and/or to reverse oxidative stress-related parameters in a mild oxidative stress context. The consumption of more than seven servings per week of tomato-based products has been associated with a 30 per cent reduction in the relative risk of CVD (Sesso *et al.*, 2003).

The health benefits of tomato seed oil (TSO) have been suggested to be related to its anti-oxidant activity, TSO was able to counteract spontaneous and H_2O_2-induced oxidative stress in human macrophages, limiting intracellular ROS production and controlling oxidative stress signaling. In particular, TSO was able to decrease the phosphorylation of the MAPK ERK1/2, JNK, and p-38, activation of the redox-sensitive NF-kB, and expression of the heat shock proteins 70 and 90. When the anti-oxidant capacity of TSO was compared with that of purified lycopene, inhibition of ROS production by TSO was remarkably higher. This was due to the high content of other antioxidants in TSO, including (5Z), (9Z), (13Z) and (15Z) lycopene isomers, β-carotene, lutein, γ-tocopherol, and α-tocopherol (Müller *et al.*, 2013).

Lycopene protects the pancreatic tissues from oxidative damage induced by cerulein, and this effect possibly involves the inhibition of neutrophil infiltration and lipid peroxidation. These results suggest that high dietary intake of tomatoes may have protective effects against acute pancreatitis (Ozkan *et al.*, 2012).

Lavelli *et al.* (2001) reported that tomato powders have multi-functional properties, which could address the prevention of oxidative degradations both in foods and *in-vivo*. Therefore, they recommended that tomato can be regarded as source of food additives for fortification and stabilization, even if it is submitted to technological processes that can cause the loss of the more labile hydrophilic antioxidants.

Therapeutic Benefits of Tomato

Consumption of processed tomato products is of significant health benefits which can be attributed to a combination of naturally occurring non-nutritional and nutritional components in tomatoes. Lycopene, the predominant carotenoid in tomato, contributes majorly to this effect (Basu and Imrhan, 2007). The ability to increase lycopene levels in tissues is one of the prerequisites for using it as a food supplement with health benefits. Moreover, tomato-based food products are the primary dietary sources of lycopene (Clinton, 1998). Epidemiologic studies also suggest that anti-oxidant capacity is improved by the consumption of tomato products, thereby decreasing the risk of the development of diseases related to oxidative stress (Burney *et al.*, 1989; Franceschi *et al.*, 1994; Giovannucci *et al.*, 1995; Parfitt *et al.*, 1994).

CVD

Consumption of tomatoes stands out for its effect on platelet anti-aggregation activity and endothelial protection both beneficial for cardiovascular health (Palomo *et al.,* 2012). The cardio-protective functions provided by tomatoes may include the reduction of low-density lipoprotein (LDL) cholesterol, homocysteine, platelet aggregation, and blood pressure (Willcox *et al.,* 2003). Tomato significantly attenuated high-fat meal-induced LDL oxidation and rise in interleukin-6, a pro-inflammatory cytokine and inflammation marker. The relevance of oxidized low-density lipoprotein and inflammation to vascular injury suggests a potentially important protective role of tomato in reducing cardiovascular disease risk (Burton-Freeman *et al.,* 2012).

A high dietary intake of tomato products significantly reduced LDL cholesterol levels, and increased LDL resistance to oxidation in healthy normo-cholesterolaemic adults. According to Maruyama *et al.* (2001) alpha-tocopherol is a major determinant in protecting LDL from oxidation, while lycopene from tomato juice may protect phospholipid in LDL, from oxidation. Thus, oral intake of lycopene might be beneficial for ameliorating atherosclerosis. These athero-protective features are associated with changes in serum lycopene, beta-carotene and gamma-carotene levels (Silaste *et al.,* 2007). Lycopene is reported to inhibit the oxidative modification of LDL as dietary supplementation of tomato's lycopene (60 mg/day) to 6 males for a 3 months period resulted in a significant 14 per cent reduction in their plasma LDL cholesterol concentrations (Fuhrman *et al.,* 1997). Data in some studies suggest an inverse association of lycopene with intima-media thickness, atherosclerosis and CVD (Rissanen *et al.,* 2000; 2001; 2002).

There is epidemiologic and clinical evidence that high intake of lycopene, vitamin E, and vitamin C may be associated with a decreased risk of CHD (Diaz *et al.,* 1997; Kohlmeier *et al.,* 1997). Bub *et al.* (2000) concluded that the additional consumption of carotenoid-rich vegetable products enhanced lipoprotein carotenoid concentrations, but only tomato juice reduced LDL oxidation in healthy men.

Owing to their lipo-philic nature, lycopene and other carotenoids are found to concentrate in low-density and very-low-density lipoprotein fractions of the serum (Clinton, 1998). Meta-analysis of intervention studies between 1955 and September 2010 suggests that lycopene in doses ≥25 mg daily is effective in reducing LDL cholesterol by about 10 per cent comparable to the effect of low doses of statins in patients with slightly elevated cholesterol levels (Ried and Fakler, 2011). Comparing the effects of lycopene, beta-carotene and (5Z)-lycopene on ROS production, cell growth and apoptosis show that lycopene and its isomer were more effective than beta-carotene in counteracting the dangerous effects of 7-KC in human macrophages suggesting that lycopene may act as a potential anti-atherogenic agent (Palozza *et al.,* 2010b).

Although the anti-oxidant property of lycopene may be one of the principal mechanisms for lycopene in the prevention of CHD, other mechanisms may also be responsible (Rao, 2002). One of the possible mechanism for its protective activities is by down-regulation of the inflammatory response through the inhibition of pivotal pro-inflammatory mediators, such as the reduction of reactive oxygen species, the inhibition of synthesis and release of pro-inflammatory cytokines, changes in the expression of cyclo-oxygenase and lipoxygenase, modifications of eicosanoid synthesis, and modulation of signal transduction pathways, including that of the inducible nitric oxide synthase via its inhibitory effects on Nuclear Factor-kB (NF-kB), Activated protein-1 (AP-1) and mitogen-activated protein kinase (MAPK) signaling (Palozza *et al.*, 2010a). Lycopene may also enhance LDL degradation. Available evidence suggests that intimal wall thickness and risk of myocardial infraction are reduced in persons with higher adipose tissue concentrations of lycopene (Arab and Steck, 2000).

Consumption of a lycopene-rich food reduced inflammation in people who are overweight or obese in a total of 106 overweight or obese female students of the Tehran University of Medical Sciences. Serum concentrations of IL-8 and TNF-α decreased significantly in the overweight group while in obese subjects, serum IL-6 concentration was decreased with no differences in IL-8 and TNF-α. Thus, increasing tomato intake may provide a useful approach for reducing the risk of inflammatory diseases such as CVD and diabetes, which are associated with obesity (Ghavipour *et al.*, 2013).

Karppi *et al.* (2012) demonstrated that high serum concentrations of lycopene, as a marker of intake of tomatoes and tomato-based products, decrease the risk of any stroke and ischemic stroke in men. Serum lycopene level changes with dietary lycopene intake irrespective of the amount of fat intake. A diet high in olive oil and rich in lycopene may decrease the risk of coronary heart disease by improving the serum lipid profile compared with a high-carbohydrate, low-fat, lycopene-rich diet (Ahuja *et al.*, 2006). Cholesterol reduction was correlated with lycopene uptake (Jacob *et al.*, 2008). A synergic antioxidant activity has been noticed between the n-3 polyunsaturated and tomato anti-oxidants upon supplementation with n-3 polyunsaturated fatty acids enriched tomato juice on the serum lipid profile through an increase in overall flow-mediated dilatation (García-Alonso *et al.*, 2012; Xaplanteris *et al.*, 2012).

Jacques *et al.* (2013) expressed that correct method of recording dietary lycopene intake is necessary to confirm its health benefits. They reported an inverse relationship between long term consumption (9-11 years) of lycopene and the risk of CVD and CHD but not stroke.

Supplementation of lycopene in low quantity has been suggested as a preventive measure for contrasting and ameliorating many aspects of CVD (Mordente *et al.* 2011). But on the other hand, there are reports that whole tomato or its products are more effective than lycopene. In a human clinical study, total anti-oxidant capacity and phenolic contents in plasma were increased after administration of fresh tomato and tomato juice, but no significant difference was found for lycopene drink consumption. Triglyceride levels and low-density lipoprotein cholesterol were decreased after administration of fresh tomato and tomato juice, and high-density lipoprotein cholesterol was increased (Shen *et al.*, 2007).

Hsu *et al.* (2008) observed that 9 per cent tomato paste reduced serum TC and LDL levels by 14.3 per cent and 11.3 per cent respectively. Hyper-lipidemic rats supplemented with tomato powder, tomato paste, or ketchup showed significant improvement in almost all the parameters studied compared to the positive control group.

Many foods are identified as hypolipidaemic as they reduce serum LDL, VLDL and triglyceride levels, but very rarely foods/food components increase HDL concentration. Interstingly, 20 mg followed by 10 mg of lycopene/kg of diet from tomato paste caused significant elevation in high-density lipoprotein cholesterol comparable to that of the negative control group (Ibrahim *et al.,* 2008). Concentrations of 3 per cent and 9 per cent of tomato paste after 8 weeks of feeding significantly increased serum HDL levels, by 19.4 per cent and 28.8 per cent respectively. The authors revealed by two dimension-gel electrophoresis (2-DE) analysis that carbonic anhydrase III (CAIII) and adenylate kinase 2 (AK2) may be two important regulators involved in the anti-lipid and anti-oxidant effects of tomato paste in hamsters. Cuevas-Ramos *et al.* (2013) also found raw tomato consumption produced a favourable effect on HDL-C levels in overweight women.

Women who consumed more than 10 servings of tomato products per week showed around 30 per cent reductions in the risk of CVD (Sesso *et al.,* 2003 and 2012). But, Shidfar *et al.* (2011) observed that 200 g of raw tomato per day had a favourable effect on blood pressure and apoA-I thus to be beneficial for reducing cardiovascular risk associated with type 2 diabetes. Similarly, tomato extract brought about a clinically significant reduction of BP by more than 10 mmHg systolic and more than 5 mmHg diastolic pressure in patients treated with low doses of ACE inhibition, calcium channel blockers or their combination with low dose diuretics (Paran *et al.,* 2009). Sixty days of tomato supplementation in 30 hypertensive subjects showed a significant improvement in the levels of serum enzymes involved in anti-oxidant activities and decreased lipid peroxidation rate, but there were no significant changes in lipid profile (Bose and Agarwal, 2007).

It has been observed that tomato inhibits platelet-aggregation induced by ADP and collagen *in-vitro* and *in-vivo* (Yamamoto *et al.,* 2003; Dutta Roy *et al.,* 2001). Anti-platelet activity is inversely related to tomato ripening and the increase in the concentration of lycopene (Yamamoto *et al.,* 2003) and aqueous and methanol extracts maintained their platelet anti-aggregation activity under various temperatures (22, 60 and 100°C), (Fuentes *et al.,* 2012; 2013). Tomato extract (20-50 µl of 100 per cent juice) inhibited both ADP- and collagen-induced aggregation by up to 70 per cent but could not inhibit arachidonic acid-induced platelet aggregation and concomitant thromboxane synthesis under similar experimental conditions.

The anti-platelet components in tomatoes are water soluble, heat stable and are concentrated in the yellow fluid around the seeds. Tomatoes contain anti-platelet compounds in addition to adenosine. Unlike aspirin, the tomato-derived compounds inhibit thrombin-induced platelet aggregation. Tomatoes might be beneficial both as a preventive and therapeutic regime for cardiovascular disease (Dutta-Roy *et al.,* 2001). The mechanism by which the tomato extract inhibits platelet aggregation appears to be through inhibition of biochemical events in the phospholipase C pathway upstream of cyclooxygenase, rather than through increased cAMP levels (Lazarus and Garg, 2004). Significant reductions in *ex-vivo* platelet aggregation induced by ADP and collagen were observed 3 h. after supplementation with doses of tomato extract. Aqueous tomato extract was found to inhibit platelet function the most in subjects with the highest plasma homo-cysteine and C-reactive protein concentrations.

Thus modulation of platelet function is considered one of the cardio-protective mechanisms of tomatoes (O'Kennedy *et al.,* 2006a; b).

New varieties of tomatoes have been developed using genetic engineering. The Food and Drug Administration approved that the new tomato is as safe as tomatoes bred by conventional means. Further, metabolic engineering of resveratrol (a stilbene phytoalexin) has been achieved in tomato plants in order to improve their nutritional value. The total anti-oxidant capability and ascorbate content in transformed fruits was increased explaining the higher capability to counteract the pro-inflammatory effects of phorbol ester in monocyte-macrophages *via* the inhibition of induced cyclo-oxygenase-2 enzyme (D'Introno *et al.,* 2009).

Diabetes

Besides being an anti-oxidant, lycopene has been found to possess considerable therapeutic potential as an anti-diabetic agent too. Upritchard *et al.* (2000) reported that short term supplementation of tomato juice for 4 weeks increased plasma lycopene levels and the intrinsic resistance of LDL to oxidation almost as effectively as supplementation with a high dose of vitamin E which in addition decreased plasma levels of CRP, a risk factor for myocardial infarction, and thus reducing risk of myocardial infarction in patients with diabetes.

Administration of lycopene at 90 mg/kg body weight decreased glucose level, increased insulin concentration, decreased H_2O_2 and TBARS levels, increased total anti-oxidant status with increased anti-oxidant enzyme activities (CAT, SOD and GPx) and improvement in serum lipid profile (Ali and Agha, 2009). A systematic review and meta-analysis by Valero *et al.* (2011) revealed that intervention with lycopene increased the plasma levels of lycopene and decreased the malonyl-dialdehyde and lipid peroxidation. The non-provitamin A/provitamin A carotenoids ratio was found to be negatively associated with the risk for suffering from diabetic retinopathy.

Long term tomato supplementation in 40 diabetic patients significantly improved the anti-oxidant enzymes and decreased lipid peroxidation rate though there were no significant changes in lipid profile and glycated haemoglobin HbA1c levels (Bose and Agrawal, 2006).

Cancer

High intake of tomatoes was linked with decreased cancer risks and a 50 per cent reduction in mortality from cancers at all sites. (Colditz *et al.,* 1985; Djuric and Powell, 2001). The mechanisms underlying the inhibitory effects of lycopene on carcinogenesis could involve ROS scavenging, upregulation of detoxification systems, interference with cell proliferation, induction of gap-junctional communication, inhibition of cell cycle progression and modulation of signal transduction pathways (Bhuvaneswari and Nagini, 2005). Serum and tissue lycopene levels have also been inversely related with the chronic disease risk. Although the anti-oxidant properties of lycopene are thought to be primarily responsible for its beneficial properties, evidence is accumulating to suggest other mechanisms such as modulation of intercellular gap junction communication, hormonal and immune system and metabolic pathways may also be involved (Rao and Agarwal, 2000; Agarwal and Rao, 2000).

Phenolics in fresh as well as heated tomatoes contribute to the suppression of COX-2 expression in KB cells. Cell studies showed that phenolic extracts of heated tomatoes resulted in increased suppression of COX-2 expression compared with that of fresh tomato. Non-condensed tannin containing fraction of fresh tomato greatly suppressed COX-2 expression that compared to the negative control, but both non-condensed tannin containing and condensed tannin containing fractions of heated tomatoes showed suppression on COX-2 expression (Shen *et al.,* 2008).

Prospective and retrospective epidemiological studies indicate an inverse relationship between lycopene intake risk of cervical and prostate cancer risk in a intra-epithelial neoplasia case-control study (Van Eewyck *et al.,* 1991). The role of lycopene in prostate cancer has been extensively studied.

Prostate Cancer

Numerous potentially beneficial compounds are present in tomatoes, and, conceivably, complex interactions among multiple components may contribute to the anti-cancer properties of tomatoes (Giovannucci, 1999). Dietary intake of tomatoes and tomato products containing lycopene have been shown in cell culture, animal, epidemiologic and case-control studies to be inversely associated with the risk of prostate cancer. Its unique structural and biologic properties enable lycopene to prevent free-radical damage to cells caused by reactive oxygen species, thus acting as a potent antioxidant (Ansari and Ansari, 2005). The estimated intake of lycopene from various tomato products, and not any other carotenoid, was inversely related to prostate cancer risk reduction of almost 35 per cent on 10 or more servings of tomato products per week, and the protective effects were even stronger for more advanced or aggressive prostate cancer (Giovannucci *et al.,* 1995).

In-vitro and *in-vivo* experiments show that oral lycopene is bioavailable, accumulates in prostate tissue and is localized to the nucleus of prostate epithelial cells. In addition to anti-oxidant activity, *in-vitro* experiments indicate other mechanisms of chemo-prevention by lycopene including induction of apoptosis and anti-proliferation in cancer cells, anti-metastatic activity, and the up-regulation of the anti-oxidant response element leading to the synthesis of cyto-protective enzymes (van Breemen and Pajkovic, 2008). Based on a review of *in-vitro* data and human clinical trials on lycopene, Dahan *et al.* (2008) reported enough evidence to warrant use of lycopene in phase I and II clinical trials to examine its safety and efficacy as a potential chemo-preventive agent for prostate cancer.

Consumption of tomato sauce, the primary source of bioavailable lycopene, was associated with an even greater reduction in prostate cancer risk especially for extra-prostatic cancers (Giovannucci *et al.,* 2002). Contrary to this Boileau *et al.* (2003) reported that the consumption of tomato powder but not lycopene inhibited prostate carcinogenesis, suggesting that tomato products contain compounds in addition to lycopene that modify prostate carcinogenesis.

Diet restriction also reduced the risk of prostate cancer. Tomato phytochemicals and diet restriction may act by independent mechanisms. An overview on the efficacy of supplementation with tomatoes, tomato products and lycopene on appropriate surrogate end point biomarkers, provided increasing evidence suggesting that a single serving of tomatoes or tomato products ingested daily may contribute to protect from DNA damage. As DNA damage seems to be involved in the pathogenesis of prostate cancer (Ellinger *et al.,* 2006). Basu and Imrhan (2007) reviewing the effects tomato product supplementation, found the mechanisms of action involve protection of plasma lipoproteins, lymphocyte DNA and serum proteins against oxidative damage, and anti-carcinogenic effects, including reduction of prostate-specific antigen, up-regulation of connexin expression and overall decrease in prostate tumor aggressiveness.

Serum and tissue levels of lycopene were inversely associated with prostate cancer risk in recent case-control and cohort studies (Gann *et al.,* 1999; Rao *et al.,* 1999). It has also been demonstrated that an inverse association exists between dietary lycopene and the risk of cervical intra-epithelial neoplasia in a case-control study (Van Eewyck *et al.,* 1991). Plasma lycopene levels increased by two fold after supplementation with tomato lycopene extract. The plasma concentration of insulin-like growth factor-I decreased significantly by about 25 per cent after tomato lycopene extract supplementation as compared with the placebo-treated group. While no significant change was observed in insulin-like growth factor-I-binding protein-3 or insulin-like growth factor-II, the insulin-like growth factor-I/insulin-like growth factor-I-binding protein-3 molar ratio decreased significantly. Given that high plasma levels of insulin-like growth factor-I have been suggested as a risk factor for various types of cancer including colon cancer, the results support the view that tomato lycopene extract has a role in the prevention of colon and possibly other types of cancer (Walfisch *et al.,* 2007).

Newly diagnosed prostate cancer patients who received a tomato oleoresin extract containing 30 mg of lycopene for 3 weeks before radical prostatectomy had smaller tumors (80 per cent vs 45 per cent, less than 4 ml), less involvement of surgical margins and/or extra-prostatic tissues with cancer (73 per cent vs 18 per cent, organ-confined disease), and less diffuse involvement of the prostate by high-grade prostatic intraepithelial neoplasia (33 per cent vs 0 per cent, focal involvement) compared with subjects in the control group. Mean plasma prostate-specific antigen levels were lower in the intervention group compared with the control group (Kucuk *et al.,* 2002). Tomato polyphenols, specifically quercetin, kaempferol, and naringenin, individually decreased cancer cell growth *in-vitro* and that the combination of these polyphenols, which are present in whole foods, may have additive effects in decreasing cancer proliferation (Campbell *et al.,* 2003). Freeze-dried whole-tomato powder or lycopene alone could reduce the growth of prostate tumors in rats (Campbell *et al.,* 2004). Lo *et al.* (2007) demonstrated that lycopene inhibits PDGF-BB-induced signaling, proliferation and migration in rat A10 and aortic smooth muscle cells (SMCs). The mechanism of action is that lycopene is capable of binding PDGF-BB and inhibiting its interaction with SMC, which is quite different from those previously developed PDGFR-beta antagonists.

The lycopene/ciprofloxacin group also showed a statistically significant decrease in bacterial growth and improvement in prostatic inflammation compared with the ciprofloxacin group too suggesting that lycopene may have an additional (synergistic) effect with ciprofloxacin in the treatment of chronic bacterial prostatitis (Han *et al.*, 2008).

In contrast to the pharmacologic approach with pure lycopene, many nutritional scientists direct their attention upon the diverse array of tomato products as a complex mixture of biologically active phytochemicals that together may have anti-prostate cancer benefits beyond those of any single constituent (Hadley *et al.*, 2002).

Other Cancers

The anti-cancer activity of lycopene has been demonstrated both in *in-vitro* and *in-vivo* tumour models. Ozkan *et al.* (2012) observed that lycopene protects the pancreatic tissues from oxidative damage induced by cerulein, and this effect possibly involves the inhibition of neutrophil infiltration and lipid peroxidation.

In an animal study, reduction in tumour incidence (42.05 per cent), tumour burden (1.39) and tumour multiplicity (3.42) was observed upon lycopene pre-treatment to female Balb/c mice with N-diethylnitrosamine-induced experimental hepato-carcinogenesis (Gupta *et al.*, 2013). Consistently lower risk of cancer for a variety of anatomic sites is associated with higher consumption of tomatoes and tomato-based products. A case-control study from the Breast Cancer Serum Bank in Columbia, Missouri, reported that only serum lycopene and none of the other anti-oxidants showed a significant inverse relationship with breast cancer risk (*Dorgan et al., 1998*).

Small amounts of tomato puree added to the diet over a short period can increase carotenoid concentrations and the resistance of lymphocytes to oxidative stress (Porrini and Riso, 2000). In a double-blind, cross-over study, twenty-six healthy subjects consumed 250 ml of the drink daily, providing about 6 mg lycopene, 4 mg phytoene, 3 mg phytofluene, 1 mg beta-carotene and 1.8 mg alpha-tocopherol, or a placebo drink. After 26 d of consumption of the drink, plasma carotenoid levels and lymphocyte carotenoid concentrations also increased significantly. The alpha-tocopherol concentration remained nearly constant. The intake of the tomato drink significantly reduced DNA damage in lymphocytes subjected to oxidative stress by about 42 per cent. Porrini *et al.* (2005) thus suggest that a low intake of carotenoids from tomato products improves cell anti-oxidant protection. Polívková *et al.* (2010) showed that lycopene has anti-mutagenic effects, although the effects are lower than that of tomato purée, which contains a complex mixture of bioactive phytochemicals. The anti-mutagenic effect is connected with the chemo-protective role of lycopene, tomatoes, and tomato products in the prevention of carcinogenesis.

Tomato paste containing lycopene provides protection against ultraviolet radiation (UVR)-induced erythema and potentially longer-term aspects of photo damage (Rizwan *et al.*, 2011). The first reported biological activities of lycopene *in-vivo* were protection against radiation and development of certain types of ascites tumors (Forssberg *et al.*, 1959). Due to its anti-oxidant and anti-inflammatory properties, lycopene may reduce or prevent the side effects of chemotherapy (Sahin *et al.*, 2010) and protect from carcinogenesis (Sengupta and Das, 1999).

Active components in tomato, such as kaempferol and chlorogenic acid, have anti-mutagenic activities and lycopene, the most active oxygen quencher (Di Mascio *et al.,* 1989) with potential chemo-preventive activities (Sengupta *et al.,* 2003) is effective in protecting blood lymphocytes from NO_2 radical damage (Bohm *et al.,* 1995). In a cohort study, serum lycopene levels were found to be inversely related to the risk of bladder cancer (Helzlsouer *et al.,* 1989).

Other Benefits

Epidemiological data and *in-vitro* and *in-vivo* data reveal the anti-oxidant as well anti-inflammatory effects of lycopene showing promising preventive mechanisms of lycopene in several other clinical conditions such as arthritis, depression, kidney diseases etc. (Böhm, 2012).

Suppression of oxidative stress, reduction of plasma TNF-alpha and NO levels, and the down-regulation of TNF-alpha in lungs contribute to the alleviation of pulmonary fibrosis in rats administered lycopene (Zhou *et al.,* 2008).

Lycopene administration ameliorated biochemical indices of cisplatin-induced nephro-toxicity in both plasma and kidney tissues. Pre-treatment with lycopene was more effective (Atessahin *et al.,* 2005). Significant increase in the levels of plasma creatinine and urea observed in the adriamycin-induced heart and kidney toxicity in rats were normalized by lycopene treatment (Yilmaz *et al.,* 2006). Administration of gentamicin to rats induced a marked renal failure, characterized by a significant increase in plasma creatinine, urea and MDA concentrations, and decrease in GSH-Px and CAT activities without affecting GSH concentrations. But pre-treatment with lycopene produced amelioration in the same biochemical indices of nephro-toxicity (Karahan *et al.,* 2005).

Tomato reduces heavy metal (mercury, lead and cadmium) accumulation in the liver while enhancing the elimination of these metals in a time dependent manner. The highest hepato-protective effect was to cadmium followed by mercury and least to lead (Nwokocha *et al.,* 2012). In addition, tomato paste was found to significantly lower the adverse effects of lead exposure on the kidney as well as lead induced oxidative stress (Salawu *et al.,* 2009).

Dietary lycopene restriction resulted in significantly decreased serum lycopene, lutein/zeaxanthin, and α -/β –carotene. GPx, lipid and protein oxidation increased though not significant, while CAT and SOD were significantly depressed respectively coinciding with significant increase in NTx. Therefore, Mackinnon *et al.* (2011b) suggested that the daily consumption of lycopene may be important as it acts as an anti-oxidant to decrease bone resorption in post-menopausal women and may therefore be beneficial in reducing the risk of osteoporosis. Moreover, Cheong and Chang (2009) reported that fermented milk supplemented with tomato and taurine might improve bone health in post-menopausal osteoporotic rats.

Lycopene prevented cigarette smoke extract -induced IL-8 production through a mechanism involving an inactivation of NF-kB. NF-kB inactivation was accompanied by an inhibition of redox signalling and an activation of PPARγ signalling. The ability of lycopene in inhibiting IL-8 production, NF-kB/p65 nuclear translocation, and redox signalling and in increasing PPARγ expression was also found in isolated rat alveolar macrophages exposed to cigarette smoke extract (Simone *et al.*, 2011). Matrix metalloproteinase-9 (MMP-9) has been implicated in both inflammation and fibrosis. Palozza *et al.* (2012) observed that lycopene may inhibit cigarette smoke exposure mediated MMP-9 induction, primarily by blocking prenylation of Ras in a signaling pathway, in which MEK1/2-ERK1/2 and NF-κB are involved.

Niu *et al.* (2013) demonstrated that a tomato-rich diet is independently related to lower prevalence of depressive symptoms suggesting that it may have a beneficial effect on the prevention of depressive symptoms.

Bodet *et al.* (2008) clearly indicated that naringenin is a potent inhibitor of the pro-inflammatory cytokine response induced by lipo-polysaccharide in both macrophages and in whole blood. Naringenin markedly inhibited the phosphorylation on serines 63 and 73 of Jun proto-oncogene-encoded AP-1 transcription factor in lipo-polysaccharide-stimulated macrophages. Pre-administration of naringenin significantly reduced the severity of colitis and resulted in down-regulation of pro-inflammatory mediators (inducible NO synthase (iNOS), intercellular adhesion molecule-1 (ICAM-1), monocyte chemo-attractant protein-1 (MCP-1), cyclo-oxygenase-2 (Cox2), TNF-α and IL-6 mRNA) in the colon mucosa. The decline in the production of pro-inflammatory cytokines, specifically TNF-α and IL-6, correlated with a decrease in mucosal Toll-like receptor 4 (TLR4) mRNA and protein. Phospho-NF-κB p65 protein was significantly decreased, which correlated with a similar decrease in phospho-IκBα protein (Dou *et al.*, 2013).

Conclusion

Though the various health benefits of tomatoes are primarily attributed to lycopene, consumption of whole tomatoes with skin and seeds would be more beneficial. It has also been recommended by several researchers perhaps due to the less efficient absorption of pure lycopene. Moreover, according to Cohen (2002) the absorption was poor through grain-based diet than casein-based diet. Hence, it is interesting to study the influence of other ingredients in a recipe as well as cooking methods on the absorption of lycopene. New varieties of tomatoes have been developed using genetic engineering which are declared as safe for consumption by the Food and Drug Administration.

References

Agarwal S, Rao AV (2000). Tomato lycopene and its role in human health and chronic diseases. *CMAJ*. 163 (6): 739–44.

Ahuja KD, Pittaway JK, Ball MJ (2006). Effects of olive oil and tomato lycopene combination on serum lycopene, lipid profile, and lipid oxidation. *Nutrition*. 22 (3): 259–65.

Ali MM, Agha FG (2009). Amelioration of streptozotocin-induced diabetes mellitus, oxidative stress and dyslipidemia in rats by tomato extract lycopene. *Scand. J. Clin. Lab. Invest.* 69 (3): 371–79.

Ansari MS, Ansari S (2005). Lycopene and prostate cancer. *Future Oncol.* 1 (3): 425–30.

Arab L, Steck S (2000). Lycopene and cardiovascular disease. *Am. J. Clin. Nutr.* 71: 1691S–95S.

Atessahin A, Yilmaz S, Karahan I, Ceribasi AO, Karaoglu A (2005). Effects of lycopene against cisplatin-induced nephrotoxicity and oxidative stress in rats. *Toxicology.* 212 (2-3): 116–23.

Basu A, Imrhan V (2007). Tomatoes versus lycopene in oxidative stress and carcinogenesis: conclusions from clinical trials. *Eur. J. Clin. Nutr.* 61 (3): 295–303.

Beecher GR (1998). Nutrient content of tomatoes and tomato products. *Proc. Soc. Exp. Biol. Med.* 218 (2): 98–100.

Bhuvaneswari V, Nagini S (2005). Lycopene: a review of its potential as an anti-cancer agent. *Curr. Med. Chem. Anti-cancer Agents.* 5 (6): 627–35.

Bodet C, La VD, Epifano F, Grenier D (2008). Naringenin has anti-inflammatory properties in macrophage and *ex-vivo* human whole-blood models. *J. Periodontal. Res.* 43 (4): 400–407.

Böhm V (2012). Lycopene and heart health. *Mol. Nutr. Food Res.* 56 (2): 296–303.

Bohm F, Tinkler JH. and Truscott, TG (1995). Carotenoids protect against cell membrane damage by nitrogen dioxide radical. *Nat. Med.* 1: 98–99.

Böhm V, Bitsch R (1999). Intestinal absorption of lycopene from different matrices and interactions to other carotenoids, the lipid status, and the anti-oxidant capacity of human plasma. *Eur. J. Nutr.* 38 (3): 118–25.

Boileau TW, Liao Z, Kim S, Lemeshow S, Erdman JW Jr, Clinton SK (2003). Prostate carcinogenesis in N-methyl-N-nitrosourea (NMU)-testosterone-treated rats fed tomato powder, lycopene, or energy-restricted diets. *J. Natl. Cancer Inst.* 95 (21): 1578–86.

Bose KS, Agrawal BK (2006). Effect of long term supplementation of tomatoes (cooked) on levels of antioxidant enzymes, lipid peroxidation rate, lipid profile and glycated haemoglobin in Type 2 diabetes mellitus. *West Indian Med J.* 55 (4): 274–78.

Bose KS, Agrawal BK (2007). Effect of lycopene from tomatoes (cooked) on plasma anti-oxidant enzymes, lipid peroxidation rate and lipid profile in grade-I hypertension. *Ann. Nutr. Metab.* 51 (5): 477–81.

Bub A, Watzl B, Abrahamse L, Delincee H, Adam S, Wever J, Muller H, Rechkemmer G (2000). Moderate intervention with carotenoid-rich vegetable products reduces lipid peroxidation in men. *J. Nutr.* 130: 2200–06.

Burney PGJ, Comstock GW. Morris JS (1989). Serologic precursors of cancer: serum micro-nutrients and the subsequent risk of pancreatic cancer. *Am. J. Clin. Nutr.* 49: 895–900.

Burton-Freeman B, Talbot J, Park E, Krishnankutty S, Edirisinghe I (2012). Protective activity of processed tomato products on postprandial oxidation and inflammation: a clinical trial in healthy weight men and women. *Mol. Nutr. Food Res.* 56 (4): 622–31.

Campbell JK, King JK, Lila MA, Erdman JW Jr. (2003) Anti-proliferation effects of tomato polyphenols in Hepa1c1c7 and LNCaP cell lines. *J. Nutr.* 133: 3858S–59S.

Campbell JK, Canene-Adams K, Lindshield BL, Boileau TW, Clinton SK, Erdman JW Jr. (2004). Tomato phytochemicals and prostate cancer risk. *J. Nutr.* 134 (12 Suppl): 3486S–92S.

Cheong SH, Chang KJ (2009). The preventive effect of fermented milk supplement containing tomato (*Lycopersion esculentum*) and taurine on bone loss in ovariectomized rats. *Adv. Exp. Med. Biol.* 643: 333–40.

Clinton SK (1998). Lycopene: chemistry, biology, and implications for human health and disease. *Nutr. Rev.* 56: 35–51.

Cohen LA (2002). A review of animal model studies of tomato carotenoids, lycopene, and cancer chemoprevention. *Exp. Biol. Med. (Maywood).* 227 (10): 864–68.

Colditz GA, Branch LG, Lipnic RJ (1985). Increased green and yellow vegetables intake and lowered cancer death in an elderly population. *Am. J. Clin. Nutr.* 41: 32 –36.

Cuevas-Ramos D, Almeda-Valdés P, Chávez-Manzanera E, Meza-Arana CE, Brito-Córdova G, Mehta R, Pérez-Méndez O, Gómez-Pérez FJ (2013). Effect of tomato consumption on high-density lipoprotein cholesterol level: a randomized, single-blinded, controlled clinical trial. *Diabetes Metab. Syndr. Obes.* 6: 263–73.

Dahan K, Fennal M, Kumar NB (2008). Lycopene in the prevention of prostate cancer. *J. Soc. Integr. Oncol.* 6 (1): 29–36.

Diaz MN, Frei B, Vita JA, Keaney JF (1997). Anti-oxidants and atherosclerotic heart disease. *N. Engl. J. Med.* 337: 408–416.

Di Mascio P, Kaiser S, Sies H (1989). Lycopene as the most efficient biological carotenoid singlet oxygen quencher. *Arch. Biochem. Biophys.* 274: 532–38.

D'Introno A, Paradiso A, Scoditti E, D'Amico L, De Paolis A, Carluccio MA, Nicoletti I, DeGara L, Santino A,Giovinazzo G (2009). Anti-oxidant and anti-inflammatory properties of tomato fruits synthesizing different amounts of stilbenes. *Plant Biotechnol. J.* 7 (5): 422–29.

Djuric Z, Powell LC (2001). Anti-oxidant capacity of lycopene-containing foods. *Int. J. Food Sci. Nutr.* 52 (2): 143–49.

Dorgan JF, Sowell A, Swanson CA, Potischman N, Miller R, Schussler N, Stephenson HE Jr. (1998). Relationship of serum carotenoids, retinol, α-tocopherol, and selenium with breast cancer risk: results from a prospective study in Columbia, Missouri (United States). *Cancer Causes Control.* 9: 89–97.

Dou W, Zhang J, Sun A, Zhang E, Ding L, Mukherjee S, Wei X, Chou G, Wang ZT, Mani S (2013). Protective effect of naringenin against experimental colitis via suppression of Toll-like receptor 4/NF-κB signalling. *Br. J. Nutr.* 110 (4): 599–608.

Dutta-Roy AK, Crosbie L, Gordon MJ (2001). Effects of tomato extract on human platelet aggregation *in-vitro*. *Platelets*. 12 (4): 218–27.

Ellinger S, Ellinger J, Stehle P (2006). Tomatoes, tomato products and lycopene in the prevention and treatment of prostate cancer: do we have the evidence from intervention studies? *Curr. Opin. Clin. Nutr. Metab. Care.* 9 (6): 722–27.

Forrsberg A, Lingen C, Ernster L, Lindberg O (1959). Modification of the x-irradiation syndrome by lycopene. *Exp. Cell. Res.* 16 (1): 7–14.

Franceschi S, Bidoli E, La Vecchia C, Talamini R, D'Avanzo, B, Negri E (1994). Tomatoes and risk of digestive-tract cancers. *Int. J. Cancer.* 59: 181–84.

Fuentes E, Astudillo L, Gutiérrez M, Contreras S, *et al.* (2012). Fractions of aqueous and methanolic extracts from tomato (*Solanum lycopersicum* L.) present platelet anti-aggregant activity. *Blood Coagulation and Fibrinolysis.* 23: 109–17.

Fuentes E, Carle R, Astudillo L, Guzmán L, Gutiérrez M, Carrasco G, Palomo I (2013). Anti-oxidant and anti-platelet activities in extracts from green and fully ripe tomato fruits (*Solanum lycopersicum*) and pomace from industrial tomato processing. Evid. Based Complement. *Alternat. Med.* 2013: 867578.

Fuhrman B, Elis A, Aviram M (1997). Hypo-cholesterolemic effect of lycopene and beta-carotene is related to suppression of cholesterol synthesis and augmentation of LDL receptor activity in macrophages. *Biochem. Biophys. Res. Commun.* 233: 658–62.

Gann P, Ma J, Giovannucci E, Willett W, Sacks FM, Hennekens CH, Stampfer MJ (1999). Lower prostate cancer risk in men with elevated plasma lycopene levels: results of a prospective analysis. *Cancer Res.* 59: 1225–30.

García-Alonso FJ, Bravo S, Casas J, Pérez-Conesa D, Jacob K, Periago MJ (2009). Changes in anti-oxidant compounds during the shelf life of commercial tomato juices in different packaging materials. *J. Agric. Food Chem.* 57 (15): 6815–22.

García-Alonso FJ, Jorge-Vidal V, Ros G, Periago MJ (2012). Effect of consumption of tomato juice enriched with n-3 poly-unsaturated fatty acids on the lipid profile, anti-oxidant biomarker status, and cardiovascular disease risk in healthy women. *Eur. J. Nutr.* 51 (4): 415–24.

Gartner C, Stahl W, Sies, H (1997). Lycopene is more bioavailable from tomato paste than from fresh tomatoes. *Am. J. Clin. Nutr.* 66: 116–22.

Ghavipour M, Saedisomeolia A, Djalali M, Sotoudeh G, Eshraghyan MR, Moghadam AM, Wood LG (2013). Tomato juice consumption reduces systemic inflammation in overweight and obese females. *Br. J. Nutr.* 109 (11): 2031–35.

Giovannucci E (1999). Tomatoes, tomato-based products, lycopene, and cancer: review of the epidemiologic literature. *J. Natl. Cancer Inst.* 91 (4): 317–31.

Giovannucci E, Rimm EB, Liu Y, Stampfer MJ, Willett WC (2002). A prospective study of tomato products, lycopene, and prostate cancer risk. *J. Natl. Cancer. Inst.* 94 (5): 391–98.

Giovannucci E, Ascherio A, Rimm EB, Stampfer MJ, Colditz GA, Willett WC (1995). Intake of carotenoids and retinol in relation to risk of prostate cancer. *J. Natl. Cancer Inst.* 87: 1767–76.

Gitenay D, Lyan B, Rambeau M, Mazur A, Rock E (2007). Comparison of lycopene and tomato effects on biomarkers of oxidative stress in vitamin E deficient rats. *Eur. J. Nutr.* 46 (8): 468–75.

Gupta P, Bansal MP, Koul A (2013). Spectroscopic characterization of lycopene extract from *Lycopersicum esculentum* (Tomato) and its evaluation as a chemo-preventive agent against experimental hepato-carcinogenesis in mice. *Phytother. Res.* 27 (3): 448–56.

Hadley CW, Miller EC, Schwartz SJ, Clinton SK (2002). Tomatoes, lycopene, and prostate cancer: progress and promise. *Exp. Biol. Med. (Maywood).* 227 (10): 869–80.

Han CH, Yang CH, Sohn DW, Kim SW, Kang SH, Cho YH (2008). Synergistic effect between lycopene and ciprofloxacin on a chronic bacterial prostatitis rat model. *Int. J. Anti-microb.* Agents. 31 (Suppl 1): S102–07.

Helzlsouer KJ, Comstock GW, Morris JS (1989). Selenium, lycopene, α-tocopherol, β-carotene, retinol, and subsequent bladder cancer. *Cancer Res.* 49: 6144–48.

Hsu YM, Lai CH, Chang CY, Fan CT, Chen CT, Wu CH (2008). Characterizing the lipid-lowering effects and anti-oxidant mechanisms of tomato paste. *Biosci. Biotechnol. Biochem.* 72 (3): 677–85.

Ibrahim HS, Ahmed LA, El-din MM (2008). The functional role of some tomato products on lipid profile and liver function in adult rats. *J. Med. Food.* 11 (3): 551–59.

Jacob K, Periago MJ, Böhm V, Berruezo GR (2008). Influence of lycopene and vitamin C from tomato juice on biomarkers of oxidative stress and inflammation. *Br. J. Nutr.* 99 (1): 137–46.

Jacques PF, Lyass A, Massaro JM, Vasan RS, D'Agostino Sr. RB (2013). Relationship of lycopene intake and consumption of tomato products to incident CVD. *Br. J. Nutr.* 15: 1-7.

Karahan I, Ate°°ahin A, Yilmaz S, Ceriba°i AO, Sakin F (2005). Protective effect of lycopene on gentamicin-induced oxidative stress and nephro-toxicity in rats. *Toxicology.* 215 (3): 198–204.

Karppi J, Laukkanen JA, Sivenius J, Ronkainen K, Kurl S (2012). Serum lycopene decreases the risk of stroke in men: a population-based follow-up study. *Neurology.* 79 (15): 1540–47.

Kohlmeier L, Kark JD, Gomez-Garcia E, Martin BC, Steck SE, Kardinaal AFM, Ringstad J, Thamm M, Masaev V, Riemersma R, Martin-Moreno JM, Huttunen JK, Kok FJ

(1997). Lycopene and myocardial infarction risk in the EURAMIC Study. *Am. J. Epidemiol.* 146: 618–26.

Kucuk O, Sarkar FH, Djuric Z, Sakr W, Pollak MN, Khachik F, Banerjee M, Bertram JS, Wood DP Jr. (2002). Effects of lycopene supplementation in patients with localized prostate cancer. *Exp. Biol. Med. (Maywood).* 227 (10): 881–85.

Lavelli V, Hippeli S, Dornisch K, Peri C, Elstner EF (2001). Properties of tomato powders as additives for food fortification and stabilization. *J. Agric. Food Chem.* 49 (4): 2037–42.

Lazarus SA, Garg ML (2004). Tomato extract inhibits human platelet aggregation *in-vitro* without increasing basal cAMP levels. *Int. J. Food Sci. Nutr.* 55 (3): 249–56.

Li H, Deng Z, Liu R, Young JC, Zhu H, Loewen S, Tsao R (2011). Characterization of phytochemicals and anti-oxidant activities of a purple tomato (*Solanum lycopersicum* L.). *J. Agric. Food Chem.* 59 (21): 11803–11.

Lo HM, Hung CF, Tseng YL, Chen BH, Jian JS, Wu WB (2007). Lycopene binds PDGF-BB and inhibits PDGF-BB-induced intra-cellular signaling transduction pathway in rat smooth muscle cells. *Biochem. Pharmacol.* 74 (1): 54–63.

Mackinnon ES, Rao AV, Josse RG, Rao LG (2011a). Supplementation with the anti-oxidant lycopene significantly decreases oxidative stress parameters and the bone resorption marker N-telopeptide of type I collagen in post-menopausal women. *Osteoporos. Int.* 22 (4): 1091–01.

Mackinnon ES, Rao AV, Rao LG (2011b). Dietary restriction of lycopene for a period of one month resulted in significantly increased biomarkers of oxidative stress and bone resorption in post-menopausal women. *J. Nutr. Health. Aging.* 15 (2): 133–38.

Maruyama C, Imamura K, Oshima S, Suzukawa M, Egami S, Tonomoto M, Baba N, Harada M, Ayaori M, Inakuma T, Ishikawa T (2001). Effects of tomato juice consumption on plasma and lipoprotein carotenoid concentrations and the susceptibility of low density lipoprotein to oxidative modification. *J. Nutr. Sci. Vitaminol. (Tokyo).* 47 (3): 213–21.

Miller NJ, Sampson J, Candeias LP, Bramley PM, Rice-Evans CA (1996). Anti-oxidant activities of carotenes and xanthophylls. *FEBS Lett* 384: 240–46.

Mordente A, Guantario B, Meucci E, Silvestrini A, Lombardi E, Martorana GE, Giardina B, Böhm V (2011). Lycopene and cardio-vascular diseases: an update. *Curr. Med. Chem.* 18 (8): 1146–63.

Mortensen A, Skibsted LH (1997). Relative stability of carotenoid radical cations and homologue tocopheroxyl radicals. A real time kinetic study of anti-oxidant hierarchy. *FEBS. Lett.* 417: 261 -66.

Müller L, Catalano A, Simone R, Cittadini A, Fröhlich K, Böhm V, Palozza P (2013). Anti-oxidant capacity of tomato seed oil in solution and its redox properties in cultured macrophages. *J. Agric. Food Chem.* 61 (2): 346–54.

Niu K, Guo H, Kakizaki M, Cui Y, Ohmori-Matsuda K, Guan L, Hozawa A, Kuriyama S, Tsuboya T, Ohrui T,Furukawa K, Arai H, Tsuji I, Nagatomi R (2013). A tomato-rich diet is related to depressive symptoms among an elderly population aged 70 years and over: a population-based, cross-sectional analysis. *J. Affect. Disord.* 144 (1-2): 165–70.

Nwokocha CR, Nwokocha MI, Aneto I, Obi J, Udekweleze DC, Olatunde B, Owu DU, Iwuala MO (2012). Comparative analysis on the effect of *Lycopersicon esculentum* (tomato) in reducing cadmium, mercury and lead accumulation in liver. *Food Chem. Toxicol.* 50 (6): 2070–73.

O'Kennedy N, Crosbie L, van Lieshout M, Broom JI, Webb DJ, Duttaroy AK (2006a). Effects of anti-platelet components of tomato extract on platelet function *in-vitro* and *ex-vivo*: a time-course cannulation study in healthy humans. *Am. J. Clin.Nutr.* 84 (3): 570–79.

O'Kennedy N, Crosbie L, Whelan S, Luther V, Horgan G, Broom JI, Webb DJ, Duttaroy AK (2006b). Effects of tomato extract on platelet function: a double-blinded crossover study in healthy humans. *Am. J. Clin. Nutr.* 84 (3): 561–69.

Ozkan E, Akyüz C, Dulundu E, Topaloðlu U, Sehirli AÖ, Ercan F, Sener G (2012). Protective effects of lycopene on cerulein-induced experimental acute pancreatitis in rats. *J. Surg. Res.* 176 (1): 232–38.

Palomo I, Fuentes E, Padró T, Badimon L (2012). Platelets and atherogenesis: Platelet anti-aggregation activity and endothelial protection from tomatoes (*Solanum lycopersicum* L.). *Exp. Ther. Med.* 3 (4): 577–84.

Palomo I, Gutiérrez M, Astudillo L, Rivera C, *et al.* (2009). Efecto anti-oxidante de frutas y hortalizas de la zona central de Chile. Revista. Chilena. De Nutrición. 36: 152–58.

Palozza P, Parrone N, Catalano A, Simone R (2010a). Tomato lycopene and inflammatory cascade: basic interactions and clinical implications. *Curr. Med. Chem.* 17 (23): 2547–63.

Palozza P, Simone R, Catalano A, Boninsegna A, Böhm V, Fröhlich K, Mele MC, Monego G, Ranelletti FO (2010b). Lycopene prevents 7-ketocholesterol-induced oxidative stress, cell cycle arrest and apoptosis in human macrophages. *J. Nutr. Biochem.* 21 (1): 34–46.

Palozza P, Simone RE, Catalano A, Saraceni F, Celleno L, Mele MC, Monego G, Cittadini A. (2012). Modulation of MMP-9 pathway by lycopene in macrophages and fibroblasts exposed to cigarette smoke. *Inflamm. Allergy. Drug Targets.* 11 (1): 36–47.

Paran E, Novack V, Engelhard YN, Hazan-Halevy I (2009). The effects of natural anti-oxidants from tomato extract in treated but uncontrolled hyper-tensive patients. Cardiovasc. *Drugs Ther.* 23 (2): 145–51.

Parfitt VJ, Rubba P, Bolton C, Marotta G, Hartog M, Mancini M (1994). A comparison of anti-oxidant status and free radical peroxidation of plasma lipoproteins in healthy young persons from Naples and Bristol. *Eur. Heart J.* 15: 871–76.

Park HY, Kim GY, Choi YH (2012). Naringenin attenuates the release of pro-inflammatory mediators from lipo-polysaccharide-stimulated BV2 microglia by inactivating nuclear factor-κB and inhibiting mitogen-activated protein kinases. *Int. J. Mol. Med.* 30: 204–10.

Polívková Z, Šmerák P, Demová H, Houška M (2010). Anti-mutagenic effects of lycopene and tomato purée. *J. Med. Food.* 13 (6): 1443–50.

Porrini M, Riso P (2000). Lymphocyte lycopene concentration and DNA protection from oxidative damage is increased in women after a short period of tomato consumption. *J. Nutr.* 130 (2): 189–92.

Porrini M, Riso P, Brusamolino A, Berti C, Guarnieri S, Visioli F (2005). Daily intake of a formulated tomato drink affects carotenoid plasma and lymphocyte concentrations and improves cellular antioxidant protection. *Br. J. Nutr.* 93 (1): 93–99.

Rafi MM, Yadav PN, Reyes M (2007). Lycopene inhibits LPS-induced pro-inflammatory mediator inducible nitric oxide synthase in mouse macrophage cells. *J. Food Sci.* 72 (1): S069–74.

Ragab AS, Van Fleet J, Jankowski B, Park JH, Bobzin SC (2006). Detection and quantitation of resveratrol in tomato fruit (*Lycopersicon esculentum* Mill.). *J. Agric. Food Chem.* 54 (19): 7175–79.

Rao AV (2002). Lycopene, tomatoes, and the prevention of coronary heart disease. *Exp. Biol. Med. (Maywood).* 227 (10): 908–13.

Rao AV (2004). Processed tomato products as a source of dietary lycopene: bioavailability and anti-oxidant properties. *Can. J. Diet. Pract. Res. Winter.* 65 (4): 161–65.

Rao AV, Agarwal S (2000). Role of anti-oxidant lycopene in cancer and heart disease. *J. Am. Coll. Nutr.* 19 (5): 563–69.

Rao AV, Fleshner N, Agarwal S (1999). Serum and tissue lycopene and biomarkers of oxidation in prostate cancer patients: a case control study. *Nutr. Cancer.* 33: 159–64.

Ried K, Fakler P (2011). Protective effect of lycopene on serum cholesterol and blood pressure: Meta-analyses of intervention trials. *Maturitas.* 68 (4): 299–310.

Riso P, Visioli F, Erba D, Testolin G, Porrini M (2004). Lycopene and vitamin C concentrations increase in plasma and lymphocytes after tomato intake. Effects on cellular anti-oxidant protection. *Eur. J. Clin. Nutr.* 58 (10): 1350–58.

Rissanen T, Voutilainen S, Nyyssonen K, Salonen R, Salonen JT (2000). Low plasma lycopene concentration is associated with increased intima-media thickness of the carotid artery wall. *Arterioscler. Thromb. Vasc. Biol.* 20: 2677–81.

Rissanen T, Voutilainen S, Nyyssonen K, Lakka TA, Sivenius J, Salonen R, Kaplan GA, Salonen JT (2001). Low serum lycopene concentration is associated with an excess incidence of acute coronary events and stroke: the Kuopio Ischaemic Heart Disease Risk Factor Study. *Br. J. Nutr.* 85: 749–54.

Rissanen T, Voutilainen S, Nyyssönen K, Salonen JT (2002). Lycopene, atherosclerosis, and coronary heart disease. *Exp. Biol. Med. (Maywood).* 227 (10): 900–07.

Rizwan M, Rodriguez-Blanco I, Harbottle A, Birch-Machin MA, Watson RE, Rhodes LE (2011). Tomato paste rich in lycopene protects against cutaneous photodamage in humans in vivo: a randomized controlled trial. *Br. J. Dermatol.* 2011 Jan; 164 (1): 154–62.

Sahin K, Sahin N, Kucuk O (2010). Lycopene and chemotherapy toxicity. *Nutr. Cancer.* 62 (7): 988–95.

Sakamoto H, Mori H, Ojima F, Ishiguro Y, Arimoto S, Imae Y, Nanba T, Ogawa M, Fukuba H (1994). Elevation of serum carotenoids after continual ingestion of tomato juice. *J. Jpn. Soc. Nutr. Food Sci.* 47: 93–99.

Salawu EO, Adeeyo OA, Falokun OP, Yusuf UA, Oyerinde A, Adeleke AA (2009). Tomato (*Lycopersicon esculentum*) prevents lead-induced testicular toxicity. *J. Hum. Reprod. Sci.* 2 (1): 30–34.

Sengupta A, Das S (1999). The anti-carcinogenic role of lycopene, abundantly present in tomato. *Eur. J. Cancer Prev.* 8 (4): 325–30.

Sengupta A, Ghosh S, Das S (2003). Tomato and garlic can modulate azoxymethane-induced colon carcinogenesis in rats. *Eur. J. Cancer. Prev.* 12 (3): 195–200.

Sesso HD, Liu S, Gaziano JM, Buring JE (2003). Dietary lycopene, tomato-based food products and cardiovascular disease in women. *J. Nutr.* 133: 2336–41.

Sesso HD, Wang L, Ridker PM, Buring JE (2012). Tomato-based food products are related to clinically modest improvements in selected coronary biomarkers in women. *J. Nutr.* 142 (2): 326–33.

Shen YC, Chen SL, Zhuang SR, Wang CK (2008). Contribution of tomato phenolics to suppression of COX-2 expression in KB cells. *J. Food Sci.* 73 (1): C1–10.

Shen YC, Chen SL, Wang CK (2007). Contribution of tomato phenolics to anti-oxidation and down-regulation of blood lipids. *J. Agric. Food Chem.* 55 (16): 6475–81.

Shi J, Le Maguer M (2000). Lycopene in tomatoes: chemical and physical properties affected by food processing. *Crit. Rev. Biotechnol.* 20 (4): 293–334.

Shidfar F, Froghifar N, Vafa M, Rajab A, Hosseini S, Shidfar S, Gohari M (2011). The effects of tomato consumption on serum glucose, apolipoprotein B, apolipoprotein A-I, homocysteine and blood pressure in type 2 diabetic patients. *Int. J. Food Sci. Nutr.* 62 (3): 289–94.

Silaste ML, Alfthan G, Aro A, Kesäniemi YA, Hörkkö S (2007). Tomato juice decreases LDL cholesterol levels and increases LDL resistance to oxidation. *Br. J. Nutr.* 98 (6): 1251–58.

Simone RE, Russo M, Catalano A, Monego G, Froehlich K, Boehm V, Palozza P (2011). Lycopene inhibits NF-kB-mediated IL-8 expression and changes redox and PPARγ signalling in cigarette smoke-stimulated macrophages. *PLoS. One.* 6 (5): e19652.

Story EN, Kopec RE, Schwartz SJ, Harris GK (2010). An update on the health effects of tomato lycopene. *Annu. Rev. Food Sci. Technol.* 1: 189–210.

Takeoka GR, Dao L, Flessa S, Gillespie DM, Jewell WT, Huebner B, Bertow D, Ebeler SE (2001). Processing effects on lycopene content and anti-oxidant activity of tomatoes. *J. Agr. Food Chem.* 49: 3713–17.

Tyssandier V, Feillet-Coudray C, Caris-Veyrat C, Guilland JC, Coudray C, Bureau S, Reich M, Amiot-Carlin MJ, Bouteloup-Demange C, Boirie Y, Borel P (2004). Effect of tomato product consumption on the plasma status of anti-oxidant micro-constituents and on the plasma total antioxidant capacity in healthy subjects. *J. Am Coll Nutr.* 23 (2): 148–56.

Upritchard JE, Sutherland WH, Mann JI (2000). Effect of supplementation with tomato juice, vitamin E, and vitamin C on LDL oxidation and products of inflammatory activity in type 2 diabetes. *Diabetes Care.* 23 (6): 733–38.

USDA, National Nutrition Data Base for standard reference, Release-26, NDB No. 11529.

Valero MA, Vidal A, Burgos R, Calvo FL, Martínez C, Luengo LM, Cuerda C (2011). [Meta-analysis on the role of lycopene in type 2 diabetes mellitus]. *Nutr. Hosp.* 26 (6): 1236–41. [Article in Spanish].

van Breemen RB, Pajkovic N (2008). Multi-targeted therapy of cancer by lycopene. *Cancer Lett.* 269 (2): 339–51.

Van Eewyck J, Davis FG, Bowen PE (1991). Dietary and serum carotenoids and cervical intra-epithelial neoplasia. *Int. J. Cancer.* 48: 34–38.

Walfisch S, Walfisch Y, Kirilov E, Linde N, Mnitentag H, Agbaria R, Sharoni Y, Levy J (2007). Tomato lycopene extract supplementation decreases insulin-like growth factor-I levels in colon cancer patients. *Eur. J. Cancer Prev.* 16 (4): 298–303.

Walfisch Y, Walfisch S, Agbaria R, Levy J, Sharoni Y (2003). Lycopene in serum, skin and adipose tissues after tomato-oleoresin supplementation in patients undergoing haemorrhoidectomy or peri-anal fistulotomy. *Br. J. Nutr.* 90 (4): 759–66.

Weisburger JH (1998). Evaluation of the evidence on the role of tomato products in disease prevention. *Proc. Soc. Exp. Biol. Med.* 218 (2): 140–43.

Willcox JK, Catignani GL, Lazarus S (2003). Tomatoes and cardio-vascular health. *Crit. Rev. Food Sci. Nutr.* 43 (1): 1–18.

Woodall AA, Lee SWM, Weesie RJ, Jackson MJ, Britton G (1997). Oxidation of carotenoids by free radicals: relationship between structure and reactivity. *Biochim Biophys. Acta.* 1336: 33–42.

Xaplanteris P, Vlachopoulos C, Pietri P, Terentes-Printzios D, Kardara D, Alexopoulos N, Aznaouridis K,Miliou A, Stefanadis C (2012). Tomato paste supplementation improves endothelial dynamics and reduces plasma total oxidative status in healthy subjects. *Nutr. Res.* 32 (5): 390–94.

Yamamoto J, Taka T, Yamada K, Ijiri Y, Murakami M, Hirata Y, Naemura A, Hashimoto M, Yamashita T, Oiwa K, Seki J, Suganuma H, Inakuma T, Yoshida T (2003). Tomatoes have natural anti-thrombotic effects. *Br. J. Nutr.* 90: 1031–38.

Yilmaz S, Atessahin A, Sahna E, Karahan I, Ozer S (2006). Protective effect of lycopene on adriamycin-induced cardio-toxicity and nephro-toxicity. *Toxicology.* 218 (2-3): 164–71.

Zhou C, Han W, Zhang P, Cai M, Wei D, Zhang C (2008). Lycopene from tomatoes partially alleviates the bleomycin-induced experimental pulmonary fibrosis in rats. *Nutr. Res.* 28 (z2): 122–30.

Glossary

AAPH–2,2'-Azobis(2-amidinopropane) dihydrochloride (AAPH)–Is a chemical compound used to study the chemistry of the oxidation of drugs. It is a free radical-generating azo compound. It is gaining prominence as a model oxidant in small molecule and protein therapeutics for its ability to initiate oxidation reactions via both nucleophilic and free radical mechanisms.

ABROGATED–Abolished or avoided.

ADAPTOGENIC–Pharmacological concept whereby administration results in stabilization of physiological processes and promotion of homeostasis.

ADP–Adenosine Diphosphate. Is the oldest and one of the most important agonists of platelet activation. ADP induces platelet shape change, exposure of fibrinogen binding sites, aggregation, and influx and intracellular mobilization of Ca^{2+}. ADP-induced platelet aggregation is important for maintaining normal hemostasis, but aberrant platelet aggregation manifests itself pathophysiologically in myocardial ischemia, stroke, and atherosclerosis.

AIN76A–Purified rodent diet extensively used in research.

AML 14.3D10–Human acute myeloid leukaemia cells.

ANTITUSSIVES–(Cough suppressants) Capable of relieving or suppressing coughing.

BAP INDUCED DNA DAMAGE–Metabolism of benzo[a] pyrene (BaP), a well-established carcinogen and ubiquitous environmental contaminant, can result in either detoxification or bioactivation to its genotoxic forms.

Bax: Bcl-2 ratio–Determines the Susceptibility of Human Melanoma Cells to CD95/Fas-Mediated Apoptosis.

CATHARTICS–Sorbitol, magnesium citrate, magnesium sulfate, orsodium sulfate were previously used as a form of gastro-intestinal decontamination following poisoning *via* ingestion. They are no longer routinely recommended for poisonings.

CD-11b cells–Macrophage-1 antigen (or integrin $\alpha_M\beta_2$)- Is a complement receptor ("CR3") consisting of CD11b (integrin α_M) and CD18(integrin β_2).

CD-14 cells–Is a human gene. The protein encoded by this gene is a component of the innate immune system.

CHAFFED SKIN–Chafing refers to the irritation of skin caused by repetitive friction, usually generated through skin to skin contact of multiple bodyparts.

CHOLAGOGUE–Cholagogue–a drug or other substance that promotes the flow of bile from the gall bladder into the duodenum.

CISPLATIN–The most potent chemo-therapeutic anti-tumor drug which induces Oxidative stress.

COULOMBIC INTERACTION- Interactions of charged particles associated with the Coulomb forces they exert on one another. Also known as electrostatic interactions.

CYCLOOXYGENASE–COX is officially known as prostaglandin-endoperoxide synthase (PTGS), is an enzyme (EC 1.14.99.1) that is responsible for formation of important biological mediators called prostanoids, including prostaglandins, prostacyclin and thromboxane.

COX-1–Appears to be responsible for the production of prostaglandins (PG) that are important for homeostatic functions, such as maintaining the integrity of the gastric mucosa, mediating normal platelet function, and regulating renal blood flow.

COX-2 expression–COX-2 is dramatically upregulated during inflammation. For example, synovial tissues in patients with rheumatoid arthritis (RA) express increased levels of COX-2. In animal models of inflammatory arthritis.

CROCIN–is a natural carotenoid chemical compound found in the flowers crocus. It is the diester formed from the disaccharide gentiobiose and the dicarboxylic acid crocetin.

Cy-3-glc–A flavonoid, cyanidin-3-glucoside.

CYP activity–The cytochrome P450 superfamily of monooxygenases (officially abbreviated as CYP) is a large and diverse group of enzymes that catalyze the oxidation oforganic substances. The substrates of CYP enzymes include metabolicintermediates such as lipids and steroidal hormones, as well as xenobioticsubstances such as drugs and other toxic chemicals. CYPs are the major enzymes involved in drug metabolism and bioactivation, accounting for about 75 per cent of the total number of different metabolic reactions.

CYP1A1 and CYP1B1–Cytochromes.

DLA–Glucagon-like peptide-1 (GLP-1) is an incretin hormone with antidiabetic action through its ability to stimulate insulin secretion, increase beta cell neogenesis, inhibit beta cell apoptosis, inhibit glucagon secretion, delay gastric emptying and induce satiety.

DPP4–dipeptidyl peptidase-4 (DPP-4), an enzyme that inhibits GLP-1.

DU145 cells–Prostatic Cancer Cell Line.

EAC cells–Ehrlich Ascites Carcinoma (EAC) tumor cell lines.

EC APOPTOSIS–Endothelial cell apoptosis.

EGCG–Epigallocatechin-3-gallate, the main and most significant polyphenol in green tea.

EMOLLIENT–Moisturizers.

FEBRIFUGE–Any drug or agent that lowers body temperature to prevent or alleviate fever.

FE-NTA–Ferric Nitrilotriacetate.

FRAP–An optical technique capable of quantifying the two dimensional lateral diffusion of a molecularly thin film containing fluorescently labeled probes, or to examine single cells.

GADD153–Also known as chop/growth arrest- and DNA damage-inducible gene is a highly stress-inducible gene that is robustly expressed following disruption of homeostasis in the endoplasmic reticulum (ER) (so-called ER stress).

GYPENOSIDES–Derived from *Gynostemma pentaphyllum.* Also called jiaogulan is a dioecious, herbaceous climbing vine of the Cucurbitaceae family (cucumber or gourd family) indigenous to the southern reaches of China, northern Vietnam, southern Korea, and Japan. Best known as an herbal medicine reputed to have powerful antioxidant and adaptogenic effects purported to increase longevity. Pharmacological research has indicated a number of therapeutic qualities of Jiaogulan, such as lowering cholesterol and highblood pressure, strengthening immunity, and inhibiting cancer growth.

HBeAg–Stands for hepatitis B "e" antigen. This antigen is a protein from the hepatitis B virus that circulates in infected blood when the virus is actively replicating. Also a marker of contagious infection.

HBsAg–Is the surface antigen of the hepatitis B virus (HBV). It indicates current hepatitis B infection.

HepG2–Hep G2 is a perpetual cell line which was derived from the liver tissue of a 15-year-old Caucasian American male with a well-differentiated hepatocellular carcinoma.

HNSCC–Head and Neck Squamous Cell Carcinomas.

HOMA-IR- The homeostatic model assessment (HOMA) is a method used to quantify insulin resistance and beta-cell function. It was first described under the name HOMA by Matthews *et al.,* in 1985. The formula to calculate IR is as follows.

HT29–Cell Line.

iNOS–Intracellular reactive oxygen species.

ISO–Isoproterenol induced myocardial injury is widely used as an experimental animal model.

IκB-α- Immunogen. synthetic peptide corresponding to the C-terminus of human *IκBα* (amino acids 297-317 with N-terminally added lysine) conjugated to KLH.

LACTADHERIN–a major glycoprotein of the human fat globule membrane, is abundant in human breast milk and expressed in human breast carcinomas.

LANGMUIR ADSORPTION ISOTHERM- The Langmuir equation (also known as the Langmuir isotherm, Langmuir adsorption equation or Hill-Langmuir equation) relates the coverage oradsorption of molecules on a solid surface to gas pressure or concentration of a medium above the solid surface at a fixed temperature. The equation was developed by Irving Langmuir in 1916. The equation is stated as:.

MANGANISM–It is manganese poisoning, a toxic condition resulting from chronic exposure to manganese and first identified in 1837 by James Couper. Manganism or manganese poisoning is a toxic condition resulting from chronic exposure to manganese and first identified in 1837 by James Couper.

MAPK ERK1/2, JNK–Mitogen-activated protein kinases also known as MAP kinases areserine/threonine/tyrosine-specific protein kinases belonging to the CMGC (CDK/MAPK/GSK3/CLK) kinase group.

MCF-7–Human breast cancer cell line.

MCF-7/ADR cells–Human breast adenocarcinoma.

MDA-MB468 cells–Human breast cancer cell line.

MELANOGENESIS–formation of melanin.

MELANOSOMOGENESIS–Biogenesis of Melanosomes in Kupffer Cells.

MELIOIDOSIS–Melioidosis is an infectious disease caused by a Gram-negative bacterium,*Burkholderia pseudomallei*, found in soil and water.

MITOCANS–emerging class of drugs called "mitocans", a term that reflects their mitochondrial targeting and anti-cancer roles.

MMP-13–Collagenase 3 is an enzyme that in humans is encoded by the MMP13 gene. It is a member of the matrix metalloproteinase family.

MMP-3–Stromelysin-1 also known as matrix metalloproteinase-3 (MMP-3) is an enzyme that in humans is encoded by the MMP3 gene. The MMP3 gene is part of a cluster of MMP genes which localize to chromosome 11q22.3.

NF-Kb- NF-κB (nuclear factor kappa-light-chain-enhancer of activated B cells) is a protein complex that controls transcription of DNA. NF-κB is found in almost all animal cell types and is involved in cellular responses to stimuli such as stress, cytokines, free radicals, ultraviolet irradiation, oxidized LDL, and bacterial or viralantigens.

NOCICEPTION–Encoding and processing in the nervous system of noxious stimuli.

NOOTROPIC–Also referred to as smart drugs, memory enhancers, neuro enhancers, cognitive enhancers, and intelligence enhancers, are drugs, supplements, nutraceuticals, and functional foods that purportedly improve mental functions such as cognition, memory, intelligence, motivation, attention, and concentration.

Nrf2–The transcription factor Nrf2 (NF-E2-related factor 2) binds to the antioxidant responsive element (ARE) of α9-nAChR and cyclin D3 expression.

ORAC-Oxygen radical absorbance capacity (ORAC)–Is a method of measuring antioxidant capacities in biological samples *in vitro*.

OTOTOXICITY- Ototoxicity is damage to the hearing or balance functions of the ear by drugs or chemicals.

OUABAIN–A toxic compound obtained from certain trees, used as a very rapid cardiac stimulant. It is a polycyclic glycoside.

PC-3 cells–Human prostate cancer cells.

PETIT MAL AND GRAND MAL SEIZURES–The 1970 classification the distinction was whether the symptoms involved elementary sensory or motor functions (simple) or whether "higher functions" were involved (complex). Prior to this, terms such as *petit mal, grand mal, Jacksonian, psychomotor* and *temporal-lobe* seizures were used.

PHASE G1–The first gap phase (G1) of the cell cycle.

POULTICE–A soft, moist mass of material, typically consisting of bran, flour, herbs, etc., applied to the body to relieve soreness and inflammation and kept in place with a cloth.

PURGATIVES–Are foods, compounds and/or drugs that facilitate or increase bowel movements. They are most often used to treat constipation, causing evacuation of the bowels; cathartic.

QUAIL–Is a collective name for several genera of mid-sized birds generally considered in the order Galliformes.

RAGE- The Receptor for Advanced Glycation Endproducts is a 35kD transmembrane receptor of the immunoglobulin super family.

RESPIRATORY CATARRHS–An acute inflammatory disease involving the mucous membranes of the respiratory tract (rhinitis, pharyngitis, laryngotracheitis, bronchitis).

RULEIN–Ceruletide (INN)-Also known as cerulein or caerulein, is a ten amino acidoligopeptide that stimulates smooth muscle and increases digestive secretions. Ceruletide is similar in action and composition to cholecystokinin.

SCLEROSANT–An injectable irritant that is used in the treatment of varicose veins and that causes inflammation and subsequent fibrosis, thus obliterating the lumen of the vein.

SIN-1–*in-vitro* systems.

STILBENE- May refer to one of the two isomers of 1,2-diphenylethene: (E)-stilbene (trans-isomer), (Z)-stilbene (cis-isomer).

SW480–Colon cancer cells.

T(MAX)–The time after administration of a drug when the maximum plasma concentration is reached; when the rate of absorption equals the rate of elimination.

THELARCHE–Is the onset of secondary (postnatal) breast development, usually occurring at the beginning of puberty in girls.

TRAP–Total Reactive Antioxidant Potential (TRAP).

UBIQUITINATION–Is a post-translational modification (an addition to a protein after it has been made) where ubiquitin is attached to a substrate protein. The addition of ubiquitin can affect proteins in many ways: It can signal for their degradation via the proteasome, alter their cellular location, affect their activity, and promote or prevent protein interactions.

UGT activity–glucuronyltransferases (UGTs).

VEGF–*Vascular endothelial growth factor*.

WESSEL CRITERIA–Infantile colic refers to a behavioural syndrome occurring during the first 3 months of life. Excessive crying is common. The most widely used quantitative definition of colic is the one proposed by Wessel *et al.,* known as the rule of threes. Infants are considered to have colic if they cry for more than 3 hour a day, for more than 3 days a week, and for more than 3 weeks.

Index

www.ingramcontent.com/pod-product-compliance
Lightning Source LLC
Chambersburg PA
CBHW050512190326
41458CB00005B/1513